AF593591

Application of FRACTURE MECHANICS for Selection of Metallic Structural Materials

Edited by

JAMES E. CAMPBELL
Metallurgical Consultant
Columbus, Ohio

WILLIAM W. GERBERICH
Chairman, Department of Chemical Engineering and Material Science
University of Minnesota

JOHN H. UNDERWOOD
Research Materials Engineer
U. S. Army Benet Weapons Laboratory

American Society for Metals
Metals Park, Ohio 44073

Library of Congress Catalog Card No.: 82-71791

ISBN: 0-87170-136-7

SAN: 204-7586

PRINTED IN THE UNITED STATES OF AMERICA

Preface

This monograph is the product of several years of effort by ASM technical committees. The starting point of the work was the article "Toughness and Fracture Mechanics" in *Metals Handbook,* Vol 10, Eighth Edition, prepared by a committee chaired by the late Dr. George E. Pellissier. In the present effort, state-of-the-art information has been compiled on the fracture mechanics of structural metals and on the application of fracture mechanics data to selection of metals for critical structural components. In the past 20 years, the concepts of fracture mechanics have developed into a broadly based technology that can be applied to the selection of materials and to the design of structures to obtain improved service performance. This development has been fostered by thousands of publications covering research and applications that have indicated the usefulness and the limitations of fracture mechanics technology.

At the present time, it is generally recognized that the requirements for the application of fracture mechanics in design are expanding. Furthermore, the value of the fracture mechanics approach in supplementing well-established design procedures is being demonstrated in many applications, from aerospace components to agricultural equipment. Because of the rapid growth of the technology and because of the need for education of engineering students and other technical people in the fundamentals and applications of the technology, a summary of the available information on the fracture mechanics of metals is needed in a basic text. This monograph is intended to meet these needs.

With regards to the level of expertise required for effective use of the information in this text, a Bachelor's degree is sufficient. This would include all disciplines in the physical sciences, although mechanical, materials and metallurgical engineers should find this book most useful. Although in monograph form, this compilation of fracture mechanics approaches and data could serve as reference material for an upper-division undergraduate or first-year graduate course in fracture mechanics or materials selection.

The first three chapters present, respectively, a brief historical background on the development of fracture mechanics, explanations of fracture mechanics concepts, and reviews of standard fracture mechanics testing methods. The basic concepts of fracture mechanics are unlike other concepts related to the mechanical properties of metals. For the reader who is unfamiliar with these concepts and with standard testing methods, Chapters 2 and 3 will provide the fundamental information that is prerequisite to an understanding of the chapters on fracture mechanics properties of structural metals. For the reader who is familiar with these concepts and testing methods, the information presented in the later chapters on fracture mechanics properties of structural metals and on materials selection will be most useful.

Chapters 4 through 8 provide information on fracture mechanics properties of alloy steels, stainless steels, aluminum alloys, titanium alloys and superalloys. The effects of testing and service environments on fracture properties are considered, including the effects of chemical environments, temperature, and rate and duration of loading. The properties are given in the SI system of units, with alternate values given in the customary English system of units in most cases.

Considerable information on the metallurgy of each class of metals also is presented, so that the reader with a limited background in metallurgy will have an opportunity to acquire information needed for an understanding of the effects of microstructure, heat treatment and other variables on fracture mechanics properties of materials. More information is presented on the metallurgies of the less common materials than on those of the more common materials.

The fracture mechanics information in each chapter is intended to present typical or average values of various fracture mechanics parameters, including effects of variations in heat treatment, orientation, material thickness and environmental effects. The data are not design data, but they are useful in determining what alloy or alloys would be preferred for certain critical applications.

Chapter 9 presents examples and case histories that illustrate how the data presented in the earlier chapters may be used in calculating critical flaw sizes, fatigue crack growth rates and other requirements for safe service lives of structural components. Other examples show how fracture mechanics information may be used in the design of critical structures and in analyzing fractures in failed components.

Those who served on one or both of the two ASM committees which were responsible for this monograph are:

Dr. William W. Gerberich, Chairman, ASM Committee on Fracture Toughness of Stainless Steels, Superalloys and Non-Ferrous Alloys; Director of Material Science, Department of Chemical Engineering and Material Science, University of Minnesota, Minneapolis, MN

Mr. John H. Underwood, Chairman, ASM Committee on Fracture Toughness of Steels; Research Materials Engineer, U. S. Army Armament Research and Development Command, Benet Weapons Laboratory, Watervliet, NY

Mr. James E. Campbell, Secretary of the Committees; Metallurgical Consultant, Columbus, OH

Dr. Steven D. Antolovich, Professor, Department of Material Science and Metallurgical Engineering, University of Cincinnati, Cincinnati, OH

Mr. James C. Chesnutt, Member Technical Staff, Rockwell International Science Center, Thousand Oaks, CA

Mr. Allan W. Gunderson,Materials Engineer, Air Force Materials Laboratory, Wright-Patterson Air Force Base, OH

Mr. J. G. Kaufman, Director of Engineering and Research, Anaconda Aluminum, Louisville, KY

Dr. Joseph M. Krafft, Head, Mechanics of Materials Branch, Ocean Technology Division, Naval Research Laboratory, Washington, DC

Dr. Harold Margolin, Professor, Department of Physical and Engineering Metallurgy, Polytechnic Institute of New York, Brooklyn, NY

Mr. Philip O. Metz, Research Metallurgist, Armco Steel Corporation, Middletown, OH

Dr. Harry W. Rosenberg, Manager, Aerospace Market Development, TIMET, Pittsburgh, PA

Mr. Joseph S. Santner, Materials Research Laboratories, Inc., Glenwood, IL

Dr. William R. Tyson, Research Scientist, Canadian Department of Energy, Mines and Resources, Ottawa, Ontario

Dr. Robert H. Van Stone, Metallurgist, General Electric Company, Aircraft Engine Group, Evendale, OH

Dr. William Wallace, Metallurgy Group, National Aeronautical Establishment, National Research Council of Canada, Ottawa, Ontario

Mr. Alexander D. Wilson, Research Engineer, Lukens Steel Company, Coatesville, PA

Dedication

This monograph is dedicated to the memory of
Dr. George E. Pellissier.

Contents

1 **Historical Development of Fracture Mechanics** **1**
J. E. Campbell

References 2

2 **Concepts of Fracture Mechanics** **5**
J. H. Underwood and W. W. Gerberich

Basic Quantities 5
Energy-Release Rate 5
Assumptions 9
Material Flaws and Nondestructive Inspection 9
Stress-Intensity Factor 10
Fracture Toughness 12
Plastic Zone 14
Additional Applications and Limitations 16
Elastic-Plastic Crack Growth 16
Fatigue Crack Growth 17
Effects of Testing Environment on Crack Growth 19
References 20

3 **Fracture Test Methods** **23**
J. H. Underwood

Plane-Strain Fracture Toughness, K_{Ic} 23
ASTM Method E-399 23
New K_{Ic} Tests 27

Non-Plane-Strain Toughness Tests . . . 29
R-Curve Determination . . . 29
J_{Ic} Fracture Toughness . . . 30
COD Methods . . . 32
Correlation Fracture Tests . . . 32
Fatigue Crack Growth Rate Test Methods . . . 34
Sustained-Loading Crack Growth Tests . . . 35
Use of Fracture Test Methods . . . 36
References . . . 38

4 Fracture Properties of Carbon and Alloy Steels . . . 41
R. H. Van Stone and A. W. Gunderson

Introduction . . . 41
Metallurgy of Steels . . . 42
Phase Relationships and Chemistries . . . 42
Special Embrittlement Considerations . . . 46
Fracture Toughness . . . 46
Microstructural Fracture Modes . . . 47
Effects of Alloy Chemistry . . . 49
Effects of Microstructure and Heat Treatment . . . 56
Temperature and Strain Rate Variables . . . 62
Fatigue Crack Propagation . . . 66
Microstructural Fracture Modes . . . 66
Effects of Microstructure and Heat Treatment . . . 68
Loading, Temperature and Frequency Variables . . . 73
Aggressive Environments . . . 78
Sustained-Load Crack Propagation . . . 86
Microstructural Fracture Modes . . . 88
Effects of Alloy Chemistry . . . 88
Effects of Microstructure and Heat Treatment . . . 92
Effects of Temperature and Environment . . . 93
References . . . 96

5 Fracture Properties of Wrought Stainless Steels . . . 105
J. E. Campbell

Introduction . . . 105
Compositions of Wrought Stainless Steels . . . 106
Austenitic Stainless Steels . . . 106
Ferritic Stainless Steels . . . 107

Martensitic Stainless Steels 107
Precipitation Hardening Stainless Steels 110
Fracture Mechanics Data for Stainless Steels 111
Fracture Properties of Austenitic Stainless Steels 112
Fracture Properties of Martensitic Stainless Steels 142
Fracture Properties of Precipitation Hardening Stainless Steels 149
References 163

6 Fracture Properties of Aluminum Alloys 169
J. G. Kaufman and J. S. Santner

Introduction 169
Major Alloy Systems 169
Fracture Toughness 171
Effects of Microstructure, Composition and Heat Treatment 173
Effects of Mechanical Processing and Orientation 184
Effect of Temperature 187
Fatigue Crack Propagation 189
Effects of Composition, Microstructure and Thermal Treatments 190
Effects of Product Form and Orientation 192
Effects of Exposure Temperature and Humidity 192
Effect of Load-Time History 193
Effect of Load Ratio, R 195
Sustained-Load Crack Growth in Aggressive Environments 196
Effects of Composition, Microstructure and Thermal Processing 200
Effects of Product Form and Orientation 204
Effects of Temperature and Environment 204
References 208

7 Fracture Properties of Titanium Alloys 213
H. W. Rosenberg, J. C. Chesnutt and H. Margolin

Introduction 213
Metallurgy of Titanium Alloys 214
Phase Relationships and Chemistry 214
Metallurgy of Ti-6Al-4V 214
Metallurgy of High-Strength Alpha-Beta Alloys 222

Metallurgy of "Super Alpha" Alloys......... 223
Tensile Properties............................ 224
Fracture Toughness............................ 224
Effects of Alloy Chemistry................ 225
Effects of Microstructure................. 227
Effects of Texture........................ 230
Effects of Environment.................... 231
Fatigue Crack Propagation..................... 232
Effects of Alloy Chemistry................ 235
Effects of Microstructure................. 237
Effects of Crystallographic Texture....... 238
Environmental Effects..................... 238
Sustained-Load Crack Propagation.............. 241
Concluding Remarks............................ 244
References Cited in Text...................... 247
General References............................ 251

8 Fracture Properties of Superalloys....................253
Stephen D. Antolovich and J. E. Campbell

Introduction.................................. 253
Physical Metallurgy of Nickel-Base Superalloys... 254
Alloys Strengthened by Gamma Prime Phase.......................... 254
Alloys Strengthened by Gamma Double-Prime........................ 263
Fatigue Crack Propagation..................... 265
Empirical Models for Data Representation.... 265
Effects of Microstructure................. 265
Effects of Testing Variables: Environment, Temperature, Frequency and Waveform.... 277
Creep Crack Growth............................ 294
Applicability of Stress Intensity Parameter to Creep Crack Growth.................. 294
Mechanisms and Models for CCG.......... 299
Metallurgical Variables................... 301
Environmental Effects..................... 304
Summary....................................... 305
References.................................... 307

9 Design, Materials Selection and Failure Analysis......311
W. W. Gerberich and A. W. Gunderson

Introduction.................................. 311
Principles of the Fracture Toughness Approach.... 313
Pressure Vessel Leak-Before-Burst Concept... 314
Overpressurization of Thick-Walled Vessels... 315

General Design/Performance Model 317
Subcritical Crack Growth Approach 318
Fatigue Methodology 320
Stress Corrosion Cracking, Hydrogen Embrittlement and Corrosion Fatigue 328
Quality Control 333
Incoming Quality 333
In-Plant Control 334
Exit Quality 334
Failure Analysis 334
Materials Selection 335
Case Studies Using K_{Ic} Considerations 336
Case Studies Using ΔK Considerations 347
Case Studies Using K_{Iscc} Considerations 354
References 363

INDEX 367

Chapter 1

Historical Development of Fracture Mechanics

J. E. Campbell

The technology of linear-elastic fracture mechanics has been developed over the past three decades to provide quantitative methods for evaluating the load-carrying capacities of flawed structures and components. When fracture mechanics is applied to the selection of metallic materials for structural components, the materials engineer, metallurgist or designer has assumed that cracks or cracklike flaws will be present in the structure at some point in its lifetime. Past experience in design, fabrication, testing and failure analysis of many structures has indicated that cracklike flaws may be present in as-fabricated structures as well as in these structures after they have been exposed to service stresses and service environments. Fracture mechanics technology provides the design procedures required to minimize catastrophic service failures originating from such flaws and cracks.

Historically, reference is often made to catastrophic failures in the British Comet aircraft (Ref 1.1), in welded World War II ships (Ref 1.2), in certain bridges (Ref 1.3), and in a molasses tank (Ref 1.4) that failed during cold weather, to illustrate problems of brittle fracturing. However, the technology of fracture mechanics was first applied to metallic structures by Irwin and coworkers at the Naval Research Laboratory when analyzing fractures in welded high-strength alloy steel Polaris missile cases which had failed on proof testing (Ref 1.5). Following the fracture mechanics analysis of a number of failures, the fabrication techniques were modified, inspection techniques were improved, and materials and heat treatment specifications were revised to improve fracture resistance and alleviate the fracture problem.

Several successful applications of failure analysis followed the Polaris missile case work. Fracture during proof testing of a 260-in.-diam motor case fabricated from maraging steel was traced to a flaw that was estimated to be of critical size at the bursting pressure (Ref 1.6). Laboratory burst tests on preflawed pressure vessels made of steels, aluminum alloys and titanium alloys (Ref 1.7 and 1.8) have confirmed the validity of fracture mechanics analysis. Recently, fracture mechanics theory has been applied to more complex structures. As a good example, the design of the B-1 supersonic bomber involved the application of fracture mechanics to analysis of critical components. These components were produced for the prototypes from high-strength alloy steels, precipitation-hardening stainless steels, high-strength aluminum alloys, and titanium alloys (Ref 1.9). In this monograph, special emphasis is directed toward the application of fracture mechanics data for selecting appropriate alloys for specific components from alloy steels, stainless steels, aluminum alloys, titanium alloys, and superalloys. For further insights into the magnitude of the fracture problem in critical components, the reader is referred to two other ASM publications: *Source Book in Failure Analysis* (Ref 1.10) and Volume 10 of the 8th Edition of *Metals Handbook: "Failure Analysis and Prevention"* (Ref 1.11).

Originally, the technology of fracture mechanics was limited to relatively high-strength materials that could be tested in sizes that met certain requirements for linear-elastic displacement during slow testing of specimens of certain configurations. More recent advancements in the state of the art, such as R-curve and J-integral tests, have extended the use of fracture mechanics to elastic-plastic conditions which are associated with lower-strength materials and smaller section sizes. Furthermore, the technology has been applied to fatigue crack growth rate assessment under various conditions, environmental and stress-corrosion problems, dynamic fracturing, and determinations of the effects of elevated and cryogenic testing temperatures. These developments, which have occurred in the past 20 years, have led to broad use of fracture mechanics and to greater confidence in the design of fracture-critical structures.

REFERENCES

1.1. Fatigue and the Comet Disasters, by T. Bishop: *Metal Progress,* Vol 67, No. 5, May 1955, p 79-85

1.2. Investigation of Structural Failures of Welded Ships, by M. L. Williams and G. A. Ellinger: *Welding Journal,* Vol 32, No. 10, Oct 1953, p 498s-528s

1.3. Metallurgical Aspects of the Failure of the Pt. Pleasant Bridge, by J. A. Bennett and H. Mindlin: *Journal of Testing and Evaluation,* Vol 1, No. 2, March 1973, p 152-161

1.4. Brittle Failure of Steel Structures – A Brief History, by M. E. Shank: *Metal Progress,* Vol 66, No. 3, Sept 1954, p 83-88

1.5. Fracture Mechanics Theory and Its Application to Analysis of Fractures of Rocket Chambers, by G. R. Irwin: paper presented at the Symposium on the Testing and Evaluation of Materials for Solid Rocket Motor Casings, MAB, National Research Council, Washington, Feb 2-3, 1959, p 24-58

1.6. Investigation of Hydrotest Failure of Thiokol Chemical Corporation 260-Inch-Diameter SL-1 Motor Case, by J. E. Srawley and J. B. Esgar: NASA TMX-1194, Jan 1966

1.7. Stable and Unstable Crack Growth in Pressure Vessel Models, by G. C. Smith *et al:* Report CONF-780213-3, Contract W-7415-eng-26, Oak Ridge National Laboratory, Oak Ridge, TN, Feb 26, 1978

1.8. Practical Fracture Mechanics Applications to Design of High Pressure Vessels, by T. E. Davidson and J. F. Throop: in *Application of Fracture Mechanics to Design,* edited by J. J. Burke and V. Weiss, Plenum Publishing Corp., New York, 1979

1.9. Fracture Mechanics Evaluation of B-1 Materials (Vol I, Text; Vol II, Fatigue Crack Growth Data), by R. R. Ferguson and R. C. Berryman: Report AFML-TR-76-137 (I and II), Contract F33657-70-0800, Rockwell International Corporation, B-1 Division, Los Angeles, Oct 1976
1.10. *Source Book in Failure Analysis:* Metal Progress Bookshelf Series, American Society for Metals, Metals Park, OH, 1974
1.11. *Failure Analysis and Prevention:* Vol 10 of 8th Edition of *Metals Handbook,* American Society for Metals, Metals Park, OH, 1975

Chapter 2

Concepts of Fracture Mechanics

J. H. Underwood and W. W. Gerberich

2.1. BASIC QUANTITIES

The concepts of fracture mechanics are basic ideas for developing methods of predicting the load-carrying capabilities of structures and components containing cracks. The concepts deal with basic quantities or parameters of fracture mechanics. These quantities can be discussed in relation to a simple example: a center crack in a plate remotely loaded by a uniform tensile stress (See Fig. 2.1). When the half-crack length, a, is less than 10% of the total plate width, the relationship among stress-intensity factor, K, applied stress, σ, and half-crack length, a, is very close to the relationship for a crack in an infinitely wide plate, which is:

$$K = \sigma\sqrt{\pi a} \qquad \text{(Eq 2.1)}$$

The stress applied to the component, the length of the crack, and the stress-intensity factor in the loaded component with a crack are the basic quantities of fracture mechanics. The example in Fig. 2.1 also provides a simple explanation for the units of stress-intensity factor—i.e., the product of stress and square root of length. But more important is the concept that the stress-intensity factor, K, is a single parameter which includes both the effect of the stress applied to a sample and the effect of a crack of a given size in a sample. Still using the example of Fig. 2.1, if the combination of σ and a in Eq 2.1 were to exceed a critical value of K, then the fracture strength of the plate would be exceeded and the crack would be expected to propagate by one of the several mechanisms mentioned in this chapter and discussed in detail in the following chapters.

2.2. ENERGY-RELEASE RATE

The origins of modern-day fracture mechanics may be traced to Griffith (Ref 2.1), who established an energy-release-rate criterion for brittle materials.

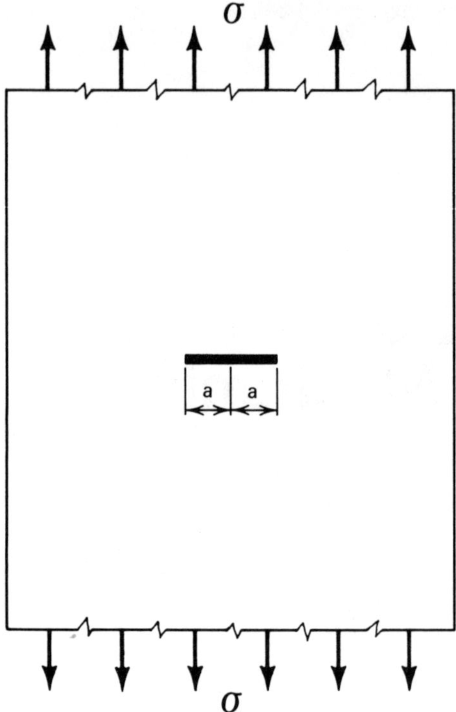

σ is remotely applied uniform tensile stress; a is half-crack length.

Fig. 2.1. Schematic illustration of a center crack in a wide plate

Observations of the fracture strength of glass rods had shown that the longer the rod, the lower the strength. Thus the idea of a distribution of flaw sizes evolved, and it was discovered that the longer the rod, the larger the chance of finding a large natural flaw. This physical insight led to an instability criterion which involved the elastic energy released in a solid at the time a flaw grew catastrophically under an applied stress.

From the theory of elasticity comes the concept that the strain energy contained in an elastic body per unit volume is simply the area under the stress-strain curve, or:

$$U_0 = \frac{\sigma^2}{2E} \qquad \text{(Eq 2.2)}$$

where σ is the applied stress and E is Young's modulus. However, there is a reduction (that is, a release) of energy in an elastic body containing a flaw or a crack because of the inability of the unloaded crack surfaces to support a load. We shall assume that the volume of material whose energy is released is the area of an elliptical region around the crack (as shown in Fig. 2.2) times the plate thickness, B; the volume is $\pi(2a)(a)B$. This is based on the area of an ellipse being $\pi r_a r_b$, where r_a and r_b are the major and minor radii of the ellipse. Then, the total energy released from the body due to the crack is the energy per unit volume times the volume, which is:

$$U = \pi(2a)\,(a)B\,\frac{\sigma^2}{2E} = \frac{\pi\sigma^2 a^2 B}{E} \qquad \text{(Eq 2.3)}$$

In ideally brittle solids, the released energy can be offset only by the surface energy absorbed, which is:

$$W = (2aB)\,(2\gamma_s) = 4aB\gamma_s \qquad \text{(Eq 2.4)}$$

where 2aB is the area of the crack and $2\gamma_s$ is twice the surface energy per unit area (because there are two crack surfaces).

Griffith's energy-balance criterion, in the simplest sense, is that crack growth will occur when the amount of energy released due to an increment of crack advance is larger than the amount of energy absorbed:

$$\frac{dU}{da} \geq \frac{dW}{da} \qquad \text{(Eq 2.5)}$$

Performing the derivatives indicated in Eq 2.5 and rearranging gives the Griffith criterion for crack growth:

$$\sigma\sqrt{\pi a} = \sqrt{2E\gamma_s} \qquad \text{(Eq 2.6)}$$

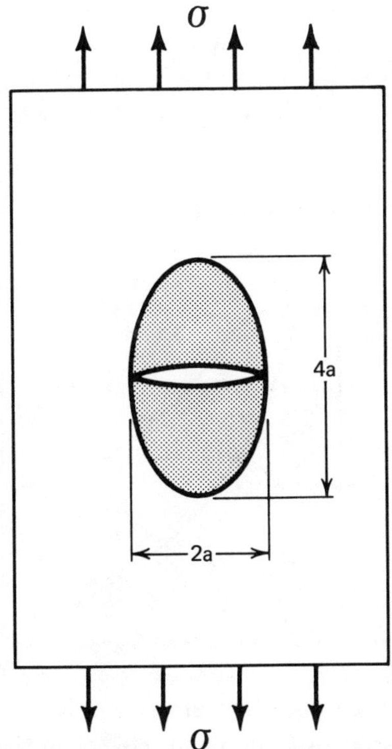

Fig. 2.2. Schematic illustration of the concept of energy release around a center crack in a loaded plate

Fracture theory was built upon this criterion in the early 1940's by considering that the critical strain energy release rate, G_c, required for crack growth was equal to twice an effective surface energy, γ_{eff}:

$$G_c = 2\gamma_{eff} \quad \text{(Eq 2.7)}$$

This γ_{eff} is predominantly the plastic energy absorption around the crack tip, with only a small part due to the surface energy of the crack surfaces. Then, with the development of complex variable and numerical techniques to define the stress fields near cracks, this energy view was supplemented by stress concepts—that is, the stress-intensity factor, K, and a critical value of K for crack growth, K_c. Replacing γ_s with γ_{eff} in Eq 2.6 and noting that the energy and stress concepts are essentially identical (that is, $K = \sqrt{EG}$) gives:

$$K_c = \sqrt{EG_c} = \sigma\sqrt{\pi a} \quad \text{(Eq 2.8)}$$

which is the crack-growth-criterion equivalent of Eq 2.1. Thus, K_c is the critical value of K which, when it is exceeded by a combination of applied stress and crack length, will lead to crack growth. For thick-plate plane-strain conditions, this critical value became known as the plane-strain fracture toughness, K_{Ic}, and any combination of applied stress and crack length that exceeds this value could produce unstable crack growth, as indicated schematically in Fig. 2.3(a) (linear-elastic).

In work with tougher, lower-strength materials, it was later noted that stable slow crack growth could occur even though accompanied by considerable plastic deformation. Such phenomena led to the nonlinear J-integral and R-curve concepts which could be used to predict the onset of stable slow crack growth and final

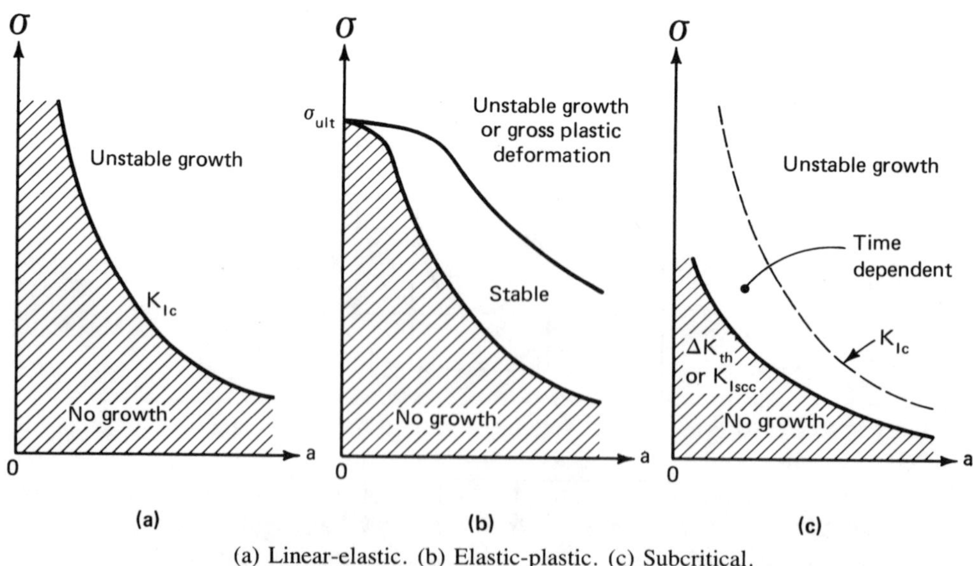

(a) Linear-elastic. (b) Elastic-plastic. (c) Subcritical.

Fig. 2.3. Relationships between stress and crack length, showing regions and types of crack growth

instability under elastic-plastic conditions, as noted in Fig. 2.3(b). Finally, the fracture mechanics approach was applied to characterize subcritical crack growth phenomena where time-dependent slow crack growth, da/dt, or cyclic crack growth, da/dN, may be induced by special environments or fatigue loading. For combinations of stress and crack length above some environmental threshold, K_{Iscc}, or fatigue threshold, ΔK_{th}, subcritical growth occurs, as indicated in Fig. 2.3(c).

2.3. ASSUMPTIONS

Additional concepts of fracture mechanics have evolved which are related to the assumptions made in using linear-elastic fracture mechanics. Three assumptions basic to fracture mechanics analysis are as follows:

1. Cracks exist in the body to be analyzed. Indeed, cracks, or flaws with cracklike behavior, are frequently present in real structural materials.
2. A crack can be represented by a flat, free surface in a linear-elastic, homogeneous, isotropic continuum.
3. A characteristic stress field surrounds any crack in a loaded body. The magnitude of this field, the K value, at the onset of crack extension is a material property which is independent of specimen size and geometry for many conditions of loading and environment.

2.3.1. Material Flaws and Nondestructive Inspection

A great many failure investigations have confirmed the first assumption given above—that is, that cracks exist in structural materials. In engineering design, materials are often considered to be free of cracklike flaws, but in reality flaws of some sort are always present in structural materials. The flaws may be so small or so unlike cracks that they will not become the sources of crack growth. However, the usefulness of the fracture mechanics approach is that it can predict what would happen if a cracklike flaw were to exist and to cause crack growth at a critical location of a component. Then, appropriate steps can be taken in design, manufacture and inspection to make sure that the flaw would be detected before the component is placed in service or that the undetected flaw would have no significant effect on the performance or lifetime of the component.

Nondestructive inspection methods are often used to detect flaws—particularly in critical, highly stressed locations of structural components. There is always a chance with any inspection method that certain small or unfavorably oriented flaws will go undetected. This topic is discussed in detail on pages 414 to 424 in Volume 11 of the 8th Edition of *Metals Handbook*. In general, however, the nondestructive testing method that is used should have the capability of detecting any flaw which could lead to failure during the lifetime of the component. The flawed component can then be withheld from service until its flaws are repaired.

An example of a material flaw which was detected by nondestructive inspection and which led to crack growth and component failure during simulated service testing is shown in Fig. 2.4, which shows the fracture surface of a forged AISI 4335 steel pressure vessel which was subjected to cyclic, internal pressure

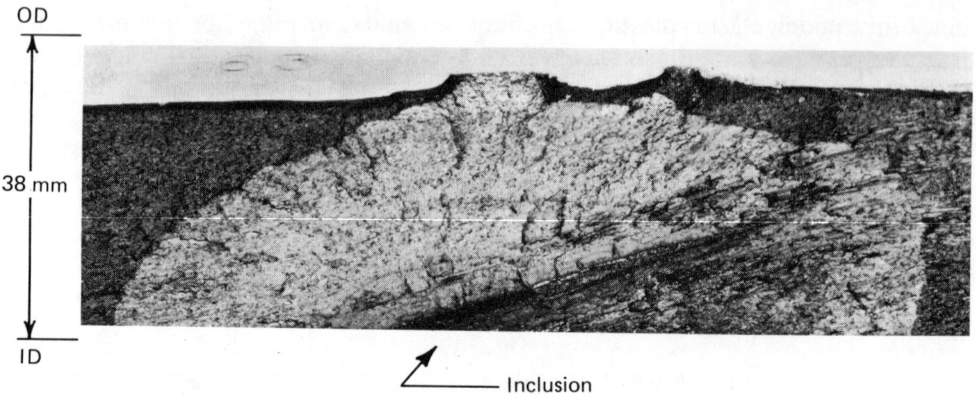

The vessel was a 4335 steel forging that had been hardened and tempered at 620°C (1150°F) to a yield strength of 1050 MPa (152 ksi).

Fig. 2.4. Fracture surface of a steel pressure vessel, showing fatigue crack growth from a flaw

in the laboratory. The unusually large nonmetallic inclusion at the ID of the tube was clearly the initiation point of the failure. In this case an ultrasonic nondestructive inspection method had shown the location and approximate size of the flaw, so failure initiation at this point was expected. A fatigue crack, shown as the lighter area in Fig. 2.4, grew from the initiation point through to the OD of the vessel. At this point, leakage occurred, and the test was stopped.

In the example described above, the flaw was much larger than those commonly observed in structural materials. Regardless of the types and sizes of flaws which can be present in structural components, the important point is that nondestructive inspection methods can provide an indication of the flaw sizes. Then the maximum flaw size can be used to calculate the value of K, using an expression such as Eq 2.1, and a quantitative description of the conditions for crack growth and failure can be made.

2.3.2. Stress-Intensity Factor

The second assumption given above makes possible a mathematical description of the stresses in the vicinity of the crack tip. This assumption of a crack in a linear-elastic solid at first appears contradictory to what is known about fracture of metals, because some plastic deformation is always found to accompany fracture. However, when the region of plastic deformation around the crack is small compared with the size of the crack, which often is true with large structures and with high-strength materials, this is a good assumption. Using linear-elastic theory and referring to Fig. 2.5, the stress at a point P near the crack tip can be expressed as:

$$\sigma_y = \frac{K_I}{\sqrt{2\pi r}} \cos\frac{\theta}{2}\left(1 + \sin\frac{\theta}{2}\sin\frac{3\theta}{2}\right) \qquad \text{(Eq 2.9)}$$

where σ_y is the stress perpendicular to the crack plane, r is the distance from the crack tip to point P, θ is the angle between the crack plane and the line from point P to the crack tip, and K_I is the applied opening mode stress-intensity factor. In Eq 2.9, as the distance from the crack tip, r, approaches zero, the stress, σ_y, approaches infinity. This $1/\sqrt{r}$ singularity can not actually occur, because plastic deformation relieves elastic stresses very near the crack tip. Nevertheless, in many cases the $1/\sqrt{r}$ singularity provides an adequate over-all description of the stresses near the crack tip and therefore a good description of the conditions for crack growth.

The important implications of Eq 2.9 are: (*a*) a crack in a loaded component or specimen generates its own intensified stress field near the crack tip, a stress field that differs from another crack-tip stress field only by the scaling factor represented by K; and (*b*) the factor K expresses how much the stress intensifies at the crack tip, and thereby allows the loading and geometry factors that influence crack growth in a specimen to be described on a uniform basis using a single parameter.

The stress-intensity factor, K, can have a simple relation to applied stress and crack length, as in Eq 2.1. But, more often, the K relation is of greater complexity because of complex loading, various configurations of real structural components, or variations in crack shapes. The K relations for many different types of loading and specimen and crack geometries have been obtained by various experimental and analytical methods. Two handbooks that give a variety of K relations are listed as Ref 2.2 and 2.3. These handbooks give K relations for the three basic types of

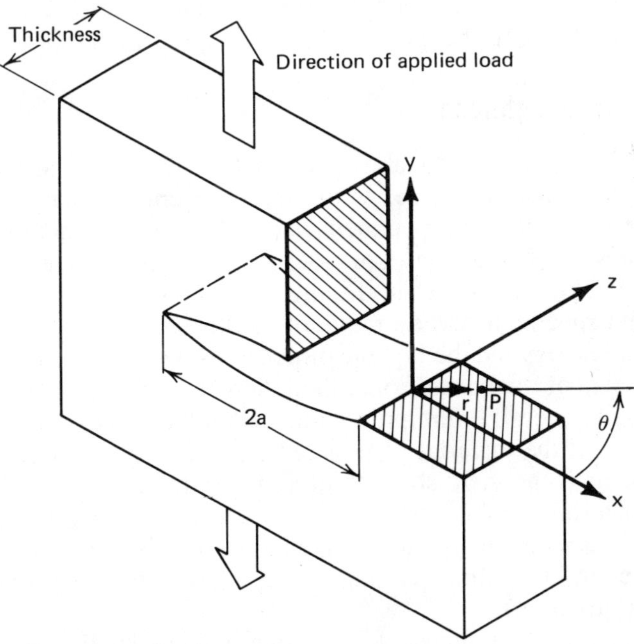

x is the direction perpendicular to the crack tip and in the plane of the crack; y is the direction perpendicular to the plane of the crack; z is the direction parallel to the crack tip; P is a point near the crack tip; r is the distance from the crack tip to point P; θ is the angle between the plane of the crack and a line from point P to the crack tip; a is half-crack length.

Fig. 2.5. Schematic illustration of a through-thickness crack

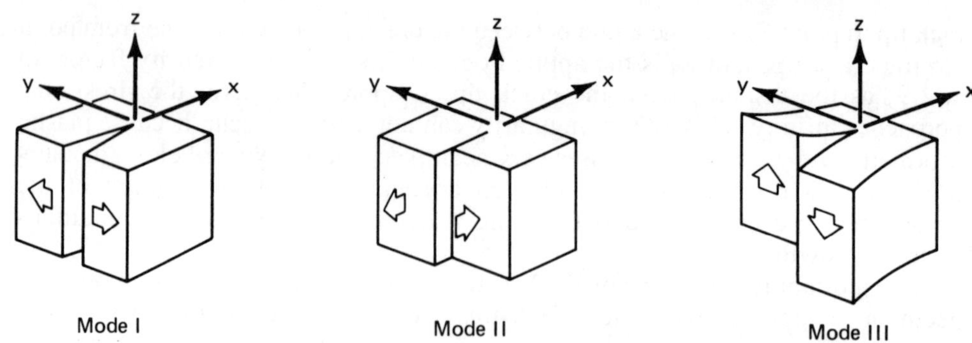

Mode I—opening mode: tension stress in y direction (perpendicular to crack surfaces). Mode II—edge-sliding mode: shear stress in x direction (perpendicular to crack tip). Mode III—screw-sliding mode: shear stress in z direction (parallel to crack tip).

Fig. 2.6. Modes of crack deformation

crack-face displacement shown in Fig. 2.6. Nearly all crack-related fracture processes of practical significance for metals involve mode I, opening mode deformation, in which the displacement of the crack faces is in a direction perpendicular to the crack plane. There are cases in which shear deformation by modes II and III accompanies opening mode deformation, but shear deformation often has little significant effect on the over-all, macroscopic fracture process.

2.3.3. Fracture Toughness

The third assumption given above states that for many test conditions the magnitude of the stress field, K, at the onset of crack extension is a material constant. Tests on precracked specimens of a wide variety of materials have shown that the critical K value at the onset of crack extension approaches a constant value as specimen thickness increases. Figure 2.7 shows this effect in tests with AISI 4340 steel specimens over a range of thickness (Ref 2.4). In general, when the specimen thickness and the inplane dimensions near the crack are large enough relative to the size of the plastic zone, then the value of K at which growth begins is a constant and generally minimum value called the plane-strain fracture toughness factor, K_{Ic}, of the material. The parameter K_{Ic} is a true material property in the same sense as is the yield strength of a material. The value of K_{Ic} determined for a given material is unaffected by specimen dimensions or type of loading, provided that the specimen dimensions are large enough relative to the plastic zone to ensure plane-strain conditions around the crack tip (strain is zero in the through-thickness or z-direction).

Plane-strain fracture toughness, K_{Ic}, is directly related to the energy required for the onset of crack propagation by the formula

$$K_{Ic} = \sqrt{\frac{EG_{Ic}}{1 - \nu^2}} \qquad \text{(Eq 2.10)}$$

where E is the elastic modulus (in Mpa or psi), ν is Poisson's ratio (dimensionless), and G_{Ic} is the critical plane-strain energy release rate for crack extension (in kJ/m^2 or in.-lb/in.^{2}). In simplified concept, G_{Ic} is the critical amount of strain energy that is released from the elastic stress field of the specimen per unit area of new cracked surface for the first small increment of crack extension. The concepts of K_{Ic} and G_{Ic} are essentially interchangeable; K_{Ic} is generally preferred because it is more easily associated with the stress or load applied to a specimen. The value of K_{Ic} for a given material can be measured directly using ASTM Standard Test Method E-399 (Ref 2.5), which is discussed in Chapter 3.

Plane-strain fracture toughness, K_{Ic}, is particularly pertinent in materials selection because, unlike other measures of toughness, it is independent of specimen configuration. For comparison, the notch toughness of a material, which is most commonly measured by Charpy testing, does depend on the configuration of the specimen. Changes in the size of the specimen or in the root radius of the notch will affect the amount of energy absorbed in a Charpy test. The main reason for this is that the total energy required for initiation of the crack from the notch, for propagation of the crack across the specimen, and for complete fracture of the specimen is measured in a Charpy test. In contrast, a K_{Ic} test measures only the critical load required for a small extension of a pre-existing crack. Even though K_{Ic} is more difficult to measure than notch toughness, because of the requirements of a pre-existing crack and a specimen large enough for plane-strain conditions, it is a constant material property and can be more generally applied to materials selection.

In selection of structural materials, the single most important characteristic of K_{Ic} for nearly all materials is that it varies *inversely* with yield strength. This is shown schematically in Fig. 2.8 using typical values from the published literature for high-strength steels. Specific values will be shown in Chapter 4. These data

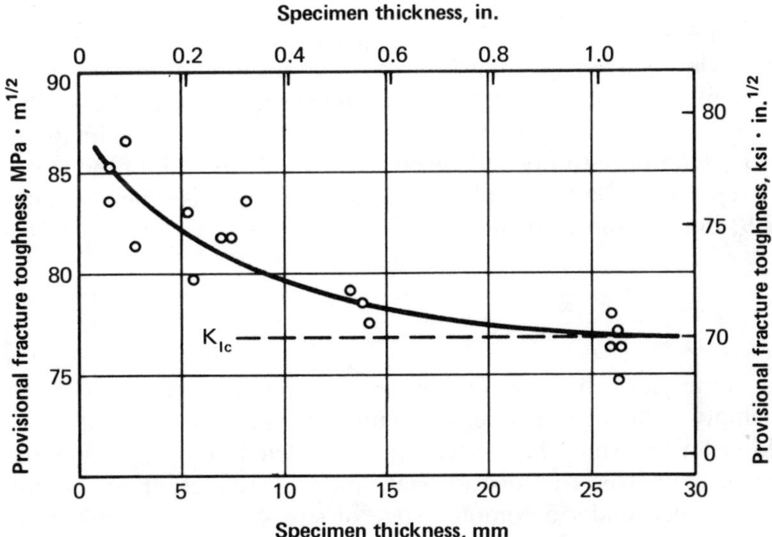

The material was 4340 steel plate that had been hardened and tempered at 400°C (750°F) to a yield strength of 1470 MPa (213 ksi).

Fig. 2.7. Effect of specimen thickness on the critical K value for crack extension in steel specimens (Ref 2.4)

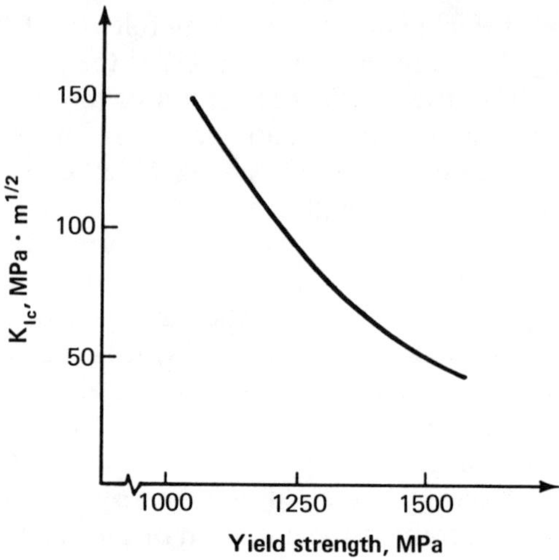

Fig. 2.8. Schematic illustration of the variation of K_{Ic} with yield strength for high-strength steels

for steels are representative of the significant decrease in fracture toughness which occurs with an increase in yield strength for nearly all structural materials. It is easy to see that if materials for structural components are selected with only yield strength in mind, then the materials selector may be inviting the occurrence of a brittle, catastrophic fracture.

2.4. PLASTIC ZONE

As mentioned in section 2.3.2, when the stresses get sufficiently high near the crack tip, plastic deformation must occur. A description of the size of the plastic zone is essential to the understanding of how section size affects the fracture behavior of a specimen or component. The simplest concept is to imagine the point at which σ_y, according to Eq 2.9, reaches the yield strength of the material. This value of r represents the radius of the plastic zone. This is schematically depicted in Fig. 2.9; in equation form it is:

$$r_p\big|_{\theta=0^\circ} = \frac{K_I^2}{2\pi\sigma^2_{ys}} \qquad \text{(Eq 2.11)}$$

where σ_{ys} is the yield strength. Of course, material at the tip of a crack does not respond simply to the σ_y stress but responds to the complex triaxial state of stress. One yield criterion which has successfully modeled complex states of stress in metals is the von Mises or distortional strain energy criterion. Simply stated, yielding is reached under a complex state of stress when the sum of the squares of the principal stress differences is equal to two times the uniaxial tensile yield squared, or

$$2\sigma^2_{ys} = (\sigma_1 - \sigma_2)^2 + (\sigma_2 - \sigma_3)^2 + (\sigma_3 - \sigma_1)^2 \qquad \text{(Eq 2.12)}$$

From linear elastic theory, all of the values of principal stress may be determined, which for mode I are:

$$\sigma_1 = \frac{K_I}{\sqrt{2\pi r}} \cos\frac{\theta}{2}\left[1 + \sin\frac{\theta}{2}\right] \qquad \text{(Eq 2.13a)}$$

$$\sigma_2 = \frac{K_I}{\sqrt{2\pi r}} \cos\frac{\theta}{2}\left[1 - \sin\frac{\theta}{2}\right] \qquad \text{(Eq 2.13b)}$$

$$\sigma_3 = 2\nu\frac{K_I}{\sqrt{2\pi r}} \cos\frac{\theta}{2} \text{ (plane strain)} \qquad \text{(Eq 2.13c)}$$

$$\sigma_3 = 0 \text{ (plane stress)} \qquad \text{(Eq 2.13d)}$$

where ν is Poisson's ratio. Combining Eq 2.12 and 2.13 gives the elastic-plastic boundary by the inverse method. Although only a first approximation, this approach illustrates the differences between thin-section and thick-section behavior. For plane stress, which would apply to relatively thin sections having $\sigma_3 = 0$ (stress is zero in the through-thickness direction),

$$r_p \Big|_{\text{plane stress}} \simeq \frac{K_I^2}{2\pi\sigma^2_{ys}} \cos^2\left(\frac{\theta}{2}\right)\left\{1 + 3\sin^2\left(\frac{\theta}{2}\right)\right\} \approx \frac{K_I^2}{2\pi\sigma^2_{ys}} \qquad \text{(Eq 2.14)}$$

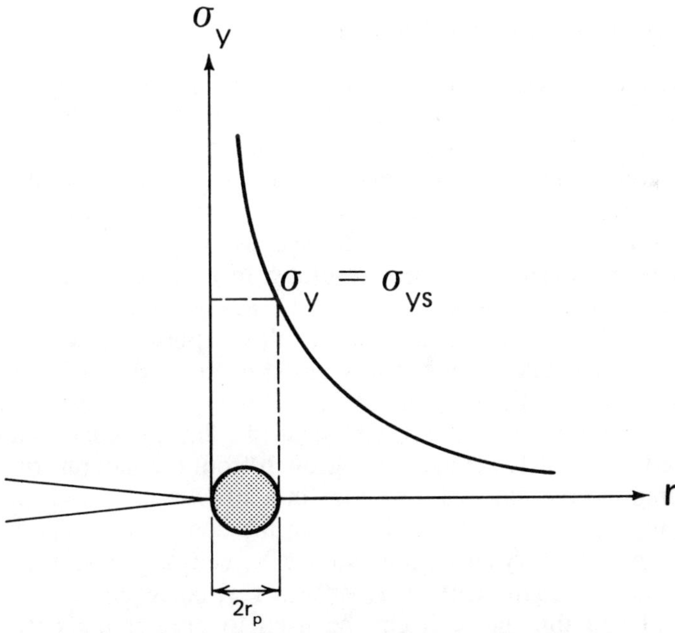

σ_y is the stress perpendicular to the crack plane; σ_{ys} is the yield strength; r is the distance from the crack tip; r_p is the radius of the plastic zone.

Fig. 2.9. Schematic illustration of plastic zone and stress distribution at a crack tip

which is equivalent to Eq 2.11 for $\theta = 0°$. However, for thick plates where plane-strain conditions prevail, σ_3 is as given in Eq 2.13, and now Eq 2.12 and 2.13 give:

$$r_p \Big|_{\text{plane strain}} \simeq \frac{K_I^2}{2\pi\sigma_{ys}^2} \cos^2\left(\frac{\theta}{2}\right)\left\{1 + 3\sin^2\left(\frac{\theta}{2}\right) - 4\nu(1-\nu)\right\} \approx \frac{K_I^2}{6\pi^2\sigma_{ys}} \quad \text{(Eq 2.15)}$$

For a Poisson's ratio of $\nu = 0.3$, the average plane-strain plastic-zone size is about ⅓ the plane-stress value of Eq 2.14, considering all values of θ. The important point is that yielding in thick plates is more difficult, the plastic zones are smaller, and hence the energy absorbed around the crack is less.

2.5. ADDITIONAL APPLICATIONS AND LIMITATIONS

Fracture mechanics is generally applied to describe the crack-growth process for a number of test and service conditions in addition to those of plane-strain, K_{Ic}-type fracture. The main additional categories of crack growth, as discussed in Chapter 1, are elastic-plastic crack growth, fatigue crack growth, and crack growth as affected by testing environment in the general sense—that is, chemical environment, temperature and loading rate. The application of fracture mechanics to describe crack growth under these conditions and the limitations involved are summarized in this section.

2.5.1. Elastic-Plastic Crack Growth

For cracked and loaded specimens of relatively low-strength materials and relatively small section sizes, the size of the plastic zone at the crack tip can be significant relative to the section size. When this occurs, the assumption that the elastic stress around the crack controls the fracture process is no longer valid, and the conditions around the crack represent plane stress (strain is nonzero in the z-direction). Furthermore, crack growth generally occurs with a shear-lip type of deformation near external surfaces, resulting in a critical value of K which is higher than that for elastic-stress-controlled crack growth.

A standardized method for describing elastic-plastic fracture involves the resistance-curve or R-curve concept discussed in Ref 2.6 and 2.7 and in Chapter 3. Briefly, the resistance-curve concept involves measurement of the K values at which various amounts of crack growth occur in a thin-plate laboratory specimen. Then a plotted curve of K versus crack growth from the laboratory specimen can be used to predict crack-growth behavior in a structural component of the same material. Limitations of the method are that the component must have the same thickness as the laboratory specimen and that K relations must be known for both component and specimen. However, once a resistance curve is obtained for a given material and thickness, it can be used to predict the crack-growth and crack-instability behavior of other components of the same material.

Another concept for use in the analysis of elastic-plastic fracture is the J-integral concept (Ref 2.8). In brief statement, J is a nonlinear generalization of G, the elastic strain energy release rate. So J can be thought of as the amount of elastic-plastic strain energy per unit area of crack growth which is applied toward extending the crack in a specimen under load. A critical value of J, called J_{Ic}, is the value required for the start of crack extension from a pre-existing crack. For material having a sufficiently high yield strength or for specimens of sufficient size, elastic stresses control the crack extension, and J_{Ic} is equal to G_{Ic}.

The great advantage of the J_{Ic} approach is that it makes possible the prediction of the failure load of a cracked component or the measurement of the fracture toughness of a material even when there is significant plastic deformation present in the component or material sample. This is an important advantage because, in recent years, high-toughness, medium-strength alloys (with more tendency toward plastic deformation) have been used in place of high-strength, low-toughness alloys for many fracture-critical applications. Details of the J_{Ic} method are described in Ref 2.9 and in Chapter 3. As with K_{Ic}, the J_{Ic} method uses precracked specimens of fixed geometry and of a minimum required size. The size requirement allows the use of smaller specimens than for the K_{Ic} test, so fracture toughness measurements are possible using J_{Ic} for materials and sizes which are outside the range of the K_{Ic} method.

2.5.2. Fatigue Crack Growth

Crack growth can occur at K levels much below K_{Ic} in any structural alloy when cyclic loading is applied. In simplified concept, it is the accumulation of damage from the cyclic plastic deformation in a small zone at the crack tip which accounts for fatigue crack growth at K levels much below K_{Ic}. The concentration of damage is increased further when strain hardening occurs and causes the zone of reversed plastic deformation to be smaller than the zone formed during one application of load.

The general nature of fatigue crack growth and its description using fracture mechanics can be briefly summarized by the example data shown in Fig. 2.10. This figure, based on the work of Paris *et al* (Ref 2.10) shows a logarithmic plot of the crack growth per cycle, da/dN, versus the stress-intensity factor range, ΔK, corresponding to the load cycle applied to a sample. The da/dN versus ΔK plot shown is from five specimens of ASTM A533 B-1 steel tested at 24 °C (75 °F). A plot of similar shape is expected with most structural alloys; the absolute values of da/dN and ΔK are dependent on the material. Results of fatigue crack growth rate tests for nearly all metallic structural materials have shown that the da/dN versus ΔK curves have the following characteristics: (*a*) a region at low values of da/dN and ΔK in which fatigue cracks grow extremely slowly or not at all below a lower limit of ΔK called the threshold of ΔK, ΔK_{th}; (*b*) an intermediate region of power-law behavior described by the Paris equation (Ref 2.11):

$$\frac{da}{dN} = C(\Delta K)^n \qquad \text{(Eq 2.16)}$$

where C and n are material constants; and (*c*) an upper region of rapid, unstable crack growth with an upper limit of ΔK which corresponds either to K_{Ic} or to gross plastic deformation of the specimen.

The fracture mechanics description of fatigue crack growth as summarized in Fig. 2.10—particularly the power-law description of behavior—is broadly applied to calculate growth rate and fatigue life of components and structures. The

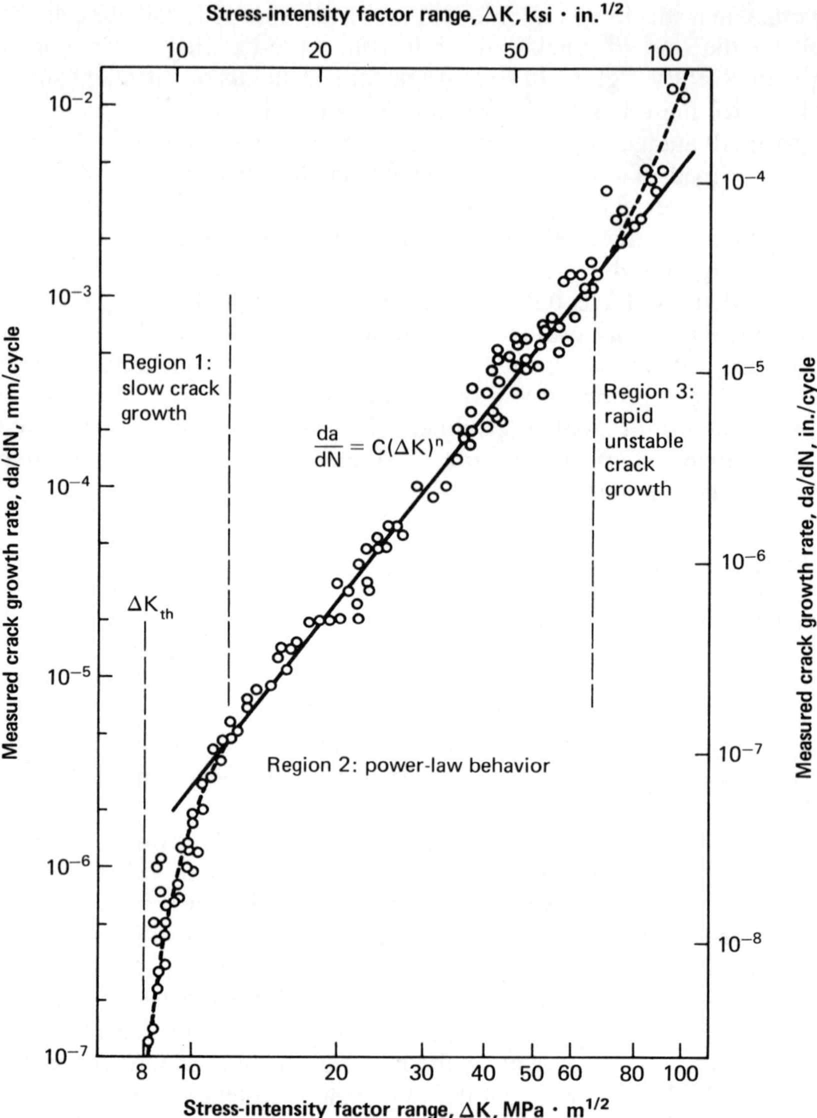

The material was ASTM A533 B-1 steel, with a yield strength of 470 MPa (70 ksi). Test conditions: R = 0.10; ambient room air; 24°C (75°F).

Fig. 2.10. Fatigue crack growth behavior of A533 steel (Ref 2.10)

general limitation of this application is the obvious one, that the power-law representation applies only in the intermediate region of da/dN and ΔK. It is particularly important that the power-law relation not be extended to higher values of da/dN and ΔK than those supported by laboratory test data, because an underestimate of da/dN, and thus an overestimate of fatigue life, would result. To be sure that the power-law relation is not extended to excessive values of ΔK, both K_{Ic}-type failure and gross yielding failure of the component must be considered.

Generally, for small components and for low-strength materials, gross yielding of the component will limit the range of a power-law description of da/dN; for large components and for high-strength materials, plane-strain K_{Ic}-type fracture will limit the range of a power-law description.

A standardized method for measuring fatigue crack growth rates is described in ASTM Test Method E-647 (Ref 2.12), which is discussed in Chapter 3.

2.5.3. Effects of Testing Environment on Crack Growth

There are countless combinations of physical and chemical testing environments which can affect crack growth in structural alloys. For a few conditions, an increase in the K value is required for crack growth at a designated rate, but generally the K value is decreased when specimens are exposed to service environments other than normal laboratory air. The mechanisms of crack growth that can occur with various testing environments are too numerous and complex for discussion here. However, the general variations in testing environment can include variations in testing temperature, rate of loading, and chemical environment. For these three types of variations, fracture tests generally can be performed using laboratory specimens of the same or similar types as those used for testing under normal conditions. (For example, refer to the specimens used for measuring K_{Ic} and fatigue crack growth rates described in Chapter 3.)

As with fatigue crack propagation, results of sustained loading may be shown in logarithmic plots of crack growth rate versus applied K_I—except with da/dt being plotted instead of da/dN. Results of sustained-load tests on specimens of aluminum alloy 7075-T6 alternately immersed in an aqueous solution of 3.5% NaCl to develop stress-corrosion cracking are shown in Fig. 2.11. Both crack-mouth-opening-displacement and traveling-microscope techniques were used to measure crack growth rate, because the relatively low velocities permitted direct measurement of crack length during the out-of-solution period. These results are typical, and they show that, as K_I is increased from some threshold value, the subcritical crack velocity increases by several orders of magnitude. A plateau region is then reached where velocity is relatively independent of K_I but dependent on temperature. As might be expected, temperature generally enhances the kinetics of environmental attack and diffusion of embrittling species, so growth rates increase at higher temperatures. Finally, if growth rates had been measured at K_I values closer to K_{Ic}, the crack velocities would have increased dramatically as K_{Ic} was approached.

The major difficulty and limitation of measuring crack growth as affected by testing environment is in accurately producing and maintaining the specified test conditions during the laboratory test. The proper use of high and low loading rates, high and low temperature chambers, and chemical environment chambers can greatly increase the time and cost required to obtain laboratory test results. Nevertheless, test results for various service conditions are required in order to apply fracture mechanics to the selection of materials and thereby to avoid the catastrophic failures which can occur in components and structures which are adversely affected by service environments. Typical procedures for application of fracture mechanics to structural materials subjected to special environments are presented in Chapter 9.

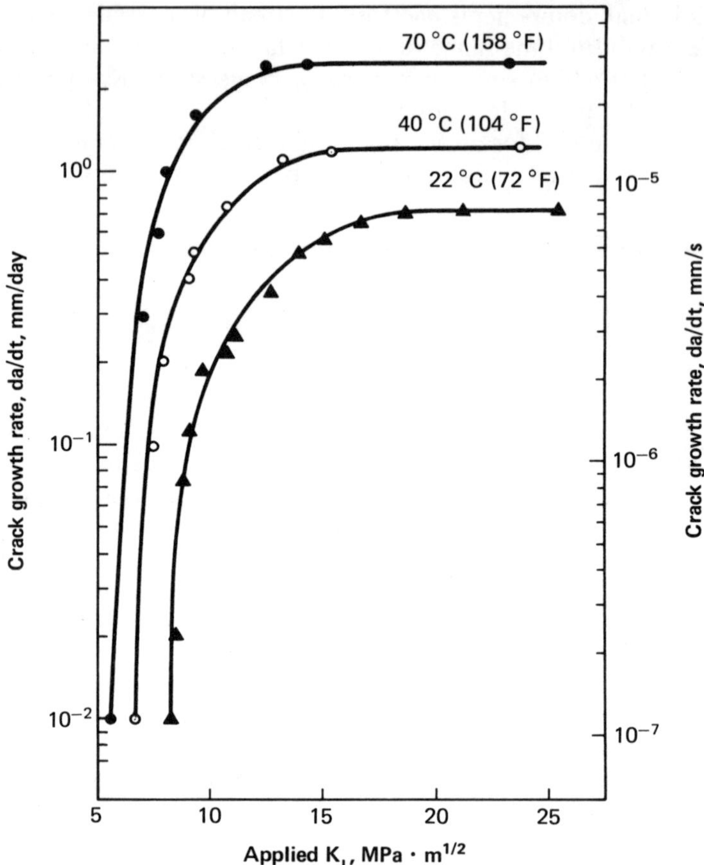

The material was aluminum alloy 7075-T6, with a yield strength of 520 MPa (75 ksi). Test conditions: alternate immersion in 3.5% aqueous solution of NaCl.

Fig. 2.11. Stress corrosion crack growth behavior of aluminum alloy 7075-T6

2.6. REFERENCES

2.1. The Phenomena of Rupture and Flow in Solids, by A. A. Griffith: *Philosophical Transactions*, Vol 221, 1921, p 163-198

2.2. *Compendium of Stress Intensity Factors*, by D. P. Rooke and D. J. Cartwright: Hillingdon Press for Her Majesty's Stationery Office, Uxbridge, Middlesex, UK, 1976

2.3. *The Stress Analysis of Cracks Handbook,* by H. Tada, P. C. Paris and G. R. Irwin: Del Research Corporation, Hellertown, PA, 1973

2.4. The Influence of Crack Length and Thickness in Plane Strain Fracture Toughness Tests, by M. H. Jones and W. F. Brown, Jr.: in *Review of Developments in Plane Strain Fracture Toughness Testing,* STP 463, American Society for Testing and Materials, Philadelphia, 1970, p 63-101

2.5. Standard Test Method for Plane-Strain Fracture Toughness of Metallic Materials:

E399-81, *1981 Annual Book of ASTM Standards,* Part 10, American Society for Testing and Materials, Philadelphia, 1981, p 588-618

2.6. *Fracture Toughness Evaluation by R-Curve Methods:* STP 527, American Society for Testing and Materials, Philadelphia, 1973

2.7. Standard Recommended Practice for R-Curve Determination: E561-81, *1981 Annual Book of ASTM Standards,* Part 10, American Society for Testing and Materials, Philadelphia, 1981, p 673-692

2.8. Recent Developments in J_{Ic} Testing, by J. D. Landes and J. A. Begley: in *Developments in Fracture Mechanics Test Methods Standardization,* edited by W. F. Brown, Jr., and J. G. Kaufman, STP 632, American Society for Testing and Materials, Philadelphia, 1977, p 57-81

2.9. A Procedure for the Determination of Ductile Fracture Toughness Values Using J Integral Techniques, by G. A. Clarke, W. R. Andrews, J. A. Begley, J. K. Donald, G. T. Embley, J. D. Landes, D. E. McCabe and J. H. Underwood: *Journal of Testing and Evaluation*, Vol 7, Jan 1979, p. 49-56

2.10. Extensive Study of Low Fatigue Crack Growth Rates in A533 and A508 Steels, by P. C. Paris, R. J. Bucci, E. J. Wessel, W. R. Clark and T. R. Mager: in *Stress Analysis and Growth of Cracks*, Proceedings of the 1971 National Symposium on Fracture Mechanics, Part I, STP 513, American Society for Testing and Materials, Philadelphia, 1972, p 141-176

2.11. A Critical Analysis of Crack Propagation Laws, by P. C. Paris and F. Erdogan: *Journal of Basic Engineering, Transactions of ASME*, Vol 85, Dec 1963, p 528-534

2.12. Standard Test Method for Constant-Load-Amplitude Fatigue Crack Growth Rates Above 10^{-8} m/cycle: E647-81, *1981 Annual Book of ASTM Standards,* Part 10, American Society for Testing and Materials, Philadelphia, 1981, p 765-783

Chapter 3

Fracture Test Methods

J. H. Underwood

3.1. PLANE-STRAIN FRACTURE TOUGHNESS, K_{Ic}

3.1.1. ASTM Method E-399

The first fracture mechanics test method to become widely accepted in the United States is ASTM Method E-399 for measuring plane-strain fracture toughness, K_{Ic}, of metals (Ref 2.5). Because of this, ASTM Method E-399 is perhaps the most important method to consider in the application of fracture mechanics to selection of structural alloys. In addition, many other test methods use some of the same test specimens and test procedures as those in E-399.

Method E-399 first appeared in the 1969 ASTM Standards, and the latest revision was in 1981. As discussed in the previous chapter, ASTM Method E-399 specifies the requirements and procedures for measuring the critical value of stress-intensity factor, K_{Ic}. The measurement corresponds to at most a 2% extension of a pre-existing fatigue crack in a specimen large enough that plane-strain conditions predominate around the crack. Three types of test specimens which can be used with the method are shown in Fig. 3.1. They are, in the order in which they were developed for use with E-399, the bend specimen, the compact specimen and the arc specimen. The specimens may be taken from plate and other product forms in any of six orientations for various crack growth directions. These are shown, using the compact specimen as an example, in Fig. 3.2, which was taken from ASTM Method E-399. For cylindrical products, a similar procedure is used, except that the circumferential and radial (C and R) directions are indicated instead of long transverse and short transverse (T and S) directions.

The critical dimensions for each of the test specimens are thickness, B, width, W, and over-all crack length, a, which includes the machined starter notch and the fatigue precrack. One of the important requirements of this test method is that the specimen thickness, B, and the crack length, a, be at least equal to the quantity 2.5 $(K_{Ic}/\sigma_{ys})^2$, where σ_{ys} is the yield strength of the material. Thus, it is desirable to determine the yield strength and to estimate the K_{Ic} value for any material to be tested by this method before the specimens are prepared. For some materials,

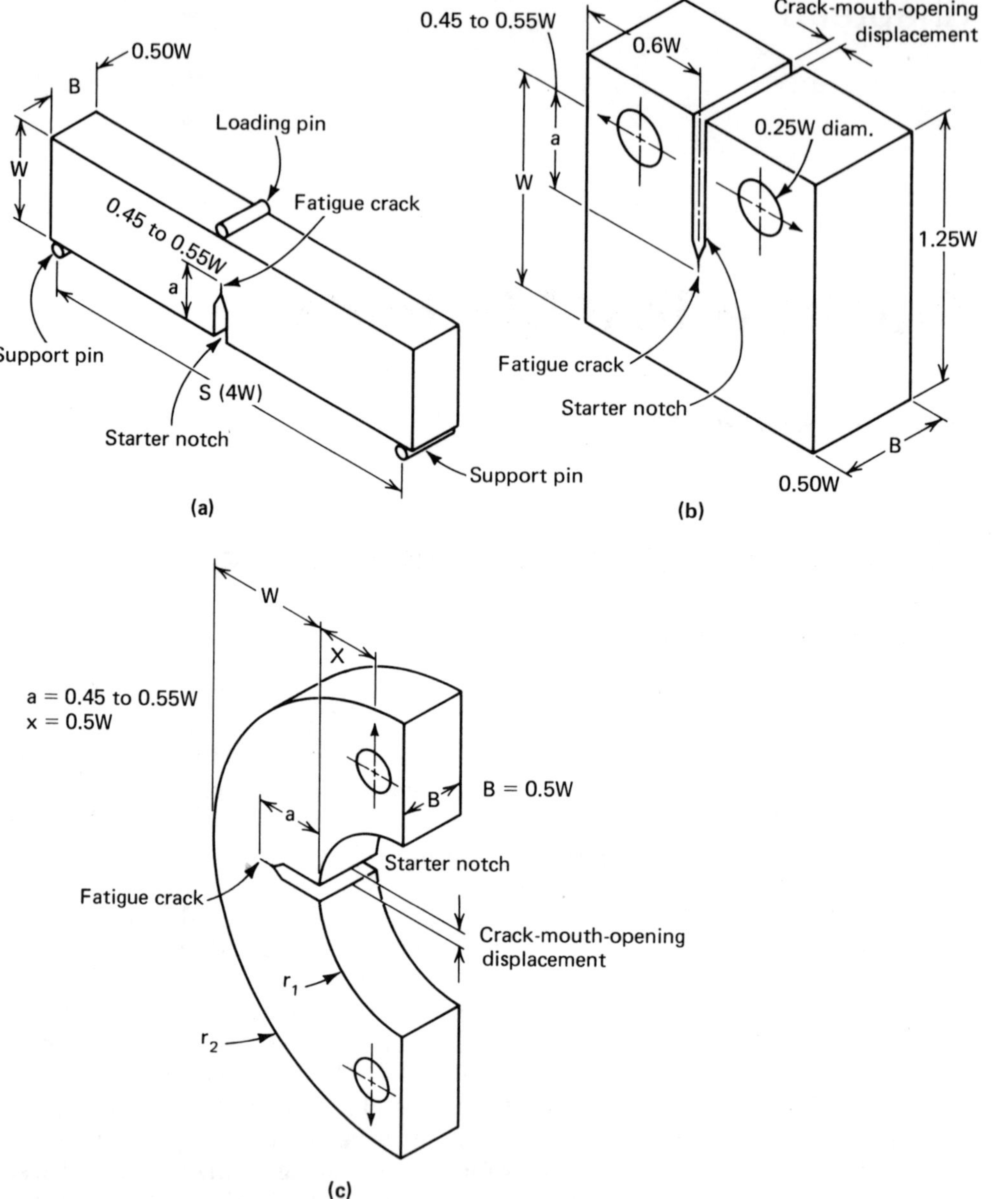

(a) Bend specimen. (b) Compact specimen. (c) Arc specimen. a is notch plus crack length; B is specimen thickness; W is specimen depth.

Fig. 3.1. Fracture toughness test specimens

often those of lower strength, the dimensions required by the relation 2.5 $(K_{Ic}/\sigma_{ys})^2$ are greater than can be obtained from the available section sizes of material. For such material and section size combinations, measurement of K_{Ic} is not possible. In these cases, alternative methods of fracture toughness measurement are necessary, as discussed in Section 3.2. When measurement of K_{Ic} *is*

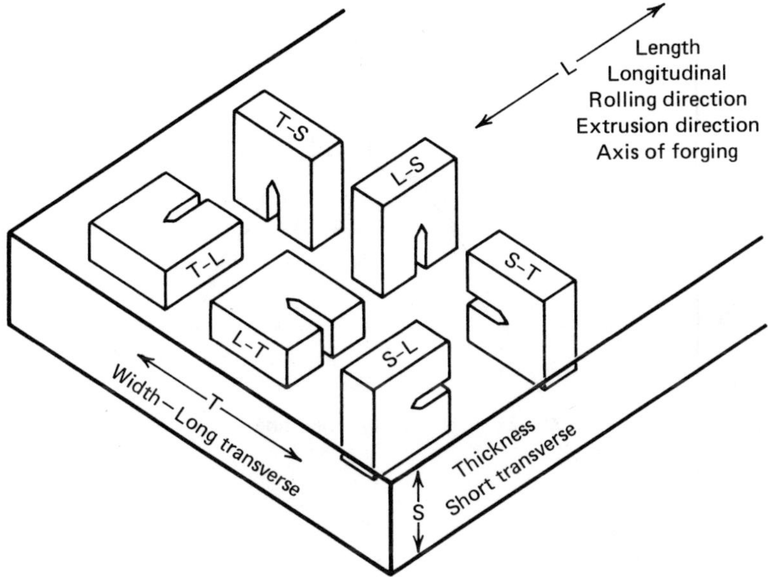

First letter designates the direction perpendicular to the crack plane; second letter designates the direction parallel to the direction of crack growth.

Fig. 3.2. Orientations of crack plane and direction for fracture specimens taken from rectangular product forms such as plate (from ASTM Method E399)

possible, the K_{Ic} specimen dimensions should be somewhat greater than the minimum requirements estimated from 2.5 $(K_{Ic}/\sigma_{ys})^2$. For a few materials, notably some aluminum alloys, recent testing (Ref 3.1) has shown that dimensions about twice as great as those normally required give more consistent K_{Ic} results.

Another important set of requirements for a valid K_{Ic} test involves the fatigue precracking process. The test results will be valid only if the length of the fatigue crack and the straightness and flatness of the crack are within the prescribed limits. Furthermore, the maximum cyclic load used in producing the fatigue crack must be held below a designated limit to restrict the size of the plastic zone ahead of the fatigue crack.

Once the test specimen has been precracked, the procedure of the K_{Ic} test itself is quite similar to the load-versus-displacement procedure of a standard tension test. The displacement used in the K_{Ic} test is the opening displacement of the notch surfaces at the notch mouth and in the direction perpendicular to the plane of the notch and crack. This displacement is called the crack-mouth-opening displacement (see Fig. 3.1). A calibrated displacement gage and autographic load-versus-displacement recording equipment are needed to measure and record the test data. Special fixtures with loading pins which are free to rotate during the test are required. This ensures that the same free-rotation loading condition used for the stress and K analysis of the specimens is also present during the test. A load-versus-displacement plot for a high-strength aluminum alloy, which is generally representative of the plots from many steel and aluminum alloys, is shown in Fig. 3.3 (Ref 3.2).

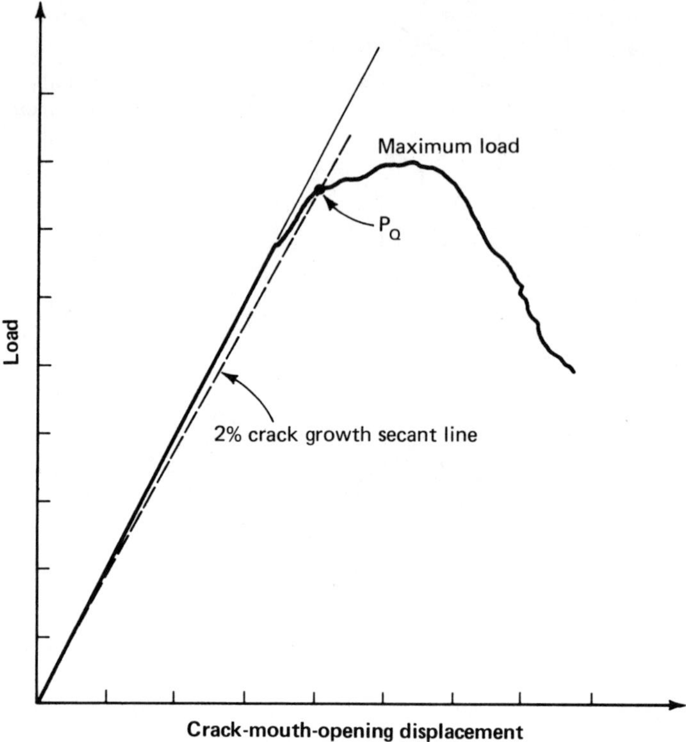

Fig. 3.3. Load vs crack-mouth-opening displacement typical of high-strength aluminum alloys (Ref 3.2)

Interpretation of the load-versus-displacement plot and calculation of K_{Ic} is described in detail for the various specimen types in ASTM E-399. The procedures are described briefly here for the compact specimen. The provisional value of the load at which onset of crack propagation occurs, P_Q, is obtained from the test record. The P_Q value is either the maximum load during the test or, more commonly and as shown in Fig. 3.3, the load corresponding to 2% crack growth as determined by the intersection of a secant line and the plot. The provisional value of the stress-intensity factor, K_Q, is determined by means of equations such as the following, for compact specimens in which K_Q is in units of MPa · $m^{1/2}$ (psi · in.$^{1/2}$):

$$K_Q = (P_Q/BW^{1/2}) \cdot f(a/W) \qquad \text{(Eq 3.1)}$$

where

$$f(a/W) = \frac{[2 + a/W][0.886 + 4.64a/W - 13.32(a/W)^2 + 14.72(a/W)^3 - 5.6(a/W)^4]}{(1 - a/W)^{3/2}}$$

and where P_Q is load, in MN (lbf); B is specimen thickness, in m (in.); W is specimen width, in m (in.); and a is crack length, in m (in.). It is important to note that, for the compact specimen, the crack length is measured from the centerline of the loading-pin holes. Equation 3.1 is considered to be accurate within 0.5% over a wide range of crack lengths, for $0.2 \leq a/W < 1$, so it can be used for a variety of fracture mechanics tests and analyses as well as in the range of interest for K_{Ic} tests, $0.45 \leq a/W \leq 0.55$.

A K_Q value calculated from Eq 3.1 must meet several criteria in order to be a valid measurement of K_{Ic}. The first, mentioned earlier, is that the specimen thickness and over-all crack length must be equal to or greater than $2.5\,(K_{Ic}/\sigma_{ys})^2$. Also, the maximum load supported by the specimen prior to complete fracture must not be more than 10% greater than P_Q. This further ensures that P_Q corresponds to the load at which crack extension takes place rather than the load at which excessive plastic deformation occurs around the crack. The fatigue crack portion of the fracture surface is examined and measured as designated in the test method, in regard to straightness of the crack front and other criteria.

Because of the complexity of the calculations and the requirements for validity, some of the laboratories in which these tests are run have developed computer programs which compute K_Q values and indicate their validity based on appropriate input data. The complexity of the K_{Ic} test method cannot be denied. Nevertheless, the method has been applied to a wide variety of metallic materials in the past ten years, and very similar methods have been adopted by many other standards organizations around the world. In spite of its complexity, ASTM Method E-399 is the preferred standard for the measurement of plane-strain fracture toughness, K_{Ic}.

3.1.2. New K_{Ic} Tests

A number of test specimens and methods have been proposed as new K_{Ic} procedures. Three new procedures are described here—two that are related primarily to new specimen geometries and a third that is primarily a new method with existing specimen geometries. These procedures are now being generally used to some extent and are in various stages of consideration by ASTM as K_{Ic} test methods.

A test specimen that is much like the compact specimen, but that is round rather than rectangular in shape, has been used in the United States since 1975. Recently, a round compact specimen and its associated K solution have been added to ASTM Method E-399 (Ref 3.3) and is referred to as the disk specimen. The specimen geometry is sketched in Fig. 3.4. The K solution is similar both in form and in value (within ±6%) to that of the rectangular compact specimen in the K_{Ic} testing range. The advantage of the disk specimen is that it can be less costly to fabricate because turning operations are often faster than milling. In addition, when round cores are cut from structures or product forms for use in K_{Ic} tests, the disk specimen has a clear advantage.

Another test specimen which has been proposed for use in K_{Ic} measurements is the short rod specimen (Ref 3.4). The specimen geometry is shown in Fig. 3.5. The significant advantage in the proposed use of this specimen is that, due to the triangular shape of the area to be cracked, a crack initiates and grows stably during a single application of a relatively low load. This could make possible K_{Ic} tests without the necessity of a fatigue precrack. However, because cracks which have been produced by a single application of load may be sufficiently different from

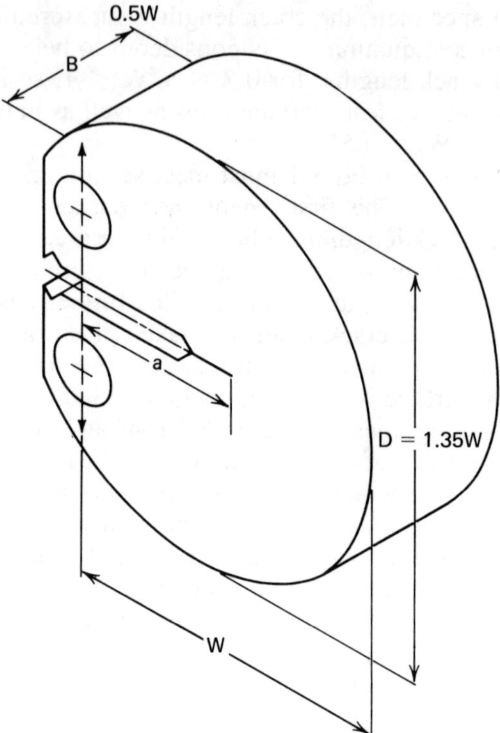

Fig. 3.4. A round compact K_{Ic} test specimen; the disk specimen

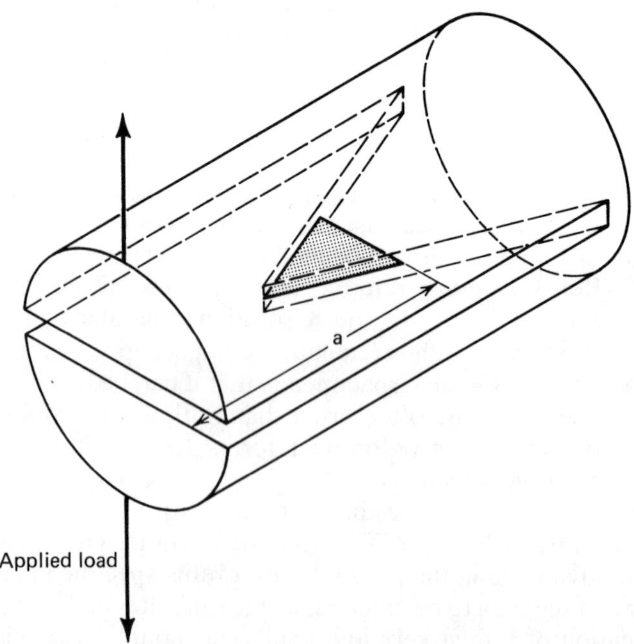

a is notch plus crack length; shaded area indicates crack growth.

Fig. 3.5. Proposed short rod fracture toughness specimen

cracks produced by fatigue loading, the resulting K values may not be consistent with those obtained on fatigue-precracked specimens. Further evaluation is needed to determine whether the short rod specimens qualify for plane-strain fracture toughness testing or for screening tests.

For many years, various laboratories have performed K_{Ic}-type tests at significantly higher loading rates than those specified in E-399. Reference 3.5 describes one example of high-loading-rate K_{Ic} tests on specimens of an alloy steel. A K_{Ic} test procedure based on E-399 but at much higher loading rates is being considered by ASTM. Interlaboratory tests have been completed using a pressure-vessel steel tested in closed-loop machines at loading rates in the range of 10^4 MPa · $m^{1/2}$/s; this is about a factor of 10^4 faster than the essentially static loading rate of E-399. Compact specimens (Fig. 3.1b) and many guidelines presented in E-399 are generally used in obtaining fracture toughness data by dynamic K_{Ic} tests. The primary application of dynamic K_{Ic} tests is with the low-to-medium-strength steels, which are known to show loading-rate effects on mechanical properties, particularly yield strength. Problems with the development of the dynamic K_{Ic} test method are related to uncertainties in the yield strength of materials at high loading rates and uncertainties in load measurement at high loading rates.

3.2. NON-PLANE-STRAIN TOUGHNESS TESTS

3.2.1. R-Curve Determination

A consensus R-curve test method is described in ASTM Recommended Practice E-561 (Ref 2.7), first published in 1974 and last revised in 1981. Method E-561 describes the determination of the crack growth resistance of a material under relatively thin-sheet plane-stress conditions which result in a significant amount of plastic deformation at the crack tip. The practice describes the use of center-cracked tensile specimens of the same general type shown in Fig. 2.1, and compact-type specimens similar to those in Fig. 3.1(b) except generally much thinner. The basic procedures of the R-curve method can be well described by referring to a sketch of typical measurements and calculations from the method using a center-cracked sheet specimen as the example. This sketch, shown in Fig 3.6, has coordinates of stress-intensity factor, K, and crack length, a. The solid curve in Fig. 3.6, marked K_R, is a plot of K applied to the specimen versus the measured crack extension, Δa, produced by the applied K; thus the K_R curve represents the measured crack-extension *resistance* of the material for the particular specimen thickness of the test. The dashed curves are calculations of crack-extension *force* which is available to cause extension of the crack at a given load level. When the applied load is sufficiently high (curve P_4), the crack-extension force exceeds the crack-extension resistance of the material and the crack will grow. Moreover, since the slope of the crack-extension force curve is larger than that of the K_R curve, the crack will become unstable and grow to failure unless the load is decreased. The K value at this point, K_c, is the plane-stress fracture toughness, at which unstable crack growth occurs for this particular combination of material, thickness and other specimen dimensions. The K_R curve is unique to the material and specimen thickness, whereas K_c values may depend on other specimen dimensions in addition to thickness.

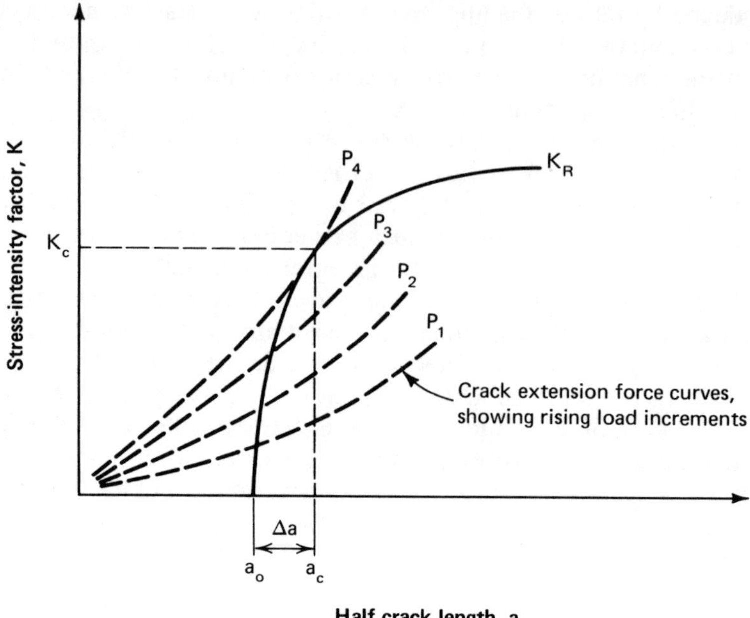

Fig. 3.6. Typical R-curve test results and analysis (from ASTM Practice E561)

The measured data required to plot the K_R curve—that is, load and crack extension—can often be obtained using the same equipment and procedures as those used in K_{Ic} tests. The calculations required to plot the crack-extension force curves are based on the K solution for the specimen, such as Eq 3.1; the only special requirement is that an adjustment is added to the actual physical crack length, a_p, to account for the effective extension of the crack due to a crack-tip plastic zone. Thus, the effective crack size is:

$$a_e = a_p + \frac{K^2}{2\pi\sigma_Y} \qquad \text{(Eq 3.2)}$$

where σ_Y is the effective yield strength—that is, the average of the yield and ultimate strengths of the material. Note that the estimate of plastic zone size from Eq 2.14 is used here.

Once a K_R curve is obtained for a given material and thickness, it can be used to predict the load at which unstable crack extension will occur in any other geometry, providing that the K solution is known so that the crack-extension curve can be calculated. The K_R curve approach accounts for crack-tip plastic deformation as discussed above, with the general limitation that the specimen must remain predominantly elastic. Additional information on R-curve methods and their use is available in Ref 2.6 as well as in ASTM Practice E-561 (Ref 2.7).

3.2.2. J_{Ic} Fracture Toughness

The development of elastic-plastic fracture toughness tests based on the J-integral concept followed soon after the analyses and estimation of J by Rice,

Paris and Merkle (Ref 3.6) for specimen geometries commonly used in fracture toughness testing. ASTM has sponsored two cooperative J_{Ic} testing programs, and, based on results from these programs, a proposed J_{Ic} test method has been described in the literature (Ref 2.9) and has been published as an ASTM standard (Ref. 3.7).

The J_{Ic} test method can be outlined and described in relation to the sketch of typical results in Fig. 3.7. Values of J are calculated from load versus load-point-displacement data obtained from bend or compact specimens using generally the same specimens, procedures and equipment as in E-399. A key requirement is that the displacement must be at the load point; this is allowed but not required in a K_{Ic} test. With this requirement, the area A under the load-versus-displacement curve is a true measure of the combined elastic and plastic strain energy input to the specimen. The J value is calculated from

$$J = \frac{2A}{B(W\text{-}a)} \qquad \text{(Eq 3.3)}$$

which applies directly for the bend specimen and with some modification for the compact specimen. The dimensions B, W, and a are as described in Fig. 3.1. At least four specimens are loaded to produce a range of crack extension, Δa, of about 1 to 2 mm. The value of J calculated from each specimen is plotted versus Δa, and a linear regression line is fitted to the data. This is the J_R curve as represented by a regression line (see Fig. 3.7). The blunting line, calculated from $J = 2\Delta a\sigma_Y$, is drawn to represent the amount of apparent crack extension associated with the crack-tip blunting which occurs before actual crack extension occurs. The value of J_{Ic} is determined at the intersection of the J_R curve regression line and the blunting line; so J_{Ic} is a measure of the fracture toughness of the material at the start of actual crack extension from the blunted crack tip.

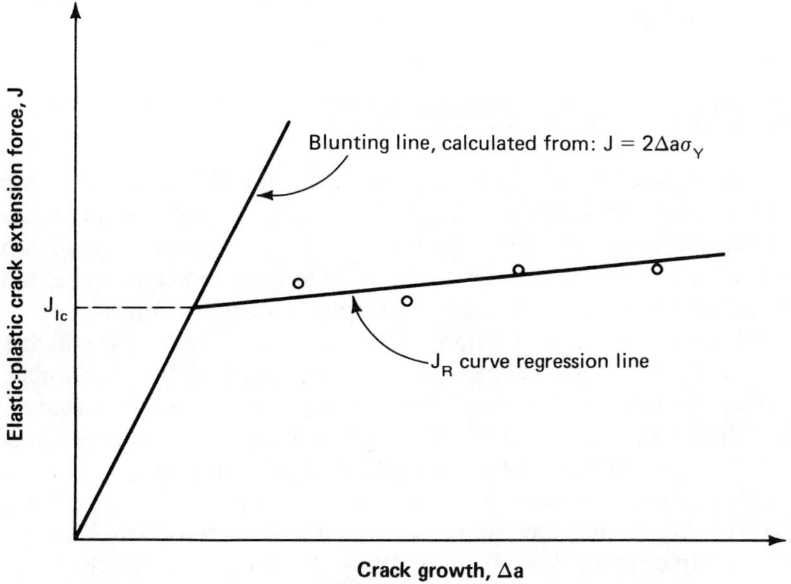

Fig. 3.7. Typical J_{Ic} test results and analysis (Ref 2.9)

An important advantage of the J_{Ic} test method is that it can accommodate a significant amount of crack-tip blunting and general plastic deformation in the specimen (see Section 2.5.1). If the amount of plastic deformation is small enough, J_{Ic} will be identical to G_{Ic}, and thus J_{Ic} can be converted to an approximately equivalent measure of K_{Ic} (see Eq 2.10). For large amounts of plastic deformation, a specimen size requirement limits the size of the specimen and, indirectly, the amount of plastic deformation which can be allowed. The specimen size requirement allows a significantly smaller specimen, often ten times smaller, to be tested with the J_{Ic} procedure than with the K_{Ic} procedure. So, although the J_{Ic} test is relatively time consuming due to multiple tests, it can be used over a wider range of material properties and specimen sizes than the K_{Ic} test. In addition, single-specimen J_{Ic} test procedures, such as incremental unloading methods, can reduce both testing time and the required number of specimens in obtaining J_{Ic} test data.

3.2.3. COD Methods

A British test method, "Methods for Crack Opening Displacement (COD) Testing" (Ref 3.8), is generally similar to the ASTM K_{Ic} method. In this method, a bend specimen is used as in Fig. 3.1(a) and a load versus crack-mouth-opening-displacement (CMOD) plot is obtained as in Fig. 3.3. Then, by use of results from experiments and analyses, the displacement at the crack tip, called the crack-tip-opening displacement, or CTOD, is calculated from CMOD. Critical values of CTOD are defined including values corresponding to elastic, plane-strain conditions and values corresponding to significant amounts of plastic deformation. Methods which determine CTOD have the advantage of concentrating on the area in which the actual crack extension process occurs; however, CTOD methods have the disadvantage of being indirect measurements of the parameter of interest—that is, the opening of the crack faces at the crack tip. Harrison *et al* (Ref 3.9) have described the British COD method and its application to welded structural components.

3.3. CORRELATION FRACTURE TESTS

A number of tests with notched or cracked specimens have been developed for specific engineering applications, in contrast to the tests described in Sections 3.1 and 3.2, which were developed to measure directly the fracture toughness properties of materials. These engineering application tests have been used extensively for correlation and screening in relation to fracture toughness tests.

The most long-standing and probably the most widely used test which is used for correlation with fracture toughness is ASTM Method E-23, "Notched Bar Impact Testing of Metallic Materials" (Ref 3.10), commonly known as the Charpy impact test. The Charpy specimen is a three-point bend-type specimen of one size, with B = W = 10 mm and with a length between support points of 40 mm (see Fig. 3.8). A somewhat similar test and test specimen are described in ASTM Method E-604, "Dynamic Tear Energy of Metallic Materials" (Ref 3.11), which uses a bend specimen with B = 16 mm, W = 38 mm, and a length of 165 mm.

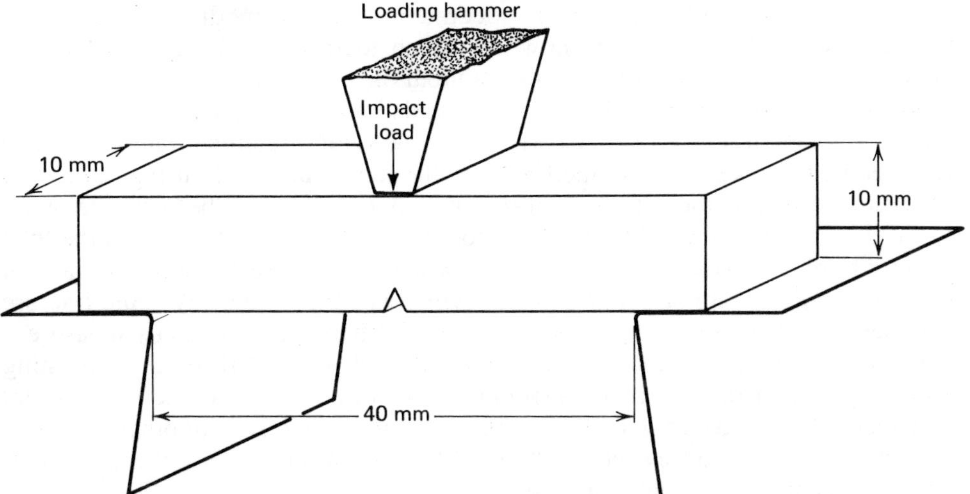

Fig. 3.8. Schematic illustration of Charpy impact specimen and test arrangement

Both of these methods measure the energy required to break the notched bend specimen by impact loading with a falling mass. As discussed previously in Section 2.3.3, the Charpy impact test, as well as the dynamic tear test, measure the total energy required to initiate and then grow a crack to complete failure of the specimen, whereas K_{Ic} is a measure of K for initial growth of a pre-existing crack. This difference, along with the differences in loading rate and specimen size between Charpy impact and K_{Ic} tests, limits the correlations between Charpy energy and K_{Ic} to certain materials and to certain ranges of Charpy energy and K_{Ic} values. However, the use of Charpy impact tests to correlate with K_{Ic} has resulted in significant time and cost savings, particularly with medium-to-high strength steels. Many structural components are purchased and inspected to a Charpy energy specification which has been shown to correspond in laboratory testing to a certain required level of K_{Ic}. However, errors could result if the correlation between Charpy energy and K_{Ic} is extended beyond the range of laboratory data for which it was established.

Two tests which are used for correlation with K_{Ic}, generally with high-strength aluminum alloys, are ASTM Method E-338, "Sharp-Notch Tension Testing of High-Strength Sheet Materials" (Ref 3.12), and ASTM Method E-602, "Sharp-Notch Tension Testing with Cylindrical Specimens" (Ref 3.13). Method E-338 describes tests on center-notched and edge-notched sheet specimens; Method E-602 describes tests on circumferentially notched cylindrical specimens. Both methods require a maximum radius of 0.018 mm (0.0007 in.) at the notch, and this effectively limits these methods to aluminum and magnesium alloys, which are readily machinable. The sharp-notch tensile-strength parameters are used extensively in the aluminum industry to screen materials in respect to their ability to meet corresponding K_{Ic} requirements. The test results are very sensitive to notch sharpness, and thus machining and inspection of the notch must be done carefully to ensure proper correlation.

Two tests that employ the Charpy specimen (Fig. 3.8) modified by a fatigue precrack ahead of the notch are now being considered by ASTM. They are the proposed "Method for Nominal Crack Strength of Slow-Bend Precracked Charpy Specimens of High-Strength Metallic Materials" and the proposed "Method for Impact Testing of Precracked Charpy Specimens of Metallic Materials." The slow-bend precracked Charpy method uses the maximum load during the test to calculate a nominal crack strength of the specimen which may be correlated with K_{Ic}. The impact precracked Charpy method uses an instrumented tup on the load hammer to obtain a load-versus-time plot which, if it meets certain criteria, can be used to calculate a maximum K value which is a measure of dynamic fracture toughness. Both of these proposed precracked Charpy test methods measure a maximum stress or K value associated with crack growth from a pre-existing crack; this type of measurement is similar to that of a K_{Ic} test, so the results from these methods should correlate better with K_{Ic} than do the results of other methods. This advantage is offset by the increased difficulty in precracking the specimens and in performing and analyzing the tests.

A review of the correlation fracture tests described above, as well as others, is given in a report of the National Materials Advisory Board (Ref 3.14).

3.4. FATIGUE CRACK GROWTH RATE TEST METHODS

Testing procedures for measuring fatigue crack growth rates are described in ASTM Method E-647 (Ref 2.12). This method applies to medium-to-high crack growth rates, that is, above 10^{-8} m/cycle (3.9×10^{-7} in./cycle). Procedures for growth rates below 10^{-8} m/cycle are under consideration by ASTM. For applications involving fatigue lives of up to about 10^6 load cycles, the procedures of E-647 can be used. Fatigue lives greater than about 10^6 cycles correspond to growth rates below 10^{-8} m/cycle, and these require special testing procedures, which are related to the threshold of fatigue growth described in Section 2.5.2 and illustrated in Fig. 2.10.

ASTM Method E-647 describes the use of center-cracked specimens (Fig. 2.1) and compact specimens (Fig. 3.1b). The specimen orientation designation of Fig. 3.2 is used. The specimen thickness-to-width ratio, B/W, is smaller than the 0.5 value for K_{Ic} tests; the maximum B/W values for center-cracked and compact specimens are 0.125 and 0.25, respectively. With the thinner specimens, it is feasible to use crack length measurements on the sides of the specimens as representations of through-thickness crack growth behavior. The specimens are loaded in the same general manner as for K_{Ic} testing. For tension-tension fatigue loading, the K_{Ic} loading fixtures often can be used. For this type of loading, both the maximum and minimum loads are tensile and the load ratio, $R = P_{min}/P_{max}$, is in the range $0 < R < 1$. A ratio of $R = 0.1$ is commonly used. Tension-compression loading can be performed with the compact specimen, but it is a more complex type of loading and requires more care.

Testing normally is performed in laboratory air at room temperature; however, any gaseous or liquid environment and temperature of interest may be used in order to determine the effect of corrosion or other chemical reaction on cyclic loading. Cyclic loading may involve various wave forms for constant amplitude loading, spectrum loading or random loading.

For constant amplitude loading, a set of crack-length versus elapsed-cycle data (a versus N) is collected, with the specimen loading, P_{max} and P_{min}, generally held constant. The minimum crack length increment, Δa, between data points is required by ASTM Method E-647 to be larger than a certain value. This prevents the measurement of erroneous growth rates from a group of data points which are too closely spaced relative to the precision of data measurement and relative to the scatter of the data. The growth rates may be calculated by either of two methods. The secant method is simply the slope of the straight line connecting two adjacent data points. This method, although simpler, results in more scatter in measured crack growth rate. The polynomial method fits a second order polynomial expression (parabola) to typically 5 to 7 adjacent points, and the slope of this expression is the growth rate. The polynomial method, particularly when used with a large number of adjacent points, eliminates some of the scatter in growth rate which is inherent in fatigue testing. The measured values of growth rate typically are plotted as in Fig. 2.10, where ΔK is calculated from $\Delta P = P_{max} - P_{min}$ (for tension-tension loading) using a K expression such as Eq 3.1.

The measured growth rate data are represented by an equation of the form of the Paris equation (Eq 2.16), repeated here:

$$\frac{da}{dN} = C(\Delta K)^n \qquad \text{(Eq 3.4)}$$

where, as discussed in Section 2.5.2, the material constants C and n apply only within a certain range of da/dN and ΔK values. Other relationships based on the Paris equation, such as the commonly used Forman equation (Ref 3.15), are used to represent the variation of da/dN with other key variables, including load ratio, R, and the critical K value, K_c, at which fast fracture of the specimen occurs. The Forman equation is:

$$\frac{da}{dN} = \frac{C(\Delta K)^n}{(1 - R)K_c - \Delta K} \qquad \text{(Eq 3.5)}$$

where C and n here are material constants of the same types as those in the Paris Equation, but of different values. An advantage of the Forman equation is that it describes the type of accelerated da/dN behavior which is often observed at high values of ΔK and which is not described by the Paris equation. For example, for zero-to-tension loading (in which $R = P_{min}/P_{max} = 0$), as ΔK approaches K_c in Eq 3.5, da/dN increases rapidly, and this is often observed in tests. In addition, the Forman equation describes the often observed decrease in da/dN associated with an increase in R from zero toward one. So when it is necessary to describe the effect of ΔK approaching K_c or the effect of R on da/dN, the Forman equation can be used to represent the da/dN behavior. When only ΔK, the primary variable affecting da/dN, is involved, the less complex Paris equation may be used.

3.5. SUSTAINED-LOADING CRACK GROWTH TESTS

The two general types of sustained-loading crack growth tests are (*a*) determination of K_{Iscc}, the lowest threshold value of K_I at which stress corrosion cracking occurs, and (*b*) determination of stress-corrosion crack growth rate, da/dt, resulting from the combination of sustained loading and an aggressive environment. Similar terminology and testing procedures are used for threshold, K_{th}, and crack growth rate, da/dt, involving internal or external embrittling species such as hydrogen or liquid metals. A standard method for measuring K_{Iscc} is now (1982) being prepared by ASTM Committee E-24 on Fracture. The method is based on a large body of research, such as in Ref 3.16, as well as on the experience and results of an ASTM interlaboratory testing program. Tests were conducted at 17 laboratories using AISI 4340 steel specimens in a 3.5% NaCl solution for times up to 7000 hours. Additional interlaboratory tests with 7075-T7651 aluminum alloy are under way in a cooperative program of ASTM Committees E-24 and G-1 on Corrosion of Metals.

In determining K_{Iscc}, a group of fatigue-precracked specimens are held at different values of K_I (calculated from load and crack length) in the environment of interest, and the lowest value of K_I at which cracking occurs is K_{Iscc}. Another procedure for determining K_{Iscc} is the rising load procedure, in which the load on a single precracked specimen is very slowly increased until crack growth occurs, and the associated value of K_I is K_{Iscc}. Rising load K_{Iscc} tests of an AISI 4340 steel in an H_2S atmosphere are described by Clark in Ref 3.17.

Measurements of stress-corrosion crack growth rate, da/dt, are performed at constant load with crack length measurements taken at various time increments depending on the value of da/dt. Direct measurements of crack length on the specimen surfaces are made if the environment and test plan allow it. Indirect measurement of crack length is often made by measuring specimen displacement—usually the crack-mouth-opening displacement (CMOD)—as shown in Fig. 3.1. The crack length can then be determined by using the known compliance relations between CMOD and crack length. The example of da/dt measurements in 7075-T6 aluminum alloy in 3.5% NaCl, described in Section 2.5.3 and illustrated in Fig. 2.11, shows the type of da/dt test data which are obtained.

Various specimen geometries are used for stress-corrosion cracking tests. Specimen orientation is designated as it is for fracture toughness specimens (Fig. 3.2). The more commonly used specimens are described here. The most frequently used is the compact specimen; it is tension-loaded through pins as in other fracture tests (Fig. 3.9a), or it is self-loaded using a bolt (Fig. 3.9b). A double-beam specimen (often called double cantilever beam), with a configuration similar to that of the compact specimen, is used particularly for da/dt tests (see Fig. 3.9c). Because the crack is liable to grow out of the intended crack plane, side grooves are usually added to the double-beam specimen. Bend specimens also are used, either in three-point bending or in cantilever loading, as shown in Fig. 3.9(d).

3.6. USE OF FRACTURE TEST RESULTS

Measured values of plane-strain fracture toughness, K_{Ic}, and non-plane-strain and correlation measures of fracture toughness (Sections 3.1 to 3.3), are used in general to calculate the critical crack size in a loaded component which will lead

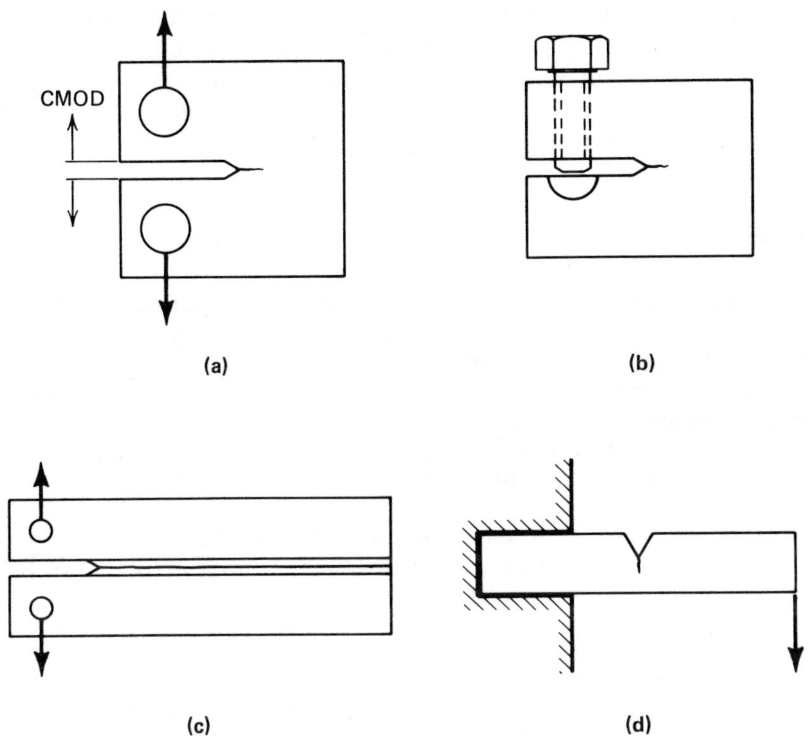

(a) Compact specimen: tension loaded; general use. (b) Modified compact specimen: bolt loaded; K_{Iscc} testing. (c) Double beam specimen: tension loaded; K_{Iscc} and da/dt testing. (d) Bend specimen: cantilever loaded; K_{Iscc} and da/dt testing.

Fig. 3.9. Schematic illustration of test specimens and loading used in sustained-load crack growth testing

to an abrupt failure. A critical-size crack may be assumed to have been present in the component as fabricated, or may have grown to critical size during service by subcritical crack growth due to fatigue loading or stress-corrosion cracking. Regardless of the cause of the crack, the measured value of toughness is used to calculate the combination of crack size and load which will cause failure.

Measured values of da/dN, K_{Iscc} and da/dt (Sections 3.4 and 3.5) are used to determine the conditions for subcritical crack growth. These test results are used to calculate the crack sizes and loading combinations which will cause cracks to grow from small initial sizes to critical sizes.

Results from the fracture tests described here are included in Chapters 4 to 8 for various structural materials. In Chapter 9, examples and case histories illustrate how such results are used to make calculations of critical crack sizes and subcritical crack growth. The calculations can be used in the design of structural components to ensure safe service lives and in the analysis of components which have failed during laboratory testing or service use.

Compilations of fracture mechanics data for specific materials have been included in various handbooks such as the *Damage Tolerant Design Handbook*

(Ref 3.18) and in reports of testing programs such as "Fracture Mechanics Evaluation of B-1 Materials" (Ref 3.19). Data from such sources may be used in a preliminary assessment of materials and their fracture properties, but when dealing with critical structures, test data should be obtained on specimens that are truly representative of the material in the structures.

REFERENCES

3.1. Experience in Plane-Strain Fracture Toughness Testing per ASTM Method E-399, by J. G. Kaufman: in *Developments in Fracture Mechanics Test Methods Standardization*, edited by W. F. Brown, Jr., and J. G. Kaufman, STP 632, American Society for Testing and Materials, Philadelphia, 1977, p 3-24

3.2. Fracture Toughness of Plain and Welded 3-In.-Thick Aluminum Plate, by F. G. Nelson and J. G. Kaufman: in *Progress in Flaw Growth and Fracture Toughness Testing*, STP 536, American Society for Testing and Materials, Philadelphia, 1973, p 350-376

3.3. A Proposed Standard Round Compact Specimen for Plane Strain Fracture Toughness Testing, by J. H. Underwood, J. C. Newman, Jr., and R. R. Seeley: *Journal of Testing and Evaluation*, Vol 8, No. 6, 1980, p 308-313

3.4. A Simplified Method for Measuring Plane Strain Fracture Toughness, by L. M. Barker: *Engineering Fracture Mechanics,* Vol 9, 1977, p 361-369

3.5. Fracture Toughness Properties of SA540 Steels for Nuclear Bolting Applications, by R. R. Seeley, W. A. Van Der Sluys and A. L. Lowe, Jr.: *Journal of Pressure Vessel Technology*, Vol 99, 1977, p 419-426

3.6. Some Further Results of J-Integral Analysis and Estimates, by J. R. Rice, P. C. Paris and J. G. Merkle: in *Progress in Flaw Growth and Fracture Toughness Testing*, STP 536, American Society for Testing and Materials, Philadelphia, 1973, p 231-245

3.7. Standard Test for J_{Ic}, A Measure of Fracture Toughness: E 813-81, *1981 Annual Book of ASTM Standards,* Part 10, American Society for Testing and Materials, Philadelphia, 1981, p 810-828

3.8. *Methods for Crack Opening Displacement (COD) Testing*: BS.5762, British Standards Institution, London, 1979

3.9. The COD Approach and Its Application to Welded Structures, by J. D. Harrison, M. G. Dawes, G. L. Archer and M. S. Kamath: in *Elastic Plastic Fracture*, STP 668, American Society for Testing and Materials, Philadelphia, 1979, p 606-631

3.10. Standard Methods for Notched Bar Impact Testing of Metallic Materials: E 23-81, *1981 Annual Book of ASTM Standards*, Part 10, American Society for Testing and Materials, Philadelphia, 1981, p 273-296

3.11. Standard Test Method for Dynamic Tear Energy of Metallic Materials: E 604-80, *1981 Annual Book of ASTM Standards*, Part 10, American Society for Testing and Materials, Philadelphia, 1981, p 702-710

3.12. Standard Method of Sharp-Notch Tension Testing of High-Strength Sheet Materials: E 338-81, *1981 Annual Book of ASTM Standards*, Part 10, American Society for Testing and Materials, Philadelphia, 1981, p 553-560

3.13. Standard Method for Sharp-Notch Tension Testing with Cylindrical Specimens: E 602-81, *1981 Annual Book of ASTM Standards*, Part 10, American Society for Testing and Materials, Philadelphia, 1981, p 693-701

3.14. *Rapid Inexpensive Tests for Determining Fracture Toughness*: National Materials Advisory Board, National Academy of Sciences, Washington, 1976

3.15. Numerical Analysis of Crack Propagation in Cyclic-Loaded Structures, by R. G. Forman, V. E. Kearney and R. M. Engle: *Journal of Basic Engineering, Transactions of ASME*, Vol 89, Sept 1967, p 459-464

3.16. *Stress Corrosion–New Approaches*, edited by H. L. Craig, Jr.: STP 610, American Society for Testing and Materials, Philadelphia, 1976

3.17. Effect of Cold Working on K_{Iscc} in a 4340 Steel, by W. G. Clark, Jr.: in *Flaw Growth and Fracture*, STP 631, American Society for Testing and Materials, Philadelphia, 1977, p 331-344

3.18. *Damage Tolerant Design Handbook*: Report MCIC-HB-01, Metals and Ceramics Information Center, Battelle Columbus Laboratories, Columbus, OH, 1975

3.19. Fracture Mechanics Evaluation of B-1 Materials, by R. R. Ferguson and R. C. Berryman: Report AFML-TR-76-137, Rockwell International, B-1 Div., Los Angeles, Contract F33657-70-C-0800, Air Force Materials Laboratory, Wright-Patterson AFB, OH, Oct 1976

Chapter 4

Fracture Properties of Carbon and Alloy Steels

W. W. Gerberich, R. H. Van Stone and A. W. Gunderson

4.1. INTRODUCTION

The steels discussed in this chapter range from the less expensive mild steels used in bridge construction to the very expensive 18Ni maraging steels used in high-strength pressure vessels and aircraft components. Such steels may differ by more than an order of magnitude in both yield strength and alloy additions as well as in cost per unit weight. It has been clear for some time that selection of ultrahigh-strength steels for critical structures requires application of fracture mechanics principles if high degrees of reliability are to be achieved. In fact, such principles have been applied to selection of steels since the early 1960's in developing rocket motor chambers with improved performance. These concepts have been adopted by the aerospace industry and have since been incorporated into standards for determining fracture toughness of metals and for design of airframes and engine components.

For example, the U. S. Air Force has established damage-tolerance requirements (MIL-A-83444) that are applied in the design, fabrication and inspection of military aircraft. This resulted because of recurring structural failures in military aircraft such as those caused by problems in material selection, fabrication and heat treatment of the D6ac steel wing carry-through structure for the F-111 aircraft. Other government agencies have included damage-tolerance-requirement specifications, and it is expected that this trend will continue. These concepts have already been extended to medium-strength rotor forging steels in the electric power industry and are gradually being applied to such problem areas as dynamic fracture toughness of roll bars for off-road vehicles and corrosion fatigue of line-pipe steel in aggressive H_2S environments.

Because these concepts are being applied to a broad spectrum of steels with extreme variations in strength, composition and microstructure, this chapter will attempt to present typical data for several classes of steels — e.g., ferrite-pearlite steels, Ni-Cr-Mo generator rotor steels, and ultrahigh-strength steels. Within the scope of this chapter, it is not possible to present sufficient data for the user to apply this as a design manual. Rather, an attempt has been made to identify those effects of alloy composition, microstructure and heat treatment which are most critical to the various types of fracture behavior encountered due to variations in loading, temperature and environment. Typical sources of data are represented by Ref 3.17 and 3.18, as noted in Chapter 3.

This chapter is organized into three major sections which address fracture toughness, fatigue crack propagation and sustained-load crack propagation. Where possible, these major sections are further subdivided to illustrate the effects of variations in alloy chemistry, microstructure, temperature, strain rate (or frequency) and environment on the various fracture toughness or crack growth rate parameters. In the following sections, the metallurgy of steel is given only a summary treatment, being followed by the major mechanical property sections dealing with static fracture toughness, K_{Ic}; dynamic fracture toughness, K_{ID}; fatigue or sustained-load crack growth rates, da/dN or da/dt; and fatigue or sustained-load thresholds, ΔK_{th} or K_{th}.

4.2. METALLURGY OF STEELS

This short section on the metallurgy of steels is intended only to give the reader some insight into several recent developments, because most of the common alloys and steel microstructures are adequately covered in textbooks or in the ASM *Metals Handbook* (Vol 1, Ninth Edition). One recent development is a class of lower-nickel maraging-type steels such as 12Ni-5Cr-3Mo and 10Ni-Cr-Mo-Co steels which are intended for high-reliability oceanspace applications. Other developments evolved from a need for high-performance, low-cost steels, which led to grain-refinement and inclusion-control procedures through controlled rolling and microalloying practices. These steels have been designated as high-strength low-alloy (HSLA) steels. More recent developments in HSLA steels include a number of "dual phase" steels which are primarily ferrite-martensite or ferrite-bainite mixtures, or various combinations of these three phases. Some of these new steels, with appropriate references, are discussed in the following sections, but most of the available information in the literature is on more standard grades.

4.2.1. Phase Relationships and Chemistries

Common classes of steels are austenitic (face-centered-cubic) steels, ferrite-pearlite or bainitic (body-centered-cubic) steels and martensitic (body-centered-tetragonal) steels. The austentic stainless steels will be discussed in Chapter 5. In this chapter, nearly all of the data will deal with ferrite-pearlite steels and steels with tempered-martensite microstructures, although some of the larger components of heat treated or slowly cooled alloys will invariably contain some bainite. The HSLA steels (Ref 4.1) contain niobium (columbium), vanadium or titanium as microalloying constituents in the form of finely dispersed carbides. The high-nickel maraging steels contain a low-carbon martensite matrix with precipitates of various compositions, after age hardening. Nominal compositions of some of these steels and of the more common grades referred to in this chapter are given in Table 4.1.

Table 4.1. Nominal compositions of carbon and alloy steels referred to in Chapter 4

Type or designation	UNS designation	Nominal composition, %									Nominal yield strength or range, MPa
		C	Mn	Si	Cr	Ni	Mo	V	Co	Other	
Ferrite-pearlite steels											
A285, grade C	K02801	0.17	0.52	0.06	...	...	...	...	...	...	240
A36	...	0.25	1.0	...	...	...	...	...	...	...	≥250
St 37-3	...	0.08	0.45	0.17	...	...	...	...	...	...	320
A588, grade A	K11430	0.15	1.1	0.20	0.5	...	...	0.05	...	0.3Cu	>290
A588, grade F	K11541	0.15	0.8	0.20	0.2	0.8	0.15	0.05	...	0.6Cu	≥345
A515, grade 70 (ASTM A212)	K03101	0.27	0.71	0.19	...	...	...	...	...	...	≥450
A302, grade B	K12022	0.20	1.29	0.18	...	0.08	0.55	...	...	...	470
A572 HSLA	...	0.26 max	1.35 max	...	...	...	...	(a)	...	(a)	350-450
API X-65	...	0.24	1.35	...	...	...	...	(b)	...	(b)	≥450
Q & T low alloy (medium strength) steels											
A514, grade F	K11576	0.15	0.80	0.25	0.5	0.9	0.5	0.05	...	0.3Cu, 0.004B	690
A533, grade B	K12539	0.25	1.25	0.20	...	0.50	0.5	...	...	...	730
A517, grade F	K11576	0.17	0.85	0.20	0.56	0.81	0.5	0.04	...	0.31Cu	805
HY 80	...	0.12	0.30	0.22	1.46	2.49	0.43	...	...	...	590
HY 100	...	0.18	0.45	0.25	1.40	2.60	0.32	0.01	...	...	730
HY 130	...	0.12	0.75	0.30	0.55	5.0	0.50	0.08	...	...	950
QT 35	...	0.18	1.2	0.20	0.90	1.1	0.45	0.07	...	...	600
API M N80	...	0.41	1.15	0.28	...	...	0.09	...	...	...	605
API N 110	...	0.43	1.19	0.26	...	...	0.09	...	...	...	920
Ni-Cr-Mo-V rotor steel	...	0.30	0.40	0.20	1.70	3.5	0.60	0.15	...	...	700
3340	...	0.40	0.45	...	1.52	3.33	...	...	...	...	890

(continued on next page)

Table 4.1 Nominal compositions of carbon and alloy steels referred to in Chapter 4 (continued)

Type or designation	UNS designation	Nominal composition, %									Nominal yield strength or range, MPa
		C	Mn	Si	Cr	Ni	Mo	V	Co	Other	
Ultrahigh-strength steels (large strength ranges due to tempering)											
4130	G41300	0.30	0.50	0.25	1.0	···	0.20	···	···	···	900-1700
4140	G41400	0.40	0.83	0.25	0.93	···	0.20	···	···	···	900-1800
4340	G43400	0.40	0.70	0.25	0.80	1.9	0.25	···	···	···	900-1800
300M	K44315	0.42	0.76	1.60	0.76	1.76	0.41	0.10	···	···	1070-1740
D6ac	K24728	0.45	0.80	0.25	1.15	0.60	1.0	0.06	···	···	900-1800
H-11	T20811	0.40	0.30	0.90	5.0	···	1.3	0.5	···	···	900-1800
HP9-4-20	K91401	0.20	0.30	0.10	0.80	9.0	1.0	0.1	4.5	···	1100-1300
HP9-4-25	···	0.27	0.21	0.02	0.39	9.48	0.51	0.08	3.9	···	1200-1400
HP9-4-45	···	0.45	0.10	0.10	0.30	8.5	0.20	0.10	3.75	···	1300-1800
10Ni-Cr-Mo-Co	···	0.11	0.2	0.10	2.20	10.0	1.0	···	8.0	···	1310
AF1410	···	0.15	0.06	0.10	2.0	10.0	1.0	···	14.0	···	1480
12Ni-5Cr-3Mo	K91890	0.02	···	···	5.0	12.1	3.2	···	···	0.25Ti, 0.30Al	···
18Ni, grade 200	K92810	0.01	0.02	0.10	···	18.0	3.0	···	8.5	0.20Ti, 0.1Al	1380
18Ni, grade 250	K92890	0.006	0.02	0.02	···	18.7	4.8	···	8.5	0.40Ti, 0.15Al	1720
18Ni, grade 300	K93120	0.013	0.02	···	···	18.6	4.8	···	8.9	0.59Ti, 0.1Al	2070
18Ni, grade 350	···	0.004	0.02	0.02	···	18.0	4.2	···	12.5	1.70Ti, 0.1Al	2400
12Cr-Mo-V	···	0.20	1.0	0.5	12.0	0.6	0.5	0.3	···	···	900-1450

(a) 0.01-0.1 V and/or 0.005-0.05 Nb. (b) 0.02 V min and/or 0.005 Nb min.

Alloying additions to low-strength ferrite-pearlite steels are minimal. However, for heavy steel sections of higher strength and for ultrahigh-strength steels, alloying additions are of primary consideration for hardenability, strength and toughness. To expand the austenite stability range, additions of elements such as boron, carbon, chromium, nickel, manganese, molybdenum and vanadium are most common whereas copper, tungsten, niobium and titanium may be added to specialty steels and silicon and aluminum are added as deoxidants. The latter element may also be added as a grain refiner, because AlN pins grain boundaries and retards ferrite grain growth. Such alloying additions, except boron and carbon, are readily dissolved in ferrite, but some tend to form strong carbides as originally outlined by Bain and Paxton (Ref 4.2) in Table 4.2.

Table 4.2. Distribution of alloying elements between ferrite and a carbide phase (Ref 3.2)

Element	Dissolved in ferrite		Combined in carbide
Nickel	Ni		
Silicon	Si		
Aluminum	Al		
Copper	Cu		
Manganese	Mn	◄——►	Mn
Chromium	Cr	◄——►	Cr
Tungsten	W	◄——►	W
Molybdenum	Mo	◄——►	Mo
Vanadium	V	◄——►	V
Titanium	Ti	◄——►	Ti

With regard to strength, ferrite-pearlite steels generally achieve their strength from carbon in the form of Fe_3C (pearlite content) or silicon and manganese as solid-solution strengtheners. The exceptions are the HSLA grades which may be strengthened by Nb, V or titanium carbides. In both instances, the standard contribution of strength from the inverse square-root grain size relationship is most important. For tempered martensitic steels, carbon is the major strengthener for low tempering temperatures whereas chromium, tungsten, molybdenum or vanadium alloy carbides contribute to secondary hardening of alloy steels tempered in the range from 500 to 600°C (930 to 1110°F).

With regard to impact energy and fracture toughness of ferritic steels, manganese or nickel is most commonly added to reduce the ductile-to-brittle transition temperature. In addition, grain refiners such as aluminum may cause net improvements in low-temperature toughness. Rare earth additions such as cerium often are made for inclusion shape control to improve toughness at higher temperatures in the upper-shelf region.

4.2.2. Special Embrittlement Considerations

In many classes of steels, two of the most important metallurgical embrittlement phenomena to be aware of are temper embrittlement and tempered martensite embrittlement. Temper embrittlement is quite common in slowly cooled heavy sections of steels tempered in the range from 400 to 560°C (750 to 1050°F). Here, segregation of very small amounts of tramp elements such as arsenic, phosphorus, sulfur, antimony, tin and tellurium may cause weakening of prior austenite grain boundaries and reductions in notch toughness. For ultrahigh-strength steels, tempered martensite embrittlement arises when tempering is done in the range from 300 to 400°C (570 to 750°F), usually because of platelike carbides being precipitated along prior austenite grain boundaries which have already been weakened by impurity segregation during the prior austenitizing treatment (Ref 4.3). These two types of embrittlement could be observed in the same steel, depending on the thermal cycle. Such a case is indicated for AISI 3340 steel in Fig. 4.1. Here, thermal heat treatment cycle 1 could produce tempered martensite embrittlement whereas either cycle 2 or cycle 3 could produce temper embrittlement.

4.3. FRACTURE TOUGHNESS

In discussing fracture toughness, it is appropriate that we should first consider the various microstructural modes and then determine how various alloying and

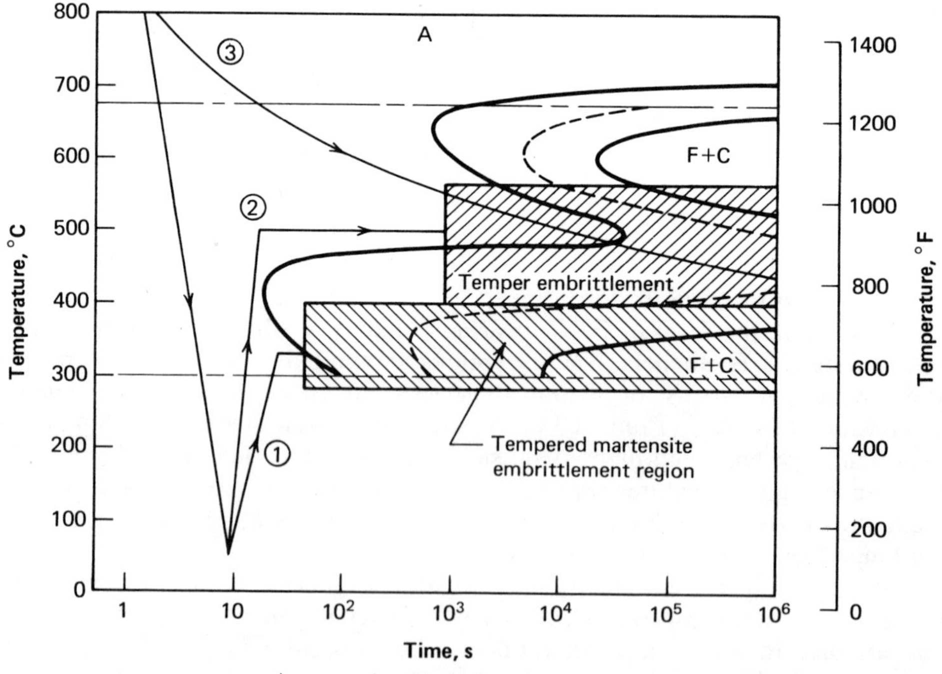

A – austenite; F – ferrite; C – cementite.

Fig. 4.1. Heat treatment cycles which could produce (1) tempered martensite embrittlement, or (2) and (3) temper embrittlement, in a 3340 steel

microstructural modifications interact to affect K_{Ic} or other measures of toughness. Finally, we will present the effects of temperature and strain rate on a number of commercial steels. It is important to note that for a given alloy chemistry or microstructural condition, the fracture toughness measured is highly reproducible for tests at different laboratories and for tests with different specimens, as shown in Table 4.3 (Ref 4.4).

4.3.1. Microstructural Fracture Modes

In steels, fractures occur by any of three basic mechanisms—cleavage, intergranular fracture and dimpled rupture—or by a mixture of these three mechanisms. The first two generally occur below the ductile-to-brittle transition temperature whereas dimpled rupture occurs above the transition temperature.

Cleavage

Cleavage is a transgranular fracture mode by which fracture occurs along crystallographic planes and after which the fracture surface appears as a series of flat planes. Figure 4.2(a) is a scanning electron microscope (SEM) fractograph of a cleavage fracture that occurred in A533B steel at −196°C (−320°F). The lines on an individual cleavage facet are known as "river patterns," which are shear steps between parallel, closely spaced cleavage planes. The critical event in cleavage fracture is the nucleation of a cleavage facet as it crosses crystallographic boundaries, causing formation of a new facet and a new set of river patterns. One way to improve the toughness of a steel which fails by cleavage is by reduction of the microstructural unit which controls the facet size. In ferritic steels, the facet size is the ferrite grain size (Ref 4.5) whereas in the multiphase structures of pearlite and bainite the fracture facet size is controlled by the prior austenite grain size (Ref 4.6).

Table 4.3. Summary of ASTM interlaboratory fracture toughness tests (Ref 4.4)

Steel	Yield strength MPa	Yield strength ksi	Number of specimens	Mean fracture toughness $MPa \cdot m^{1/2}$	Mean fracture toughness $ksi \cdot in.^{1/2}$	Coefficient of variation
Bend specimens						
4340 tempered at 260 °C (500 °F)	1640	238	35	48.5	44.4	0.045
4340 tempered at 425 °C (800 °F)	1420	206	37	87.0	79.6	0.046
18 Ni (300) maraging	1900	276	41	56.8	52.0	0.067
Compact specimens						
4340 tempered at 260 °C (500 °F)	1640	238	46	50.0	45.8	0.055
4340 tempered at 425 °C (800 °F)	1420	206	48	87.4	80.0	0.035
18 Ni (300) maraging	1900	276	45	58.6	53.6	0.041

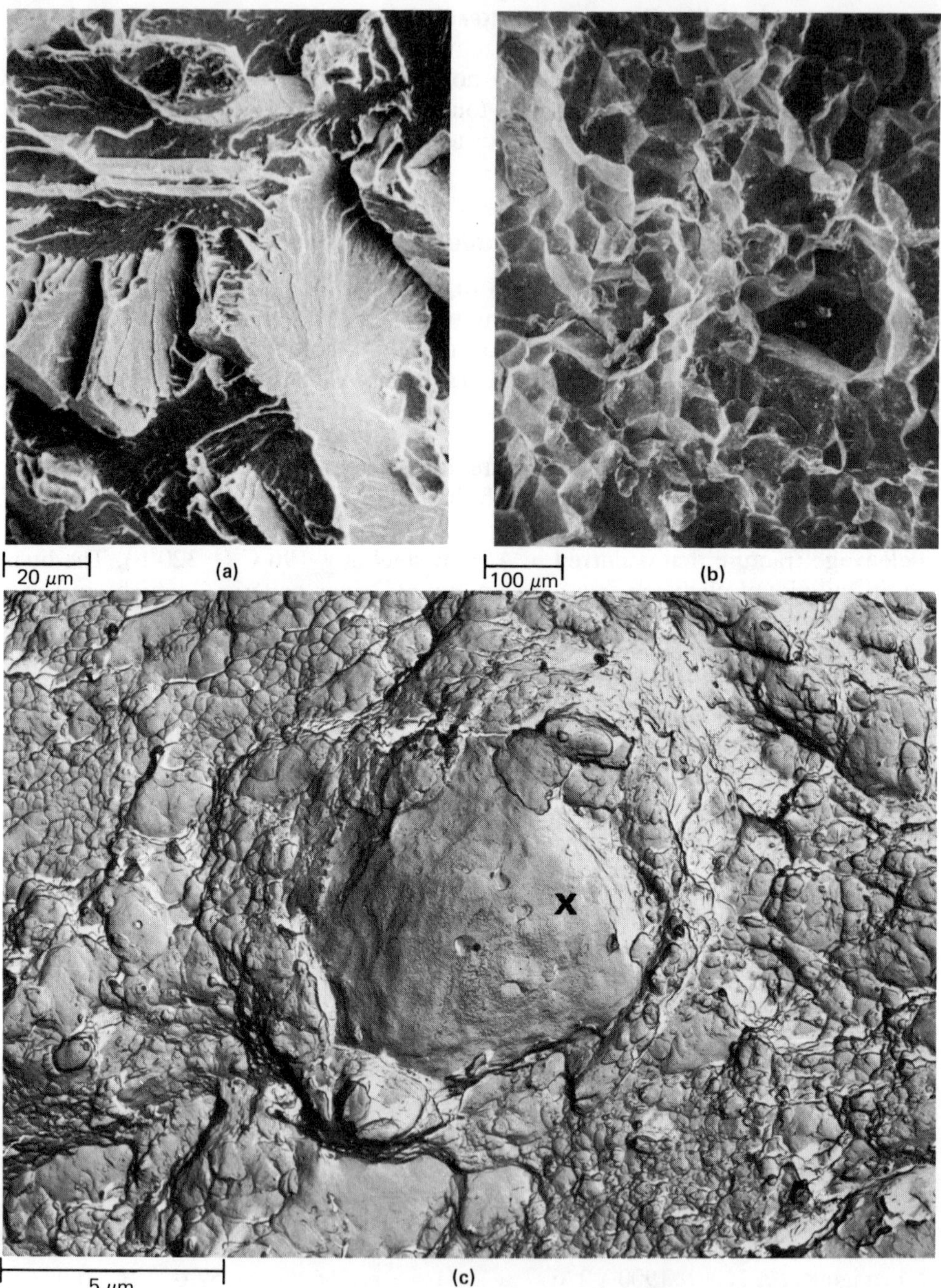

(a) SEM fractograph of cleavage in A533B steel at −196°C (−320°F). (b) SEM fractograph of intergranular fracture in HY 130 steel tempered at 480°C (900°F) for 1000 h *(Ref 4.7)*. (c) TEM fractograph of microvoid (dimple) rupture in AISI 4340 steel; "X" indicates a large dimple *(Ref 4.10)*.

Fig. 4.2. Microstructural fracture modes in steel

Low-Temperature Intergranular Fracture

Low-temperature intergranular fracture can occur by two basic mechanisms—trace-element grain-boundary segregation, or precipitation of films or finely dispersed phases along grain boundaries. Impurity-segregation-induced intergranular fracture is most often called "temper embrittlement;" fracture generally occurs along prior austenite grain boundaries. Fracture by this mode occurs most frequently in alloy steels containing nickel and chromium which have been tempered in the temperature range from 400 to 560°C (750 to 1050°F) or cooled slowly through this range. HY 130 steel tempered at 480°C (900°F) for 1000 hours fails by this fracture mode, as shown in Fig. 4.2(b) (Ref 4.7).

Intergranular fracture due to grain-boundary precipitation generally results from improper processing. An example of this is precipitation of grain-boundary films of titanium carbonitrides in maraging steels on slow cooling through the range from 1000 to 750°C (1830 to 1380°F) (Ref 4.8). Also, in low-alloy steels, high-temperature austenitization generally above 1300°C (2370°F) results in manganese sulfide precipitation along the austenite grain boundaries. These particles then act as void-nucleation sites for intergranular dimpled rupture (Ref 4.9). This phenomenon can also occur in overheated welds.

Dimpled Rupture

Dimpled rupture is a mode of fracture which occurs above the ductile-to-brittle transition temperature in low-strength steels and over a wide temperature range in high-strength steels. In this type of fracture, voids are nucleated at inclusions, carbides or other dispersed phases. With increasing plastic strain these voids grow and coalesce, resulting in a fracture surface covered with cuplets of "dimples"—hence the term "dimpled rupture." This fracture mode is also known as "microvoid coalescence" or "plastic fracture." Figure 4.2(c) shows a fractograph of quenched and tempered AISI 4340 steel with a yield strength of 1380 MPa (200 ksi) which fractured in this mode (Ref 4.10). This steel has a duplex dimple distribution with occasional large 5-micron dimples and large numbers of submicron-size dimples.

Dimpled rupture commences by void nucleation at inclusions such as manganese sulfides. Such void-nucleation events have also been observed at oxides in welds and at titanium carbonitrides in maraging steels. In some materials, such as maraging steel, final fracture occurs by growth and impingement of these voids, resulting in a relatively equiaxed dimple fracture surface. In other materials, the void-growth stage is interrupted by the formation of a sheet of voids at small dispersed-phase particles. This mechanism will be discussed further in Section 4.3.3.

4.3.2. Effects of Alloy Chemistry

The relationship between fracture toughness and yield strength for ultrahigh-strength alloy steels is shown in Fig. 4.3 (Ref 4.11). Although there is considerable scatter in the data, fracture toughness generally decreases with increasing yield strength. As shown, maraging steels represent a separate class, and data for these steels lie in a separate band. The major alloying chemistry differences here are carbon contents, which are low for the maraging steels (0.03%) but higher for

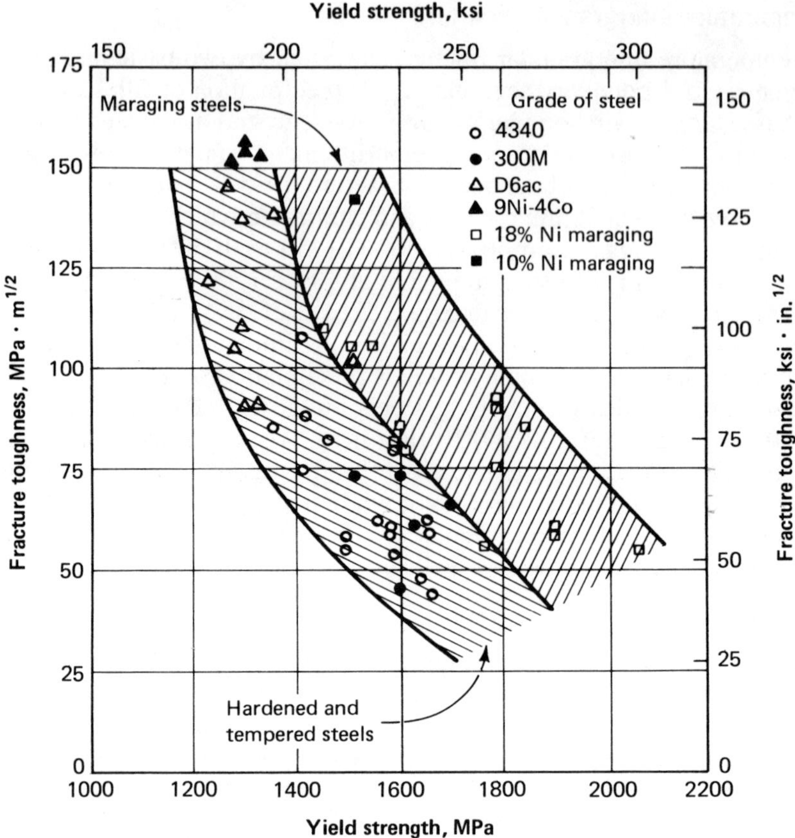

Fig. 4.3. Correlation between fracture toughness and yield strength for ultrahigh-strength steels (Ref 4.11)

the quenched and tempered steels (over 0.25%); nickel content, which is greater than 10% for maraging steels; and titanium content, which is added as a strengthener in maraging steels, being only about 0.2% for Grade 200 but over 1.0% for Grade 350. As a generalization, it may be inferred that carbon and strong carbide formers tend to decrease toughness in ultrahigh-strength steels.

Because alloy chemistry affects other characteristics such as hardenability, tempering response, strength and ductility (Ref 4.2), it is very difficult to separate the effects of chemistry on fracture toughness from the effects of other variables. In one series of experiments in which strength was held constant (Ref 4.12 and 4.13), an increase in nickel content of about 5% increased fracture toughness by about 50% in the transition temperature range, as shown in Fig. 4.4(a). Here, the yield strength was 1175 MPa (170 ksi) at room temperature. On the other hand, for lower-strength steels with yield strengths ranging from 500 to 700 MPa (73 to 102 ksi) at −100°C (−150°F), addition of 3.5% nickel increased fracture toughness by more than 100% (Ref 4.14). This is illustrated in Fig. 4.4(b). It should be noted that moderate additions of nickel mainly improve low-temperature frac-

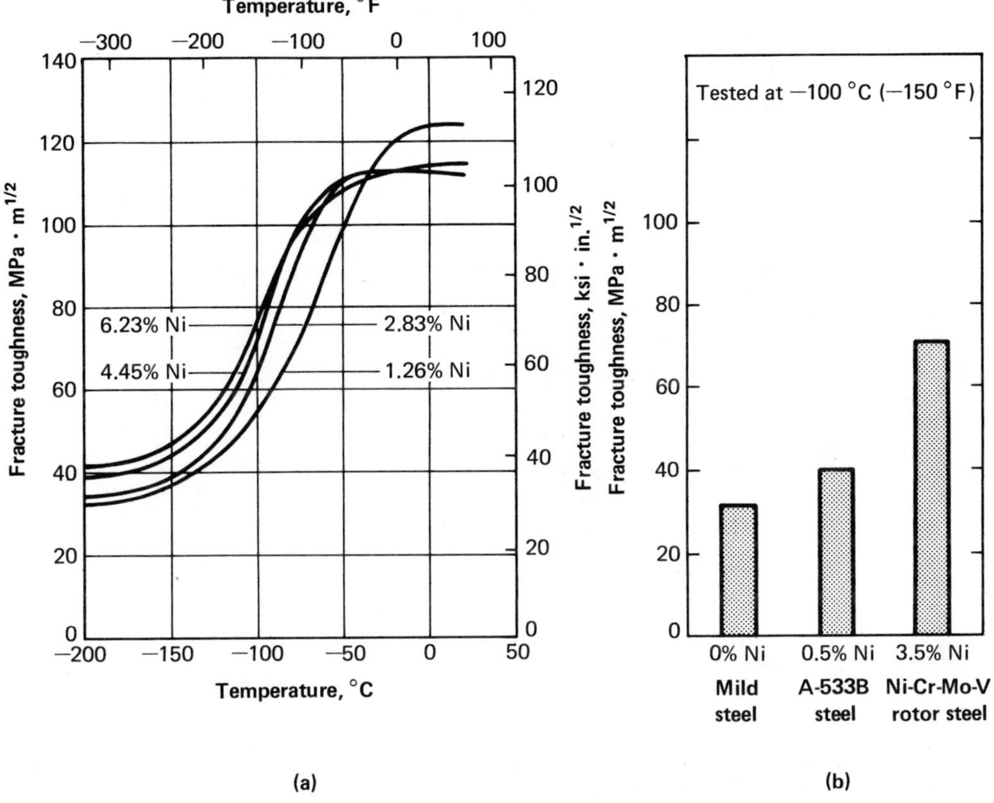

(a) High-strength steels containing 0.35 C, 0.65 Mn, 0.35 Si, 0.80 Cr, 0.30 Mo, 0.10 V and various amounts of nickel; all steels hardened to a yield strength of 1175 MPa (170 ksi) *(Ref 4.12 and 4.13)*. (b) Lower-strength steels with yield strengths ranging from 500 to 700 MPa (73 to 102 ksi) at −100°C (−150°F) *(Ref 4.14)*.

Fig. 4.4. Effect of nickel content on fracture toughness of several steels

ture toughness through resistance to cleavage, and that little effect may be seen at higher temperatures at which microvoid coalescence is the fracture mode. Such is the case in Fig. 4.4(a) at temperatures in the upper-shelf region.

It is well known that decreasing the carbon content decreases the transition temperature in ferrite-pearlite steels (*Metals Handbook*, Vol 1, Ninth Edition). This trend is less prevalent in ultrahigh-strength steels if strength is maintained at a constant level by tempering. The effect of carbon content on the fracture toughness transition curves for alloy steels containing 0.28, 0.35 and 0.41% carbon, with strength maintained at 1175 MPa (170 ksi), is illustrated in Fig. 4.5. There is little effect in the transition temperature range; however, there is a substantial effect in the upper-shelf region, where microvoid coalescence is the fracture mode. This is further discussed in Section 4.3.3.

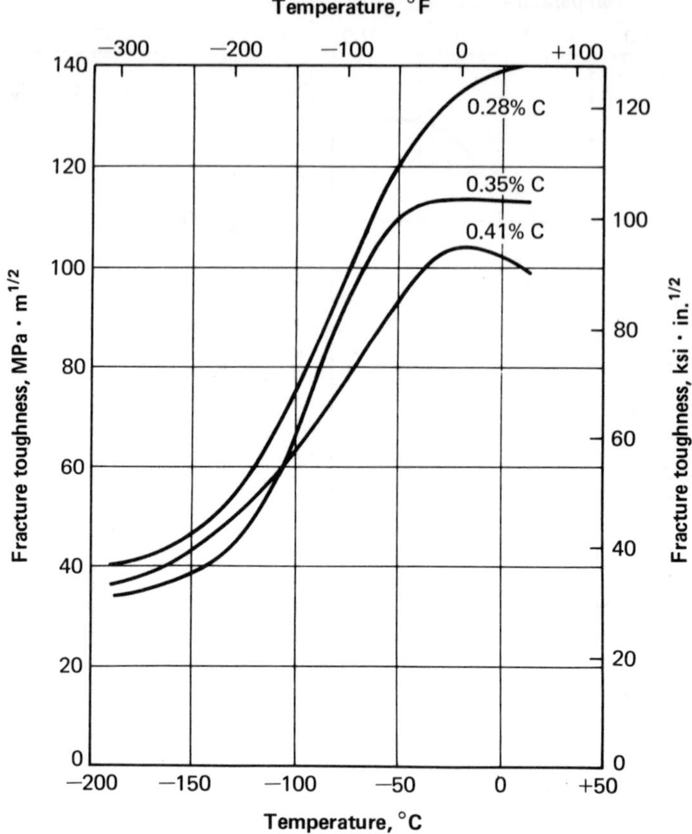

Data are for steels containing 0.65 Mn, 0.35 Si, 0.80 Cr, 3.00 Ni, 0.30 Mo, 0.10 V and various amounts of carbon; all steels were hardened and tempered to a yield strength of approximately 1175 MPa (170 ksi).

Fig. 4.5. Effect of carbon content and testing temperature on fracture toughness of alloy steels (Ref 4.12 and 4.13)

Nonmetallic inclusions (single inclusions or clusters of inclusions) reduce fracture toughness, particularly at high strength levels. The effect of sulfide inclusion content (which is a function of sulfur content) on fracture toughness of 4345 steel is shown in Fig. 4.6 (Ref 4.15). Because of this effect, special melting and processing requirements are often specified when alloy steels are selected for certain critical applications, such as aircraft landing gear. Vacuum arc remelting (VAR), vacuum induction remelting (VIR) and electroslag remelting (ESR) are three of the special processes that normally produce cleaner steels than other, more common steelmaking processes. Not only does sulfur affect ultrahigh-strength steels, but it may also affect grades of low to medium strengths (Ref 4.16). This is largely a microvoid nucleation mechanism, and, if the sulfide inclusion length is plotted versus crack tip opening displacement (CTOD), an inverse relationship is revealed, as shown in Fig. 4.7(a). As discussed in Section 3.2.2, the critical

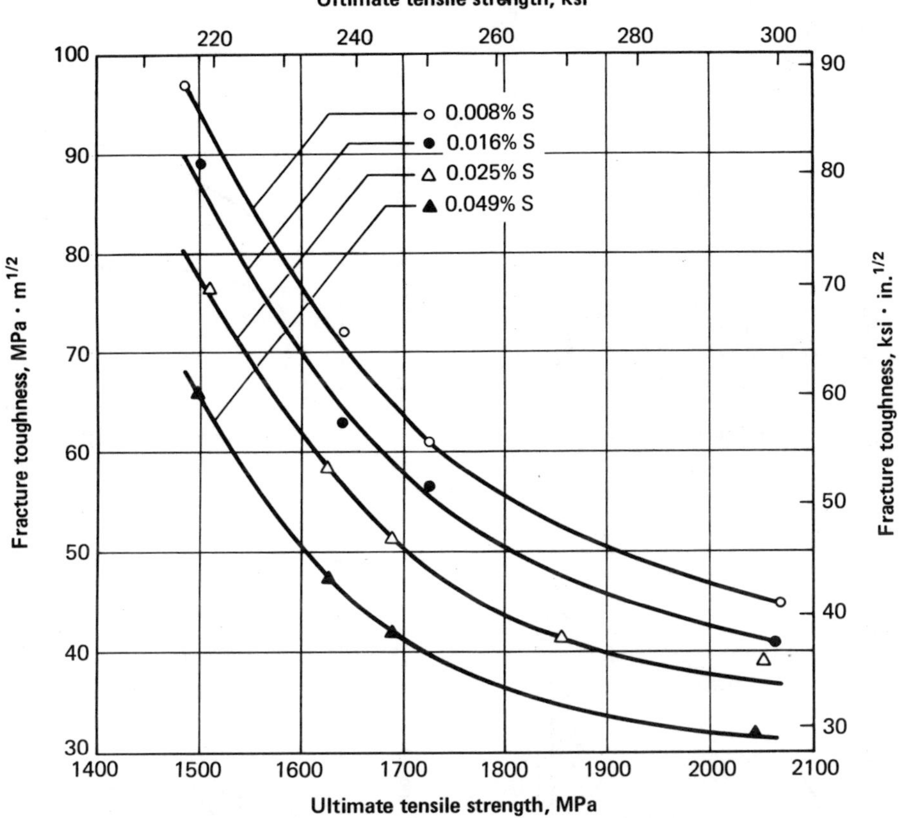

Fig. 4.6. Effect of sulfur content on fracture toughness of 4345 steel hardened and tempered to various strength levels (Ref 4.15)

crack tip opening displacement, δ_c, is a measure of fracture toughness. Under plain-strain conditions, an approximate relation between δ_c and K_{Ic} is:

$$\delta_c \simeq \frac{K_{Ic}^{\;2}}{2\sigma_{ys}E} \qquad \text{(Eq 4.1)}$$

where σ_{ys} is yield strength and E is Young's modulus. Thus, one means of improving toughness is to reduce the projected inclusion length by controlling the shape of the inclusions (Ref 4.16). This can be achieved by adding rare earths, which combine with MnS to form hard-to-deform particles comprised of sulfides, oxides and/or oxysulfides. This means of shape control greatly improves transverse notch impact properties, as shown in Fig. 4.7(b) for X65 pipeline steel (Ref 4.18). As shown in Fig. 4.7(c), notch toughness of a large heat of pipeline steel tended to increase with increasing rare earth metal additions (Ref 4.19). With regard to shape control, Luyckx *et al* (Ref 4.20) demonstrated that when there was a complete change from elongated to globular sulfides, at a cerium-to-sulfur ratio of two, the upper-shelf impact energy became a maximum. This is seen in Fig.

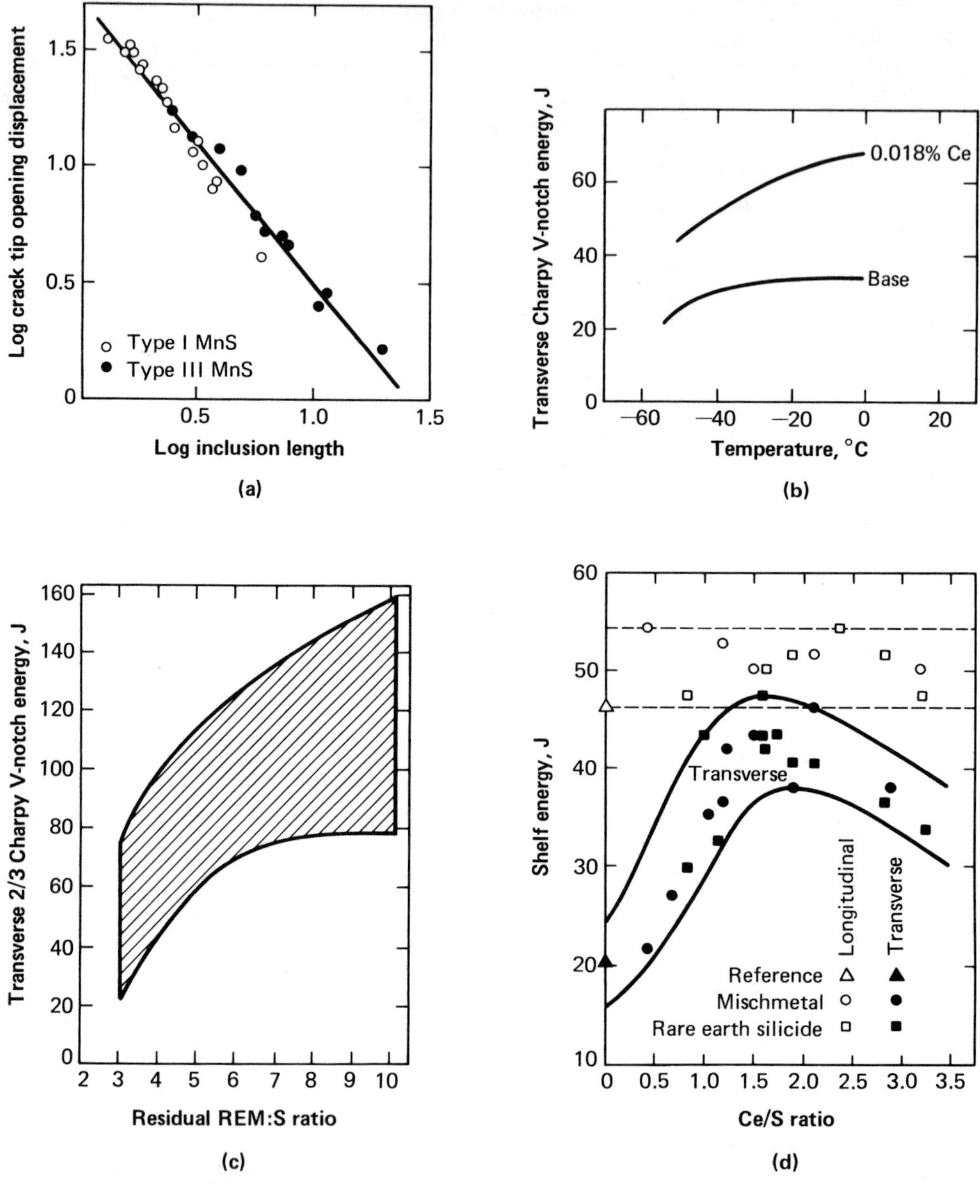

(a) Relationship between projected inclusion length per unit area and crack tip opening displacement to fracture in sulfur-bearing steels. (b) Effect of rare earth additions on impact properties of Al-Si killed X65 pipeline steel. (c) Relationship between transverse Charpy V-notch (CVN) energy and rare earth metal: sulfur ratio (⅔-size Charpy specimens) at −18°C (−1°F). (d) Relationship between shelf energy determined on longitudinal and transverse ½-size Charpy V-notch specimens and Ce:S ratio.

Fig. 4.7. Effects of sulfide inclusions on toughness of ferritic steels (Ref 4.16)

4.7(d). As might be expected, there was little effect of rare earth metal additions on longitudinal properties, because the shape of sulfides is not particularly detrimental to longitudinal fracture toughness in steels of low to medium strength.

Other alloy chemistry effects are those that cause or influence temper embrittlement. The use of Auger electron spectroscopy has shown that steels that fail by this mode have severe segregation of trace elements such as antimony, arsenic, tin and phosphorus along prior austenite grain boundaries even though the bulk concentration is in the parts per million range. This segregation exists for only a few atom layers. In Fig. 4.8, the results of an investigation (Ref 4.21) on 3340 steel indicate that the transition temperature shift was proportional to the grain boundary concentration of the embrittling trace impurity.

Temper embrittlement is a very complex phenomenon, because it is dependent not only on the trace elements present but also on the major alloying additions such as nickel and chromium. For example, in steels containing the same amounts of trace impurities, Ni-Cr alloys are more susceptible than those having either nickel or chromium separately (Ref 4.22). In either case, the degree of embrittlement can be reduced by lowering the trace element content of the steel. To avoid temper embrittlement, steels with low trace element concentrations and proper ratios of major alloying additions should be used. It is also advisable to avoid service in the embrittling temperature range. A small austenite grain size also inhibits temper embrittlement. It should be noted that this behavior is reversible—that is, it can be eliminated by heating above the temper embrittlement range provided that the

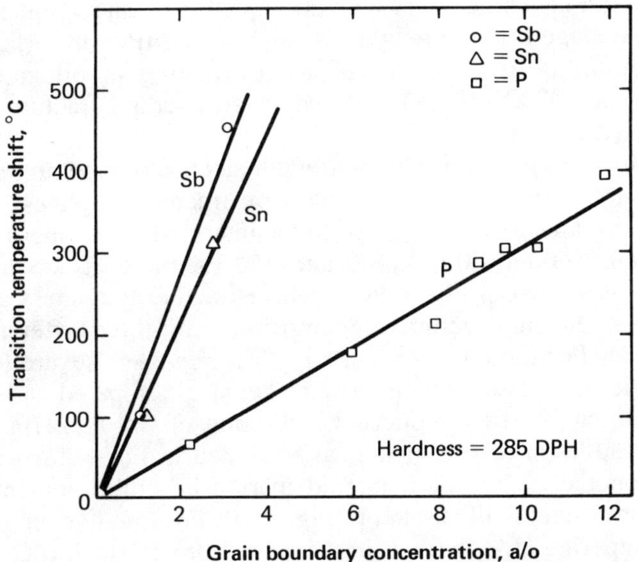

Data obtained on 3340 steel doped individually with 0.06 P, 0.06 Sn or 0.06 Sb. a/o is atomic percent. 285 DPH = 890 MPa (129 ksi) ultimate strength.

Fig. 4.8. Change in ductile-to-brittle transition temperature as a function of grain boundary impurity concentration (Ref 4.21)

material is not subsequently maintained at temperatures in the embrittlement range for extended periods of time.

4.3.3. Effects of Microstructure and Heat Treatment

Typical microstructural changes on heat treatment of high-strength steels are accomplished by changing the austenitizing temperature and time at temperature, the quenching rate, and the tempering temperature and time at temperature. Increasing the austenitizing temperature and time has a two-fold effect in that it increases the grain size and/or increases the solutionizing of alloy carbide formers. The former effect may produce mixed results in that austenitizing at 1200°C (2200°F) may increase the sharp-crack fracture toughness of an AISI 4340 steel over that obtained using conventional heat treatments; however, the large prior austenite grains may produce a lower impact toughness (Ref 4.23). As discussed below in Section 4.4.2, increased prior austenite grain size in high-strength steels may also have a detrimental effect on resistance to fatigue cracking (Ref 4.42). On the other hand, it is well known that reasonable austenitizing temperatures and times are necessary to dissolve the carbide-forming elements prior to quenching.

Variations in quenching rates also influence fracture toughness of alloy steels by causing variations in the as-quenched microstructures. Quenching alloy steels in oil to obtain nearly 100% martensite on quenching will result in higher fracture toughness in the tempered condition than will slack quenching and tempering at the same temperature. This condition is discussed by Peterman and Jones (Ref 4.24) for heat treated aircraft structures of D6ac steel. Briefly, austenitizing at 900°C (1650°F), quenching in salt at 200°C (400°F) and tempering at 200°C resulted in an average fracture toughness (K_{Ic}) of 57 MPa · $m^{1/2}$ (52 ksi · in. $^{1/2}$), whereas austenitizing at 925°C (1700°F), quenching in oil at 60°C (140°F) and tempering at 200°C (400°F) resulted in an average fracture toughness of 101 MPa · $m^{1/2}$ (92 ksi · in.$^{1/2}$).

The effects of tempering ultrahigh-strength steels at temperatures in the upper portion of the tempering range are to increase fracture toughness and to reduce yield strength, as indicated in Fig. 4.3. In the intermediate tempering temperature regime between 200 and 400°C (400 and 750°F), these effects are not always apparent. This is shown in Fig. 4.9(a), where fracture toughness at 25°C (75°F) is relatively independent of tempering temperature for three different nickel additions (Ref 4.13). For tests at −73°C (−100°F), however, severe tempered martensite embrittlement results for the low-nickel steel tempered at 300°C (570°F). This embrittlement is partly reduced by addition of 3% nickel and completely suppressed by addition of 6% nickel, as indicated in Fig. 4.9(b). The effect of carbon level on the degree of tempered martensite embrittlement in a 3% Ni ultrahigh-strength steel is illustrated in Fig. 4.10. For low-temperature service, it is clear that tempering of such steels in the range from 250 to 350°C (480 to 660°F) is to be avoided.

For lower-strength steels, useful microstructural modifications to improve low-temperature fracture toughness involve decreasing of the ferrite or prior austenite grain size. Results from two investigations presented in Fig. 4.11 show that, for temperatures above −150°C (−240°F), a substantial improvement in fracture toughness can be achieved through grain refinement (Ref 4.13 and 4.26). Here,

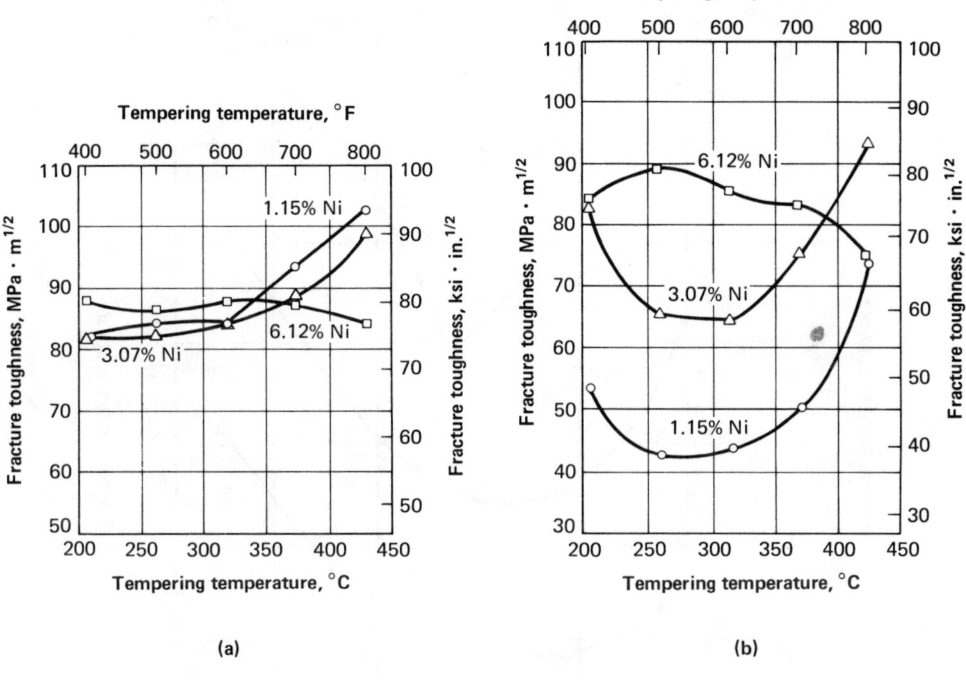

(a) Measurements made at 25°C (75°F) *(Ref 4.13)*. (b) Measurements made at −73°C (−100°F) *(Ref 4.12)*

Fig. 4.9. Variation of notch-bend fracture toughness with tempering temperature for steels containing 0.35 C, 0.65 Mn, 0.35 Si, 0.80 Cr, 0.30 Mo, 0.10 V and various amounts of nickel

it is indicated that there may be a relationship between K_{Ic} and grain size, d. This is shown for both ferrite grain size (Ref 4.27) and prior austenite grain size (Ref 4.28) in Fig. 4.12(a) and (b), respectively. Such data have been analyzed by Stonesifer and Armstrong (Ref 4.28) in terms of a critical plastic zone at the crack tip, with the grain size dependence of K_{Ic} arising from effects of grain size on yield strength. More simply, the fracture mechanism can be interpreted in terms of the plastic constraint in the triaxially stressed zone at the crack tip raising the yield stress to the cleavage fracture stress, σ_f. This has been utilized by others (Ref 4.29) and is essentially the Orowan criterion given by:

$$\text{pcf} \cdot \sigma_{ys_{(T,\dot{\varepsilon}p)}} = \sigma_f \quad \text{(Eq 4.2)}$$

where $\sigma_{ys_{(T,\dot{\varepsilon}p)}}$ is the uniaxial yield stress at the temperature and strain rate of interest and pcf is the plastic constraint factor. The simplest empirical determination of plastic constraint factor imposed by a plane-strain crack was originally proposed by Hahn and Rosenfield (Ref 4.30) to be:

$$\text{pcf} = 1 + \alpha \{K_{Ic}/\sigma_{ys_{(T,\dot{\varepsilon}p)}}\} \quad \text{(Eq 4.3)}$$

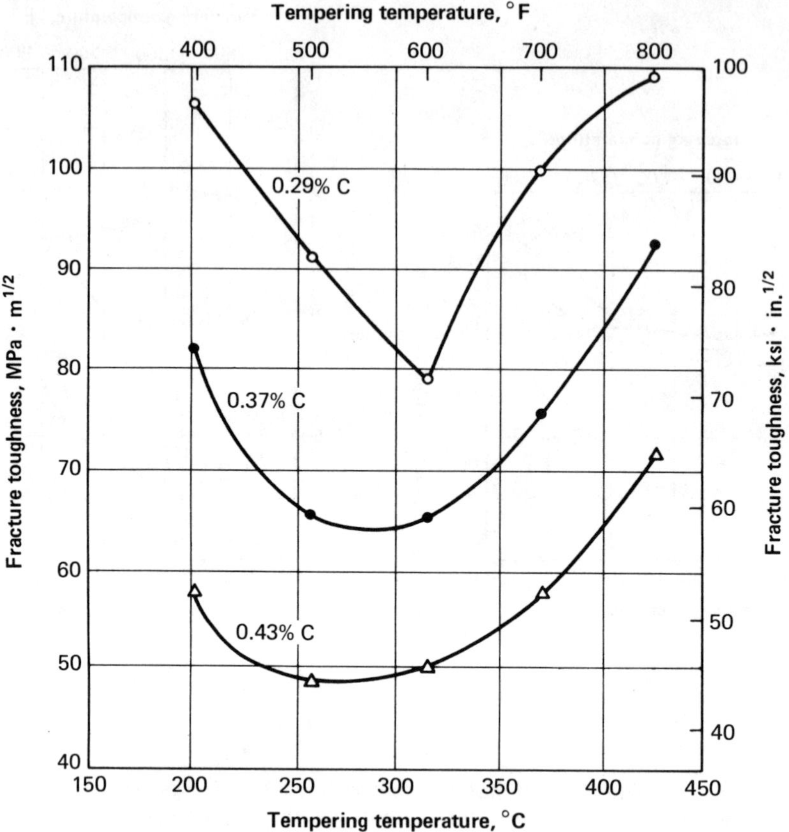

Measurements were made at −73°C (−100°F).

Fig. 4.10. Effect of carbon content and tempering temperature on fracture toughness of alloy steels containing 0.65 Mn, 0.35 Si, 0.80 Cr, 3.00 Ni, 0.30 Mo, 0.10 V and various amounts of carbon (Ref 4.12)

For strain hardening in mild steels, it may be assumed that the value of α is approximately 20 $m^{-1/2}$, because this would give plastic constraint factors on the order of 2.5 to 3.0—values that are predicted from elastic-plastic analysis (Ref 4.31). Relationships for yield and fracture stress dependence on grain size have been independently determined by Curry and Knott (Ref 4.27) and by Rosenfield, Hahn and Embury (Ref 4.32) to be, for mild steel:

$$\sigma_{ys_{-120°C}} = 210 \text{ MPa} + (0.73 \text{ MPa} \cdot m^{1/2})d^{-1/2} \qquad \text{(Eq 4.4a)}$$

and

$$\sigma_f = 343 \text{ MPa} + (3.3 \text{ MPa} \cdot m^{1/2})d^{-1/2} \qquad \text{(Eq 4.4b)}$$

By combining Eq 4.2, 4.3 and 4.4, the following equation describing the effect of grain size on K_{Ic} is obtained:

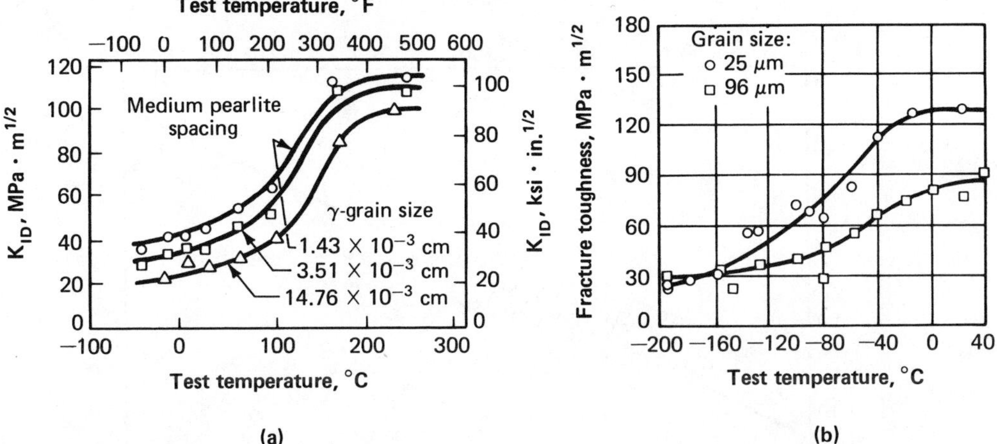

(a) Dynamic fracture toughness (K_{ID}) curves for fully pearlitic steels as a function of temperature for three prior austenite grain sizes. (b) Fracture toughness as a function of temperature for St 37-3 steel in two grain sizes.

Fig. 4.11. Effect of grain size on fracture toughness (Ref 4.13 and 4.26)

$$K_{Ic_{-120°C}} \simeq 6.6 \text{ MPa} \cdot \text{m}^{1/2} + 0.13 \text{ MPa} \cdot \text{m}(d^{-1/2}) \quad \text{(Eq 4.5)}$$

It can be seen that this equation fits the data in Fig. 4.12(a) and gives a good rationalization for the $d^{-\frac{1}{2}}$ dependence of fracture toughness. Such a relationship can also be approximated for A533B steel at room temperature, because Knott (Ref 4.33) has estimated that $\sigma_f \simeq 2600$ MPa for an A533B steel with a 15-μm grain size. This, along with Stonesifer and Armstrong's determination of the relationship between yield strength and grain size (Ref 4.28), gives, for A533B steel:

$$\sigma_{ys_{25°C}} = 572 \text{ MPa} + (0.11 \text{ MPa} \cdot \text{m}^{1/2})d^{-1/2} \quad \text{(Eq 4.6a)}$$

and

$$\sigma_f = 1750 \text{ MPa} + (3.3 \text{ MPa} \cdot \text{m}^{1/2})d^{-1/2} \quad \text{(Eq 4.6b)}$$

Following the above procedure, this leads to:

$$K_{Ic_{25°C}} \simeq 58.8 \text{ MPa} \cdot \text{m}^{1/2} + 0.16 \text{ MPa} \cdot \text{m}(d^{-1/2}) \quad \text{(Eq 4.7)}$$

and gives a curve nearly identical to the line drawn through the room-temperature data in Fig. 4.12(b) for A533B steel.

It is clear, then, that one goal for improved low-temperature fracture toughness in steels of low to medium strength is a refined grain size such as has been achieved in HSLA steels. The development of fine grain size in these steels

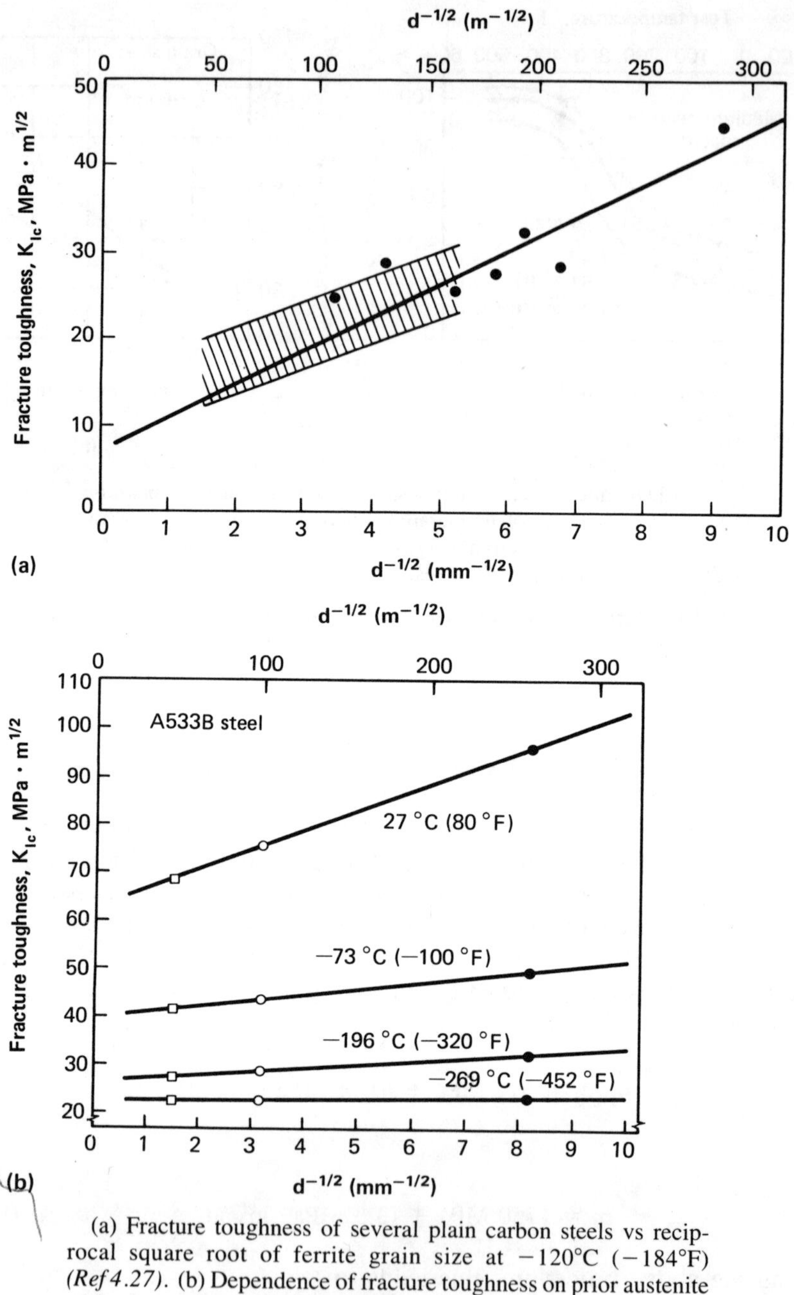

(a) Fracture toughness of several plain carbon steels vs reciprocal square root of ferrite grain size at −120°C (−184°F) *(Ref 4.27)*. (b) Dependence of fracture toughness on prior austenite grain size at four different temperatures *(Ref 4.28)*.

Fig. 4.12. Relationship of fracture toughness to inverse square root of grain size

through the use of minor alloying additions and lower hot rolling finishing temperatures has resulted in large improvements in fracture toughness (Ref 4.34).

With regard to reduced ductility and toughness due to particle nucleation of microvoids, this is most often observed in high-strength steels. Figure 4.13(a)

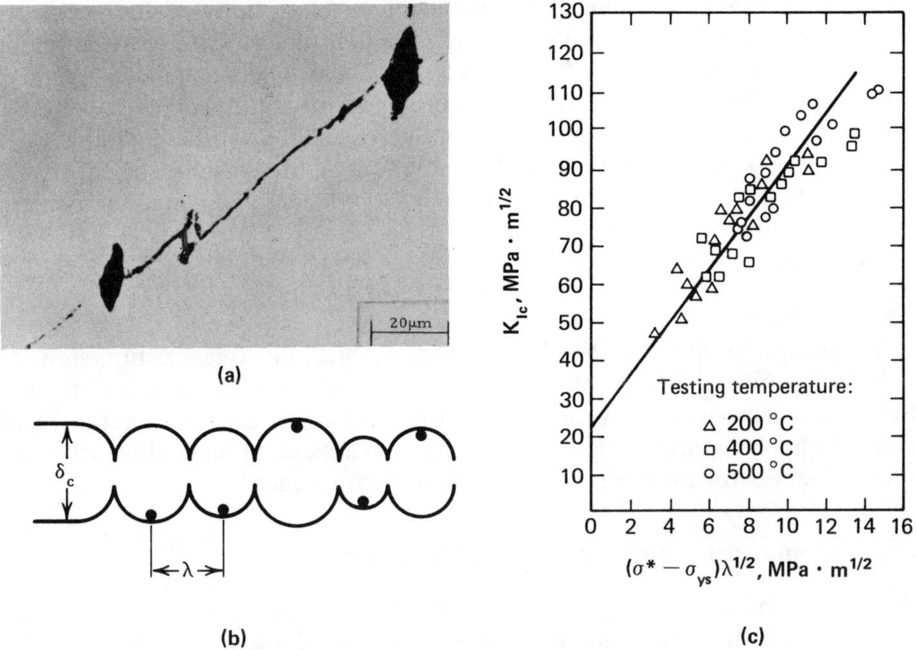

(a) Void sheets linking three inclusions in a sectioned tensile sample of 4340 steel. (b) Schematic relationship between crack tip opening displacement and inclusion spacing *(Ref 4.35)*. (c) K_{Ic} for a martensitic 0.45C-Ni-Cr-Mo-V steel as a function of inclusion spacing and yield strength *(Ref 4.36)*.

Fig. 4.13. Effect of particles on fracture toughness

shows a metallographic section of a specimen of AISI 4340 steel strained just short of fracture, exhibiting a "void sheet" between large voids which nucleated at lower strains at manganese sulfides. In this case, the smaller voids nucleated at the cementite particles which formed during tempering. Void sheet formation resulted in the duplex dimple size distributions previously shown in Fig. 4.2(c). This process aborts further void growth and tends to reduce fracture toughness. Cox and Low (Ref 4.10) concluded that void sheet formation is easier in materials with larger-size strengthening precipitates. This is supported by the absence of void sheets in maraging steel having a yield strength of 1380 MPa (200 ksi) and containing precipitates a few hundred angstroms in length and by the presence of void sheets in 4340 steel containing cementite particles 0.17 μm in length. They also suggest that this is why maraging steels are tougher than quenched and tempered steels of similar strength levels, as shown in Fig. 4.3.

The key event in dimpled rupture is the void nucleation event, which can be delayed to larger plastic strains by reductions in inclusion size (Ref 4.10). Toughness can also be improved by increasing the spacing between inclusions so that voids must grow to larger sizes during the fracture process. Reductions in impurities which occur as inclusions will reduce the inclusion volume fraction and increase the inclusion spacing, thereby improving K_{Ic}.

The effect of particles has been treated at great length in the literature (Ref 4.35), but one of the simplest quantitative relationships which has been derived is that of fracture toughness as a function of inclusion spacing. In Schwalbe's review of microstructural effects (Ref 4.35), he suggests that the critical crack tip opening displacement, δ_c, should be proportional to the distance between inclusions, λ, as shown schematically in Fig. 4.13(b). According to Eq 4.1, this would result in K_{Ic} being proportional to the square root of λ. Such a semi-empirical relationship has been found by Priest (Ref 4.36) for a 0.45C-Ni-Cr-Mo-V steel similar to 4340 steel but with somewhat higher chromium, molybdenum and vanadium contents. This relationship is given by:

$$K_{Ic} = 23\ \text{MPa} \cdot \text{m}^{1/2} + 7(\sigma^* - \sigma_{ys})(\lambda)^{1/2} \qquad \text{(Eq 4.8)}$$

with σ^* equal to 2000 MPa (290 ksi). In Fig. 4.13(c), this relationship, shown by the dashed curve, is seen to fit the data for a large variation in λ and for three different test temperatures where dimpled rupture is the microstructural fracture mode. Such observations demonstrate the advantage of providing inclusion-controlled steels for high fracture toughness performance.

4.3.4. Temperature and Strain Rate Variables

The effects of testing temperature and loading rate on fracture toughness vary from one grade of steel to another. As seen in Fig. 4.14, three types of turbine generator rotor forging steels exhibit distinct toughness transitions in the range of testing temperatures utilized (Ref 4.37). The thickness, B, required to ensure valid plane-strain behavior is given by:

$$B \geq 2.5 \left(\frac{K_{Ic}}{\sigma_{ys}}\right)^2 \qquad \text{(Eq 4.9)}$$

and is described in Section 3.1.1. For these rotor forging steels with room-temperature yield strengths on the order of 700 to 800 MPa (100 to 116 ksi), this would indicate a validity cutoff near 65 MPa $\cdot$ $\text{m}^{1/2}$ (59 ksi $\cdot$ $\text{in.}^{1/2}$) for 1T specimens (specimens 2.5 cm, or 1 in., thick).

Not only does the fracture toughness temperature transition change with alloy composition and yield strength, but it also changes with loading rate. This is extremely important in low- and medium-strength steels which are strain-rate sensitive. Such a result is shown in Fig. 4.15, where it is seen that A36 steel undergoes a temperature transition shift as the strain rate increases from 10^{-5} to 10^{-3} to 10 s^{-1}. Obviously, it would be very dangerous to select such a steel for dynamic conditions based on static fracture toughness unless the temperature shift could be adequately described. A series of studies (Ref 4.38, 4.39 and 4.40) has shown that the magnitude of this shift is dependent on yield strength. Barsom (Ref 4.41) has described this relationship as:

$$T_{shift}\ (\text{in } ^\circ\text{C}) = 119 - 0.12\sigma_{ys} \qquad \text{(Eq 4.10)}$$

for steels with strengths given by 250 MPa $< \sigma_{ys} <$ 965 MPa (36 ksi $< \sigma_{ys} <$ 139 ksi). As can be seen in Fig. 4.16, there is no temperature transition shift for steels with yield strengths greater than about 965 MPa

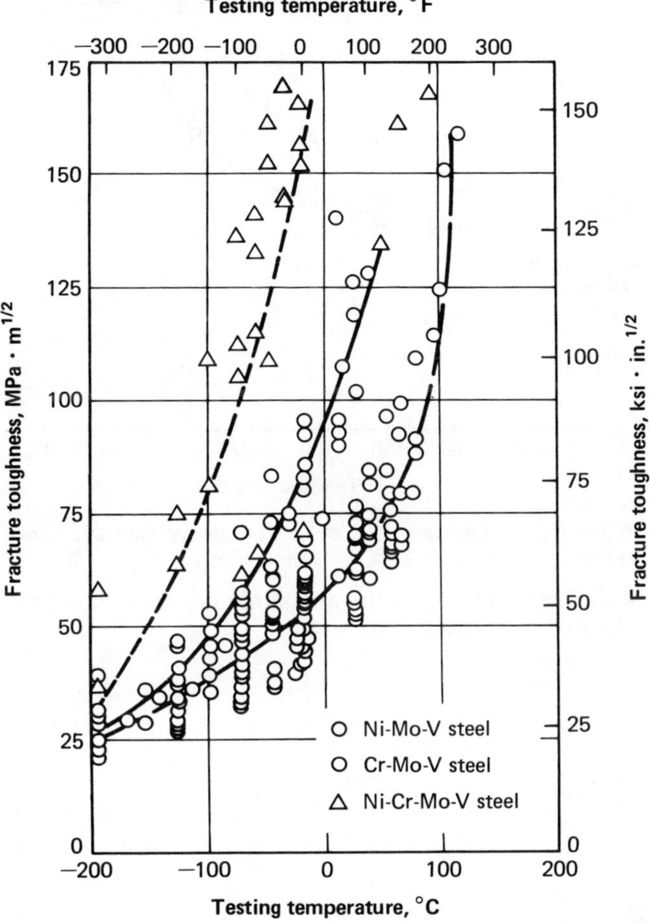

Fig. 4.14. Effect of temperature on fracture toughness of three alloy steels (Ref 4.37)

(139 ksi). This has also been verified by Priest (Ref 4.14) in a large number of tests of low-to-medium-strength steels, some of which are illustrated in Fig. 4.17. Here, up to six test temperatures and six orders of magnitude of loading rate (in K, MPa · $m^{1/2}s^{-1}$) were used to determine fracture toughness. It can be seen that there are substantial shifts for low-strength steels. However, for the steel with a yield strength of 970 MPa (140 ksi) in Fig. 4.17(d), although there is a change in fracture toughness with test temperature, there is no shift with loading rate. For steels with strengths less than 965 MPa (139 ksi), Barsom (Ref 4.41) has described the combined effect of yield strength and strain rate in the range of $10^{-3}\,s^{-1} \leq \dot{\varepsilon} \leq 10\,s^{-1}$ by:

$$T_{shift}\,(\text{in °C}) = (83 - 0.08\sigma_{ys})\dot{\varepsilon}^{0.17} \qquad \text{(Eq 4.11)}$$

where $\dot{\varepsilon}$ is in s^{-1} units.

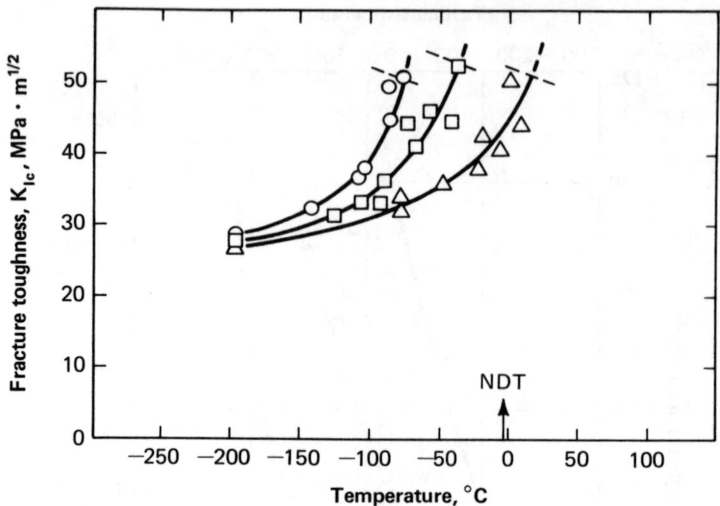

○ – K_{Ic} loading ($\dot{\varepsilon} \simeq 10^{-5}s^{-1}$); □ – Intermediate strain rate loading ($\dot{\varepsilon} \simeq 10^{-3}s^{-1}$); △ – Dynamic loading ($\dot{\varepsilon} \simeq 10s^{-1}$).

Fig. 4.15. Effect of temperature and strain rate on fracture toughness of ASTM A36 steel (Ref 4.41)

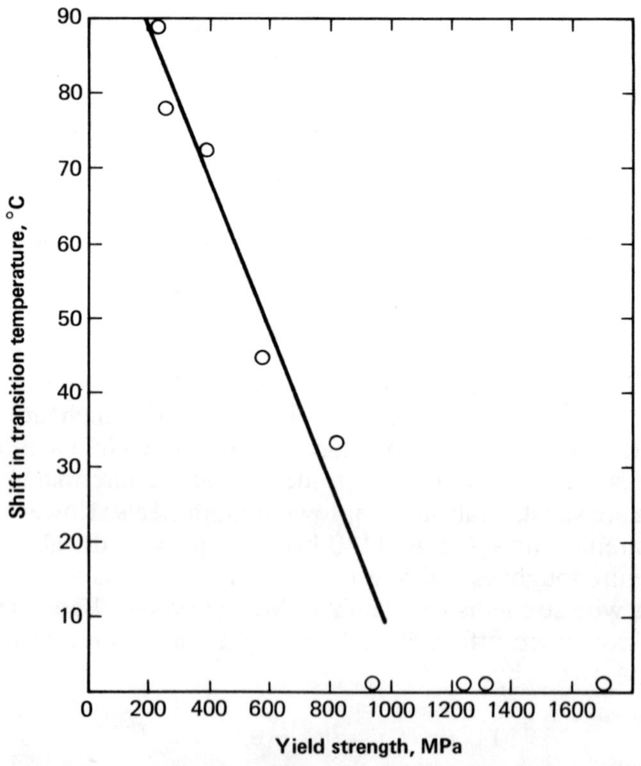

Fig. 4.16. Effect of yield strength on shift in transition temperature between impact and slow-bend K_{Ic} test loading rates (Ref 4.41)

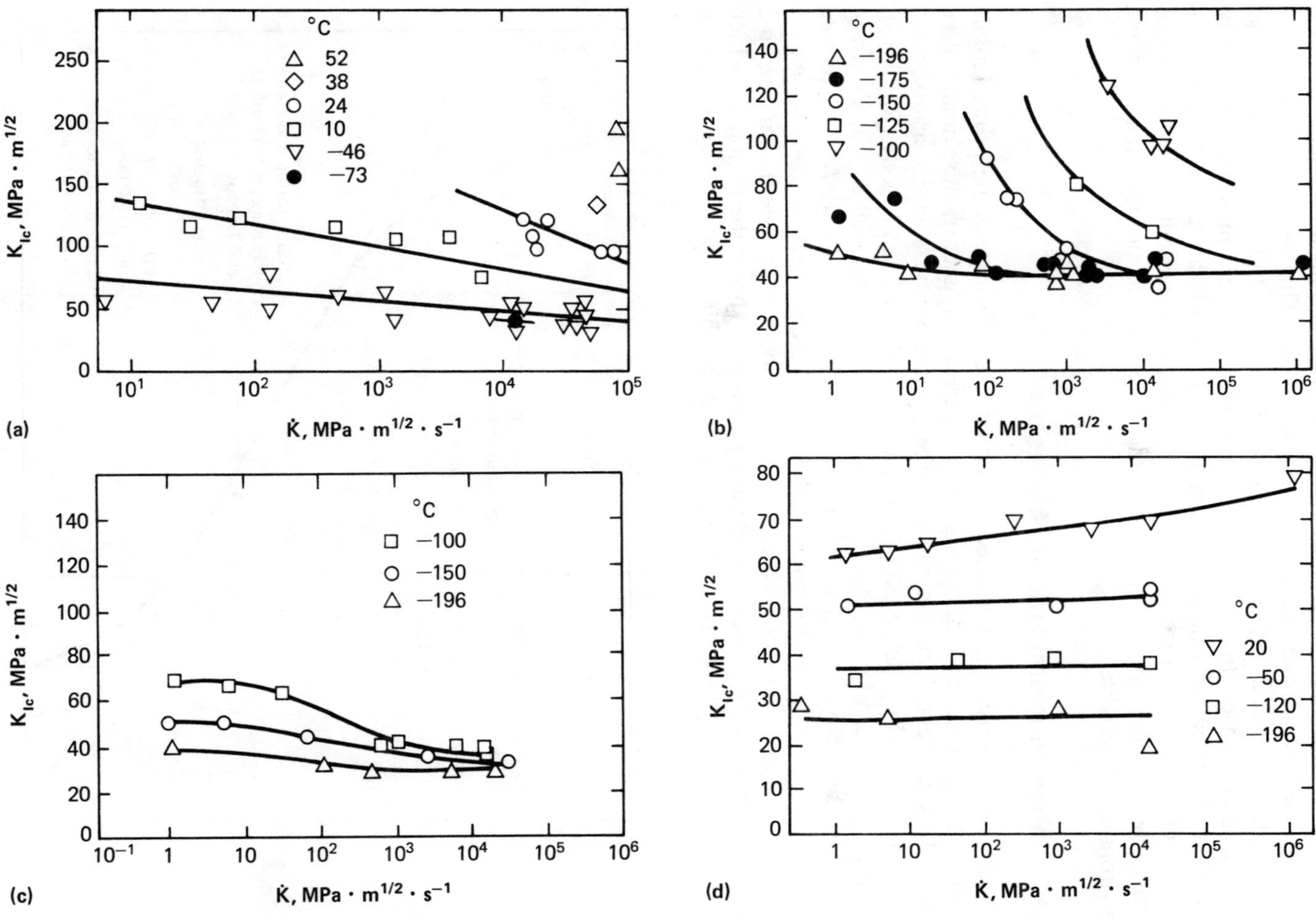

(a) A533B steel; $\sigma_{ys} \simeq 450$ MPa (65 ksi). (b) QT35 steel; $\sigma_{ys} \simeq 600$ MPa (87 ksi). (c) Ni-Cr-Mo-V rotor forging steel; $\sigma_{ys} \simeq 670$ MPa (97 ksi). (d) 0.4C-9Ni-4Co steel; $\sigma_{ys} \simeq 970$ MPa (140 ksi). (Yield strengths reported are at room temperature.)

Fig. 4.17. Influence of temperature and loading rate on fracture toughness (Ref 4.14)

4.4. FATIGUE CRACK PROPAGATION

Cyclic loading can cause crack propagation in certain alloy steels at stress intensities as low as one twentieth of the K_{Ic} value. Threshold stress intensities may range from 3 to 20 MPa · $m^{1/2}$ (2.7 to 18 ksi · $in.^{1/2}$). This demonstrates that factors other than continuum plasticity considerations are important. It will be shown that both microstructure and environment greatly affect the fatigue crack growth rate response in steels. First, the various microstructural fracture modes will be briefly considered, followed by detailed sections on heat treatment and environmental variables.

4.4.1. Microstructural Fracture Modes

Although it is usual to consider ductile fatigue striations as the prime fatigue fracture mode, it is becoming increasingly apparent that other fracture modes often occur. This has been documented by a large number of investigators, as discussed in recent reviews by Ritchie (Ref 4.42) and by Gerberich and Moody (Ref 4.43).

Crack growth can be conveniently described by graphs such as that shown in Fig. 4.18 (see also Fig. 2.10), where the crack growth rate, da/dN, is plotted against the stress-intensity range, ΔK. The typical crack growth shown in Fig. 4.18 can be divided into three regions. For most steels, the data in region 2 can be described by a power law of the type shown. This is the region for which the largest amount of data is available. These data indicate that, in general, crack

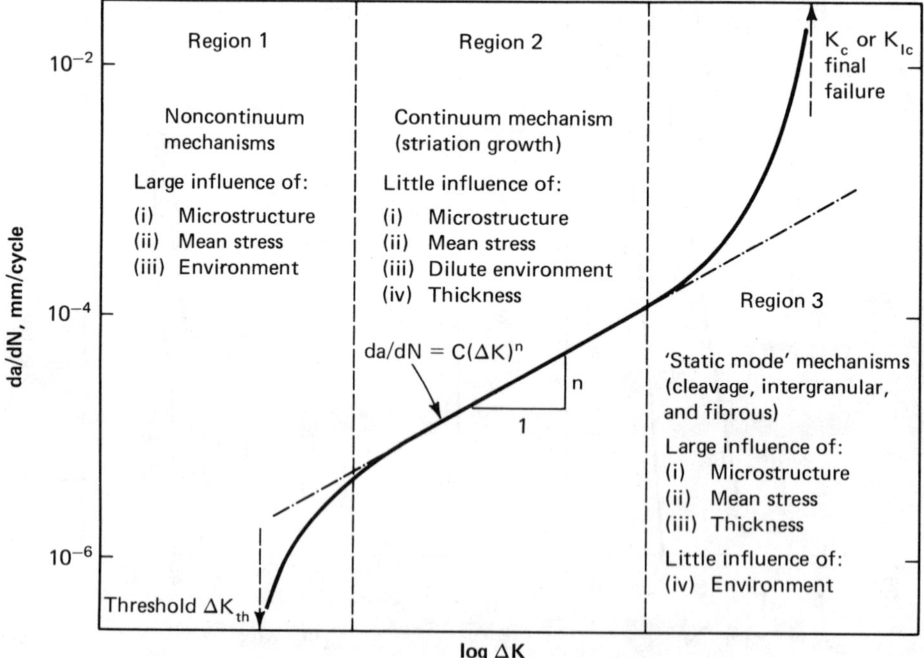

Fig. 4.18. Schematic variation of fatigue crack growth rate, da/dN, with alternating stress intensity, ΔK, in steels, showing regions of primary crack growth mechanisms (Ref 4.42)

growth in region 2 is relatively insensitive to variations in either microstructure or crack shape. In region 1, usually called the threshold region, the rate of crack growth per cycle drops off rapidly as ΔK is decreased. Behavior in this region may exhibit considerable sensitivity to the microstructure of the material. In region 3, the growth rate may become large because the maximum value of the stress-intensity factor at maximum cyclic load approaches the fracture toughness. Here the local stresses are sufficiently large to activate fracture by other microstructural modes such as intergranular fracture, cleavage fracture or dimple rupture. In Fig. 4.19, Ritchie (Ref 4.42) describes such a series of microstructural fatigue failure modes in steels. Although striations are shown at a ΔK of 30 MPa · $m^{1/2}$

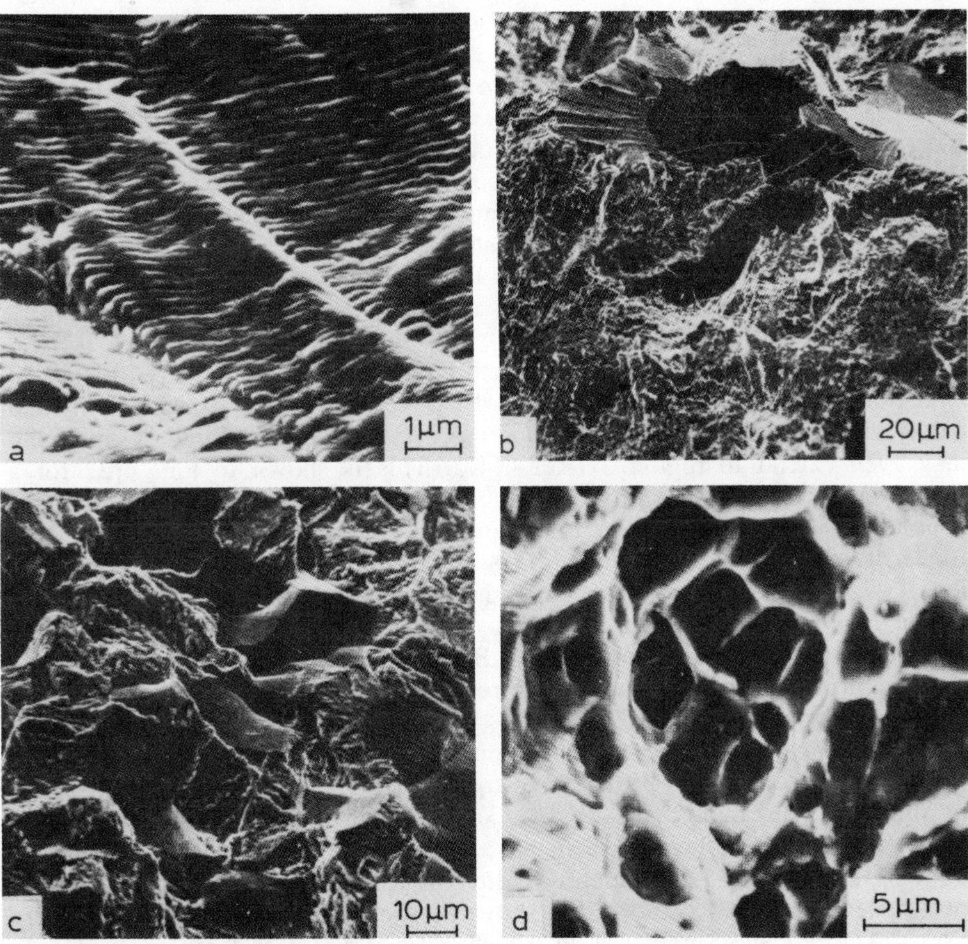

(a) Ductile striations in 9Ni-4Co steel at ΔK = 30 MPa · $m^{1/2}$. (b) Additional cleavage fracture in mild steel at ΔK = 40 MPa · $m^{1/2}$. (c) Additional intergranular fracture in 4Ni-1.5Cr steel at ΔK = 40 MPa · $m^{1/2}$. (d) Microvoid coalescence in 9Ni-4Co steel at ΔK = 70 MPa · $m^{1/2}$.

Fig. 4.19. Fractography of fatigue crack propagation at intermediate (region 2) and high (region 3) growth rates in steels tested in moist air at R = 0.1 (Ref 4.42)

(27 ksi · in.$^{1/2}$), at higher stress intensities it is seen that cleavage fracture, intergranular fracture or microvoid coalescence may occur, in addition to striation formation (see Fig. 4.19b, c and d). Such modes (Ref 4.43 and 4.44), particularly intergranular and cleavage fracture, may also appear at lower stress intensities.

Because heat treatment, alloy additions and grain size may affect the tendency towards any microstructural fracture mode, these factors must be considered in addition to yield strength and external loading variables.

4.4.2. Effects of Microstructure and Heat Treatment

In ultrahigh-strength steels, the effect of tempering is relatively small for region 2 growth rates, but the effect may be large near the threshold. An investigation of HP-9-4-30 steel quenched and tempered to yield strengths of 675, 1235 and 1400 MPa (95, 180 and 200 ksi) showed increases in crack growth rate with increasing yield strength, as illustrated in Fig. 4.20(a). On the other hand, Imhof and Barsom (Ref 4.45) evaluated AISI 4340 steel at yield strengths ranging from 895 to 1515 MPa (130 to 220 ksi) and found no significant variation in region 2. Whereas the effect on da/dN may only constitute a factor of two or three (or less) in region 2, there may be order-of-magnitude differences in growth rate near the threshold. This is shown for a 300M steel tempered at 100 to 650°C (212 to 1200°F) in Fig. 4.20(b). As discussed by Ritchie (Ref 4.42), this is a superimposed environmental effect but shows up in laboratory air with as little as 40% relative humidity.

Besides tempering effects, microstructure may also have relatively little effect on region 2 growth, as seen in Fig. 4.21. However, it has been found recently that modified microstructures may produce large variations in region 1 growth and may even extend to region 2 (Ref 4.47). This is shown in Fig. 4.22 for a low-carbon steel, where a mixture of about 38% martensite in ferrite produced about the same resistance to crack growth as that indicated in Fig. 4.21 for ferrite-pearlite microstructures. This is for heat treatment A, as indicated in Fig. 4.22, where the ferrite phase is continuous. If the thermal cycle is now varied to make the martensitic phase continuous, order-of-magnitude improvements are obtained at low ΔK with heat treatment B. Thus, it would appear that substantial improvements in fatigue crack growth resistance near the threshold are possible with microstructural control.

Another means of microstructural control is the grain size effect on fatigue thresholds, ΔK_{th} (Ref 4.42). One investigation (Ref 4.48) shows the grain effect on mild steel thresholds (see Fig. 4.23). Although there was no difference in region 2 growth, the threshold values varied from 5.3 to 7.0 MPa · m$^{1/2}$ (4.8 to 6.4 ksi · in.$^{1/2}$) as grain size increased from 7.8 to 55 μm. A collection of data from two reviews (Ref 4.42 and 4.43) shows this effect very well. One of these (Ref 4.42) indicates, in Fig. 4.24, the beneficial effect of large grain size on thresholds for low-strength steels and the detrimental effect of large grain size for high-strength steels. The former is considered to be related to a cyclic slip or microstructurally sensitive crack path that is controlled by the grain diameter whereas, for the high-strength steel, the effect is considered to be environmentally related (Ref 4.42). For the low-to-medium-strength steels, it is useful to compare the grain size with the reversed plastic zone size, $R_{p\pm}$, because many studies

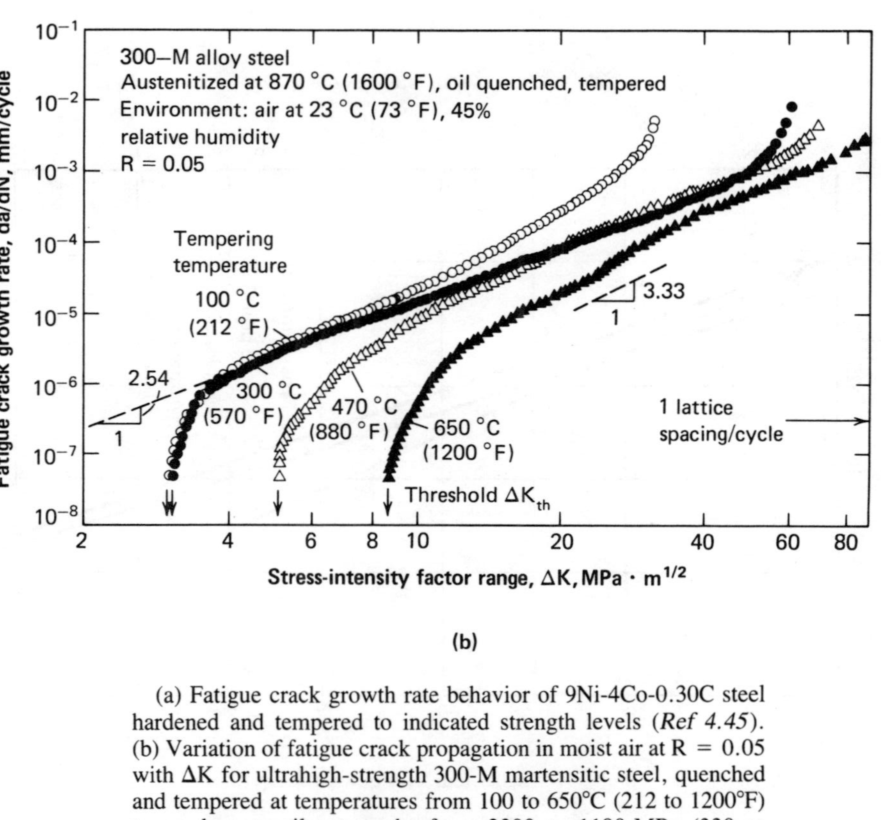

(a) Fatigue crack growth rate behavior of 9Ni-4Co-0.30C steel hardened and tempered to indicated strength levels (*Ref 4.45*). (b) Variation of fatigue crack propagation in moist air at R = 0.05 with ΔK for ultrahigh-strength 300-M martensitic steel, quenched and tempered at temperatures from 100 to 650°C (212 to 1200°F) to produce tensile strengths from 2300 to 1190 MPa (330 to 275 ksi), respectively *(Ref 4.42)*.

Fig. 4.20. Effect of strength level on fatigue crack growth rates

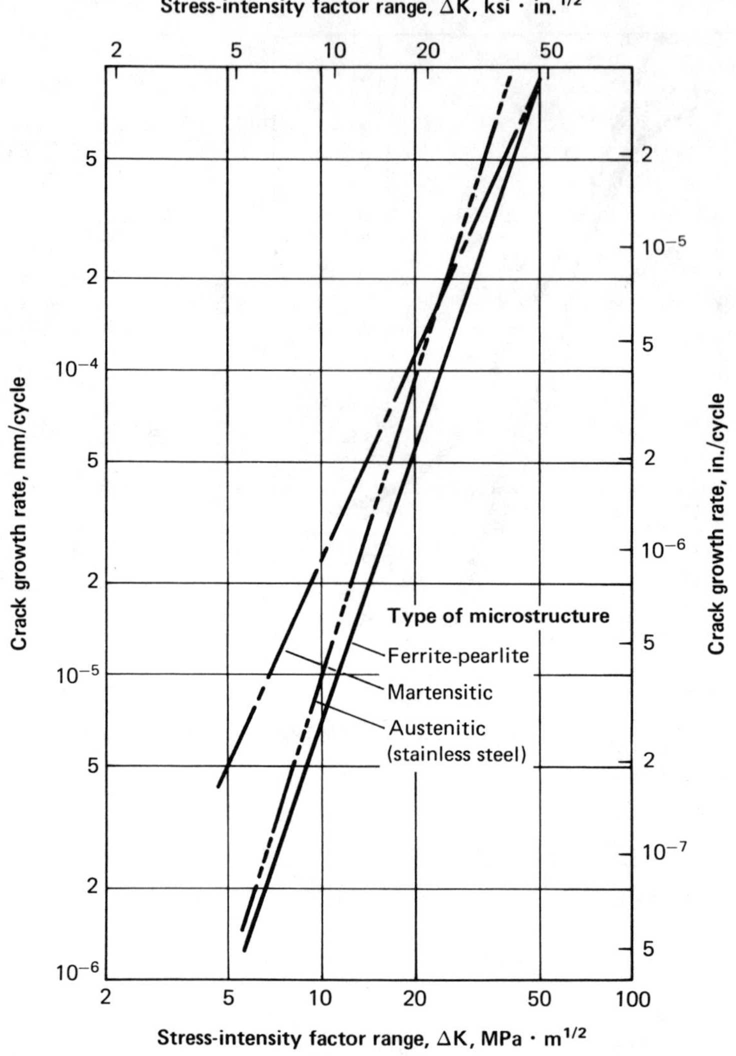

Type of microstructure	Yield strength MPa	ksi	Tensile strength MPa	ksi	Strain-hardening exponent
Austenitic (stainless steel)	205–345	30–50	515–655	75–95	>0.30
Ferrite–pearlite	205–550	30–80	345–755	50–110	0.15–0.30
Martensitic	>480	>70	>620	>90	<0.15

Fig. 4.21. Upper limits of fatigue crack growth rates for three types of steel microstructures (Ref 4.46)

(Ref 4.49, 4.50 and 4.51) have claimed that this is the controlling microstructural unit. Using twice the plane-strain plastic zone radius (from Eq 2.15), and twice the yield stress due to the stress reversal, gives:

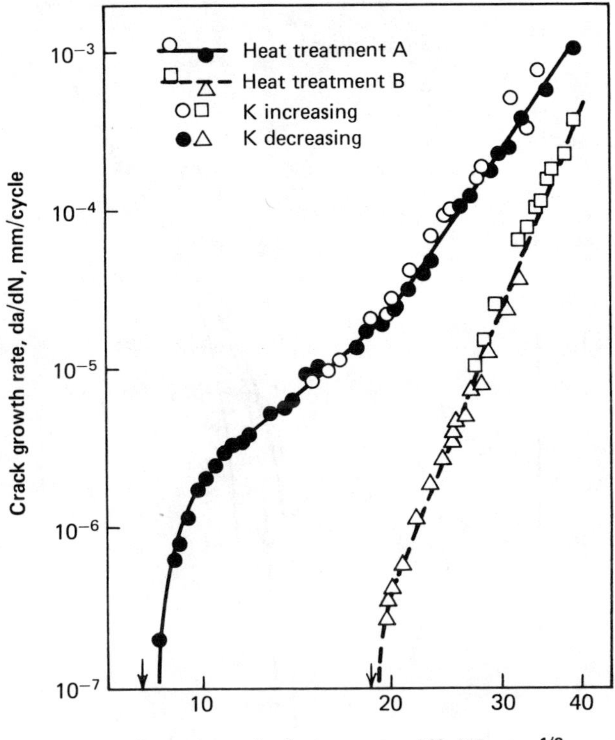

Property	Heat treatment A (ferritic)	Heat treatment B (martensitic)
0.2% proof stress, MPa (ksi)	293 (43)	452 (66)
Ultimate tensile strength, MPa (ksi)	543 (79)	750 (109)
Elongation, % in 12.7 mm	...	9.9
Reduction in area, %	32.6	...

Fig. 4.22. Rate of crack growth vs ΔK for two microstructures in low-carbon steel (0.15-0.20 C, 0.60-0.90 Mn, 0.04 max P, 0.04 max S) (Ref 4.47)

$$R_{p_\pm} \simeq \frac{2\Delta K^2}{6\pi(2\sigma_{ys})^2} = \frac{\Delta K^2}{12\pi(\sigma_{ys})^2} \qquad \text{(Eq 4.12)}$$

One way of understanding this is to consider that general cyclic slip will not proceed if the grain size is greater than the reversed plastic zone size. Thus, a description of the fatigue threshold could be obtained by substitution of $d = R_{p_\pm}$ at $\Delta K = \Delta K_{th}$ in Eq 4.12. This gives:

$$K_{th} = \sigma_{ys}(12\pi d)^{1/2} = 6.14\sigma_{ys}(d)^{1/2} \qquad \text{(Eq 4.13)}$$

If threshold values were normalized by grain size and plotted against yield strength, the slope of this curve could then be compared with Eq 4.13. A collection of data in Fig. 4.25 plotted in this manner predicts a least-squares slope of 5.0

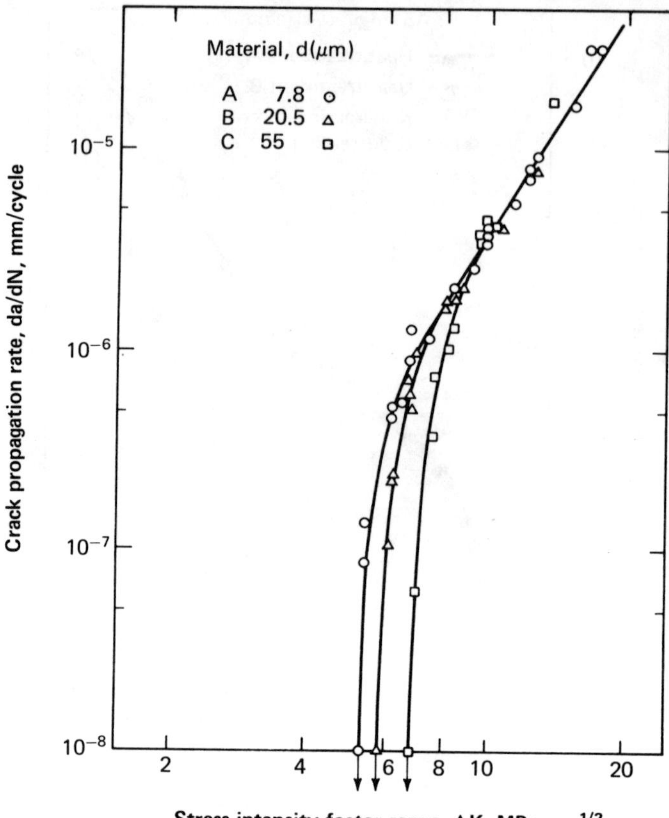

Steels are ferrite-pearlite mixtures containing 0.2 C, 0.92 Mn, 0.26 Si, 0.11 P, 0.15 S and 0.009 N, with ferrite grain sizes of 7.8, 20.5 and 55 μm.

Fig. 4.23. Relationship between crack propagation rate and stress-intensity factor range for R = −1 (Ref 4.48)

instead of 6.14. However, if the curve is forced through the origin, the best fit gives a slope of 6.06, which verifies Eq 4.13 and demonstrates that both ferrite grain size and yield strength have strong effects on low-strength threshold stress intensities.

This *does not* necessarily mean that a designer should select large-grained steels for performance. In fact, on the same steel for which Taira *et al* (Ref 4.48) measured thresholds (Fig. 4.23), they also measured smooth-bar endurance limits. As shown in Fig. 4.26, although the number of cycles to initiate slip bands was about the same, the number of cycles to produce a crack which leads to failure depends heavily on grain size, with the finest grains giving the best performance. Thus, fine grains appear to be best for resistance to fatigue initiation and coarse grains appear to be best for resistance to crack propagation for long cracks. In fact, it may be shown that there is a crossover in endurance limit for very short cracks versus long cracks. The short-crack endurance limit is highest for fine-grained steel and the long-crack endurance limit is highest for coarse-grained steel.

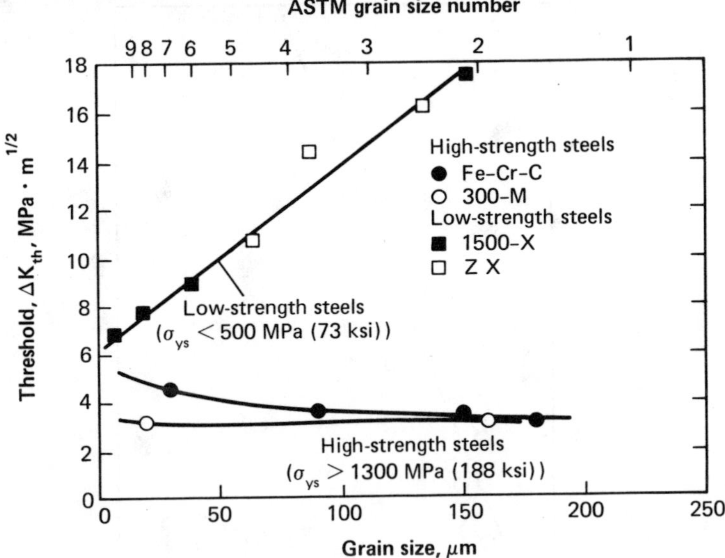

For ferritic-pearlitic low-strength steels, grain size refers to ferritic grain size; for martensitic high-strength steels, grain size refers to prior austenite grain size.

Fig. 4.24. Variation of threshold ΔK_{th} with grain size for steels at R = 0.05 (Ref 4.42)

A similar crossover has been postulated in high-strength steels, except now the critical parameter is yield strength. This is shown in Fig. 4.27 for a 300M steel quenched and tempered to wide variations in strength level. Whereas the low-strength material would offer the best performance for long cracks, the high-strength material would be superior for short cracks. As discussed by Fine and Ritchie (Ref 4.54), this becomes important because the optimization of microstructure will depend on whether the structural engineer is designing for crack initiation or crack propagation from a pre-existing flaw.

Crack growth resistance also may be different for cracks that propagate along different directions measured relative to the rolling direction. Figure 4.28 presents data on crack growth rate for specimens of conventionally melted ASTM A533, Type B, steel tested for each of six orientations. In this study, the effects of three processing techniques—conventional melting, calcium treatment and electroslag remelting—were investigated. The electroslag remelted material produced the lowest crack growth rates and the least sensitivity to orientation.

4.4.3. Loading, Temperature and Frequency Variables

In both fatigue endurance and fatigue crack propagation testing, the mean stress intensity is an important variable. Because K is proportional to stress, the standard stress nomenclature in Fig. 4.29 used in fatigue testing is also appropriate to fatigue cracking with K's substituted for σ's. The various parameters are interrelated, so that only one variable other than ΔK need be considered. The stress

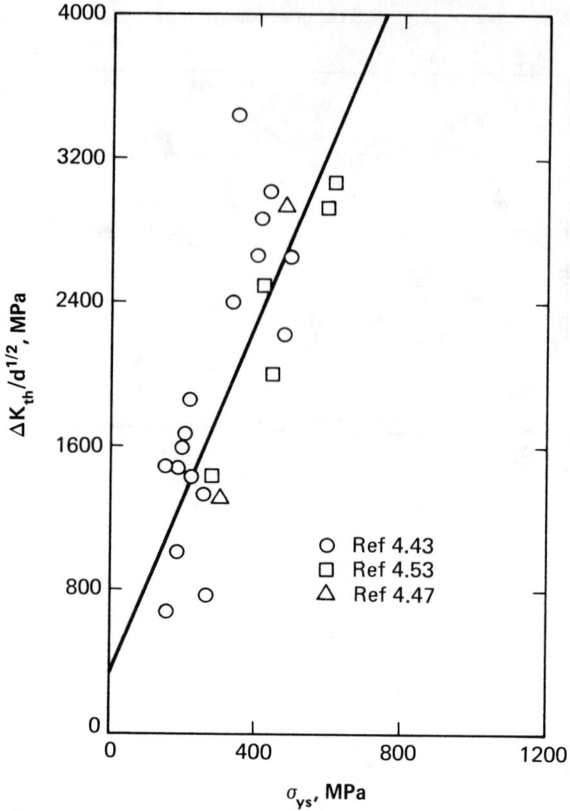

Fig. 4.25. Effect of yield strength on normalized threshold stress-intensity range

ratio, R, is the variable most used in experiments on fatigue crack growth. Figure 4.30 illustrates the effects of variations in R on the fatigue behavior of an ASTM A533, Type B, Class 1, steel in laboratory air. The R ratio has a large effect in region 1 but little effect in region 2. In a class of low- to-medium-strength steels, the effect of stress ratio on the threshold stress-intensity range, ΔK_{th}, apparently is similar for seven different materials, as shown in Fig. 4.31. For values of R greater than 0.1, ΔK_{th} can be given by:

$$\Delta K_{th} = C_1(1 - 0.85R) \qquad \text{(Eq 4.14)}$$

where C_1 is a constant whose value is 7.0 MPa · $m^{1/2}$ (6.4 ksi · in.$^{1/2}$). Because of the narrow grouping of data in Fig. 4.31, it must be assumed that these were relatively fine-grained steels without the microstructural influences discussed above. Where greater influences of strength, grain size, microstructure or environment are encountered, the constant C_1 would necessarily change and the R effect might even be qualitatively different. For example, in vacuum test conditions, the threshold may be independent of R (Ref 4.42).

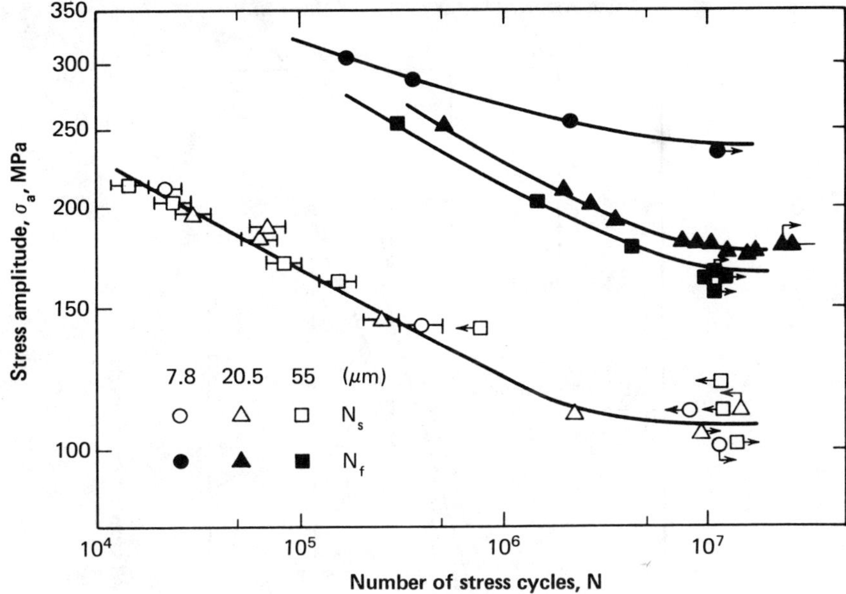

Fig. 4.26. S-N curve for slip band initiation, N_s, and total failure, N_f (Ref 4.48)

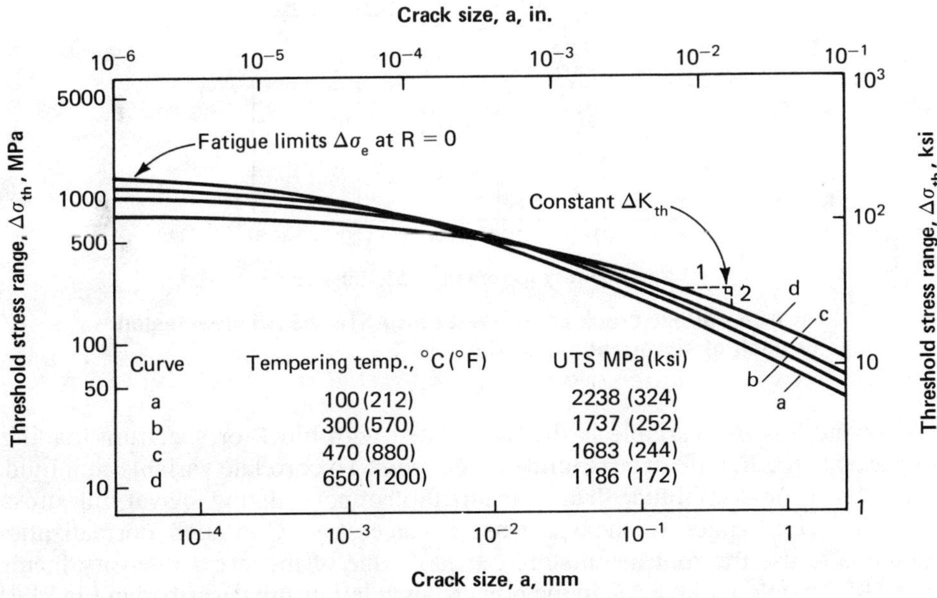

Based on data for 300M ultrahigh-strength steel tempered at temperatures from 100 to 650°C (212 to 1200°F) to produce a variety of tensile strengths.

Fig. 4.27. Predicted variation of threshold stress $\Delta\sigma_{th}$ at R = 0 with crack size a (Ref 4.42)

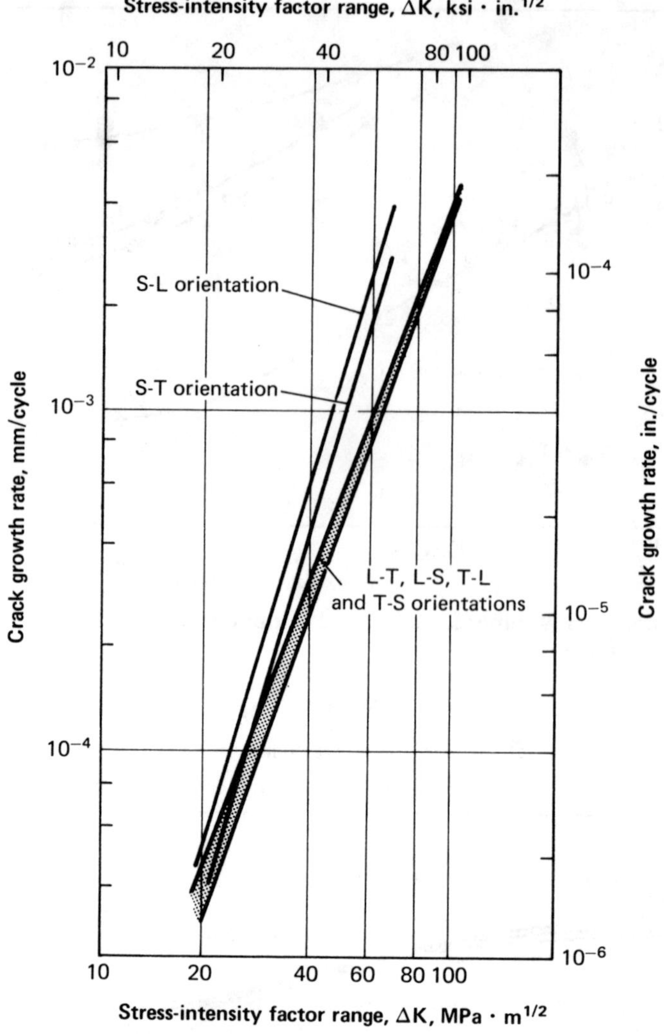

Fig. 4.28. Fatigue crack growth rates for ASTM A533B steel tested for each of six orientations (Ref 4.55)

A second loading variable is that associated with block or spectrum loading, often used to predict life of structural components. To correlate variable-amplitude data with constant-amplitude data, some method of normalizing the varying stress-intensity factor ranges in the spectrum is necessary. One such normalization scheme is to use the root mean square (rms) value of the stress-intensity factor. Here ΔK_{rms} would replace ΔK in the power-law relationship described in Fig. 4.18 (and Eq 2.16): ΔK_{rms} is given by:

$$\Delta K_{rms} = \sqrt{\frac{\sum_{i=1}^{N} (\Delta K_i)^2}{N}} \qquad \text{(Eq 4.15)}$$

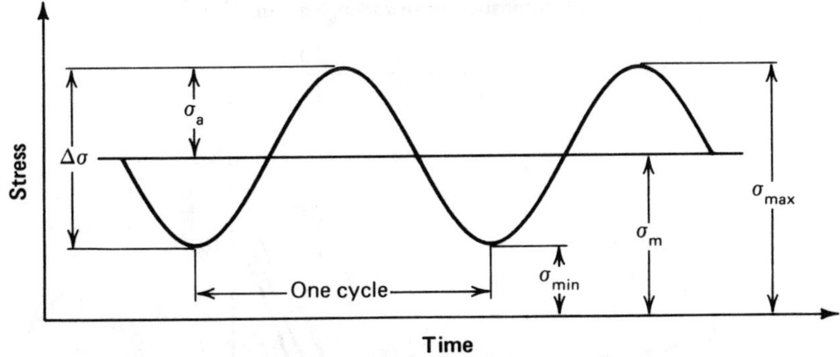

$\Delta\sigma$ = stress range; σ_a = stress amplitude; σ_{min} = minimum stress; σ_m = mean stress; σ_{max} = maximum stress; R = stress ratio ($\sigma_{min}/\sigma_{max}$); $\sigma_m = 1/2(\sigma_{max} + \sigma_{min})$.

Fig. 4.29. Nomenclature used in fatigue testing

Barsom (Ref 4.46) used the loading sequences shown in Fig. 4.32 to establish a data base for evaluation of the ΔK_{rms} normalization procedure. The results shown in Fig. 4.33 show an excellent correlation between the constant-amplitude data (ΔK) and the variable-amplitude data (ΔK_{rms}). It should be emphasized that such an independence of loading sequence may not apply to region 1 crack growth or to situations involving aggressive environments.

Temperature can affect crack growth rates in several ways. At high temperatures, ease of dislocation motion, dynamic strain aging, or oxidation can be sufficiently enhanced so that creep-fatigue or environmental-fatigue interactions result. McHenry and Pense (Ref 4.56) found that the crack growth behavior of HP 9-4-35 steel was not appreciably affected by temperatures as high as 350°C (650°F) or changes in frequency from 0.02 to 10 Hz. However, their experiments on both ASTM A212, Grade B (now ASTM A515, Grade 70), and A517, Grade F, steels showed that fatigue crack growth rates increased with increasing testing temperature as illustrated in Fig. 4.34. Decreasing the frequency of loading for specimens of A517, Grade F, steel also increased the rate of crack growth.

At low temperatures, the yield strength may be sufficiently increased so that cleavage is promoted at the fatigue crack tip. In Fig. 4.35, Stonesifer (Ref 4.57) shows that very low temperatures produce very high crack growth rates at low ΔK levels in A533B steel. It is clear that the slope of the −196°C (−320°F) curve is such that n in the power-law relationship, $da/dN = C(\Delta K)^n$ (Eq 2.16) is an order of magnitude larger for the −196°C data than it is for the 20°C (68°F) data. This effect of testing temperature on the fatigue crack propagation exponent was previously reviewed (Ref 4.43), and for a large number of iron-base alloys and steels the results in Fig. 4.36 demonstrate the increase in the value of n with decreasing testing temperature. From these results, one should not assume that crack growth rates are going to be higher at any temperature below room temperature. In fact, at temperatures only slightly below room temperature, improved resistance to fatigue crack propagation may be observed if severe microcleavage can be avoided. This implies that there is some low temperature where there is a crossover in resistance; this has been discussed elsewhere for iron-base systems (Ref 4.58).

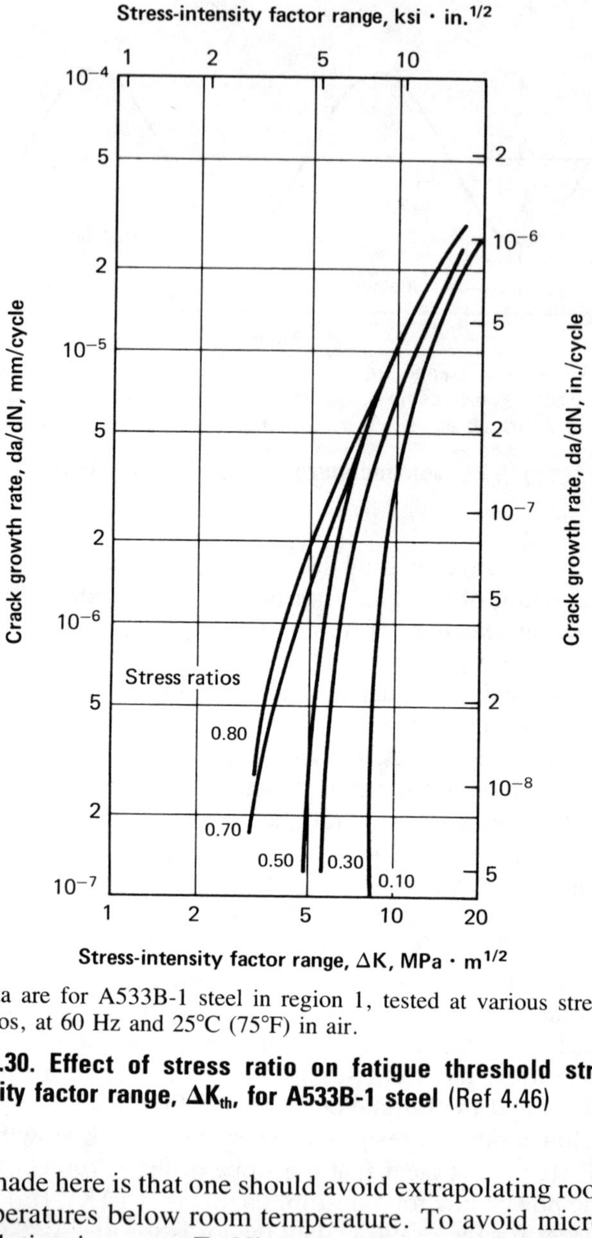

Data are for A533B-1 steel in region 1, tested at various stress ratios, at 60 Hz and 25°C (75°F) in air.

Fig. 4.30. Effect of stress ratio on fatigue threshold stress-intensity factor range, ΔK_{th}, for A533B-1 steel (Ref 4.46)

The point to be made here is that one should avoid extrapolating room-temperature behavior to temperatures below room temperature. To avoid microcleavage phenomena, one solution is to use Fe-Ni steels. The nickel additions decrease the ductile-to-brittle transition and result in low fatigue crack growth exponents even at −196°C (−320°F), as shown in Fig. 4.36.

4.4.4. Aggressive Environments

Aggressive environments affect fatigue crack growth rates in all regions of growth but most often in regions 1 and 2 where exposure times are much greater. Three patterns of corrosion fatigue behavior which might be observed in steels are

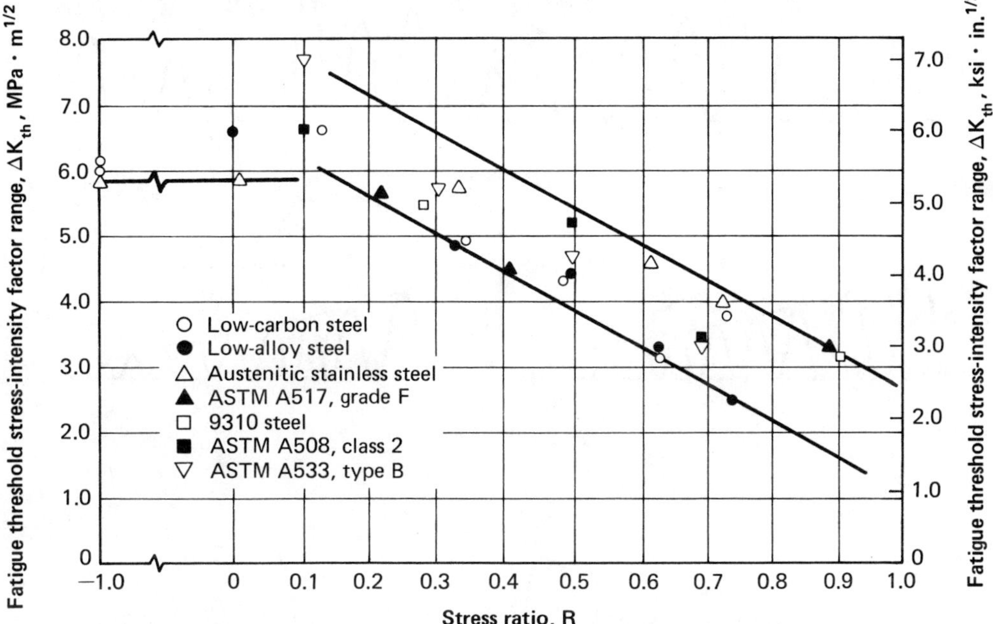

Fig. 4.31. Effect of stress ratio on fatigue threshold stress-intensity factor range, ΔK_{th}, for several steels (Ref 4.46)

indicated in Fig. 4.37. Figure 4.38(a) illustrates the corrosion fatigue behavior of AISI 4340 steel quenched and tempered to a yield strength of 1700 MPa (245 ksi) (Ref 4.59). In distilled water, a 1000-fold increase in the crack growth rate is produced by a 1000-fold decrease in the cyclic frequency. This behavior falls somewhere between types B and C behavior in Fig. 4.37. Similar behavior is observed for X-65 line pipe steel tested in salt water with a superimposed cathodic potential (Ref 4.60). However, as shown in Fig. 4.38(b), the same change in frequency does not produce quite as large a change in fatigue crack growth rate. In addition, the cathodic potential increases the growth rates substantially over the open-circuit case for X-65 steel. Both the effects in Fig. 4.38 have been associated with hydrogen entry, and in fact most corrosion fatigue of high- and ultrahigh-strength steels is considered to occur by a combined mechanism of hydrogen embrittlement and fatigue.

Effects of alloying additions are shown in Fig. 4.39 and 4.40. First, it can be seen in Fig. 4.39 that for the identical test frequency, the susceptibility of a 12Ni-5Cr-3Mo maraging steel to corrosion fatigue in a 3% aqueous sodium chloride solution is nearly identical to that of 4340M steel in distilled water. It should be noted that the K_{Iscc} value for this alloy is 60 MPa · $m^{1/2}$ (55 ksi · in.$^{1/2}$), so that nearly all of the environmental effects are showing up well below the static stress corrosion threshold, K_{Iscc}. This corresponds most closely to type A behavior in Fig. 4.37. The same data for 12Ni-5Cr-3Mo at 0.1 Hz are compared with those for two other alloys in Fig. 4.40. The lowest corrosion-fatigue crack growth rates were achieved in specimens of the 10Ni-Cr-Mo-Co steel.

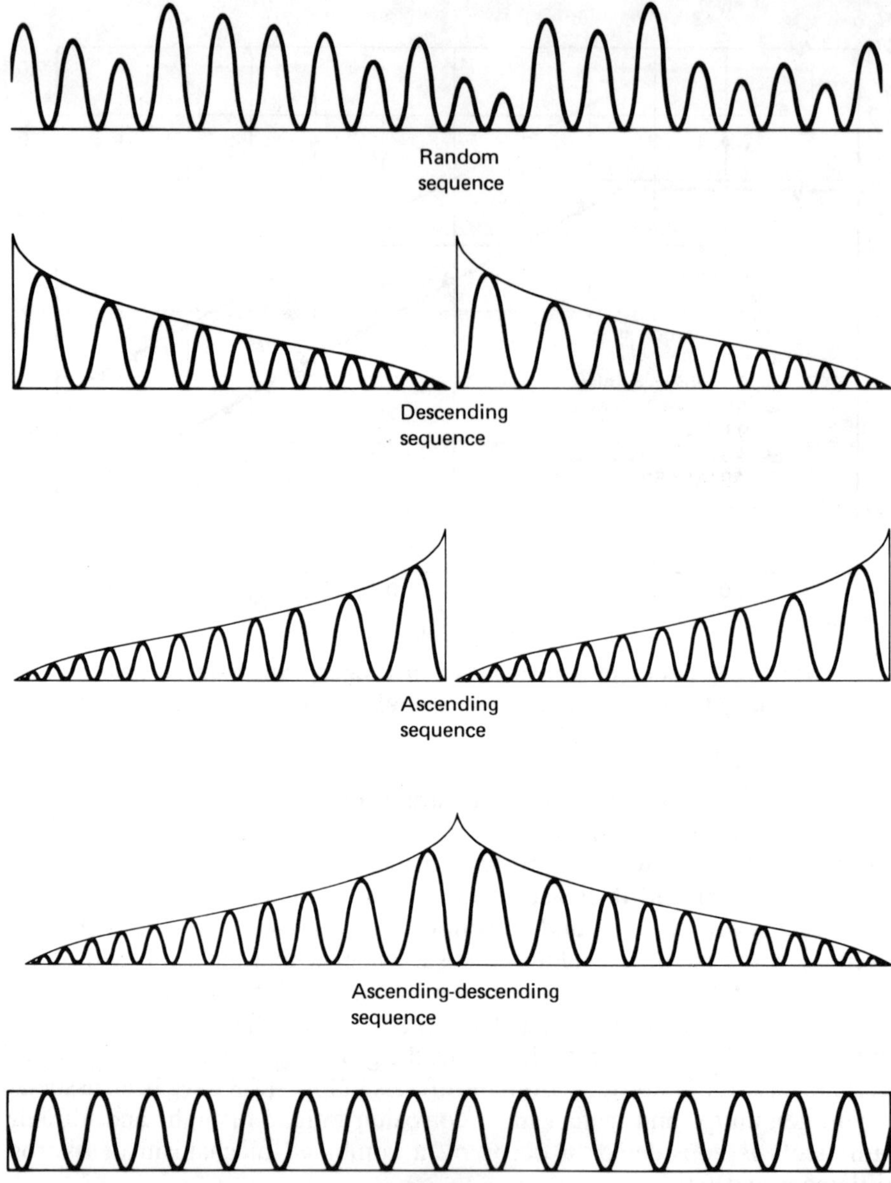

Fig. 4.32. Schematic representation of load sequences for fatigue crack growth rate testing (Ref 4.46)

For lower-strength steels tested in a 3% solution of aqueous sodium chloride, only slight effects were observed in A36, A588 Grade A and A514 Grade F steels at frequencies of 0.2 and 1 Hz (Ref 4.61). However, this was for relatively high growth rates in region 2. It is not clear what effects would occur in region 1 or at

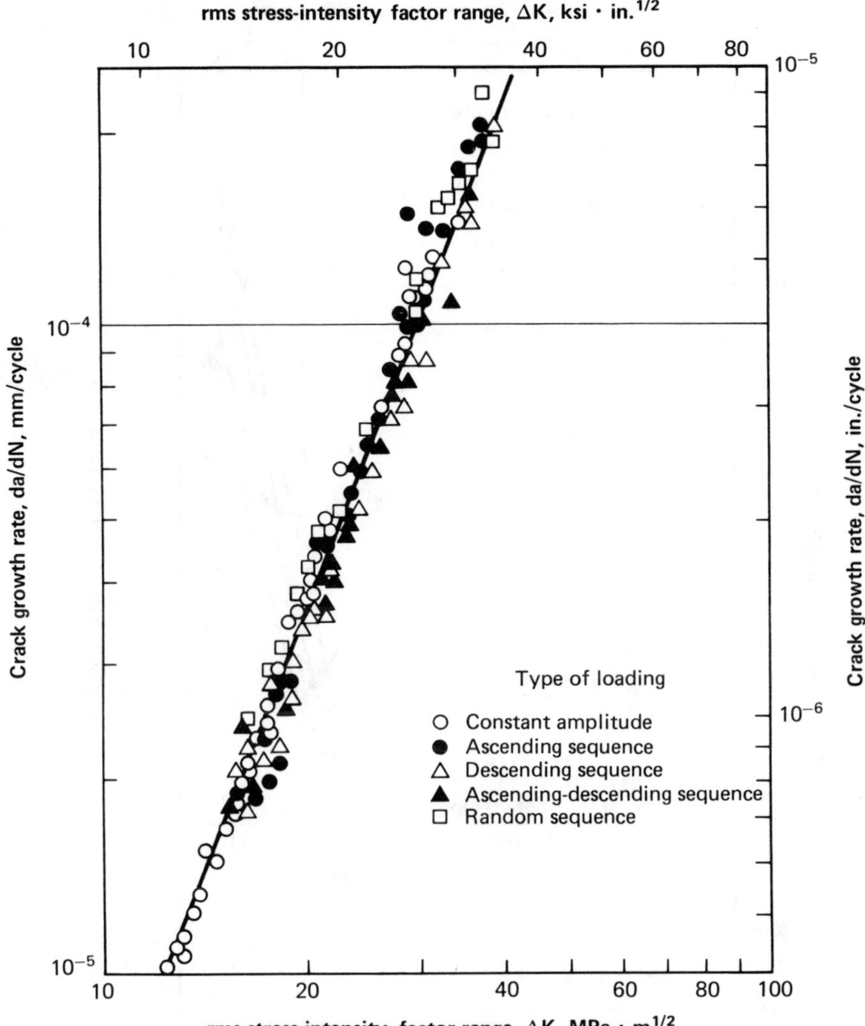

The loading patterns shown in Fig. 4.32 were imposed on specimens of A514B steel; rationalizing the stress-intensity factor range by a root-mean-square method caused the data to fall within a single band, independent of loading pattern.

Fig. 4.33. Effect of loading sequence on fatigue crack growth rate (Ref 4.46)

lower cyclic frequencies. Results of fatigue crack growth rate tests on specimens of A508-2, A533B, A516 and SA333 in reactor grade water indicate that the growth rate curves tend to lie in three basic regimes depending on test conditions (Ref 4.62).

Corrosion fatigue data for several alloy steels in a variety of environments are given in Table 4.4. In general, corrosion fatigue depends on cyclic frequency and

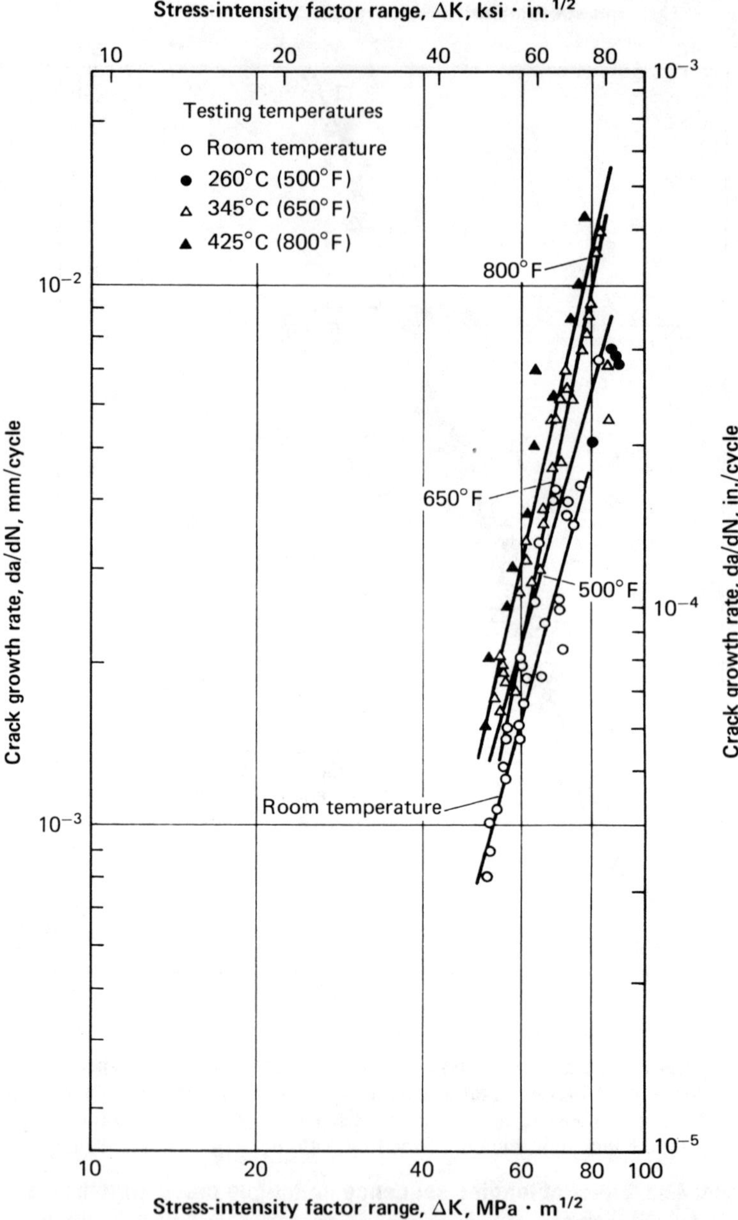

Fig. 4.34. Effect of elevated testing temperature on fatigue crack growth rates in ASTM A212B and A517F steels (Ref 4.56)

testing temperature and may occur over a large stress-intensity range and in solutions of any pH value.

With regard to inhibiting corrosion fatigue, there is some evidence that inhibitors provide effective passive films or modify the pH at the crack tip

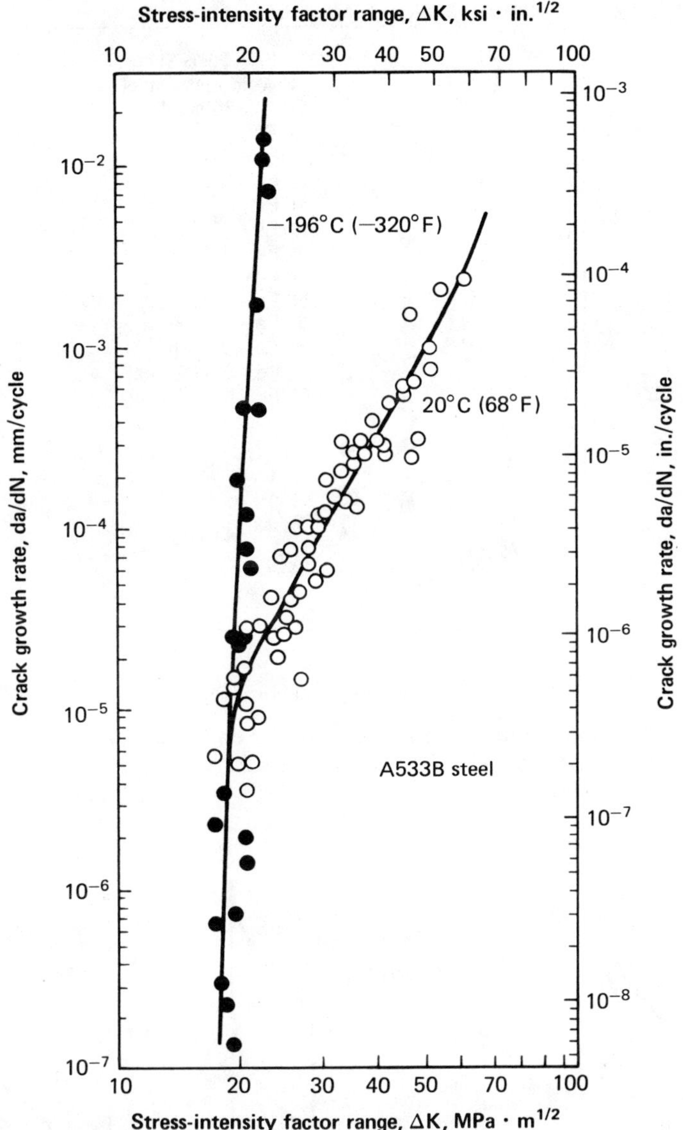

Fig. 4.35. Effect of low testing temperature on fatigue crack growth rates in A533B steel (Ref 4.57)

(Ref 4.68). The result in high-strength AISI 4340 steel, shown in Fig. 4.41, is that growth rates in air at 90% relative humidity are decreased by a factor of about three at a cyclic frequency of 0.167 Hz. These inhibitors were added to the humidified test chamber by means of an organic complex wherein the inorganic inhibitors were incorporated into a quaternary ammonium salt. The result is that the amount of intergranular fracture decreases and the environmental susceptibility to hydrogen embrittlement is reduced.

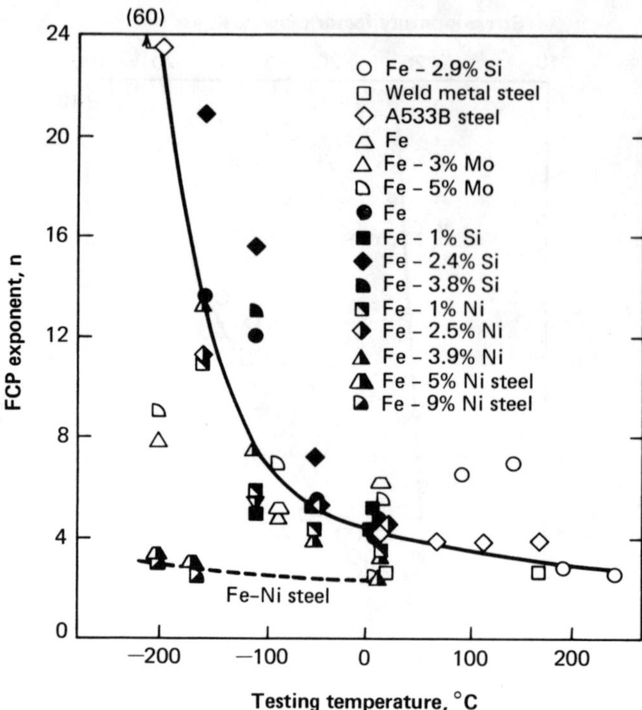

Fig. 4.36. Influence of testing temperature on fatigue crack propagation exponent for iron-base alloys (Ref 4.43)

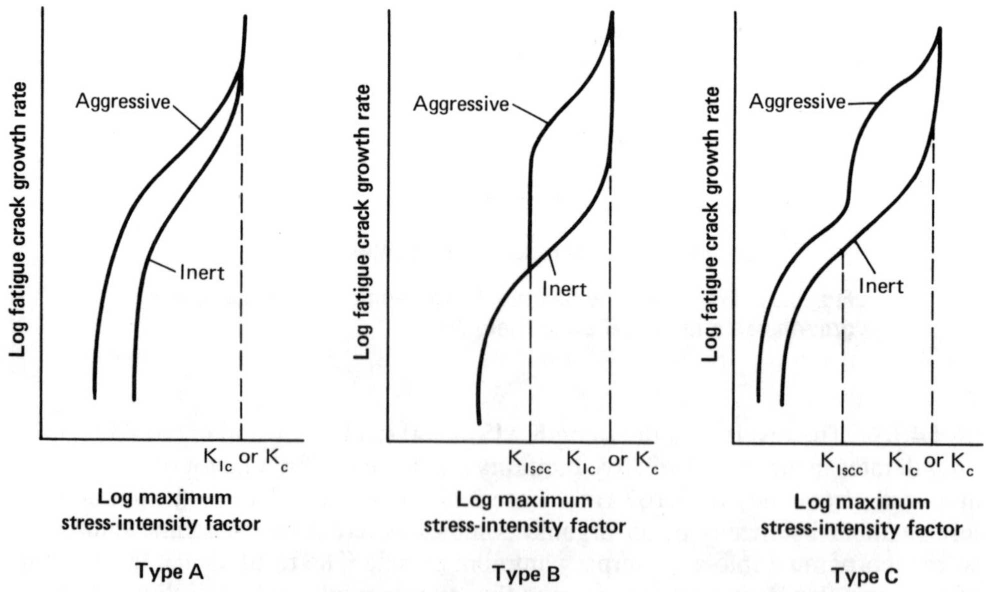

Fig. 4.37. Schematic diagrams showing three types of corrosion fatigue behavior (Ref 4.86)

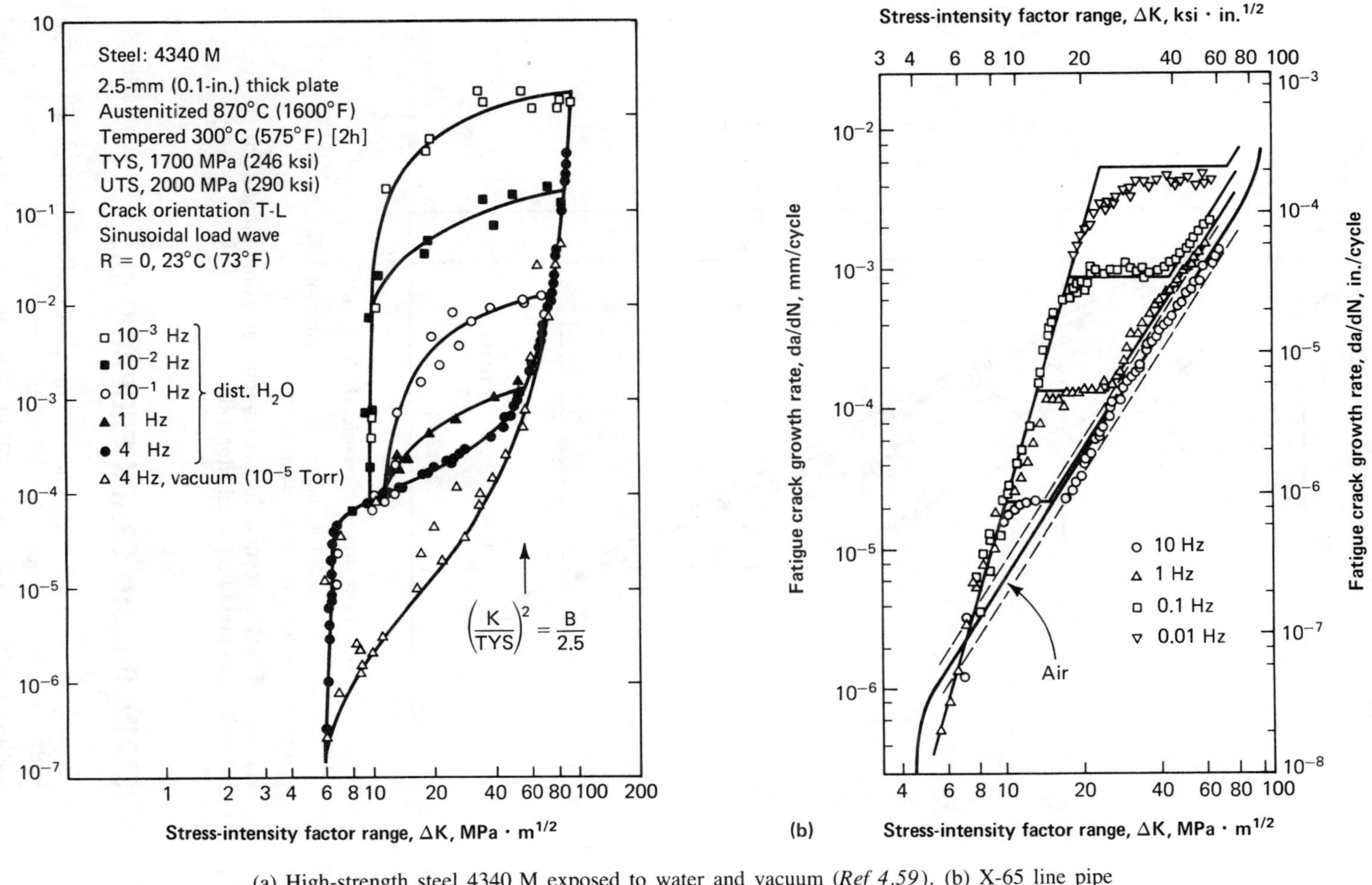

(a) High-strength steel 4340 M exposed to water and vacuum (*Ref 4.59*). (b) X-65 line pipe steel exposed to air and salt water with a superimposed cathodic potential (*Ref 4.60*).

Fig. 4.38. Effect of cyclic frequency on corrosion fatigue

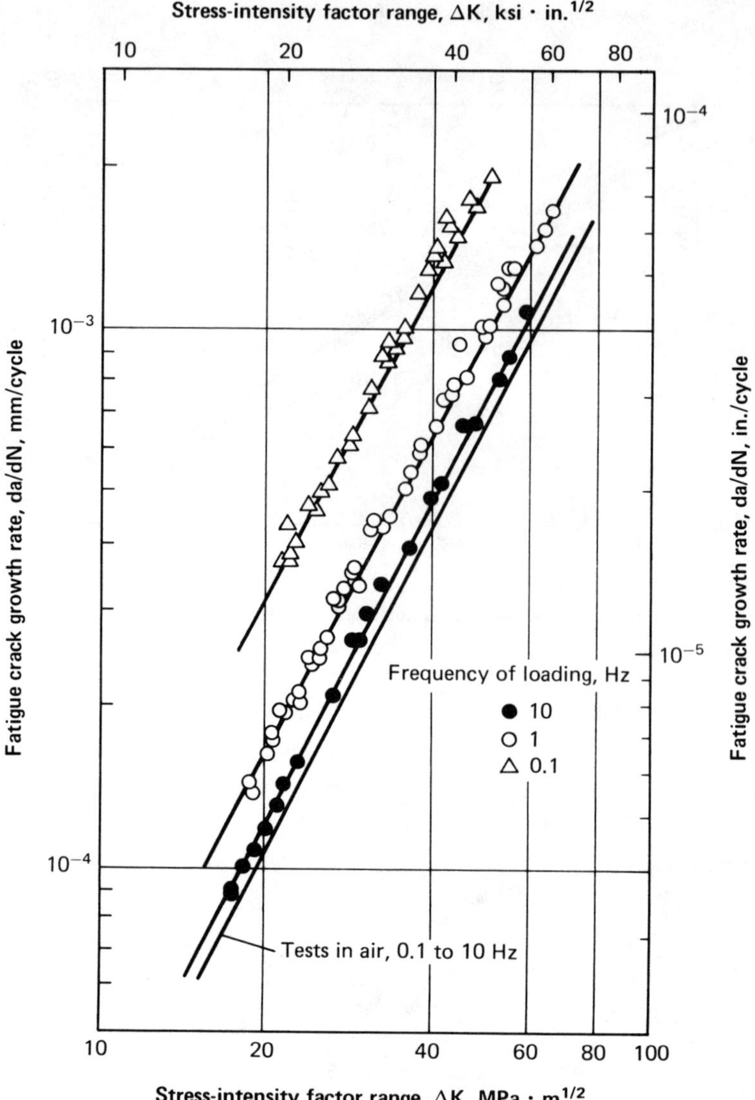

The steel was tested in air and in a 3% aqueous solution of sodium chloride with sinusoidal loading.

Fig. 4.39. Effect of cyclic frequency on corrosion fatigue for 12Ni-5Cr-3Mo maraging steel (Ref 4.46)

4.5. SUSTAINED-LOAD CRACK PROPAGATION

Environmentally assisted cracking may occur in any class of steel, given the right combination of environment, tensile stress, temperature, and presence or absence of a superimposed electrical potential. For a given combination, delayed failure may occur within minutes, as in the case of ultrahigh-strength steel fas-

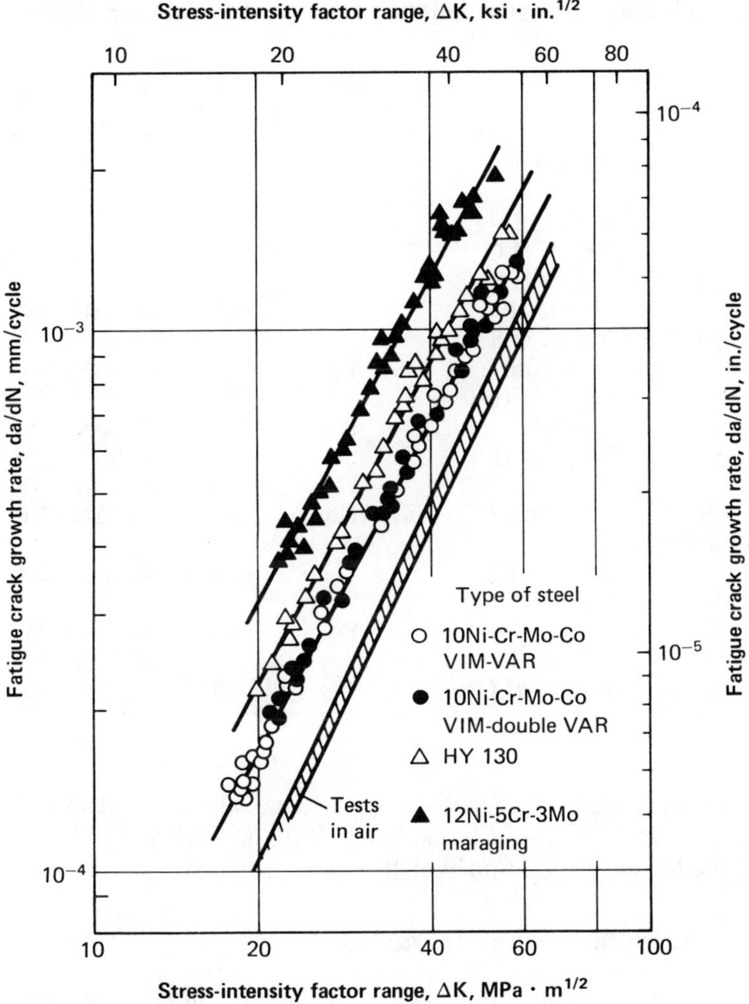

The steels were tested in air and in a 3% aqueous solution of sodium chloride at 0.1 Hz.

Fig. 4.40. Effect of composition on corrosion fatigue crack growth rates for three different high-strength steels (Ref 4.46)

teners coupled to aluminum in the presence of water, or it may take years, as in the case of steam turbine disks exposed to high-purity steam.

Typical results for static loading of precracked specimens of four high-strength steels tested in distilled water are shown in Fig. 4.42 (Ref 4.69). Each point represents a test of one specimen loaded to a particular level of stress intensity. Failure time periods depend on composition and strength of the steel and on stress intensity, provided that K_I is above a certain threshold level. The threshold stress intensity here is for tests of relatively thin sheet (plane stress). For considerably thicker plate test conditions (plane strain), the threshold value would be the critical value of K_{Iscc}, which occurs at a significantly lower K level than for thin sheet.

Table 4.4. Selected corrosion fatigue crack growth data for steels

Steel	Solution	Frequency dependent?	$\frac{(da/dN)_{sol}}{(da/dN)_{air}}$	Stress-intensity range, $MPa \cdot m^{1/2}$	Reference
A212B	Liquid Na	Yes	3	$50<\Delta K<80$	4.56
0.5Cr-0.5Mo-0.25V	Distilled H_2O	?	10	$12<\Delta K<24$	4.63
A302B	200 °C H_2O	Yes	4	$50<\Delta K<100$	4.64
X-65 line pipe	1 ppm H_2S crude oil	No	1	$10<\Delta K<25$	4.60
X-65 line pipe	1 ppm H_2S crude oil	Yes	5	$25<\Delta K<70$	4.60
X-65 line pipe	4700 ppm H_2S crude	No	20	$10<\Delta K<25$	4.60
X-65 line pipe	4700 ppm H_2S crude	Yes	20	$25<\Delta K<70$	4.60
9Ni-4Co-0.25C	3½% NaCl	No	1	$50<\Delta K<100$	4.65
12Ni (1240 MPa) maraging	3½% NaCl	?	4	$30<\Delta K<100$	4.65
18Ni (1720 MPa) maraging	Dehumidified hydrogen	Yes	10	$20<\Delta K<45$	4.66
12Ni-5Cr-3Mo	3% NaCl	Yes	4	$20<\Delta K<80$	4.45
12Cr-8Ni-2Mo	Distilled H_2O	Yes	10	$30<\Delta K<90$	4.67

This thickness transition has been discussed in detail elsewhere (Ref 4.70), but one should be very careful to evaluate the material in the thickness to be used or have data for a thicker test condition which would provide a conservative estimate.

4.5.1. Microstructural Fracture Modes

Examinations of environmentally assisted fractures in specimens of both low- and high-strength steels indicate that cracking may proceed from intergranular fracture to transgranular cleavage to transgranular dimple rupture with increasing applied stress intensity. One or more of these modes may be absent, depending on the microstructure or the strength level of the material as well as on the aggressiveness of the environment.

4.5.2. Effects of Alloy Chemistry

It is impossible to generalize on the effects of alloying additions on sustained-load cracking because of the vast number of other variables. Nevertheless, a few cases for specific environments have been isolated (Ref 4.71 and 4.72).

For high-strength low-alloy steels (Ref 4.73) in aqueous environments, manganese may lower resistance by producing either untempered martensite or twinned martensite. Nickel may also produce undesirable effects if it causes an increase in retained austenite which leads to formation of untempered martensite

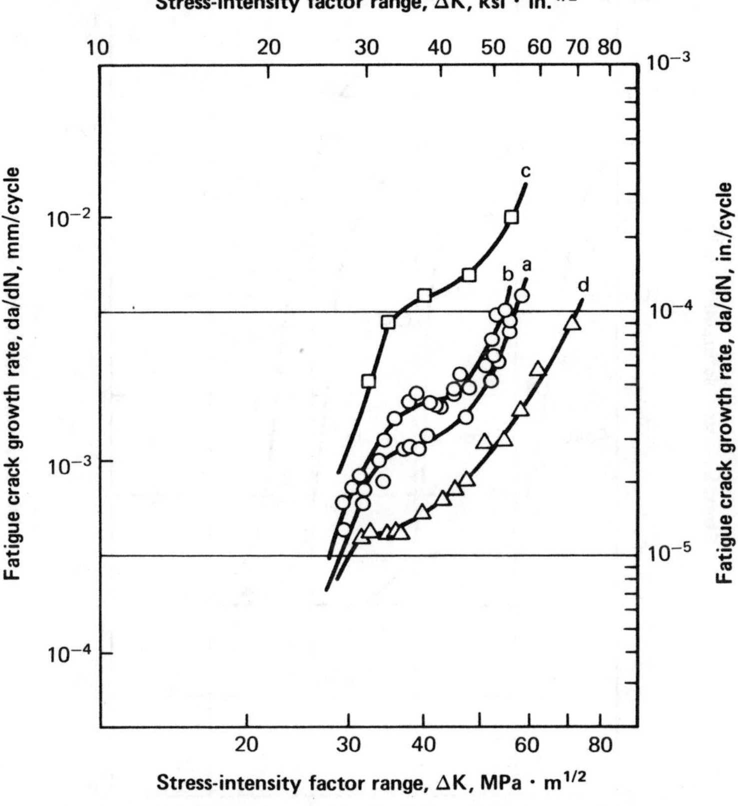

Curve a: 90% relative humidity + $Na_2Cr_2O_7$ + $NaNO_2$ + $Na_2B_4O_7$. Curve b: 90% relative humidity + $Na_2Cr_2O_7$. Curve c: 90% relative humidity only. Curve d: dry air; relative humidity 15%.

Fig. 4.41. Fatigue crack growth rate vs stress-intensity factor range, ΔK, for type 4340 steel (Ref 4.68)

after tempering. If no untempered martensite is present, nickel is considered to produce no significant effects on sustained-load cracking. Molybdenum additions on the order of 0.5% enhance resistance. For other elemental additions, either studies have not been made or results have been controversial. Effects of impurity elements are discussed at the end of this section.

For 18-Ni maraging steels, an extensive study (Ref 4.72) was conducted on 14 alloy heats with varying additions of cobalt, molybdenum, titanium and aluminum. An alloy parameter, shown as the horizontal axis in Fig. 4.43, was devised that predicted the K_{Iscc} resistance to an aqueous 3.5% sodium chloride solution with reasonable accuracy. A linear regression analysis indicated a correlation coefficient of −0.88, which means that this alloy parameter explains about 77% (r^2) of the variation in K_{Iscc}. Because the yield strengths range from 1420 to 1960 MPa (206 to 284 ksi) and there were additional possibilities of microstructural influence, it is difficult to account for all variables.

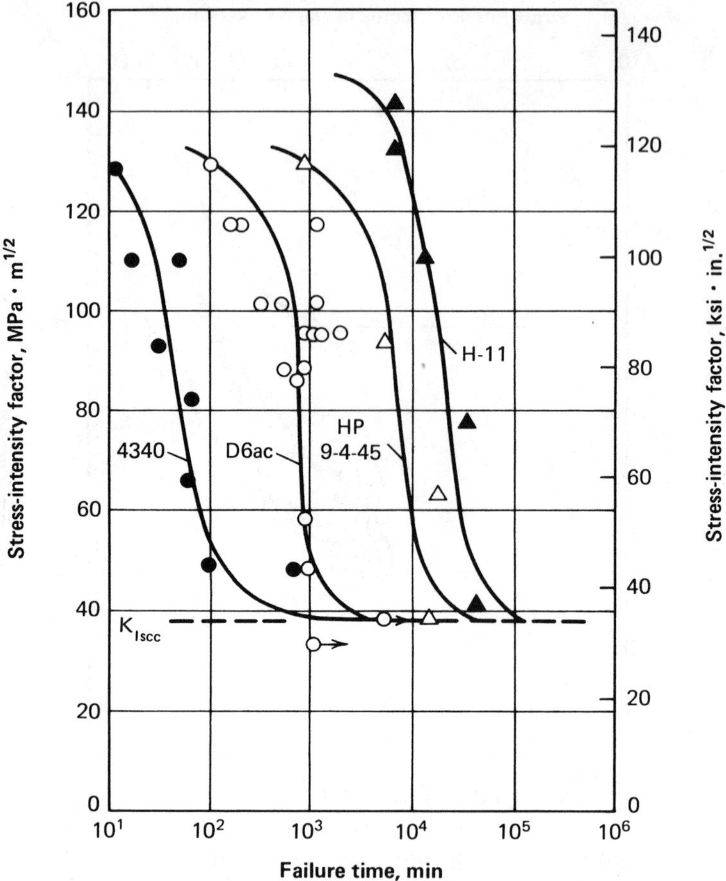

Precracked specimens of four high-strength steels were subjected to sustained loading in an environment of distilled water. Each steel had been hardened and tempered to a tensile strength of about 1650 MPa (240 ksi).

Fig. 4.42. Stress corrosion cracking in four high-strength steels (Ref 4.69)

With regard to medium-strength steels, a very extensive study (Ref 4.74) was conducted on a Mo-modified 4130 steel to determine the effect of molybdenum additions on resistance to H_2S cracking. Here, 12 heats of steel with varying molybdenum contents were tempered to a total of 56 conditions. The resulting K_{Iscc} values in an aqueous 0.5% acetic acid solution saturated with hydrogen sulfide are shown in Fig. 4.44. In this study the yield strength level was held constant at 760 MPa (110 ksi). It was observed that molybdenum additions of about 0.75% maximized resistance to cracking. Smooth-bar bend tests on the same heats and in the same solution indicated that 0.9% molybdenum maximized threshold stresses. Many possible suggestions for this improvement were offered, including the control of tramp elements by segregation of phosphorus,

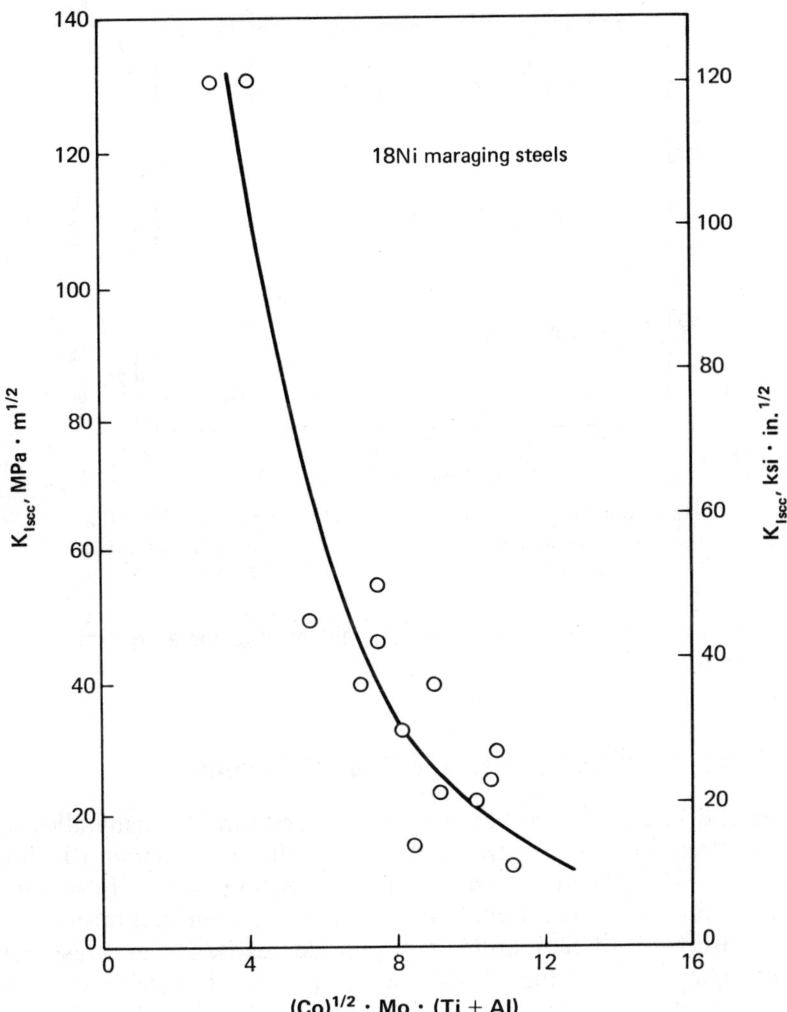

Fig. 4.43. Effect of alloying element parameter on K_{Iscc} for 18Ni maraging steels in an aqueous solution of NaCl (Ref 4.72)

arsenic, antimony and tin to carbide interfaces instead of to prior austenite grain boundaries.

The conclusion of a previous review (Ref 4.71) was that alloy chemistry probably has little effect on stress corrosion cracking resistance except for its influence on microstructure and strength. This is possibly true, with the one major exception that tramp elements segregated to prior austenite grain boundaries seriously decrease threshold stress-intensity factors. Just as it has been shown that phosphorus, sulfur, tin, antimony, arsenic and tellurium lower fracture toughness (see Section 4.3.2), combined effects of temper embrittlement and hydrogen embrittlement during stress corrosion cracking can be very detrimental (Ref 4.7).

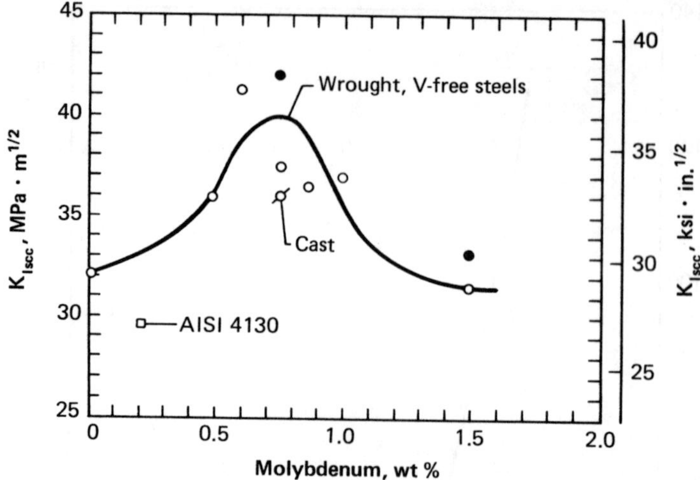

These data are for steel with a yield strength of 760 MPa (110 ksi) in an aqueous 0.5% acetic acid solution saturated with hydrogen sulfide. Solid circles denote heats with vanadium additions.

Fig. 4.44. Effect of molybdenum content on K_{Iscc} for alloy steels (Ref 4.74)

4.5.3. Effects of Microstructure and Heat Treatment

The general effect of increased tempering temperatures in many alloy steels is to lower the strength and to increase the K_{Iscc} value from about 10 MPa · m$^{1/2}$ (9 ksi · in.$^{1/2}$) to 70 MPa · m$^{1/2}$ (64 ksi · in.$^{1/2}$) (see Fig. 4.45). In this instance, the behavior parallels fracture toughness, but as will be seen later this is not always the case. The effects of tempering on increased resistance to stress corrosion cracking are also shown in Fig. 4.46(a) where the effect of increased tempering temperature is to decrease strength and lower crack velocities in aqueous solutions of 3.5% sodium chloride (Ref 4.76). On the other hand, Fig. 4.46(b) shows that while increasing tempering temperature from 100 to 550°C (212 to 1020°F) generally increases resistance to crack growth in a 12% Cr steel, two exceptions are noted at 450 and 475°C (840 and 885°F) (Ref 4.77). These temperatures are coincident with a secondary hardening peak and a ductility and fracture toughness minimum near 475°C for this steel.

It should be emphasized that resistance to stress corrosion cracking does not always parallel fracture toughness. For example, 18-Ni maraging steel undergoes a continuous improvement in resistance to stress corrosion cracking with increased aging temperature while fracture toughness goes through a minimum (Ref 4.72). This is shown in Fig. 4.47.

For lower strength conditions, quenched and tempered steels, normalized steels and isothermally transformed steels respond quite differently to sulfide cracking. A series of heat treatments that produced six different types of mixtures of martensite, ferrite and untempered martensite demonstrated the superiority of

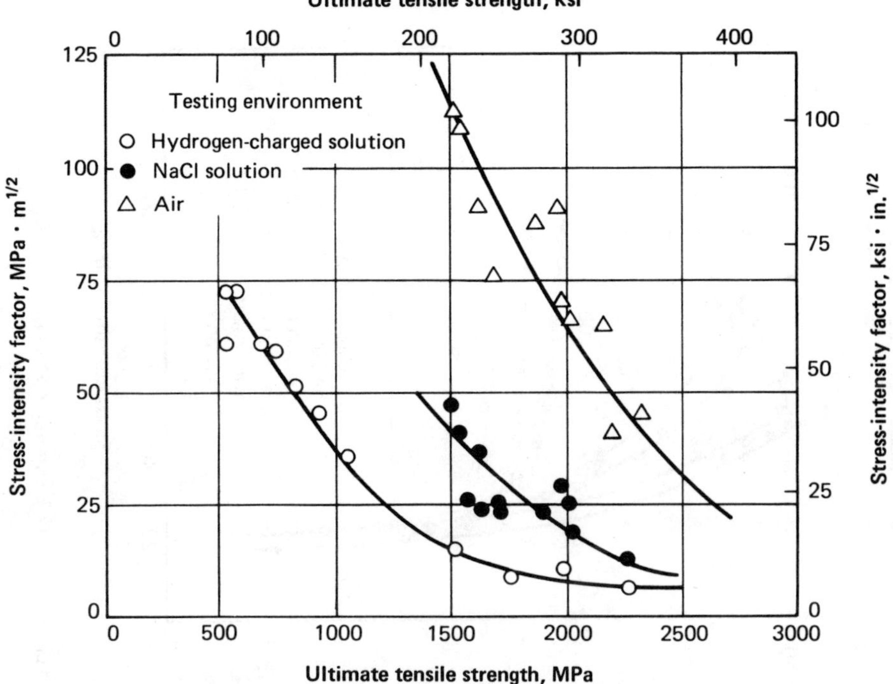

Reported values of stress-intensity factor are: K_{Ic} values for testing in air; K_{Iscc} values for testing in a 3.5% aqueous sodium chloride solution; and K values for threshold cracking of specimens electrolytically charged with hydrogen.

Fig. 4.45. Correlation between tensile strength and stress-intensity factor for crack propagation in alloy steels in several environments (Ref 4.75)

quenched and tempered microstructures, as shown in Fig. 4.48. Even though these results were for notched bars loaded in bending, the general ranking with regard to microstructural resistance to stress corrosion cracking should be the same. It was found that microstructures with fine spheroidizing carbides uniformly distributed throughout the ferrite, which are typically achieved in quenched and tempered steels, had superior resistance to sulfide cracking (Ref 4.78).

4.5.4. Effects of Temperature and Environment

Temperature and environment may produce unexpected results with regard to predictions of how a material will behave. Even though corrosion attack usually increases as temperature increases, it does not follow that stress corrosion cracking will necessarily be worse. Thus, data obtained for one combination of material, temperature, environment and loading should not be used for making predictions concerning any other combination. This is aptly illustrated in Fig. 4.49(a), where it is shown that, for AISI 4130 steel tested at one hydrogen pressure, the

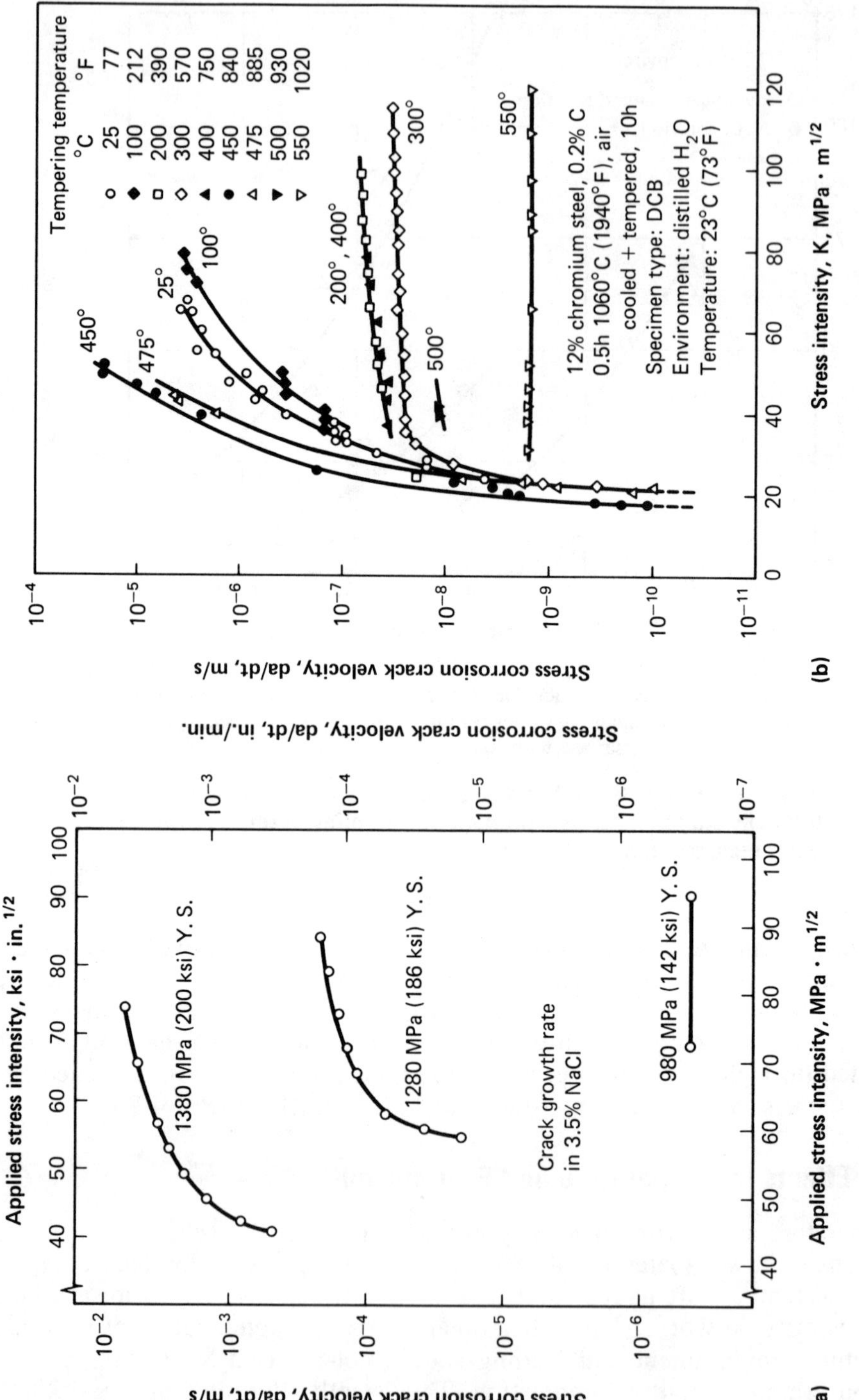

(a) Crack velocity for AISI 4340 steel (*Ref 4.76*). (b) Crack velocity for a 12% chromium steel (*Ref 4.59*).

Fig. 4.46. Effects of tempering temperature and applied stress-intensity factor on velocity of stress corrosion cracking

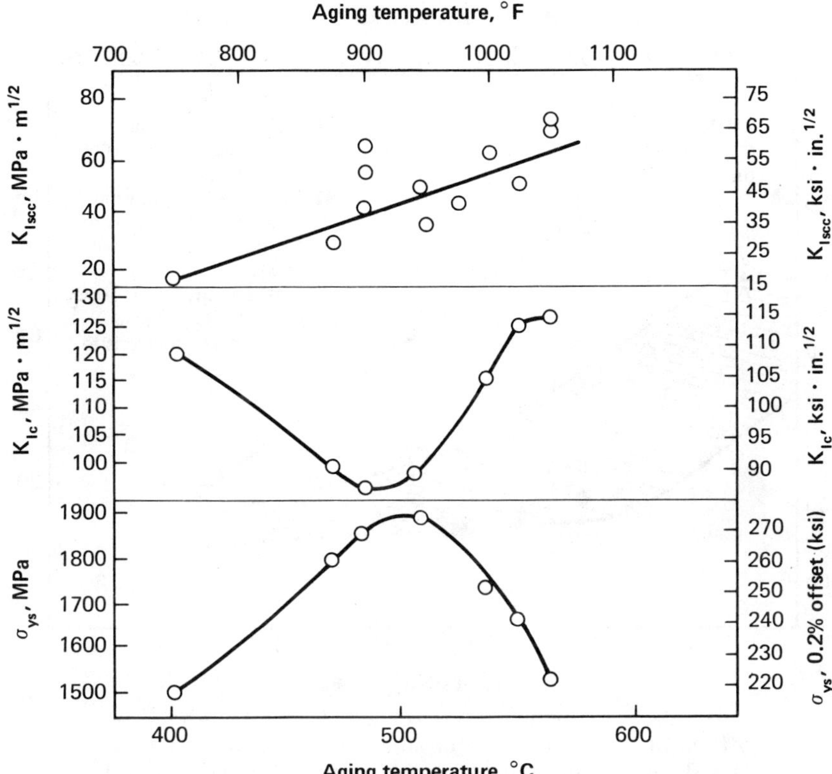

Fig. 4.47. Effects of aging temperature on K_{Iscc}, K_{Ic} and yield strength for an 18Ni (270) alloy (Ref 4.72)

lower the temperature the less resistant the material near the threshold. However, there is a crossover effect at high crack velocities, and the material is more resistant at temperatures both above and below 53°C (127°F). Thus, data obtained as a function of either temperature or stress intensity should not be extrapolated. The effect of hydrogen concentration is more orderly, as shown in Fig. 4.49(b), where an increase in hydrogen pressure causes an increase in the crack growth rate.

In general, any environmental species or electrochemical situation that promotes hydrogen entry will accelerate stress corrosion cracking in medium- and high-strength steels. For low-strength steels, there is a major categorical exception to this where stress corrosion cracking occurs by anodic dissolution in boiling solutions of nitrates, carbonates, chlorides and hydroxides as well as anhydrous ammonia. Various combinations of materials and boiling solutions that resulted in stress corrosion cracking, along with the fracture mode observed for each combination, are listed in Table 4.5 (Ref 4.79 to 4.84).

With regard to failure prevention, the effects of stress corrosion cracking can be minimized by choice of steel composition, by cathodic protection against caustic and nitrate embrittlement, or by use of protective coatings or inhibitors. Some successful applications have been (*a*) inhibited epoxy coatings on 18-Ni

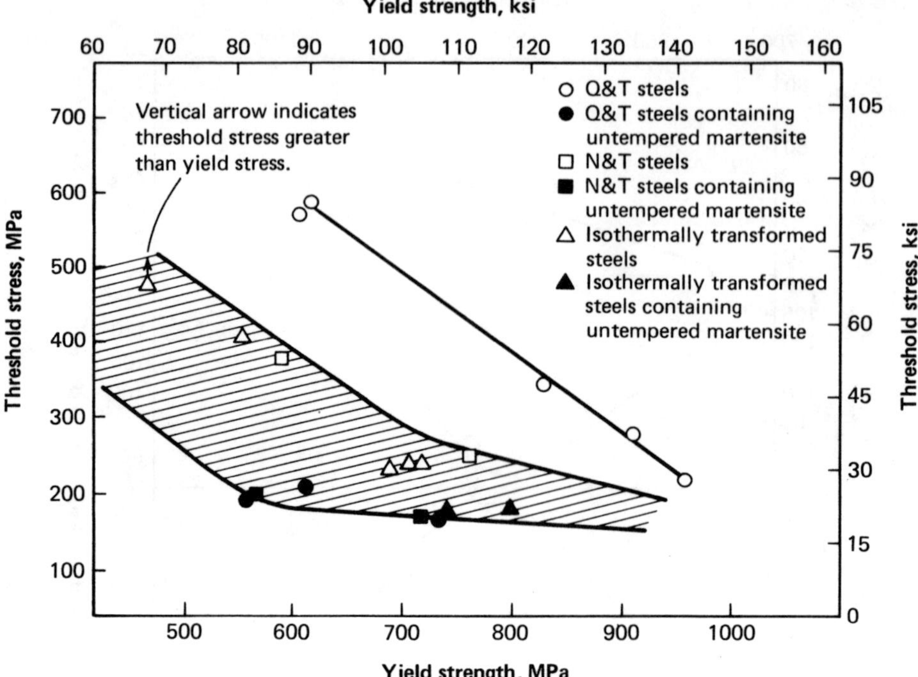

The data are for notched specimens loaded in bending in 5% NaCl−0.5% acetic acid solution saturated with H_2S. Q & T−quenched and tempered; N & T−normalized and tempered.

Fig. 4.48. Influence of microstructure on crack resistance for API N80 and AISI 4140 and 4340 steels (Ref 4.77)

maraging steel exposed to aqueous solutions containing 3% NaCl; (*b*) 0.2% water additions to vessels containing agricultural ammonia; and (*c*) film-forming amines added to hydrocrackers containing hydrogen sulfide, ammonia or hydrocyanic acid. Most of these studies were not performed on systems containing cracks, so it is not known how successful such coatings and inhibitors are for dynamic or precracked specimens. As shown in Fig. 4.50, crack velocities in high-strength 300M steel tested in distilled water can be markedly reduced by adding organic inhibitors, possibly by forming a chemisorbed double layer to prevent hydrogen entry.

REFERENCES

4.1. *Microalloying '75:* Proceedings of the International Symposium on High-Strength Low-Alloy Steels, Oct 1-3, 1975, Washington, DC, distributed by the American Society for Metals, Metals Park, OH

4.2. *Alloying Elements in Steel* (2nd Ed.), by E. C. Bain and H. W. Paxton: American Society for Metals, Metals Park, OH, 1961

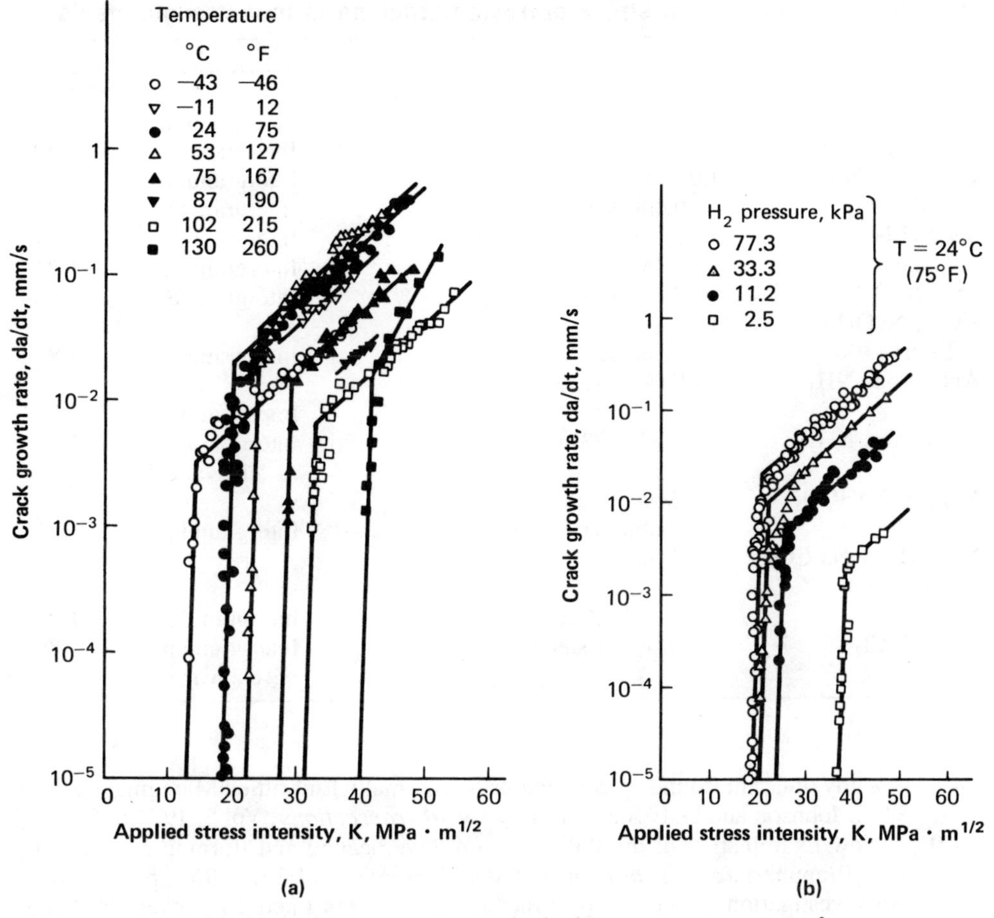

(a) At various temperatures in hydrogen at a pressure of 77.3 kPa. (b) At various hydrogen pressures at 24°C (75°F) (*Ref 4.78*).

Fig. 4.49. Dependence of crack growth rate, da/dt, on applied stress intensity, K, for 4130 steel with a yield strength of 1330 MPa (193 ksi)

4.3. Tempered Martensite Embrittlement in a High Purity Steel, by C. L. Briant and S. K. Banerji: *Metallurgical Transactions A*, Vol 10A, 1979, p 1151-1155

4.4. Evaluation of the Compact Tension Specimen for Determining Plane-Strain Fracture Toughness of High Strength Materials, by D. E. McCabe: *Journal of Materials*, Vol 7, No. 4, Dec 1972, p 449-454

4.5. The Fracture of Metals, by J. R. Low, Jr.: *Progress in Materials Science*, Vol 12, 1963, p 1-96

4.6. The Morphology of Brittle Fracture in Pearlite, Bainite and Martensite, by A. M. Turkalo: *Transactions of AIME*, Vol 218, 1960, p 24-30

4.7. The Effects of Hydrogen and Impurities on Brittle Fracture in Steel, by C. J. McMahon, Jr., C. L. Briant and S. K. Banerji: in *Proceedings of International Conference on Fracture*, Vol 4, Waterloo Press, Canada, 1977, p 363-385

Table 4.5. Observation of stress corrosion cracking in low-strength steels

Boiling solution	Steel	Fracture mode	Reference
20% NH_4NO_3	0.19%C	Intergranular	4.79
$Ca(NO_3)_2NH_4NO_3$	0.02 to 0.22%C	Intergranular	4.79
KNO_3 or $NaNO_3$	0.02 to 0.22%C	Intergranular	4.79
4N NH_4NO_3	Mild steel	Intergranular	4.80
4N NH_4NO_3	0.32C-0.5Mo-3Cr	Intergranular	4.81
9N NaOH	0.32C-0.5Mo-3Cr	Intergranular	4.81
35% NaOH	Mild steel	Intergranular	4.80
KOH, NaOH or LiOH (300 °C)	Mild steel	Intergranular	4.82
Anhydrous NH_3	ASTM A-212, A-285	Intergranular	4.79
Anhydrous NH_3	0.14C-1.73Mn	Intergranular + cleavage	4.80
Na_2CO_3-$NaHCO_3$	Mild steel (variable Ti, C)	Intergranular	4.83
NH_4CO_3 or Na_2CO_3-$NaHCO_3$	Mild steel (variable C,Si,Ni,Cr)	Intergranular	4.84
45% $MgCl_2$	1 to 6%Ni steel	Intergranular + cleavage	4.80

4.8. A Study of Grain Boundary Segregants in Thermally Embrittled Maraging Steel, by W. C. Johnson and D. F. Stein: *Metallurgical Transactions*, Vol 5, 1974, p 549-554

4.9. A New Examination of the Phenomena of Overheating and Burning of Steels, by I. S. Brammer: *Journal of Iron and Steel Institute*, Vol 201, 1963, p 752-761

4.10. An Investigation of the Plastic Fracture of AISI 4340 and 18 Nickel-200 Grade Maraging Steels, by T. B. Cox and J. R. Low, Jr.: *Metallurgical Transactions*, Vol 5, 1974, p 1457-1470

4.11. R. Van Stone, unpublished data.

4.12. Effect of Alloying Elements on Tempered Martensite Embrittlement and Fracture Toughness of Low Alloy High Strength Steels, by C. Vishnevsky: Report CR 69-18(F), Army Materials and Mechanics Research Center, Watertown, MA, Jan 1971

4.13. Influence of Alloying Elements on the Toughness of Low Alloy Martensitic High Strength Steels, by C. Vishnevsky and E. A. Steigerwald: Report CR 68-09(F), Army Materials and Mechanics Research Center, Watertown, MA, Nov 1968

4.14. Influence of Strain Rate and Temperature on the Fracture Toughness and Tensile Properties of Several Metallic Materials, by A. H. Priest: in *Dynamic Fracture Toughness*, edited by M. G. Dawes, The Welding Institute, Abington Hall, Cambridge, 1977, p 95-111

4.15. Fracture Toughness Testing in Alloy Development, by R. P. Wei: in *Fracture Toughness and Its Applications*, STP 381, American Society for Testing and Materials, Philadelphia, 1965, p 279-289

4.16. Rare Earth Additions to Steel, by P. E. Waudby: *International Metals Reviews*, Review 229, No. 2, 1978, p 74-98

4.17. *Effect of Second Phase Particles on the Mechanical Properties of Steel*, by T. J. Baker and J. A. Charles: The Iron and Steel Institute, London, 1971

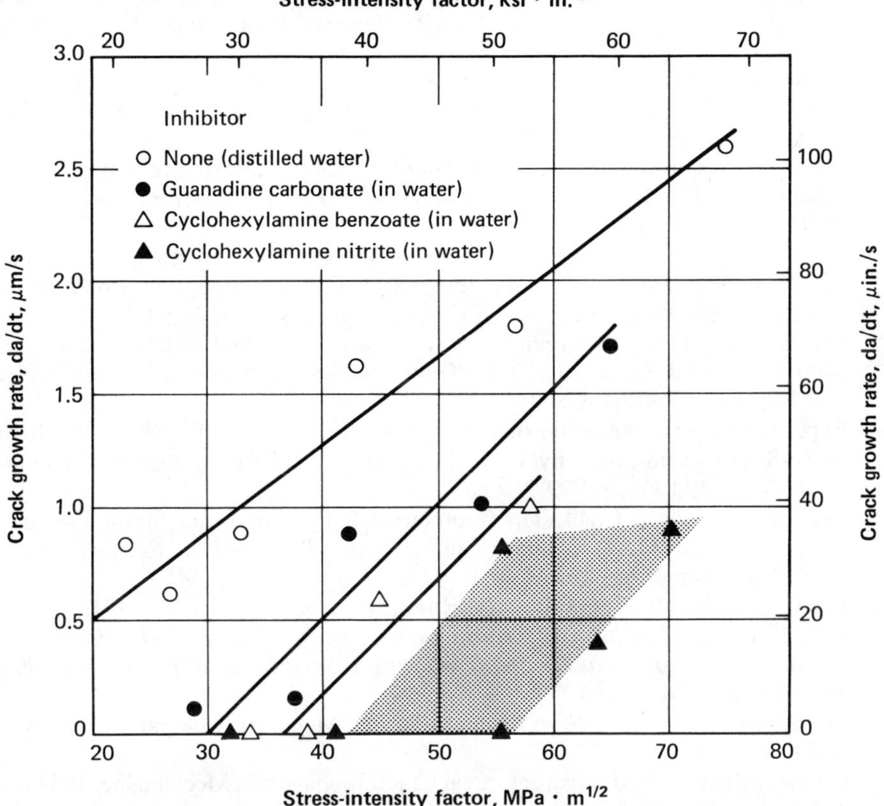

Fig. 4.50. Effect of inhibitors on crack growth rate for 300M steel (Ref 4.85)

4.18. Impact Toughness of Fuel Pipelines Demands Steel Cleanliness, by W. G. Wilson and G. J. Klems: *Industrial Heating*, Oct 1974, p 12-16

4.19. Effect of Cerium on the Properties of a Stainless Chromium Nickel Steel, by H. J. Kirsching, H. J. Hornbeck, H. Schenk and C. Carius: *Arch. Eisenhuttenwes*, Vol 34, No. 4, 1963, p 269-277

4.20. Sulfide Shape Control in High-Strength Low-Alloy Steels, by L. Luyckx, R. J. Bell, A. McLean and M. Korchynsky: *Metallurgical Transactions*, Vol 1, No. 12, 1970, p 3341-3350

4.21. Intergranular Fracture in Steels, by C. J. McMahon, Jr.: *Materials Science and Engineering*, Vol 25, 1976, p 233-239

4.22. Alloy and Impurity Effects on Temper Brittleness of Steel, by J. R. Low, Jr., D. F. Stein, A. M. Turkalo and R. P. Laforce: *Transactions of AIME*, Vol 242, 1968, p 14-24

4.23. Further Considerations on the Inconsistency in Toughness Evaluation of AISI 4340 Steel Austenitized at Increasing Temperatures, by R. O. Ritchie and R. M. Horn: *Metallurgical Transactions*, Vol 9A, No. 3, March 1978, p 331-341

4.24. Effects of Quenching Variables on Fracture Toughness of D6ac Steel Aerospace Structures, by G. L. Peterman and R. L. Jones: *Metals Engineering Quarterly*, Vol 15, No. 2, May 1975, p 59-64

4.25. The Role of Microstructure on the Strength and Toughness of Fully Pearlitic Steels, by J. M. Hyzak and I. M. Bernstein: *Metallurgical Transactions A*, Vol 7A, 1976, p 1217-1224
4.26. Influence of Precracking and Grain Size on Fracture Toughness of Structural Steels, by W. Dahl and W. B. Kretzchmann: in *Fracture 1977*, Vol 2A, edited by D. M. R. Taplin, Pergamon Press, 1977, p 17-21
4.27. The Relationship Between Fracture Toughness and Microstructure in the Cleavage Fracture of Mild Steel, by D. A. Curry and J. F. Knott: *Metal Science*, Vol 10, 1976, p 1-6
4.28. Effect of Prior Austenite Grain Size on the Fracture Toughness Properties of A 533 B Steel, by F. R. Stonsifer and R. W. Armstrong: in *Fracture 1977*, Vol 2A, edited by D. M. R. Taplin, Pergamon Press, 1977, p 1-6
4.29. On the Relationship Between Critical Tensile Stress and Fracture Toughness in Mild Steel, by R. O. Ritchie, J. F. Knott and J. F. Rice: *Journal of the Mechanics and Physics of Solids*, Vol 21, 1973, p 395-410
4.30. Experimental Determination of Plastic Constraint Ahead of a Sharp Notch Under Plane-Strain Conditions, by G. T. Hahn and A. R. Rosenfield: *Transactions of ASM*, Vol 59, 1966, p 909-919
4.31. The Role of Large Crack Tip Geometry Changes in Plane Strain Fracture, by J. R. Rice and M. B. Johnson: in *Inelastic Behavior of Solids*, edited by M. F. Kanninen *et al*, McGraw-Hill, New York, 1970, p 641-672
4.32. Fracture of Steels Containing Pearlite, by A. R. Rosenfield, G. T. Hahn and J. D. Embury: *Metallurgical Transactions*, Vol 3, 1972, p 2797-2804
4.33. *Fundamentals of Fracture Mechanics*, Revised Edition, by J. F. Knott: Butterworths, London, 1979
4.34. *Symposium, Low Alloy High Strength Steels*: Metallurg Companies, Nuremberg, BRD, May 21-23, 1970
4.35. On the Influence of Microstructure on Crack Propagation Mechanisms and Fracture Toughness of Metallic Materials, by K. H. Schwalbe: *Engineering Fracture Mechanics*, Vol 9, 1977, p 795-832
4.36. Discussion by A. H. Priest: in *Effect of Second-Phase Particles on the Mechanical Properties of Steel*, The Iron and Steel Institute, London, 1971
4.37. Fracture Toughness of Turbine Generator Rotor Forgings, by H. D. Greenberg, E. T. Wessel and W. H. Pryle: *Engineering Fracture Mechanics*, Vol 1, 1970, p 653-674
4.38. The Correlations Between K_{Ic} and Charpy V Notch Test Results in the Transition Temperature Range, by J. M. Barsom and S. T. Rolfe: in *Impact Testing of Metals*, STP 466, American Society for Testing and Materials, Philadelphia, 1970, p 281-302
4.39. Development of the AASHTO Fracture-Toughness Requirements for Bridge Steels, by J. M. Barsom: *Engineering Fracture Mechanics,* Vol 7, No. 3, 1975, p 605-618
4.40. Fracture Toughness of Bridge Steels, Phase II Report, by R. Roberts: Report FHWA-RD-74-59, Federal Highway Administration, Sept 1974
4.41. Effect of Temperature and Rate of Loading on the Fracture Behavior of Various Steels, by J. M. Barsom: in *Dynamic Fracture Toughness*, edited by M. G. Davies, The Welding Institute, Abington Hall, Cambridge 7, 1979, p 113-125
4.42. Near-Threshold Fatigue-Crack Propagation in Steels, by R. O. Ritchie: *International Metals Reviews*, Vol 24, No. 5 and 6, 1979, p 205-230
4.43. A Review of Fatigue Fracture Topology Effects on Threshold and Growth Mechanisms, by W. W. Gerberich and N. R. Moody: in *Fatigue Mechanisms*, edited by J. F. Fong, STP 675, American Society for Testing and Materials, Philadelphia, 1979, p 292-341

4.44. The Influence of Stress Intensity and Microstructure on Fatigue Crack Propagation in Ferritic Materials, by C. E. Richards and T. C. Lindley: *Engineering Fracture Mechanics*, Vol 4, 1972, p 951-978

4.45. Fatigue and Corrosion-Fatigue Crack Growth of 4340 Steel at Various Yield Strengths, by E. J. Imhoff and J. M. Barsom: in *Progress in Flaw Growth and Fracture Toughness*, STP 536, American Society for Testing and Materials, Philadelphia, 1973

4.46. *Fracture and Fatigue Control in Structures*, by S. T. Rolfe and J. M. Barsom: Prentice-Hall, Englewood Cliffs, NJ, 1977

4.47. Microstructural Effects on Fatigue Crack Growth in a Low Carbon Steel, by H. Suzuki and A. J. McEvily: *Metallurgical Transactions A*, Vol 10A, 1979, p 475-481

4.48. Grain-Size Effect on Crack Nucleation and Growth in Long-Life Fatigue of Low-Carbon Steel, by S. Taira, K. Tanaka and M. Hoshina: in *Fatigue Mechanisms*, edited by J. F. Fong, STP 675, American Society for Testing and Materials, Philadelphia, 1979, p 135-162

4.49. Fatigue Crack Growth Characteristics at Low Stress Intensities of Metals and Alloys, by C. J. Beevers: *Metal Science*, Aug/Sept 1977, p 362-367

4.50. Correlations Between Fracture Surface Appearance and Fracture Mechanics Parameters for Stage II Fatigue Crack Propagation in Ti-6Al-4V, by A. Yuen, S. W. Hopkins, G. R. Leverant and C. A. Rau: *Metallurgical Transactions*, Vol 5, 1974, p 1833-1842

4.51. The Slow Fatigue-Crack Growth and Threshold Behavior of a Medium-Carbon Alloy Steel in Air and Vacuum, by R. J. Cooke, P. E. Irving, G. S. Booth and C. J. Beevers: *Engineering Fracture Mechanics*, Vol 7, 1975, p 69-77

4.52. Fatigue Properties of Plain Carbon Steels, by J. Masounave and J. P. Bailon: in *Proceedings of the 2nd International Conference on Mechanical Behavior of Materials*, ICM II, Federation of Materials Societies, Boston, 1976, p 636-641

4.53. Effect of Microstructure on Fatigue in Threshold Region in Low-Alloy Steel, by J. P. Benson and D. V. Edmonds: *Metal Science*, Vol 12, No. 5, 1978, p 223-232

4.54. Fatigue-Crack Initiation and Near Threshold Crack Growth, by M. E. Fine and R. O. Ritchie: in *Fatigue and Microstructure*, edited by M. Meshii, American Society for Metals, Metals Park, OH, 1979, p 245-279

4.55. Fatigue Crack Propagation in A533B Steel, by A. D. Wilson: *Transactions of ASME, Journal of Pressure Vessel Technology*, Vol 99, 1977, p 459-469

4.56. Fatigue Crack Propagation in Steel Alloys at Elevated Temperatures, by H. I. McHenry and A. W. Pense: STP 520, American Society for Testing and Materials, Philadelphia, 1973, p 345-354

4.57. Effects of Grain Size and Temperature on Sub-Critical Crack Growth in A533 Steel, by F. R. Stonsifer: NRL Memorandum Report 3400, Naval Research Laboratory, Washington, Nov 1976

4.58. Fatigue Crack Propagation in Iron and Two Iron Binary Alloys at Low Temperatures, by N. R. Moody and W. W. Gerberich: *Materials Science and Engineering*, Vol 41, 1979, p 271-280

4.59. Corrosion Fatigue in Fe-Ni-Cr Alloys, by M. O. Speidel: in *Stress Corrosion Cracking and Hydrogen Embrittlement of Iron Base Alloys*, NACE-5, National Association of Corrosion Engineers, Houston, 1977, p 1071-1094

4.60. Fatigue-Crack Growth in an X-65 Line-Pipe Steel at Low Cyclic Frequencies in Aqueous Environments, by O. Vosikovsky: *Transactions of ASME*, Series H, Vol 97, 1975, p 298-305

4.61. Effect of Cyclic-Stress Form on Corrosion-Fatigue Crack Propagation Below K_{Iscc}, in a High-Yield-Strength Steel, by J. M. Barsom: in *Corrosion Fatigue*, NACE-2,

National Association of Corrosion Engineers, Houston, 1972, p 424-436

4.62. A Review of Fatigue Crack Growth of Pressure Vessel and Piping Steels in High-Temperature, Pressurized Reactor-Grade Water, by W. H. Cullen and K. Torronen: NRL Memorandum Report 4298, Naval Research Laboratory, Washington, Sept 19, 1980

4.63. Fatigue Crack Propagation in Low Alloy Steel in a De-aerated Distilled Water Environment, by T. Misawa, N. Ringshall and J. F. Knott: *Corrosion Science*, Vol 16, No. 11, Nov 1976, p 805-818

4.64. Corrosion Fatigue of ASTM A-302 B Steel in High Temperature Water, the Simulated Nuclear Reactor Environment, by T. Kondo, T. Kikuyama, H. Nakajima, M. Shindo and R. Nagasaki: in *Corrosion Fatigue*, National Association of Corrosion Engineers, Houston, 1972, p 539

4.65. The Influence of Yield Strength and Fracture Toughness on Fatigue Design Procedures for Structural Steels, by T. W. Crooker and E. Lange: in *Conference on Fatigue of Welded Structures*, Vol 2, The Welding Institute, Abington, Cambridge, 1971, p 243-256

4.66. Correlation Between Sustained Load and Fatigue Crack Growth in High Strength Steels, by R. P. Wei and J. D. Landes: *Materials Research and Standards*, Vol 9, No. 7, July 1969, p 25-27

4.67. On the Superposition Model for Environmentally Assisted Fatigue Crack Propagation, by W. W. Gerberich, J. P. Birat and V. F. Zackay: in *Corrosion Fatigue*, NACE-2, National Association of Corrosion Engineers, Houston, 1972, p 396-408

4.68. New Inhibitors for Crack Arrestment in Corrosion Fatigue of High Strength Steels, by V. S. Agarwala and J. J. DeLuccia: *Corrosion*, Vol 36, No. 4, 1980, p 208-212

4.69. Effect of Composition on the Environmentally Induced Delayed Failure of Precracked High-Strength Steel, by W. D. Benjamin and E. A. Steigerwald: *Metallurgical Transactions*, Vol 2, 1971, p 606-608

4.70. Hydrogen-Controlled Cracking – An Approach to Threshold Stress Intensity, by W. W. Gerberich and Y. T. Chen: *Metallurgical Transactions*, Vol 6A, 1975, p 271-278

4.71. Review of Stress Corrosion Cracking in Low Alloy Steels with Yield Strengths Below 150 ksi, by C. S. Carter and M. V. Hyatt: in *Stress Corrosion Cracking and Hydrogen Embrittlement of Iron Base Alloys*, NACE-5, National Association of Corrosion Engineers, Houston, 1977, p 524-600

4.72. The Stress Corrosion and Hydrogen Embrittlement Behavior of Maraging Steels, by D. P. Dautovich and S. Floreen: in *Stress Corrosion Cracking and Hydrogen Embrittlement of Iron Base Alloys*, NACE-5, National Association of Corrosion Engineers, Houston, 1977, p 798-815

4.73. The Effect of Alloying Elements on the Susceptibility to Stress Corrosion Cracking of Martensitic Steels in Salt Water, by G. Sandoz: *Metallurgical Transactions*, Vol 2, 1971, p 1055-1063

4.74. Effect of Molybdenum Content on the Sulfide Stress Cracking Resistance of AISI 4130 Steel with 0.357% Cb, by P. J. Grobner, D. L. Sponseller and D. E. Diesburg: *Corrosion*, Vol 35, No. 4, 1979, p 175-185

4.75. Effects of Metallurgy on Stress Corrosion Cracking and Hydrogen Embrittlement of Ultra-High Strength Steels, by M. Tvrdy: in *Fracture 1977*, Vol 2, edited by D. M. R. Taplin, University of Waterloo Press, Waterloo, Ontario, 1977, p 255-259

4.76. The Role of Strain Hardening Exponent in Stress Corrosion Cracking of a High Strength Steel, by V. J. Colangelo and M. S. Ferguson: *Corrosion*, Vol 25, 1969, p 509-514

4.77. Role of Composition and Microstructure in Sulfide Cracking of Steel, by E. Snape: *Corrosion*, Vol 24, 1968, p 261-282
4.78. Quantitative Observations of Hydrogen-Induced, Slow Crack Growth in a Low Alloy Steel, by H. G. Nelson and D. P. Williams: in *Stress Corrosion Cracking and Hydrogen Embrittlement of Iron Base Alloys*, NACE-5, National Association of Corrosion Engineers, Houston, 1977, p 390-404
4.79. *The Stress Corrosion of Metals*, by H. L. Logan: J. Wiley and Sons, New York, 1966
4.80. The Fractography of SCC in Carbon Steels, by B. Poulson: *Corrosion Science*, Vol 15, No. 8, Aug 1976, p 469-477
4.81. Stress Corrosion Cracking of Low Alloy Steels, by de G. Jones, J. F. Newman and R. P. Harrison: in *Proceedings of 5th International Congress on Metal Corrosion*, National Association of Corrosion Engineers, Houston, 1974, p 434-438
4.82. Preliminary Experiments on the Stress Corrosion of Mild Steel in Lithium Hydroxide Solutions, by de G. Jones and M. J. Humphries: in *Proceedings of 5th International Congress on Metal Corrosion*, National Association of Corrosion Engineers, Houston, 1974, p 410-413
4.83. A Correlation Between Electrochemical Parameters and Stress Corrosion Cracking, by R. D. Armstrong and A. C. Coates: *Corrosion Science*, Vol 16, No. 7, July 1976, p 423-433
4.84. Stress Corrosion Cracking of Mild Steel in Ammonium Carbonate Solution, by D. Hixson and H. H. Uhlig: *Corrosion*, Vol 32, No. 2, 1976, p 56-59
4.85. W. W. Gerberich, N. R. Moody and B. Miksic (unpublished results): University of Minnesota and Cortec Corp., Minneapolis
4.86. Fracture Mechanics and Corrosion Fatigue, by A. J. McEvily, Jr., and R. P. Wei: *Proceedings–International Conference on Corrosion Fatigue*, National Association of Corrosion Engineers, Houston, 1971, p 381-395

Chapter 5

Fracture Properties of Wrought Stainless Steels

J. E. Campbell

5.1. INTRODUCTION

Stainless or corrosion resistant steels represent separate groups of steels that are used for many applications in transportation, power station, refinery, chemical plant, marine, heat treating and surgical equipment because of their resistance to corrosion and to high temperature oxidation. Their corrosion resistance is attributed primarily to their high chromium contents (10 to 30%). In addition, many of these steels have good resistance to deformation (creep) at elevated temperatures. Furthermore, those stainless steels with amounts of nickel sufficient to develop microstructures of stable austenite (face centered cubic) retain excellent ductility and strength to temperatures near absolute zero.

Stainless steels are readily formed or forged on mill equipment if it is designed for working these materials. The mill products are readily fabricated and welded in the annealed or solution treated conditions. Some types of stainless steels are strengthened only by cold working; others may be strengthened by conditioning and aging treatments. High strength levels may be obtained in stainless grades from several of the groups that are discussed in this chapter. These steels are used in applications where design stresses are relatively high and where components are subjected to cyclic loading. Hydrofoils, for example, are highly stressed in service according to complex random cyclic loading in a corrosive environment of seawater. Precipitation hardening stainless steels have been selected for this application.

Tubing used in power plants and petrochemical industries is subjected to high temperatures, high stresses in high pressure service, corrosion from various liquids, and oxidation from exposure to air. In nuclear plants, the tubing also may be exposed to fast neutron fluence. Certain austenitic stainless steels have been used extensively for these applications.

Steam turbine rotors and blades are exposed to high temperature steam, high speed rotational forces, and cyclic stresses that lead to fatigue crack propagation.

Martensitic stainless steels have been used for blades and, under certain conditions, for steam turbine rotors.

Fracture mechanics concepts may be applied in the design, materials selection and failure analysis of the above components to minimize occurrence of failure in service. The information presented in this chapter is intended to aid in applying fracture mechanics concepts to selection of stainless steels for critical applications. The data presented represent state-of-the-art information from published sources. For many of the stainless steels, the available data are limited, and for some, the kind of data that are needed are nonexistent. However, for many applications, extensive experience has been gained in selecting stainless steels to meet fracture resistance requirements (damage tolerance) based on current state-of-the-art information.

5.2. COMPOSITIONS OF WROUGHT STAINLESS STEELS

Stainless steels of all types contain chromium in amounts ranging from 10 to 30%. The stainlessness and oxidation resistance of stainless steels is primarily dependent on chromium content. Other alloying elements, including silicon, nickel, molybdenum, copper and aluminum, contribute to the general corrosion resistance, but their influence is limited compared to that of chromium. Chromium and the other alloying elements have other functions, some of which are discussed in the following paragraphs. Standard compositions of stainless steels may be divided into four general groups depending on the ultimate microstructures of these steels at room temperature (see Table 5.1): these groups are austenitic, ferritic, martensitic and precipitation hardening stainless steels.

5.2.1. Austenitic Stainless Steels

The austenitic stainless steels of the UNS S3xxxx series contain relatively low percentages of carbon (generally less than 0.1%), 17% or more chromium, and sufficient amounts of nickel (usually 8% or more) to reduce the M_s temperature of the steel to below room temperature. After relatively fast cooling from the annealing temperature, the microstructures of these steels are practically 100% austenite. Austenitic stainless steels containing 6 to 10% nickel and not more than 2% manganese may be partially transformed to martensite during plastic deformation and/or exposure to temperatures in the cryogenic range. The austenite in these steels is less stable than that in stainless steels containing higher percentages of nickel. Type 310S stainless steel containing 25% chromium and 20% nickel is completely stable for all conditions of plastic deformation and cryogenic temperature exposure. The austenite stability of these steels is significant in analyzing the fracture process because of the effect of the austenite-to-martensite transformation on the plastic deformation that occurs ahead of the advancing crack.

Of the austenitic stainless steels in Table 5.1, Type 301, with the lowest nickel-plus-manganese content, has the lowest austenite stability. It has been used for cold worked sheet and cold drawn wire because of the high strength levels that may be achieved by cold working. The austenite-martensite microstructure of the cold worked Type 301 is not as corrosion resistant as the austenitic microstructure of this alloy in the annealed condition.

The most common austenitic stainless steel is Type 304, which has been used for critical elevated temperature applications such as high pressure piping in power generating stations and for very low temperature applications such as liquid helium transfer piping and storage vessels. The alloy is generally used in the annealed condition (heated at 1060°C, or 1950°F, to dissolve carbides, and quenched in water or cooled in air at a rate high enough to prevent carbide reprecipitation). Type 304 has high ductility and toughness in this condition over a wide range of service temperatures. Because of its high toughness and relatively low yield strength, plane strain conditions cannot be obtained on monotonic loading up to the point of unstable crack propagation in precracked fracture toughness specimens of reasonable size. The same limitation applies to all austenitic stainless steels in the annealed condition. Therefore, available fracture mechanics data for these alloys generally are limited to fatigue crack growth rate data (da/dN).

All of the austenitic stainless steels following Type 304 in Table 5.1 are modifications of Type 304. Increasing the nickel content increases the austenite stability and reduces the work hardening effect. Increasing the chromium content increases the resistance to deformation at elevated temperatures, and also increases the oxidation and corrosion resistance. Addition of molybdenum increases corrosion resistance. Titanium, niobium and tantalum are added to stabilize carbide formation and minimize grain boundary sensitization following welding. Up to 0.30% nitrogen and small amounts of other elements may be added to heats of stainless steel by melting under a nitrogen atmosphere and by additions of master alloys. These additions, as in Kromarc 58, increase the strength of the alloy. Compositions of other austenitic stainless steels have not been included in Table 5.1 because no fracture mechanics data have been available for them.

Of the austenitic grades, most fatigue crack growth rate data have been obtained on Types 304 and 316 because these two stainless steels have been shown to have good fatigue crack growth rate resistance in nuclear reactor environments.

5.2.2. Ferritic Stainless Steels

Ferritic stainless steels contain more than 14% chromium (but not over 27%), little or no nickel, and only limited amounts of carbon. For these steels, the elevated temperature ferrite-forming influence of chromium dominates, and the body centered cubic crystal structure is maintained at temperatures up to 1400°C (2550°F). Therefore, these steels are not hardenable by heat treatment. For some applications, ferritic stainless steels are being considered as alternative materials for the more costly austenitic grades. However, no significant fracture mechanics data have been located for the ferritic grades. Available toughness data are limited to Charpy impact data, as reported in Ref 5.2.

5.2.3. Martensitic Stainless Steels

Martensitic stainless steels contain from 11.5 to 18% chromium (in the 4xx series) and a range of carbon contents from less than 0.15 to 1.20%. Certain types in the 4xx series contain up to 3.00% nickel, 1.25% molybdenum, 1.25% tungsten, and small amounts of other alloying elements. The compositions of the martensitic stainless steels in Table 5.1 are those for which fracture mechanics data are reviewed later in this chapter. Type 403 is the basic 12% chromium grade.

Table 5.1. Compositions of stainless steels (Ref 5.1)

Type	UNS number	Composition, %(a) C	Mn	Si	Cr	Ni	P	S	Others	Yield strength(b), MPa (ksi)
Austenitic types										
301	S30100	0.15	2.00	1.00	16.0-18.0	6.0-8.0	0.045	0.03	...	205 (30) ann
304	S30400	0.08	2.00	1.00	18.0-20.0	8.0-10.5	0.045	0.03	...	205 (30) ann
304L	S30403	0.03	2.00	1.00	18.0-20.0	8.0-12.0	0.045	0.03	...	170 (25) ann
304LN	...	0.03	2.00	1.00	18.0-20.0	8.0-10.5	0.045	0.03	0.10-0.15N	205 (30) ann
308	S30800	0.08	2.00	1.00	19.0-21.0	10.0-12.0	0.045	0.03	(Welding electrode composition)	205 (30) ann
309S	S30908	0.08	2.00	1.00	22.0-24.0	12.0-15.0	0.045	0.03	...	205 (30) ann
310S	S31008	0.08	2.00	1.50	24.0-26.0	19.0-22.0	0.045	0.03	...	205 (30) ann
316	S31600	0.08	2.00	1.00	16.0-18.0	10.0-14.0	0.045	0.03	2.0-3.0Mo	205 (30) ann
316N	S31651	0.08	2.00	1.00	16.0-18.0	10.0-14.0	0.045	0.03	2.0-3.0Mo, 0.10-0.16N	240 (35) ann
321	S32100	0.08	2.00	1.00	17.0-19.0	9.0-12.0	0.045	0.03	(5 × %C) min Ti	
348	S34800	0.08	2.00	1.00	17.0-19.0	9.0-13.0	0.045	0.03	0.2Cu, (10 × %C) min (Nb + Ta) (c)	
21-6-9(d)	S21900	0.08	8.0-10.0	1.00	19.0-21.5	5.0-7.0	0.06	0.03	0.15-0.40N	
22-13-5(e)	S20910	0.06	4.0-6.0	1.00	20.5-23.5	11.5-13.5	0.04	0.03	1.5-3.0Mo, 0.1-0.3Nb,	345 (50) ann
									0.1-0.3V, 0.2-0.4N	380 (55) ann
Kromarc 58	...	0.03	9.3	0.05	15.5	23.0	0.005	0.005	2.2Mo, 0.02Al, 0.16V,	
									[illegible]	371 (54) ann

403	S40300	0.15	1.00	0.50	11.5-13.0	...	0.04	0.03	...	550 (80) ht
410	S41000	0.15	1.00	1.00	11.5-13.0	...	0.04	0.03	...	550 (80) ht
420	S42000	0.15 min	1.00	1.00	12.0-14.0	...	0.04	0.03	...	1480 (215) ht
422	S42200	0.20-0.25	1.00	0.75	11.0-13.0	0.5-1.0	0.025	0.025	0.75-1.25Mo, 0.75-1.25W, 0.15-0.30V	760 (110) ht
431	S43100	0.20	1.00	1.00	15.0-17.0	1.25-2.50	0.04	0.03	...	1030 (149) ht
Precipitation hardening types(f)										
15-5 PH	S15500	0.07	1.00	1.00	14.0-15.5	3.5-5.5	0.04	0.03	2.5-4.5Cu, 0.15-0.45 (Nb + Ta)	1000 (145) ht
17-4 PH	S17400	0.07	1.00	1.00	15.5-17.5	3.0-5.0	0.04	0.03	3.0-5.0Cu, 0.15-0.45 (Nb + Ta)	1000 (145) ht
PH 13-8 Mo	S13800	0.05	0.10	0.10	12.25-13.25	7.5-8.5	0.01	0.008	2.0-2.5Mo, 0.90-1.35Al, 0.01N	1310 (190) ht
Custom 455 (XM-16)	S45500	0.05	0.50	0.50	11.0-12.5	7.5-9.5	0.04	0.03	0.5Mo, 1.5-2.5Cu, 0.8-1.4Ti, 0.1-0.5Nb	1410 (205) ht
17-7 PH	S17700	0.09	1.00	1.00	16.0-18.0	6.5-7.75	0.04	0.03	0.75-1.5Al	1030 (150) ht
PH 15-7 Mo	S15700	0.09	1.00	1.00	14.0-16.0	6.5-7.75	0.04	0.03	2.0-3.0Mo, 0.75-1.5Al	1100 (160) ht
AM355	S35500	0.10-0.15	0.5-1.25	0.50	15.0-16.0	4.0-5.0	0.04	0.03	2.5-3.25Mo	1030 (150) ht
A-286	K66286	0.04	1.30	0.40	15.0	26.0	...	...	1.3Mo, 2.0Ti, 0.2Al, 0.3V, 0.005B	755 (110) ht

(a) Compositions are in weight percent. Single values are maximum values except for Kromarc 58 and A-286, for which nominal compositions are shown. (b) Ann–annealed, ht–heat treated; yield strengths are minimum values for annealed material and typical values for heat treated material. (c) Optional Nb+Ta. (d) Nitronic 40 (XM-10). (e) Nitronic 50 (XM-19). (f) 15-5 PH, 17-4 PH, PH 13-8 Mo and Custom 455 are martensitic precipitation hardening types; 17-7 PH, PH 15-7 Mo and AM355 are semiaustenitic precipitation hardening types; and A-286 is an austenitic precipitation hardening type.

The others have additional amounts of chromium and/or other alloying elements for improved properties for special applications. Effects of these alloying additions and the range of applications for these steels are discussed in Ref 5.3.

For certain elevated temperature applications of martensitic stainless steels, hot rolled or forged products are air cooled after hot working, tempered, and used in this condition. For better control of properties, however, reheat treatment after fabrication is preferred. To reduce the hardness of these steels after hot working, they are process annealed at subcritical temperatures in the range 650 to 760°C (1200 to 1400°F) and cooled in air. For full annealing, they are heated into the range 815 to 870°C (1500 to 1600°F), cooled in the furnace to about 540°C (1000°F) and then cooled in air. They are ferritic in this condition with pearlite in the microstructure. Hardening of components of martensitic stainless steels is accomplished by heating into the austenitic region, quenching in oil or cooling in air, and then tempering. Tempering usually is done at temperatures above 540°C (1000°F) to avoid temper embrittlement. However, required conditions for preheating and for austenitizing, as well as tempering temperatures, heating rates, holding times at temperature, and cooling rates, are dependent on composition. Transformations are relatively sluggish in these steels, and time periods for various thermal processing treatments are longer than for alloy steels with lower alloy contents (Ref 5.3).

Martensitic stainless steels are selected for certain applications in which good corrosion resistance is required along with strength and/or hardness greater than may be obtained in austenitic or ferritic grades.

5.2.4. Precipitation Hardening Stainless Steels

Precipitation hardening stainless steels have been developed to meet a need for constructional components with good corrosion resistance, with high strength, and with better toughness than that of the martensitic stainless steels discussed above. Many different types of precipitation hardening stainless steels have been developed, and many of them have been used in certain special applications. Compositions of those for which fracture mechanics data have been obtained are presented in Table 5.1.

The precipitation hardening grades may be divided into three classes. Of those shown in Table 5.1, 15-5PH, 17-4PH, PH13-8Mo and Custom 455 are martensitic; 17-7PH, PH15-7Mo and AM355 are semiaustenitic; and A-286 is austenitic. The class depends on the microstructure of the steel when it is cooled from the annealing temperature to room temperature, and this in turn depends on the relative amounts of ferrite-promoting and austenite-promoting elements in the chemical composition. Applications of the martensitic and semi-austenitic classes are usually limited to room temperature and elevated temperature service. Applications for the austenitic class (face centered cubic) include both cryogenic and elevated temperature service.

For the precipitation hardening grades in Table 5.1, carbon contents are relatively low, chromium contents range from 11.0 to 18.0%, and nickel contents range from 3.0 to 26%. Steels in the austenitic class contain the highest amounts of nickel. Other additions increase strength and influence the precipitation hardening reaction. Corrosion resistance is mainly dependent on chromium content.

Many variations of multiple-cycle heat treatments may be used in developing a range of strength and toughness properties in the precipitation hardening steels, but only the most common will be discussed here and in later sections. The steels in the martensitic class (15-5PH, 17-4PH, PH13-8Mo and Custom 455) are usually supplied by the mill in condition A—i.e., in the solution treated condition. Solution treating temperatures range from 815 to 1040°C (1500 to 1900°F) depending on composition, and aging temperatures range from 480 to 620°C (900 to 1150°F). The microstructures of these steels are martensitic after solution treating (similar to annealed 18Ni maraging steels). Aging produces precipitates of intermetallic compounds, resulting in precipitation hardening. Within the age hardening temperature range, the highest strengths are obtained after aging at 480°C (900°F). Aging at 540 or 565°C (1000 or 1050°F) results in lower strength but better ductility and fracture toughness.

Heat treatments for the semiaustenitic grades (17-7PH, PH15-7Mo and AM355) require solution annealing at 1040 to 1065°C (1900 to 1950°F) and an additional thermal conditioning step prior to aging to develop the desired properties. More detailed discussions of these treatments will be presented later in this chapter.

Heat treatment of A-286 stainless steel is not as effective in increasing its strength as for the martensitic and semiaustenitic grades because of the relatively low strength of the austenitic matrix. Mill forms of A-286 are usually obtained in the annealed condition, annealed at 980°C (1800°F). For some applications, annealed A-286 may be re-solution treated at 900°C (1650°F) and quenched in oil prior to further processing. After fabrication, components of A-286 stainless steel are aged in the range from 700 to 760°C (1300 to 1400°F). On aging, precipitates of intermetallic compounds are formed in the matrix austenite. Aging increases the yield strength from 248 MPa (36 ksi) in the annealed condition to 690 MPa (100 ksi) in the aged condition.

5.3. FRACTURE MECHANICS DATA FOR STAINLESS STEELS

As noted in the introductory chapters of this book, there are certain limitations in obtaining usable fracture mechanics data for structural alloys, depending on their strength, toughness and intended service environments. To illustrate these limitations, all of the austenitic stainless steels are too tough for obtaining valid plane strain fracture toughness (K_{Ic}) data on specimens of reasonable size even at cryogenic temperatures. However, some of the austenitic grades, particularly Types 304 and 316, are of interest for certain nuclear reactor components that are subjected to cyclic loading at elevated temperatures. In order to evaluate the fatigue crack growth rate properties of these steels under conditions which would show effects of cyclic loading, elevated temperature exposure and various environments, fatigue crack growth rate tests have been conducted in a number of laboratories on precracked specimens of Types 304, 316 and others. Results of these tests have been analyzed on the basis of fracture mechanics concepts, leading to a well documented collection of fatigue crack growth rate data for the austenitic stainless steels—particularly Type 304 (Ref 5.4).

Fatigue crack growth rate data also have been obtained on precracked specimens of Type 403 martensitic stainless steel because it is the primary steel for steam turbine buckets and for rotors in some turbines. Results of these tests show the effects of a number of variables on fatigue crack growth rates (Ref 5.5). A limited amount of plane strain fracture toughness (K_{Ic}) data also has been obtained on large specimens of Type 403 stainless steel over a range of temperatures. Results of J-integral tests also show the estimated fracture toughness, $K_{Ic}(J)$, for Type 403 at room temperature and at elevated temperatures.

Of the precipitation hardening stainless steels, the largest amount of fracture mechanics data has been obtained on PH13-8Mo because this steel was considered for certain components of the B-1 bomber during prototype production at Rockwell International. Because all major structural alloys supplied for this aircraft were subjected to fracture mechanics evaluations, a well documented backlog of fracture mechanics data is available for the PH13-8Mo alloy (Ref 5.6).

Even with the limitations on available fracture mechanics data for the stainless steels, the data presented in the following sections will show the effects on fracture of many variables in alloy processing, loading conditions and environments for the most widely used austenitic, martensitic and precipitation hardening stainless steels.

5.3.1. Fracture Properties of Austenitic Stainless Steels

Available fracture mechanics data for austenitic stainless steels are limited to fatigue crack growth rate data in air and in several other environments, and at elevated and cryogenic temperatures. The available data will be reviewed according to the sequence of alloys in Table 5.1. Effects of material variables and stressing variables are reviewed, along with effects of various environments and temperature variations.

Type 301

Fatigue crack growth rates for Type 301 stainless steel have been reviewed by Pineau and Pelloux (Ref 5.7) in the temperature range from −30 to +95°C (−22 to +203°F). The results, summarized in Fig. 5.1, were obtained on compact specimens 7 mm (0.28 in.) thick at a cyclic frequency of 20 Hz with a sinusoidal wave form at a load ratio (R) of 0.01. All specimens were tested in dry argon except one series that was tested in laboratory air. For the annealed specimens tested in argon, fatigue crack growth rates at a given ΔK value increased as the temperature increased over the testing temperature range. Fatigue crack growth rates in laboratory air at 20°C (68°F) were higher than for corresponding conditions in argon, indicating that the humidity and/or oxygen in the air influenced the growth rates.

The warm worked specimens were reduced 65% at 450 to 500°C (840 to 930°F), resulting in a substantial increase in strength. Fatigue crack growth rates for the warm worked specimens (see Fig. 5.1) indicate that the fatigue crack propagation properties of the warm worked alloy are different than those of the annealed alloy. This effect of warm working has been observed for other austenitic stainless steels. These differences are attributed to the extent of the strain-induced transformation at the crack tip. This transformation effect would be most noticeable in Type 301, because it is less stable than the other alloys in the UNS S3xxxx series.

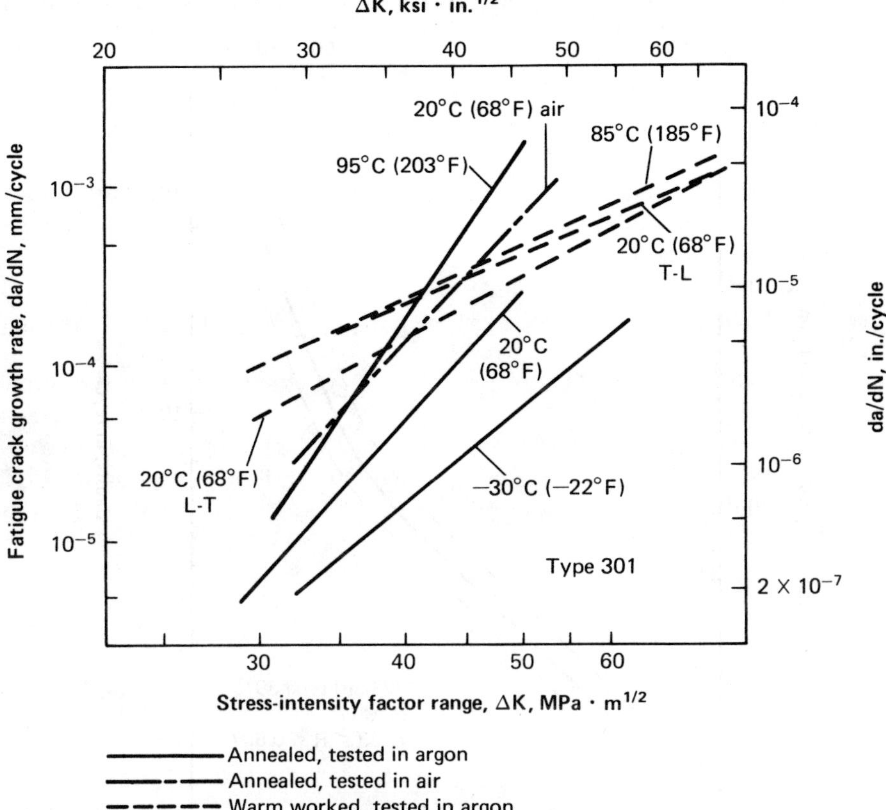

Fig. 5.1. Fatigue crack growth rates for type 301 stainless steel in the annealed and warm worked conditions, in air and argon environments, and at temperatures from −30 to +95°C (−22 to +203°F) (Ref 5.7)

Fatigue crack growth rate data reported by Walker for $\frac{1}{2}$-hard Type 301 stainless steel sheet are summarized in Fig. 5.2 (Ref 5.4 and 5.8). The data were obtained in air at room temperature over a series of load ratios (R) from 0.063 to 0.807 at a frequency of 10 Hz. These data are based on the "effective stress intensity factor", K_{eff}, rather than on ΔK, to account for the effect of the range of stress ratios. K_{eff} is defined as follows:

$$K_{eff} = K_{max}(1 - R)^m \qquad \text{(Eq 5.1)}$$

where m is determined empirically and R is the load ratio (minimum load/maximum load) on cyclic loading. The crack growth rate law then becomes:

$$da/dN = C[K_{max}(1 - R)^m]^n \qquad \text{(Eq 5.2)}$$

Results of fatigue crack growth rate tests on austenitic stainless steels have shown

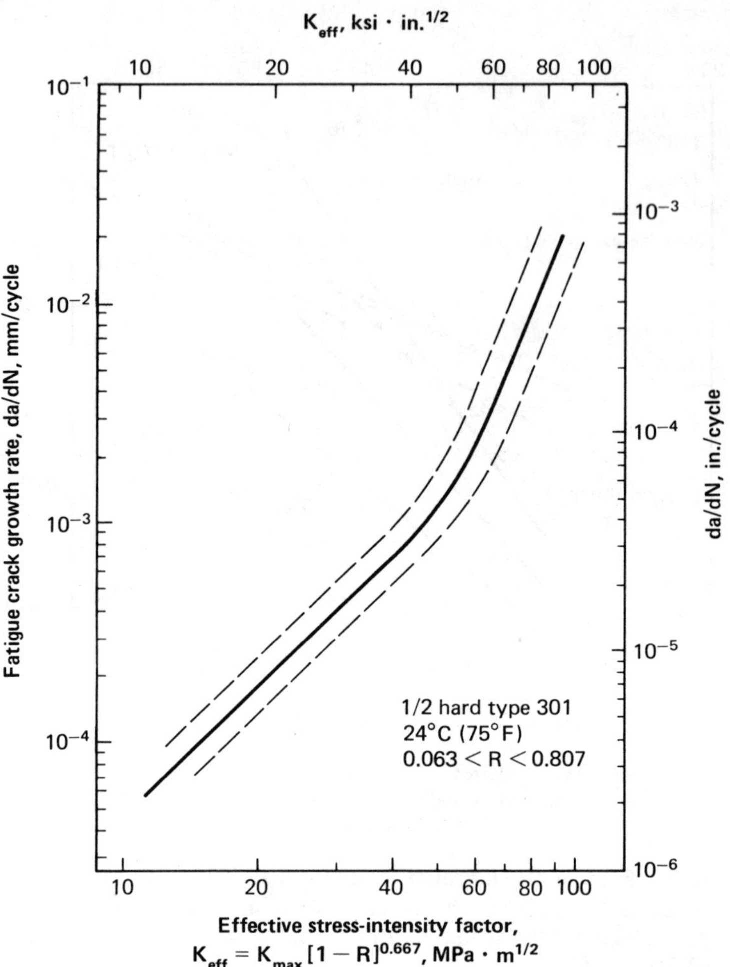

Fig. 5.2. Scatter band of fatigue crack growth rates for ½-hard type 301 stainless steel, tested at 24°C (75°F), 10 Hz, and R ratios of 0.063 to 0.807 based on effective stress-intensity factor, K_{eff} (Ref 5.4 and 5.8)

that the crack growth rate tends to increase as the R ratio is increased, when compared at given values of ΔK. If tests are made at several load ratios to determine m, then the effects of other load ratios may be estimated.

Types 304 and 304L

Fatigue crack growth rate data have been obtained for specimens of Type 304 and Type 304L stainless steels to show the effects of many of the variables associated with applications of these alloys. Types 304 and 304L are generally used in the annealed condition, but for improved strength they may be applied in the warm worked or cold worked condition (cold drawn or cold rolled). They are used extensively in construction of nuclear power station structures, and in this application they are exposed to elevated temperatures and static stressing combined with cyclic stressing over a wide range of frequencies and load ratios while being exposed to potentially corrosive environments. Exposure to fast neutron fluences must also be considered. The effects of all of these factors have been evaluated on a number of fatigue testing programs.

Fatigue Crack Growth Rates at Room Temperature and Elevated Temperatures. Results of fatigue crack growth rate tests on Types 304 and 304L stainless steels at room temperature and at elevated temperatures have been reported by James and Schwenk (Ref 5.9), by James (Ref 5.4 and 5.10) and by others (Ref 5.11 and 5.12). As shown in Fig. 5.3, increasing the exposure temperature from room temperature to 650°C (1200°F) increases the fatigue crack growth rates at any ΔK level within the range of the tests in an air environment. These data, reported by James and Schwenk, are for specimens of both the L-T and T-L orientations, for several different maximum alternating loads, for load ratios of 0 to 0.05, and for cyclic frequencies from 0.033 to 6.66 Hz for the room temperature tests and 0.067 Hz for the elevated temperature tests. Data points were omitted from the curves in Fig. 5.3 and from many of the da/dN curves in

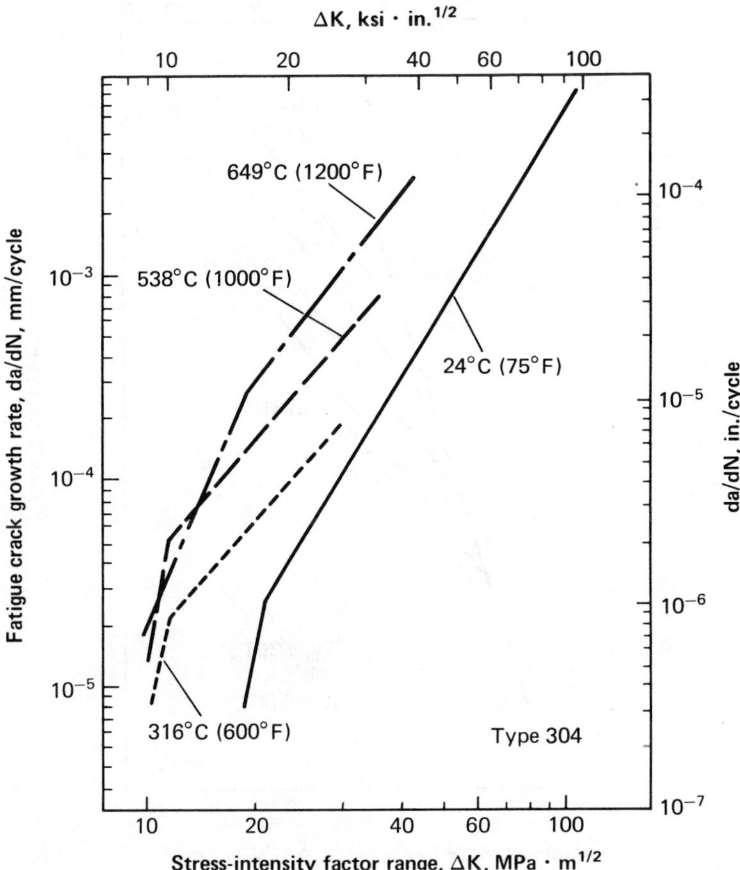

Fig. 5.3. Effect of testing temperature on fatigue crack growth rates for annealed type 304 stainless steel tested in air at 0.066 Hz and an R ratio of 0 to 0.05 (Ref 5.9)

other figures to show the trends more clearly, but the scatter in data points was relatively narrow.

For fatigue crack growth rate tests on specimens of annealed Type 304 stainless steel at elevated temperatures, increasing the cyclic frequency will decrease the crack growth rate over part of the ΔK range, as shown in Fig. 5.4 for tests at 538°C (1000°F) and at R = 0.05 (Ref 5.13 and 5.14). The data in Fig. 5.4 were obtained in tests with a sawtooth waveform. Changing from a sawtooth waveform to a waveform with a short holding period at maximum load did not influence the over-all fatigue crack growth rates according to additional data reported by James and shown in Fig. 5.5 (Ref 5.15).

Effects of holding times of 0.1 and 1.0 minute on fatigue crack growth rates for specimens of annealed Type 304 based on crack extension per unit of time, da/dt, for tests at 427 and 593°C (800 and 1100°F) have been reported by Shahinian (Ref 5.16). At 427°C (800°F), the fatigue crack growth rates were substantially

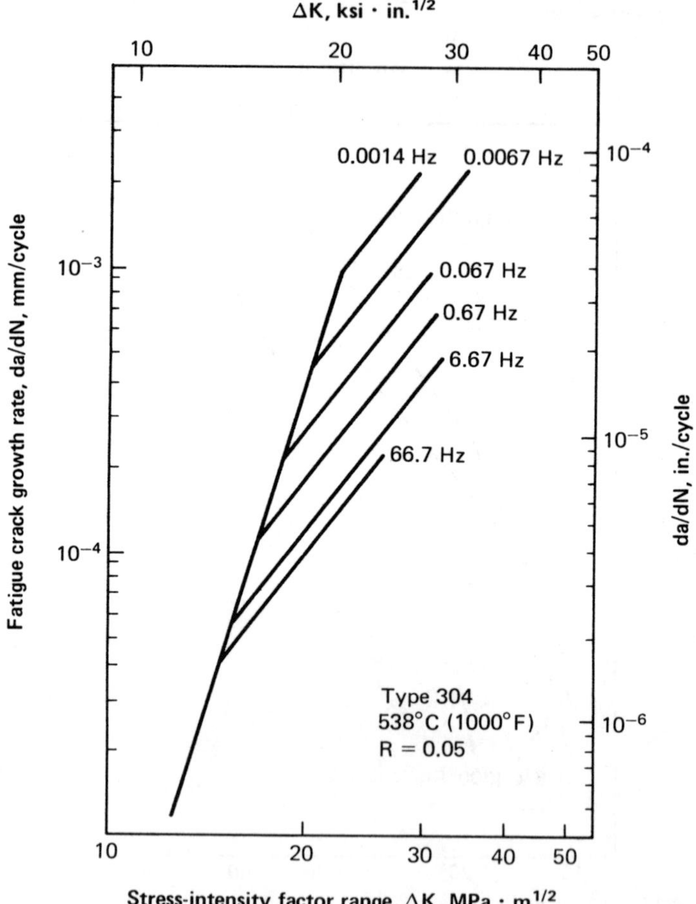

Fig. 5.4. Effect of variation in cyclic frequency on fatigue crack growth rates for annealed type 304 stainless steel at 538°C (1000°F) for an R ratio of 0.05 in air with a sawtooth waveform (Ref 5.13 and 5.14)

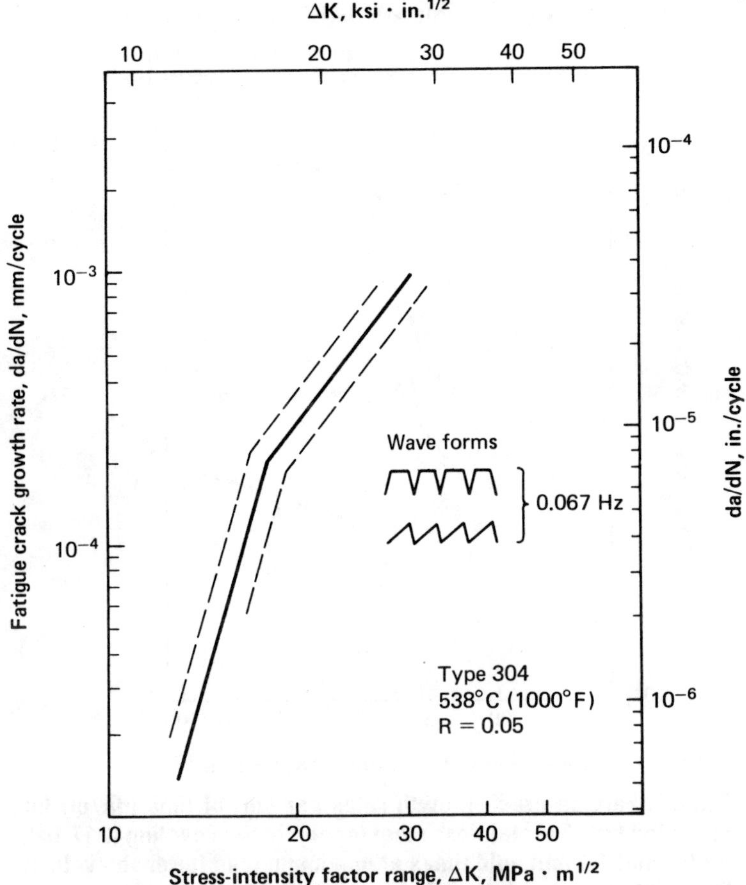

Fig. 5.5. Scatter band of fatigue crack growth rates for annealed type 304 stainless steel at 538°C (1000°F) in air at an R ratio of 0.05 with two different waveforms at 0.067 Hz (Ref 5.15)

greater for specimens tested with no holding time (continuous cycling) than for specimens held at maximum load for 0.1 or 1.0 minute per cycle. The lowest fatigue crack growth rates occurred for specimens with the longest holding time, based on da/dt. The same trend was observed for tests at 593°C (1100°F), as shown in Fig. 5.6. Therefore, cyclic loading has a more damaging effect than static loading on crack growth per unit of time.

Varying the section thickness of single edge notch cantilever specimens and compact tension loaded specimens from 7.6 to 25.4 mm (0.3 to 1.0 in.) did not affect the results of da/dN versus ΔK tests at room temperature, at 427°C (800°F) and at 593°C (1100°F), according to test data reported by Shahinian (Ref 5.17). James has also reported results of fatigue crack growth rate tests on compact specimens of annealed Type 304 plate 6.35 and 25.4 mm (0.25 and 1.00 in.) thick from the T-L orientation. Tests in air at a testing temperature of 288°C (550°F), at a frequency of 2.5 Hz and at R ratios of 0.05 and 0.50 indicated that there was no effect of specimen thickness on the da/dN data (Ref 5.18).

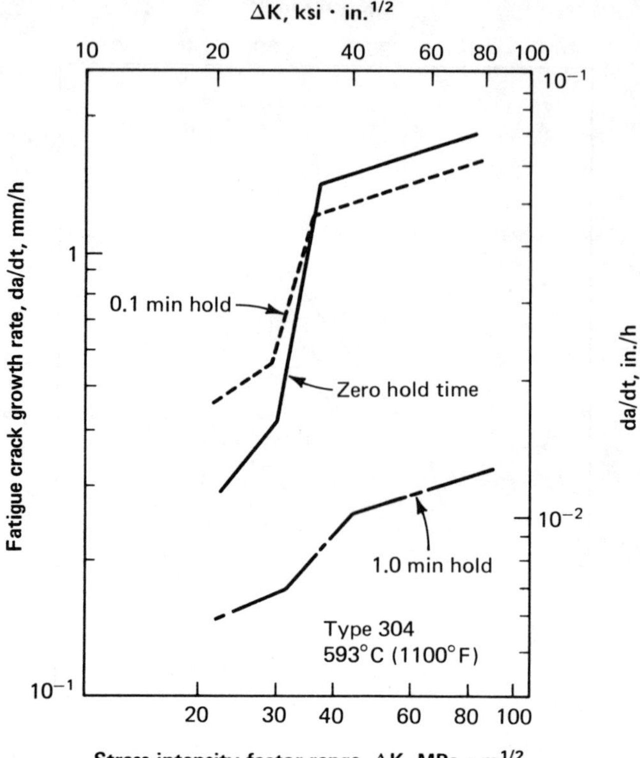

Fig. 5.6. Fatigue crack growth rates per unit of time (da/dt) for annealed type 304 stainless steel for continuous cycling (0.17 Hz), for 0.1 and 1.0-min hold times at maximum load for each cycle at 593°C (1100°F), and for an R ratio of 0 (Ref 5.16)

Results of fatigue crack growth rate tests also have been reported by James over a range of stress ratios from −0.150 to +0.750 for compact specimens of Type 304 stainless steel at 538°C (1000°F) and at a frequency of 6.67 Hz. For these tests, the parameter $K_{max}(1 - R)^m$, or K_{eff}, where $m = 0.5$ (Eq 5.1), again provided a much better correlation of results than the parameter ΔK (Ref 5.19).

In order to determine the effect of heat-to-heat variations on fatigue crack growth rates for annealed Type 304, specimens were obtained from four heats of Type 304 and one heat of Type 304L from four different sources. Each heat was annealed by the producer at temperatures from 1065 to 1120°C (1950 to 2050°F) and water quenched. Hardnesses of these heats ranged from 73 to 85 HRB, and ASTM grain sizes ranged from 3 to 5. Chromium and nickel contents were relatively consistent, but carbon contents ranged from 0.024 to 0.064%. Fatigue crack growth rate data were obtained on six specimens from one heat in air at 538°C (1000°F) at a frequency of 0.67 Hz and an R ratio of 0.05. As reported by James (Ref 5.20), the scatter band on data points plotted in the usual da/dN versus ΔK format showed a uniform spread over the ΔK range of the tests. The factor representing the variation in crack growth rates for the six specimens from one heat was 2.75.

For all data points for one specimen from each of the five heats tested under identical conditions, the scatter band again represented a uniform spread, with the factor for variation in fatigue crack growth rate being 2.78.

In addition, fatigue crack growth rate tests were made on specimens from two of the above heats at 538°C (1000°F) and R = 0.05 using a square waveform at 0.00138 Hz including a 10.8-minute holding time at maximum load on each cycle. Four specimens were tested from one heat and two from the other heat. For these tests, the factor representing the maximum spread in crack growth rate data for all specimens was 3.45. The data represent somewhat greater scatter, but there was apparently no significant difference in the behavior of the two heats. Results of these studies (Ref 5.20) indicate that, for the conditions employed, there is no effect of heat-to-heat variations on fatigue crack growth rates for commercially produced Types 304 and 304L stainless steels. Furthermore, there is no apparent effect of grain size and crack orientation (L-T vs T-L) on fatigue crack growth rates at 538°C (1000°F) for the commercial product in the annealed condition.

In some applications, Type 304 stainless steel components are fabricated in the cold worked condition to improve strength properties. A comparison of fatigue crack growth rate data by Shahinian, Watson and Smith (Ref 5.21), illustrated in Fig. 5.7, shows that the high-ΔK crack growth rates were lower for the cold worked specimens than for the annealed specimens. Crack growth rates were higher for the specimens tested at 427°C (800°F) than for corresponding specimens tested at room temperature, as noted previously.

Because the expected service lives of most components of austenitic stainless steels are many years, an evaluation of the effect of long-time aging at service temperatures is important. Results of fatigue crack growth rate tests on specimens that were tested in the unaged and aged conditions (5000 hours at 593°C, or 1100°F) are shown in Fig. 5.8 as reported by Michel and Smith (Ref 5.22). After aging for 5000 hours at this temperature, precipitation of $M_{23}C_6$ carbides is essentially complete. These results indicate that at 593°C (1100°F) there are no deleterious effects of aging on the crack growth rates of specimens that are continuously cycled. When a holding time of 0.1 or 1.0 minute is included in each loading cycle, there tends to be a slight increase in the fatigue crack growth rate at a given ΔK level.

Aging of cold worked specimens of Type 304 at 593°C (1100°F) for 5000 hours tends to increase slightly the fatigue crack growth rates of specimens that are continuously cycled at 593°C (1100°F) (Michel and Smith; Ref 5.22 and 5.23). Again, with holding times of 0.1 and 1.0 minute, the fatigue crack growth rates were increased.

The effects of humid air environments on the room temperature fatigue crack growth rates of specimens of annealed Type 304 stainless steel are shown in Fig. 5.9 for specimens cycled at 0.17 Hz with an R ratio of zero (Shahinian, Watson and Smith; Ref 5.21). At the lower end of the ΔK range, fatigue crack growth rates in humid air are substantially greater than crack growth rates in dry air. However, fatigue crack growth rates of specimens of Type 304 stainless steel tested in a pressurized water reactor environment at 260 to 315°C (500 to 600°F) with R ratios of 0.2 and 0.7 were no greater than the fatigue crack growth rates in air at the same temperature with an R ratio less than 0.1 (Bamford; Ref 5.24). However, variations in R ratios influenced the fatigue crack growth rates in the pressurized water reactor environment.

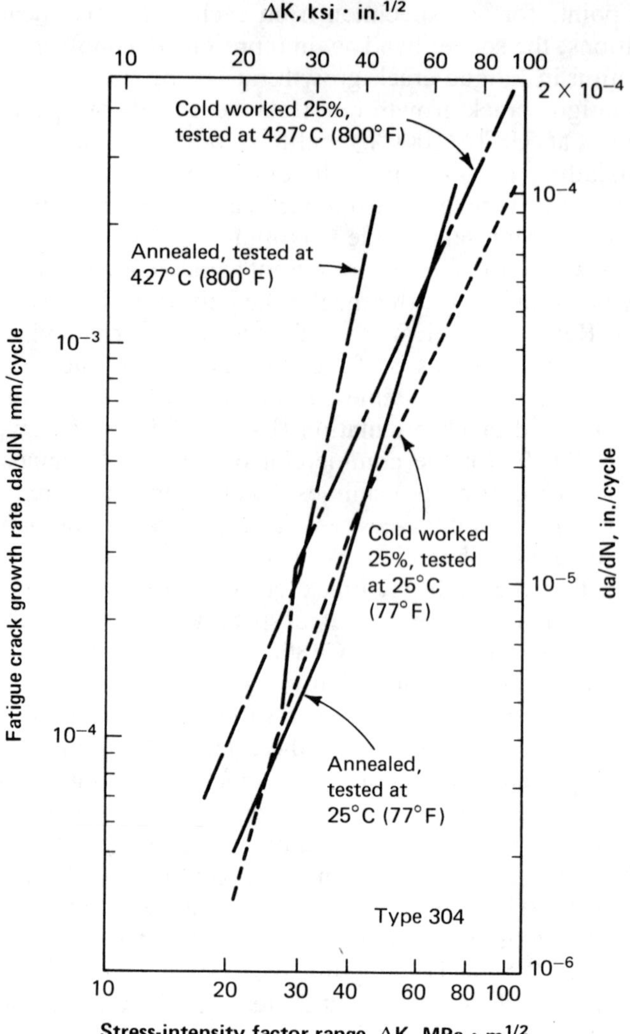

Fig. 5.7. Fatigue crack growth rates for annealed and cold worked type 304 stainless steel at 25 and 427°C (77 and 800°F), 0.17 Hz, and an R ratio of 0 (Ref 5.21)

Results of fatigue crack growth rate tests in a high pressure simulated boiling water reactor (BWR) environment at 288°C (550°F) also have been reported by Hale, Yuen and Gerber for specimens of Types 304 and 304L stainless steels (Ref 5.25). Specimens were tested as received (annealed) and after furnace sensitizing to simulate localized heating as in welding. Fatigue crack growth rates increased as the load ratio was increased and/or as the cyclic loading frequency was reduced. For these tests, the K_{eff} expression (Eq 5.1) was used to normalize the effect of variations in R ratio in the BWR environment. Sensitizing had no effect on fatigue crack growth rates.

Fatigue crack growth rate data were obtained by James and Knecht on specimens of Type 304 stainless steel at 427°C (800°F) in a vacuum and in molten

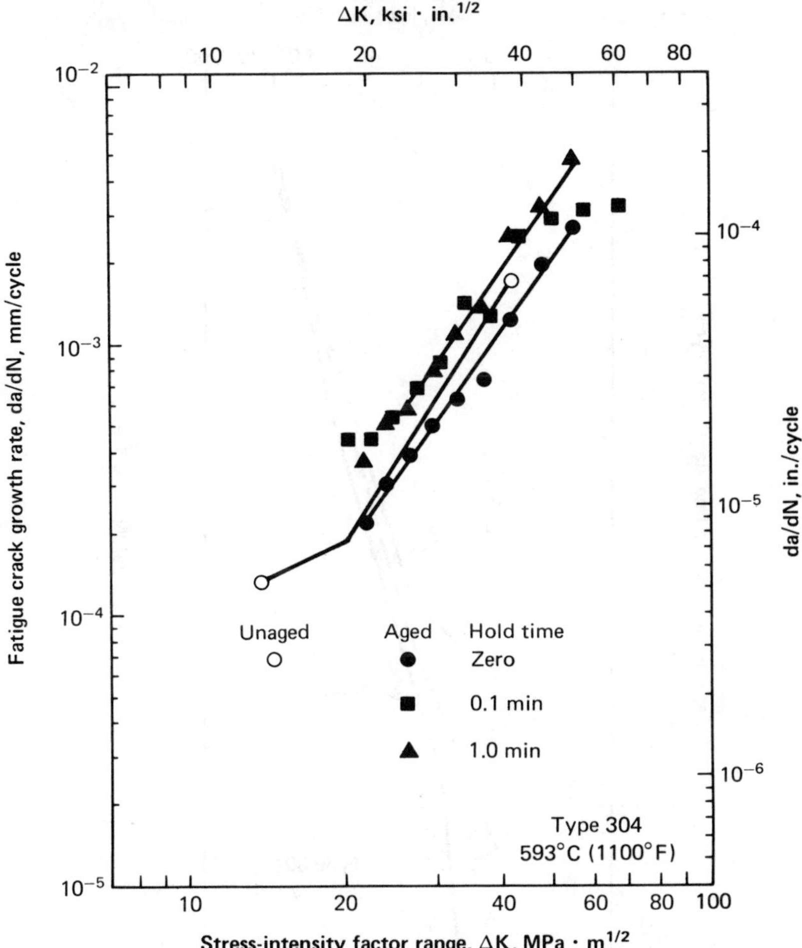

Fig. 5.8. Effect of aging at 593°C (1100°F) for 5000 h, and hold times of 0.1 and 1.0 min for each cycle, on fatigue crack growth rates of L-T oriented specimens of type 304 stainless steel tested in air at 0.17 Hz and an R ratio of 0 (Ref 5.22)

sodium environments at frequencies from 3 to 6.67 Hz and R ratios of 0.05 and 0.50 (Ref 5.26 and 5.27). Based on the K_{eff} normalization, these data show that the sodium environment is no worse than a vacuum environment under these conditions. Crack growth rates were lower than for tests at the same temperature in air. The same effect was observed for tests at 538°C (1000°F).

Effects of fast neutron irradiation at elevated temperatures on fatigue crack growth rates of specimens of Type 304 stainless steel are summarized by James in Ref 5.4, by Michel in Ref 5.28 and by Michel and Smith in Ref 5.29. For annealed and cold worked specimens of Type 304 irradiated at elevated temperatures, the effect varies depending on variations in composition, amount of cold working, irradiation temperature, testing temperature, and neutron energy spectra. As an example, fatigue crack growth rates in specimens of annealed Type 304

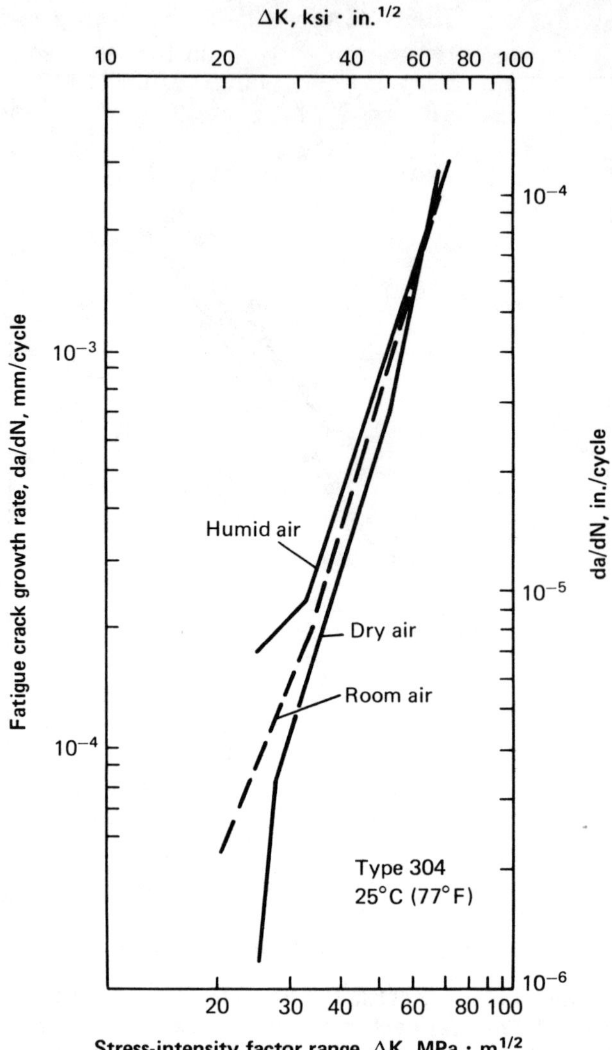

Fig. 5.9. Effect of humidity on fatigue crack growth rates for type 304 stainless steel tested at room temperature, 0.17 Hz, and an R ratio of 0 (Ref 5.21)

irradiated with fast neutrons at fluence levels of 1.3×10^{21} and 9.0×10^{21} n/cm^2 (E > 0.1 MeV) at temperatures from 454 to 477°C (850 to 890°F) were similar to fatigue crack growth rates in unirradiated specimens in tests at a temperature of 427°C (800°F), a testing frequency of 0.67 Hz and an R ratio of 0.05 (Ref 5.30). However, substantial variations may occur particularly if the testing temperature is not the same as the irradiation temperature, as observed by Shahinian, Watson and Smith (Ref 5.31). Shahinian (Ref 5.32) also has shown that irradiation of cold worked Type 304 stainless steel increases the fatigue crack growth rate substantially at 593°C (1100°F). Irradiation at higher fluences may be more damaging than irradiation at the levels for the available test data.

Fatigue Crack Growth Rates at Cryogenic Temperatures. Fatigue crack growth rate data obtained by Tobler and Reed on specimens of Types 304 and 304L stainless steels (annealed) at temperatures in the range from room temperature to liquid helium temperature (−269°C, or −452°F) are shown in Fig. 5.10 (Ref 5.33). The data for Type 304 were scattered over the range shown, while for Type 304L the data at room temperature described one curve and the data at the cryogenic temperatures described the other curve. These results indicate that cryogenic fatigue crack growth rates for Type 304 do not deviate significantly from room temperature fatigue crack growth rates over the ΔK range studied. Furthermore, if design calculations for Type 304L are based on room temperature fatigue crack growth rates, the calculations will be conservative for cryogenic exposure.

Type 304LN Base Metal and Welds. Type 304LN stainless steel is similar to Type 304L except for the added nitrogen (0.10 to 0.15%), which increases yield strength, particularly at cryogenic temperatures, without reducing ductility. Because of its increased yield strength, Type 304LN has been considered for structural applications involving exposure to cryogenic temperatures, as in super-

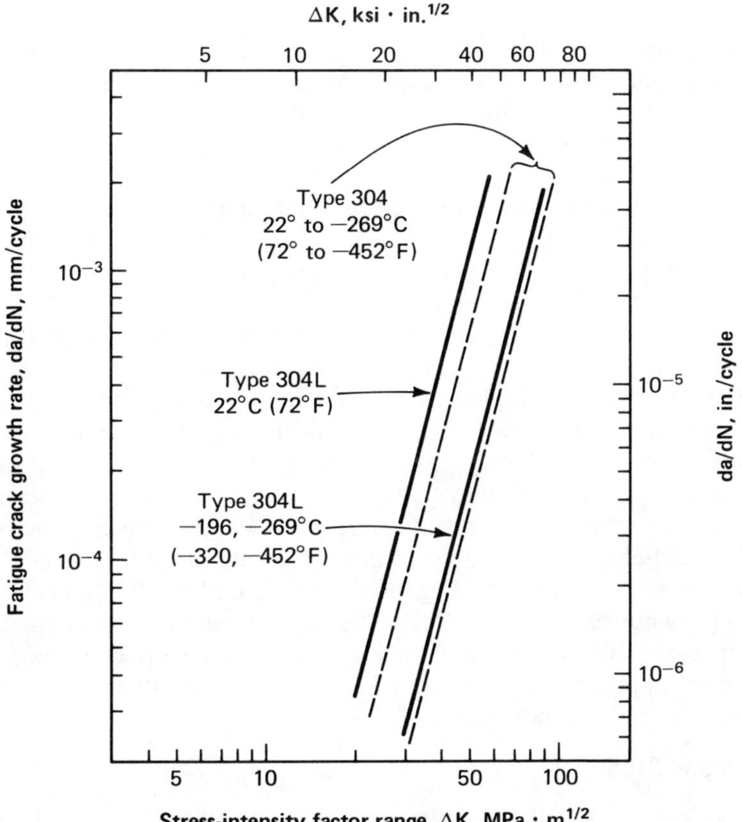

Fig. 5.10. Fatigue crack growth rates for annealed type 304 and 304L stainless steels at room and cryogenic temperatures, 20 to 28 Hz, and an R ratio of 0.1 (Ref 5.33)

conducting machinery. Results of fracture toughness tests on compact specimens of Type 304LN plate have been reported by Witherell (Ref 5.34) for tests at −269°C (−452°F). Fracture toughness data were obtained by the J-integral method. Results of these tests showed a $K_{Ic}(J)$ value of 223 MPa · $m^{1/2}$ (203 ksi · $in.^{1/2}$) for a plate 38 mm (1.5 in.) thick, and 222 MPa · $m^{1/2}$ (202 ksi · $in.^{1/2}$) for a plate 76 mm (3 in.) thick, at liquid helium temperature. Additional fracture toughness tests were made on compact specimens of Type 304LN with welds of Type 316L filler metal. The weld metal in a specimen that was welded by the submerged arc method had a fracture toughness of 111 MPa · $m^{1/2}$ (101 ksi · $in.^{1/2}$), but in the heat affected zone the $K_{Ic}(J)$ value was 144 MPa · $m^{1/2}$ (131 ksi · $in.^{1/2}$) at −269°C (−452°F). For another specimen welded with 316L filler metal deposited by the shielded metal arc method, the K_{Ic} (J) values were 145 MPa · $m^{1/2}$ (132 ksi · $in.^{1/2}$) in the weld metal and 155 MPa · $m^{1/2}$ (142 ksi · $in.^{1/2}$) in the heat affected zone at the same testing temperature. This information may be applied in estimating critical flaw sizes in thick-section structures of Type 304LN stainless steel cooled to liquid helium temperature.

Type 308 Welds in Type 304 Stainless Steel. Type 308 stainless steel is the alloy that is usually used for welding rod for weldments in Type 304 stainless steel when those weldments are to be exposed to room temperature or to elevated temperatures in service. Because service experience has shown that failures are more likely to originate in weld metal or in heat affected zones than in the base metal, it is important to have fracture information on weldments. In general, fatigue studies at elevated temperatures on specimens from Type 304 weldments have shown that the fatigue crack growth rates in the Type 308 weld metal and heat affected zones are no greater than in comparable specimens of the base metal. Fatigue crack growth rate data obtained by Shahinian for specimens of Type 304 welded with Type 308 rod by the submerged arc and shielded metal arc processes are shown in Fig. 5.11 for tests at room temperature and at 593°C (1100°F) (Ref 5.21, 5.35 and 5.36). Similar data have been obtained by James and others (Ref 5.4, 5.10, 5.29, 5.37, 5.38 and 5.39). At elevated temperatures, the fatigue crack growth rates in the weld metal are higher than at room temperature, particularly at the lower ΔK levels, but the trend is the same as that for the base metal. Aging the welded specimens at 593°C (1100°F) for 1000 hours and testing at 593°C reduced the fatigue crack growth rates in the weld metal, the same as for the base metal (Ref 5.39). However, variations in composition and testing conditions can lead to variations in results (Ref 5.4). Fatigue crack growth rate data reported by Michel and Smith indicate that welds produced by the shielded metal arc process have superior crack growth resistance at 593°C (1100°F) when exposed to fast neutron radiation (Ref 5.29). By comparison, submerged arc welds made with Type 308 filler metal had less crack growth resistance at 593°C (1100°F) when exposed to a fluence of 3.6×10^{22} n/cm^2 than shielded metal arc welds under the same conditions.

Types 309S and 310S

Types 309S and 310S stainless steels are the low-carbon versions of Types 309 and 310. They have higher chromium and nickel contents than those of Type 304 and consequently have better corrosion resistance and more stable austenite than Type 304. Fatigue crack growth rate data have been reported by Thompson for

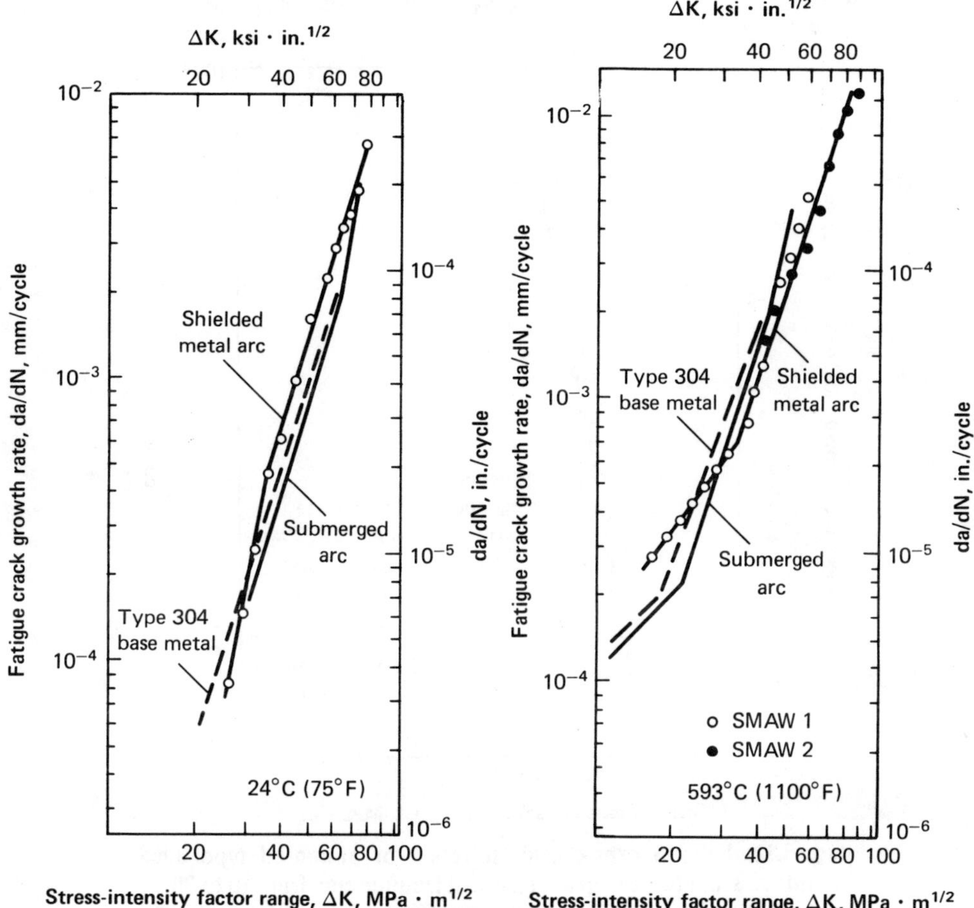

Fig. 5.11. Fatigue crack growth rates for annealed type 304 base metal and type 308 weld metal at 24 and 593°C (75 and 1100°F), 0.17 Hz, and an R ratio of 0 (Ref 5.35)

tests made at room temperature on compact specimens from plate of Type 309S in the L-T orientation after heat treating to a grain size of 45 μm in one set and 480 μm in the second set (Ref 5.40). Specimens with the smaller grain size had substantially higher yield and ultimate tensile strengths than the specimens with the larger grain size. Fatigue crack growth rates were obtained on tension-tension loading at frequencies from 10 to 30 Hz and at an R ratio of 0.05. The results are plotted in Fig. 5.12. These data provide further evidence that a wide variation in grain size, and the associated variation in strength level, does not affect the results of fatigue crack growth rate tests—at least, not in the Region 2 growth regime (see Fig. 4.18).

Because of the high nickel content of Type 310S stainless steel, it is completely stable at all cryogenic temperatures and with any amount of cold working. Therefore, it is often considered for cryogenic applications which require a high degree of austenite stability on thermal cycling and strain cycling. Fatigue crack growth

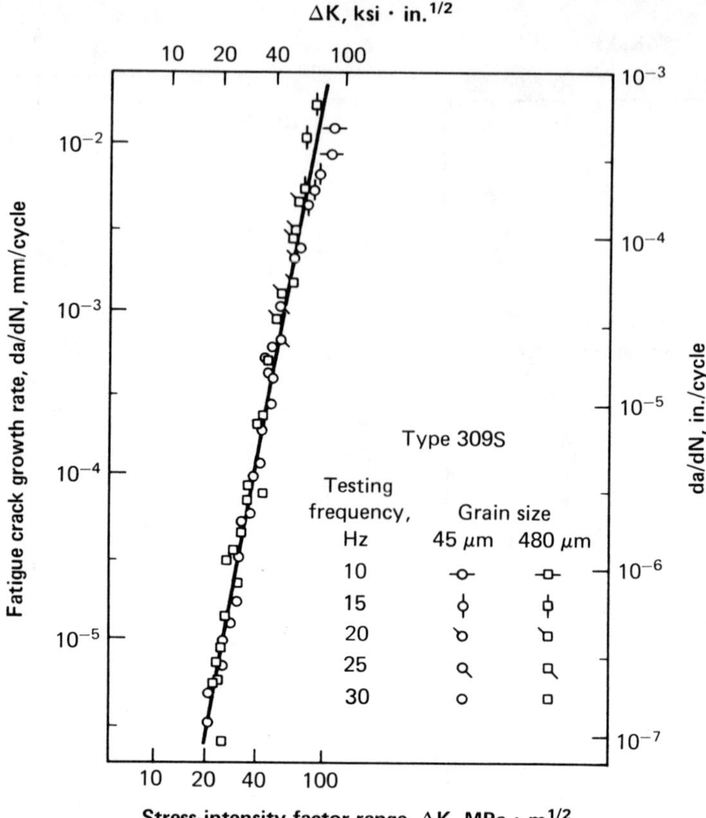

Fig. 5.12. Fatigue crack growth rates for annealed type 309S stainless steel for two grain sizes, at frequencies from 10 to 30 Hz and an R ratio of 0.05 at room temperature in air (Ref 5.40)

rate data obtained on annealed specimens of Type 310S by Tobler and Reed in the temperature range from room temperature to −269°C (−452°F) were very similar to corresponding data for Type 304L stainless steel (Fig. 5.13; Ref 5.33). These results were obtained on compact specimens tested at a frequency of 20 to 28 Hz and an R ratio of 0.1. The fatigue crack growth rates at cryogenic temperatures were lower than those at room temperature.

Additional data reported by Wells, Logsdon and Kossowsky on fatigue crack growth rates for annealed Type 310S stainless steel at cryogenic temperatures also show that the crack growth rates are lower at cryogenic temperatures than at room temperature (Ref 5.41). Results of fatigue crack growth rate tests on welded specimens of Type 310S, for specimens notched in the weld metal, also are shown in Fig. 5.13. Welds were produced by the shielded metal arc (SMAW) process using Type 310S rod. Tests were made on specimens in the as-welded condition at a frequency of 10 Hz.

Fracture toughness data also were obtained by Wells *et al* on compact specimens of 310S base metal and weld metal at −269°C (−452°F) by means of the J-integral technique (Ref 5.41). Estimated fracture toughness [K_{Ic}(J)] values were

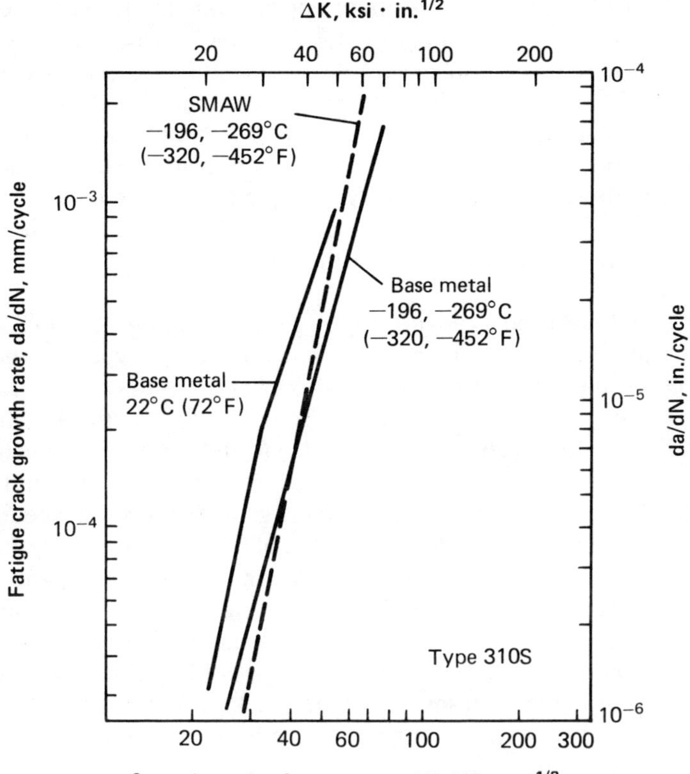

Fig. 5.13. Fatigue crack growth rates for annealed type 310S stainless steel at 22, −196 and −269°C (72, −320 and −452°F), 10 to 28 Hz, and an R ratio of 0.1, with corresponding data for SMA welds with type 316 filler metal (Ref 5.33 and 5.41)

262 MPa · m$^{1/2}$ (236 ksi · in.$^{1/2}$) for the base metal, and 118 MPa · m$^{1/2}$ (106 ksi · in.$^{1/2}$) in the weld metal, at liquid helium temperature. For these tests, fracture toughness of the weld was much lower than that of the base metal. Therefore, similar welds would be critical areas for fracture in Type 310S weldments at cryogenic temperatures.

Types 316 and 316N

Type 316 stainless steel contains 2.0 to 3.0% molybdenum in addition to chromium and nickel. The major effect of the molybdenum is to increase the tensile yield strength. Addition of 0.10 to 0.16% nitrogen to Type 316N increases both yield and ultimate tensile strengths without reducing ductility.

Most of the fatigue crack growth rate testing on Type 316 stainless steel has been oriented toward its use in components for nuclear reactors, but the data also are applicable to design of equipment for fossil fuel power stations, petrochemical refineries and chemical plants. Its improved yield strength compared with that of Type 304 stainless steel is an advantage for these applications. The austenite stability in Type 316 is greater than that in Type 304, so it is advantageous to use

Type 316 instead of Type 304 for critical applications at cryogenic temperatures.

Effects of elevated temperatures on fatigue crack growth rates for Type 316 have been reported by Shahinian, Smith and Watson and are summarized in Fig. 5.14 (Ref 5.42). The data were obtained on single-edge-notch specimens with side grooves during cantilever loading. They were obtained, at the L-T orientation, from annealed Type 316 plate. The tests were conducted in air at a frequency of 0.17 Hz according to a sawtooth waveform at an R ratio of zero. As shown by the curves in Fig. 5.14, the fatigue crack growth rates tend to increase as the testing temperature is increased, much the same as for specimens of Type 304 stainless steel.

The effects of long-time exposure (5000 hours of aging) at 593°C (1100°F) in air on the fatigue crack growth rates of specimens of Type 316 are shown in Fig. 5.15, according to data reported by Michel and Smith (Ref 5.22, 5.23, 5.43, 5.44). Aging substantially reduced the fatigue crack growth rates at ΔK levels from 18 to 55 MPa · $m^{1/2}$ (16 to 50 ksi · $in.^{1/2}$) for the continuous cycling tests and over the whole testing range for specimens cycled with 0.1- and 1.0-minute holding times for each cycle. Fatigue crack growth rates for specimens tested without prior exposure and with holding times of 0.1 and 1.0 minute for each cycle were higher than those for specimens cycled continuously under the same conditions. However, the effect of hold time was less significant for specimens that had been aged at 593°C (1100°F) before testing at the same temperature. In more recent reports by the same authors (Ref 5.45 and 5.46), holding times of up to 8 minutes did not cause significant increases in fatigue crack growth rates at 593°C (1100°F) in aged specimens. However, holding for 16 minutes caused marked increases in fatigue crack growth rates.

Fatigue crack growth rates for specimens of Type 316 stainless steel aged at 649°C (1200°F) for 6000 hours and tested at 538°C (1000°F) were either lower than, or within the scatter band for, specimens tested at 538°C without aging (James; Ref 5.10).

James has shown that the effects of variations in cyclic frequency on fatigue crack growth rates for specimens of Type 316 stainless steel do not show the same pattern as for specimens of Type 304 (Ref 5.13). For tests at frequencies in the range from 0.0067 to 6.67 Hz at 538°C (1000°F), the trend is for the crack growth rate to increase as the frequency is decreased, but there is more scatter than for Type 304. This trend is shown in Fig. 5.16.

In studying heat-to-heat variations in fatigue crack growth rates for specimens from three heats of Type 316 stainless steel, James has shown that the spread from high to low values of fatigue crack growth rates is no greater than that represented by a factor of 2.6 over the range of ΔK values studied (Ref 5.20). One heat was produced by air melting, another by vacuum arc remelting, and the third by double vacuum melting. The vacuum arc remelted heat contained sufficient nitrogen to be classed as Type 316N. The yield and ultimate tensile strengths for this heat were higher than those for the other two heats, which contained no nitrogen. ASTM grain sizes ranged from 4 to 8, and annealing temperatures ranged from 1010 to 1093°C (1850 to 2000°F). Each heat was produced at a different mill. However, the above heat-to-heat variations did not have a significant influence on the fatigue crack growth rates at 538°C (1000°F).

Fatigue crack growth rates obtained by James for specimens of cold worked Type 316 stainless steel were lower than those for comparable annealed specimens

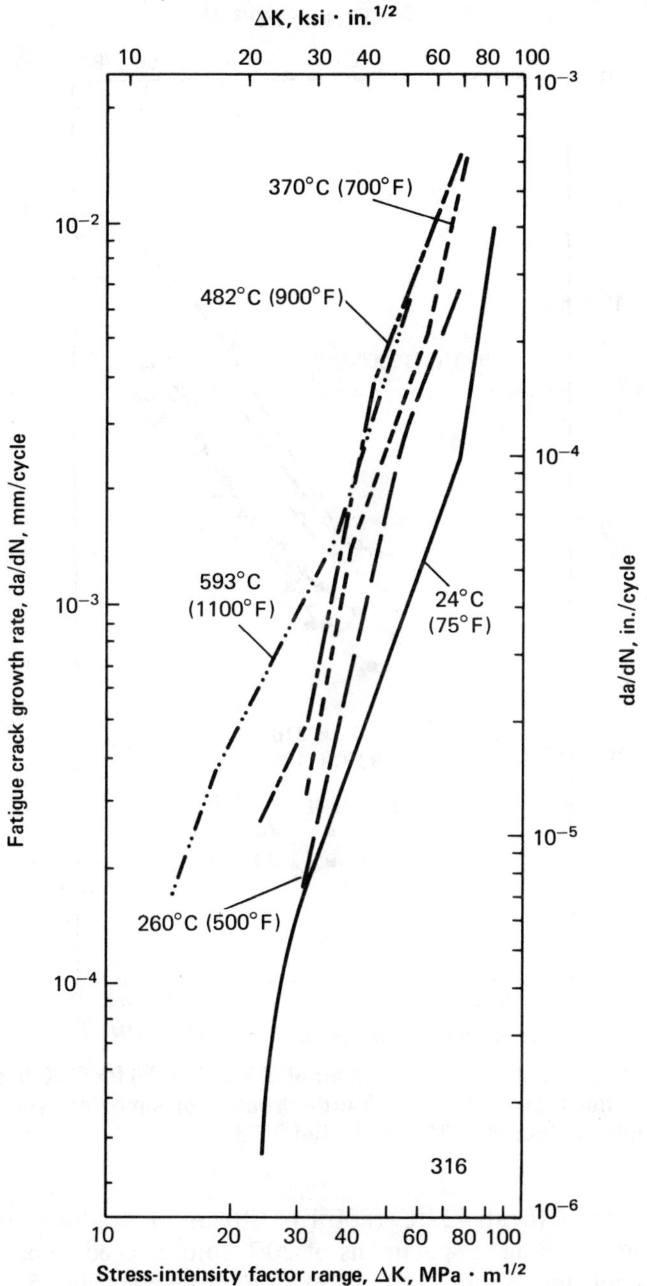

Fig. 5.14. Effect of testing temperature on fatigue crack growth rates for annealed type 316 stainless steel tested in air at 0.17 Hz and an R ratio of 0 (Ref 5.11 and 5.42)

of Type 316 at room temperature and at elevated temperatures. Results of tests on compact specimens of 20% cold worked Type 316 stainless steel at frequencies of 0.67 and 3.0 Hz and at an R ratio of 0.05 are summarized in Fig. 5.17 (Ref 5.10 and 5.47). Similar results have been reported by Shahinian (Ref 5.16) for tests on

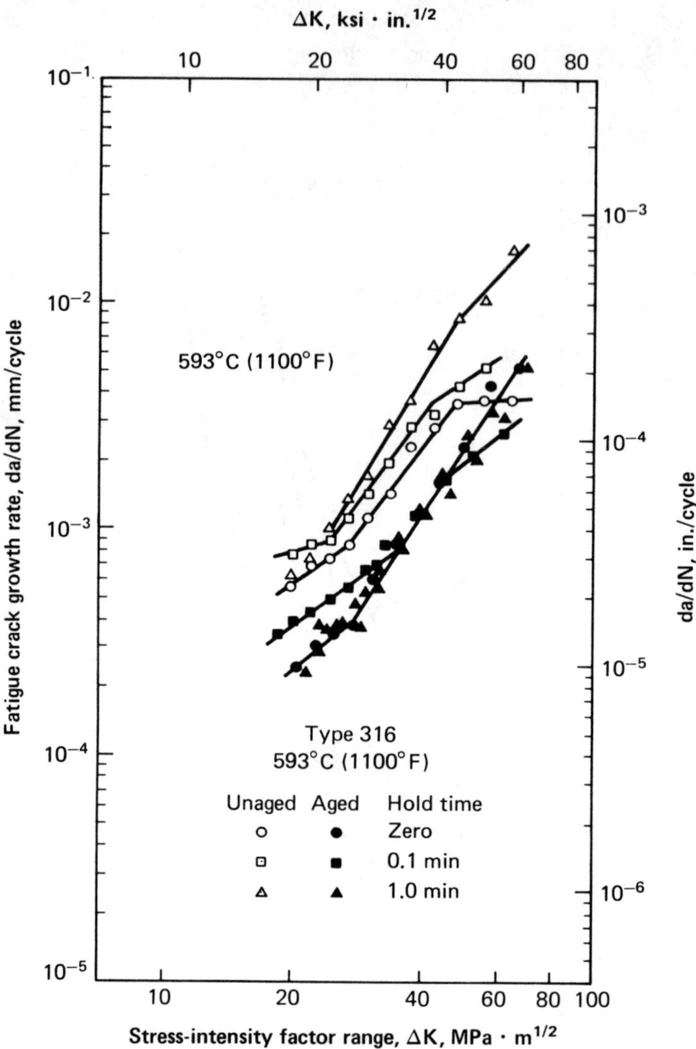

Fig. 5.15. Effect of exposure in air at 593°C (1100°F) for 5000 h, and hold times, on fatigue crack growth rates for annealed type 316 stainless steel at 593°C in air (Ref 5.22)

cold worked Type 316 at 427°C (800°F). Effects of holding times on cyclic loading of unaged and aged specimens of 20% cold worked Type 316, as determined by Michel and Smith (Ref 5.22), are shown in Fig. 5.18 for tests at 593°C (1100°F). The frequency for specimens cycled with zero holding time was 0.17 Hz, and the R ratio was zero. Aging was done for 5000 hours at 593°C (1100°F), and testing was done in air. For the unaged specimens, increasing the holding time significantly increased the fatigue crack growth rates as shown. For the aged specimens, holding at maximum load for 0.1 or 1.0 minute for each loading cycle reduced the fatigue crack growth rates over those obtained with no holding time. These data indicate that cold working and aging at 593°C (1100°F) before or during service exposure can lead to improved fatigue crack growth

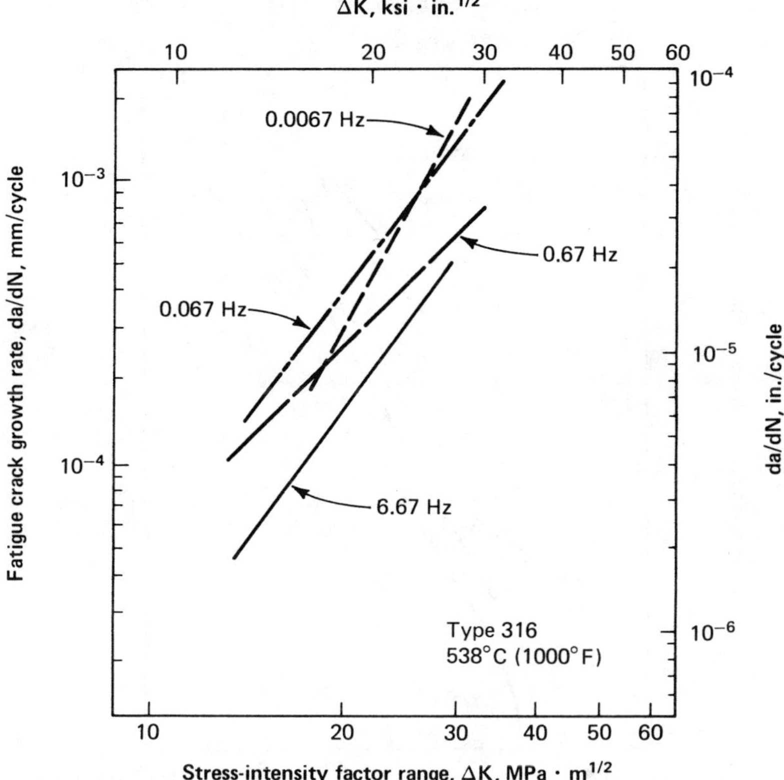

Fig. 5.16. Effect of variation in cyclic frequency on fatigue crack growth rate of annealed type 316 stainless steel in air at 538°C (1000°F) and an R ratio of 0.05 (Ref 5.13)

resistance and that short holding times at maximum load reduce fatigue crack growth rates. Short holding times (0.1 and 1.0 minute) also reduced fatigue crack growth rates in cold worked Type 316 when crack growth rate was based on a unit of time—i.e., mm of crack growth per hour of testing time (Shahinian; Ref 5.16).

Results also have been reported by James for fatigue crack growth rate tests in 20% cold worked specimens of Type 316 stainless steel which were cycled at frequencies of 0.0055 to 6.66 Hz, at 538°C (1100°F) and at an R ratio of 0.05 (Ref 5.47). Over the ΔK range studied, the fatigue crack growth rates were highest for the specimens subjected to the lowest cyclic frequency.

Results of fatigue crack growth rate tests on weldments of Type 316 stainless steel have shown that the crack growth rates in the weld metal are generally no higher than in the base metal and may be somewhat lower at elevated temperatures (Shahinian, Smith and Hawthorne; Ref 5.36). The curve shown in Fig. 5.19 for unirradiated weld metal tested at 593°C (1100°F) represents fatigue crack growth rates substantially lower than those for the unirradiated base metal at any given ΔK level (Shahinian; Ref 5.32). The weld was produced by the submerged arc method using Type 316 welding rod. Weldments were stress-relief annealed at 482°C (900°F). Specimens were single-edge-notch specimens for cantiliver loading and were tested at 0.17 Hz and at an R ratio of zero. Irradiation slightly reduced the

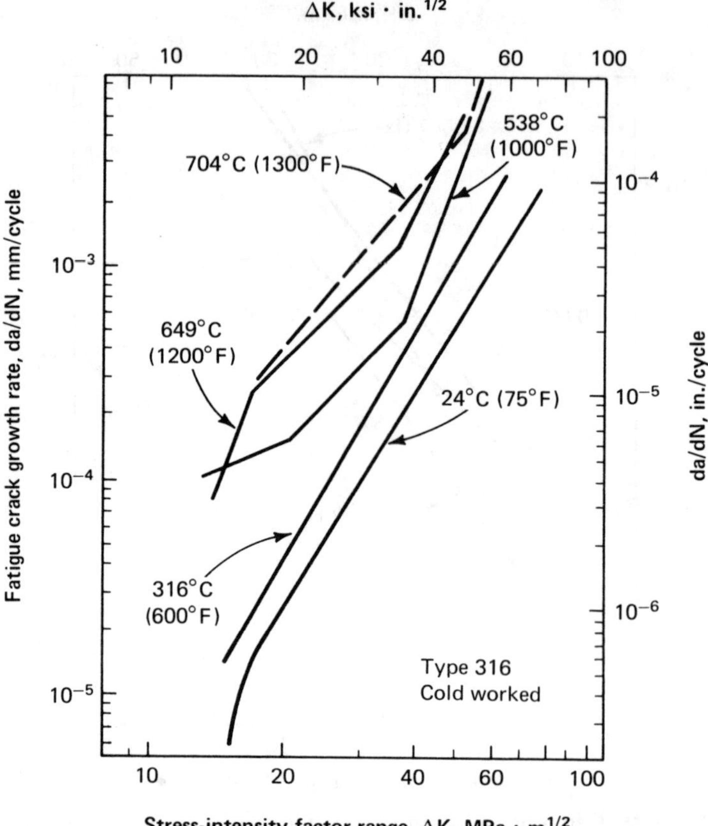

Curves are averages for L-T and T-L specimens at each temperature in air; 3 Hz at 24°C, 0.67 Hz at elevated temperatures; R = 0.05.

Fig. 5.17. Fatigue crack growth rates of 20% cold worked type 316 stainless steel for various temperatures (Ref 5.47)

fatigue crack growth resistance of the weld metal but its fatigue crack growth resistance was better than that of the unirradiated base metal.

Yuen and Copeland (Ref 5.48) also observed that the fatigue crack growth rates for Type 316 weld metal were lower than those for the base metal when tested in air at 510°C (950°F).

Fatigue crack growth rate data obtained by Hawthorne and Watson on specimens from Type 316 weldments (submerged arc welds) at 260°C (500°F) showed that substantial differences in fatigue crack growth rates were possible with various combinations of Type 316 filler metal and fluxes (Ref 5.49).

Effects of an environment of pressurized reactor water on the fatigue crack growth rates of a Type 316N stainless steel forging, of cast 316 stainless steel and of Type 316 welds at 288°C (550°F) were evaluated by Bamford (Ref 5.24). Effects of variations in load ratio, frequency, specimen orientation and heat-to-heat properties for wrought, cast and weld metal in air and in the reactor water were determined. Results of the study showed that fatigue crack growth rates in

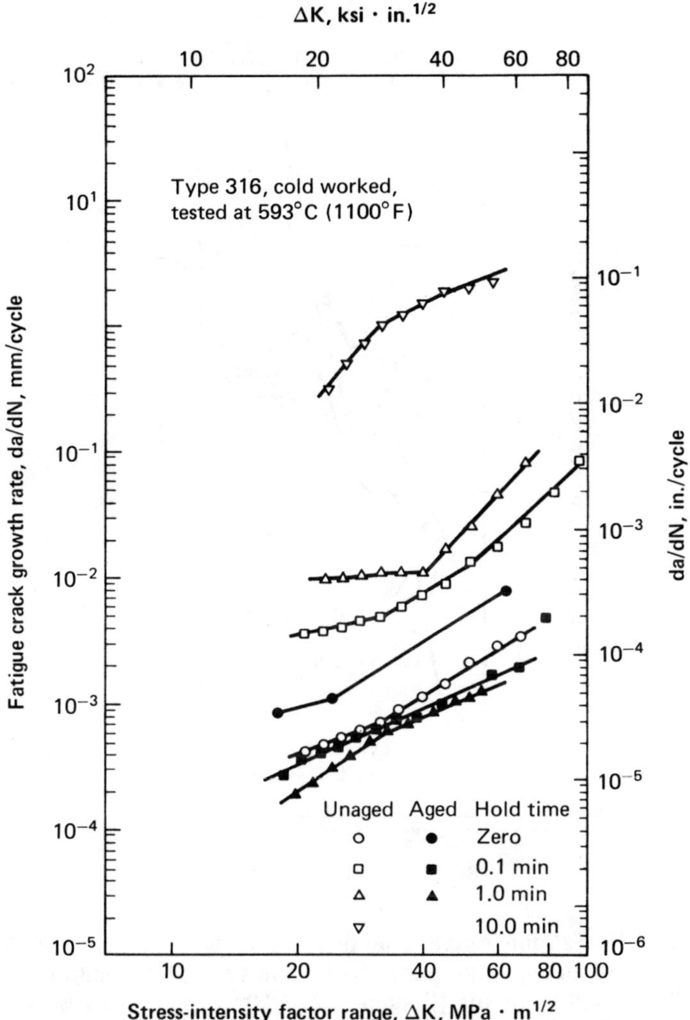

Fig. 5.18. Effect of exposure at 593°C (1100°F) for 5000 h, and hold times during cycling, on fatigue crack growth rate of 20% cold worked type 316 stainless steel at 593°C in air (Ref 5.22)

the reactor water were not significantly different from corresponding crack growth rates in air at the same load ratio and temperature. However, increasing the value of the stress ratio R resulted in increased crack growth rates at any given value of ΔK. All data points for fatigue crack growth rates at low R ratios were below the upper boundary for fatigue crack growth rates in air at R = 0.1, indicating that the corrosion effect of the reactor water was no greater than that of laboratory air.

Type 316 stainless steel is one of the few metals that has been used successfully for hip prostheses and other body implants. However, several failures have been reported in these devices. In order to obtain information on the effect of body fluids on fatigue cracking in Type 316, single-edge-notch specimens were obtained, at the L-T orientation, from annealed Type 316 plate and one was tested

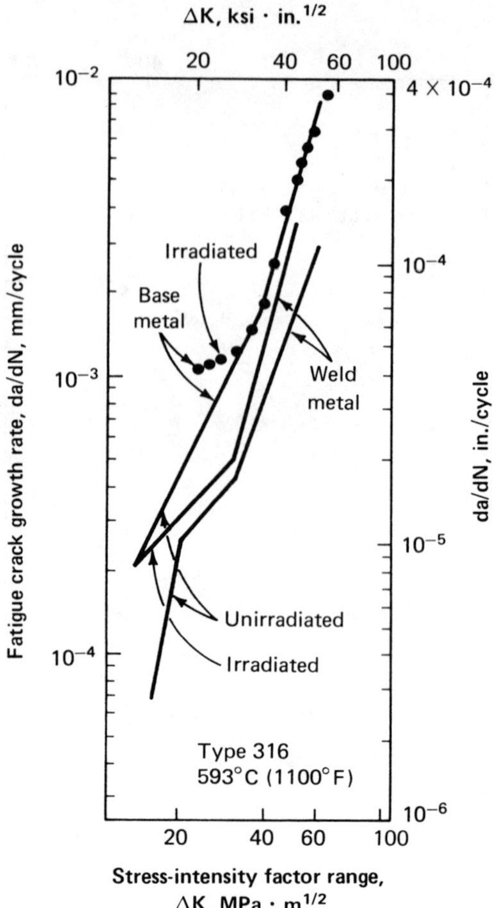

Fig. 5.19. Fatigue crack growth rates in type 316 base metal and weld metal in the unirradiated and irradiated conditions at 593°C (1100°F) in air [fluence 1.2 × 10^{22} n/cm^2, >0.1 MeV at 410°C (770°F)] (Ref 5.32)

in a saline solution (Ringer's solution) that simulates body fluids (Wheeler and James; Ref 5.50). Testing temperature was 37°C (98°F); loading frequency was 0.85 Hz, corresponding to cyclic loading during normal walking; and the load ratio was 0.04. A second specimen was tested in air. The data show that for the specimen tested in Ringer's solution, the fatigue crack growth rate was higher at a given ΔK level than for the specimen tested in laboratory air at ΔK values below 44 MPa · m$^{1/2}$ (40 ksi · in.$^{1/2}$). Therefore, it is expected that normal body fluids would affect fatigue crack propagation in Type 316 implants if fatigue cracks were to develop in these implants.

Results of an investigation of the effects of several gaseous environments on the fatigue crack growth rates in Types 316 and 321 stainless steels have been reported by Mahoney and Paton (Ref 5.51). Compact specimens were tested in fatigue loading according to a sine wave loading pattern at 5 Hz with an R ratio of 0.05 in room air, dry air, humid air, dry nitrogen, wet nitrogen and dry argon, at room temperature and at 649°C (1200°F). The results are summarized in Fig. 5.20. Fatigue crack growth rate data at 25°C (77°F) show that crack growth rates increased slightly with increased humidity when oxygen was present but that high humidity in an inert gas had no significant effect. Fatigue crack growth rates in room air at room temperature were the same for Types 316 and 321 stainless steels. Furthermore, in tests at 649°C (1200°F) in dry nitrogen, fatigue crack growth rates for Types 316 and 321 also were the same. In air, however, fatigue

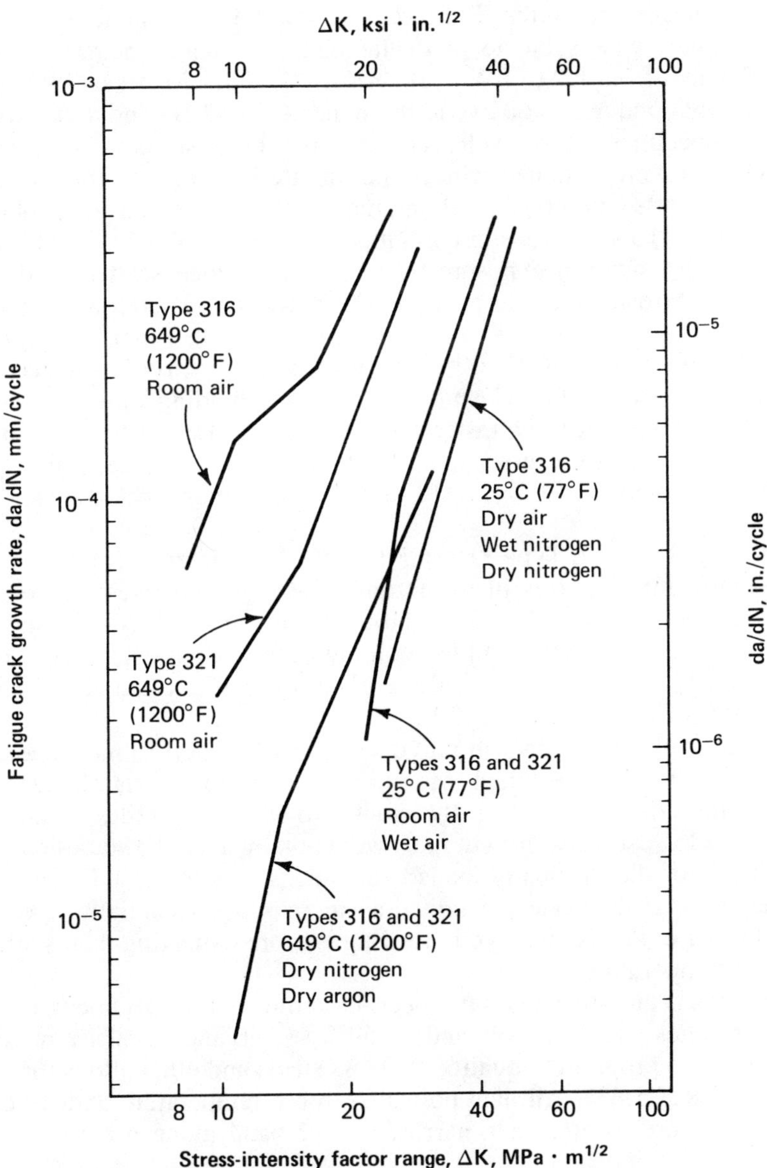

Fig. 5.20. Effect of gas environments on fatigue crack growth rates for types 316 and 321 stainless steels at 25 and 649°C (77 and 1200°F) (Ref 5.51)

crack growth rates in Type 316 specimens increased by a factor of about 22 over rates in an inert environment at the same temperature. The corresponding increase in fatigue crack growth rates for specimens of Type 321 was about 5 times that for the inert environment at 649°C (1200°F). If components of these stainless steels are exposed to inert environments instead of to air or oxygen-containing environments, fatigue crack growth rates will be substantially lower than those expected on the basis of tests in air.

Effects of fast neutron irradiation on fatigue crack growth rates have been studied by Michel and Smith and by others for specimens of Type 316 stainless steel at elevated temperatures (Ref 5.28, 5.31 and 5.52). Single-edge-notch cantilever specimens were subjected to cyclic loading tests after being exposed to fast neutron fluences from 1.2 to 1.4 × $10^{22}n/cm^2$ ($E > 0.1$ MeV) at 649°C (1200°F). The tests were conducted at a cyclic frequency of 0.17 Hz and an R ratio of zero in air. The specimens were cycled continuously in most tests, but in some tests additional data were obtained with a one-minute hold time at maximum load in each cycle. Plotted points for all fatigue crack growth rate data obtained on annealed Type 316 specimens in continuous cycling tests at 649°C (1200°F) with and without fast neutron exposure lie in a well defined scatter band with data points from corresponding tests at 593°C (1100°F). Therefore, irradiation of the annealed specimens did not change the fatigue crack growth rates at 649°C (1200°F) or at 593°C (1100°F). However, specimens of annealed Type 316 stainless steel tested at 649°C with a one-minute holding time in each loading cycle exhibited significantly faster fatigue crack growth rates after irradiation than specimens that were not irradiated. The effect was greater for specimens irradiated and tested at 649°C (1200°F) than for specimens irradiated and tested at 593°C (1100°F).

For 20% cold worked Type 316 stainless steel specimens, irradiation exposure caused significant increases in the fatigue crack growth rates for continuously cycled specimens tested at 593°C (1100°F) and 649°C (1200°F) and also for specimens cycled with a one-minute holding time in each cycle. This effect was not noted for specimens of 20% cold worked Type 316 tested at 427 and 538°C (800 and 1000°F) (James; Ref 5.30).

Static and cyclic crack growth rate data at 538°C (1000°F) have been reported by James (Ref 5.4) for specimens of cold worked Type 316 stainless steel. These data show the effect of K_{max} on the crack growth rate, da/dt, per unit of time (mm/h). The lowest crack growth rates were obtained for the static-load tests. As the frequency of the cyclically loaded specimens was increased from 0.0055 to 6.67, the crack growth rate per unit time increased substantially. These results provide further evidence that cyclic loading is more damaging than static loading at elevated temperatures.

Fatigue crack growth rates have been determined for specimens of annealed Type 316 stainless steel at −196 and −269°C (−320 and −452°F) by Tobler and Reed as part of a program to qualify stainless steels and other alloys for cryogenic applications (Ref 5.33). All data points for room temperature and for cryogenic temperatures were located in a narrow scatter band along a curve defined by:

$$da/dN = 2.1 \times 10^{-10}(\Delta K)^{3.8}, \text{ in mm/cycle.}$$

Fatigue crack growth rate data obtained at room temperature, therefore, may be used in estimating fatigue crack growth in structural components at cryogenic temperatures for components of Type 316 stainless steel.

Types 321 and 348

Types 321 and 348 stainless steels are generally similar in composition except that Type 321 contains titanium in an amount which is dependent on the carbon content required to form stable carbide particles. Type 348 contains a small amount of copper along with optional additions of niobium and tantalum to form stable carbide particles. When most of the carbon in the alloy is combined with titanium, tantalum and/or niobium as stable particles, there is less tendency for the alloy to become "sensitized" on welding and during slow cooling from other thermal processes. Sensitizing causes dispersed carbide precipitation that reduces the corrosion resistance of the alloy.

As reported by Shahinian, Smith and Watson (Ref 5.11), fatigue crack growth rate tests were made on single-edge-notch cantilever specimens of Types 321 and 348 stainless steels from the L-T orientation at 0.17 Hz with an R ratio of zero at room temperature and at elevated temperatures to 593°C (1100°F). As for Types 304 and 316, fatigue crack growth rates in air increased with increasing testing temperature. The curves in Fig. 5.21 show that, at room temperature, the fatigue crack growth rates for Types 304, 316, 321 and 348 all fall in a narrow band. For tests at 593°C (1100°F), however, specimens of Type 316 had the least fatigue crack propagation resistance, whereas specimens of Type 348 had the highest fatigue crack propagation resistance, over the ΔK range studied. Results of tests on specimens of Types 304 and 321 were nearly the same at 593°C (1100°F) in air.

As observed for Type 316 stainless steel, the presence of moisture and oxygen in the environment increased the fatigue crack growth rates substantially, and this also is true for Type 321, as shown in Fig. 5.20 (Mahoney and Paton; Ref 5.51). Similar information was not available for Type 348 stainless steel. However, an environment of humid air reduces the fatigue crack propagation resistance of these alloys at both room temperature and elevated temperatures.

Results of tests by Michel and Smith (Ref 5.22) on specimens of annealed Type 321 stainless steel that had been aged at 593°C (1100°F) for 5000 hours and then tested at 593°C have shown that long-time exposure at the service temperature does not reduce the fatigue crack propagation resistance in air. Aged specimens tested with zero holding time had lower crack growth rates than corresponding specimens that were not aged (Fig. 5.22). Fatigue cycling with holding times of 0.1 and 1.0 minute on each cycle increased the crack growth rates slightly, as shown in the figure.

Similar tests on unaged and aged specimens of Type 348 stainless steel showed that exposure at 593°C (1100°F) for 5000 hours did not change the fatigue crack propagation resistance for tests at the same temperature in air (Ref 5.22).

Alloys 21-6-9 and 22-13-5

The alloys 21-6-9 and 22-13-5 contain 9 and 5% manganese, respectively. The manganese in these steels replaces some of the nickel content that is required to yield an austenitic microstructure. They also contain small amounts of nitrogen and other additions that improve their strength. Results of fatigue crack growth rate tests by Tobler and Reed (Ref 5.33) at room temperature and at cryogenic

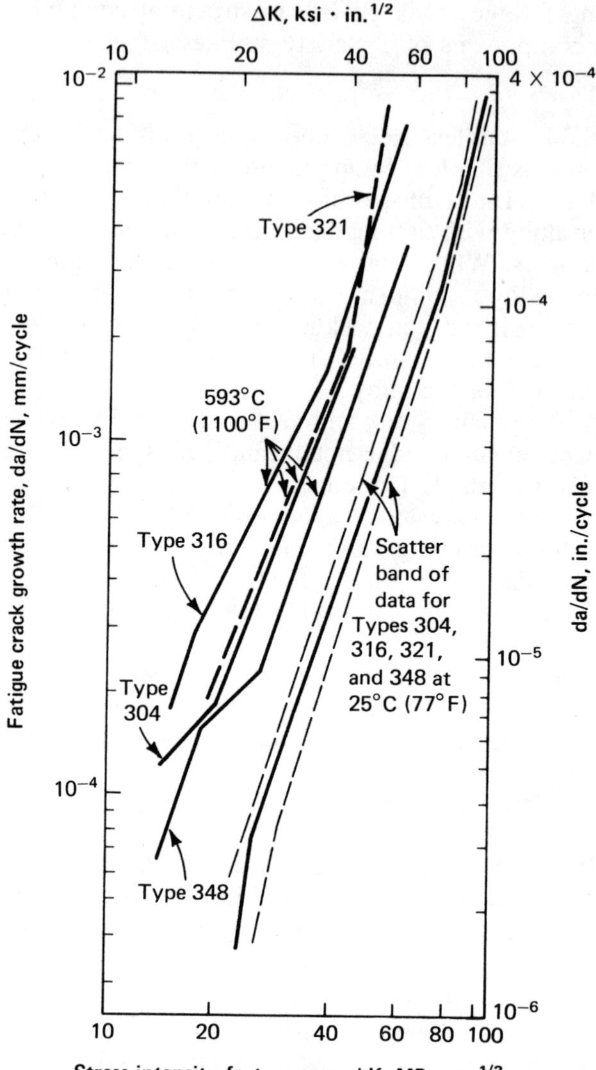

Fig. 5.21. Fatigue crack growth rates for annealed types 304, 316, 321 and 348 stainless steels in air at room temperature and 593°C (1100°F), L-T orientation, 0.17 Hz, and an R ratio of 0 (Ref 5.11)

temperatures show that the crack growth rates at −196°C (−320°F) were the same as those at room temperature, but that the fatigue crack growth rates at −269°C (−452°F) were marginally higher over the ΔK range studied for the 21-6-9 alloy (Fig. 5.23). These tests were conducted on compact specimens at frequencies of 20 and 28 Hz and at an R ratio of 0.1. Examination of the fracture surfaces after testing at −269°C (−452°F) showed cleavage-type facets, which indicate a transition in the fracture mode at this temperature. The 21-6-9 alloy is similar to Pyromet 538, Nitronic 40 and XM-10.

Compact specimens of Pyromet 538 weldments were subjected to fatigue crack growth rate tests at room temperature and at −269°C (−452°F) by Wells, Logsdon

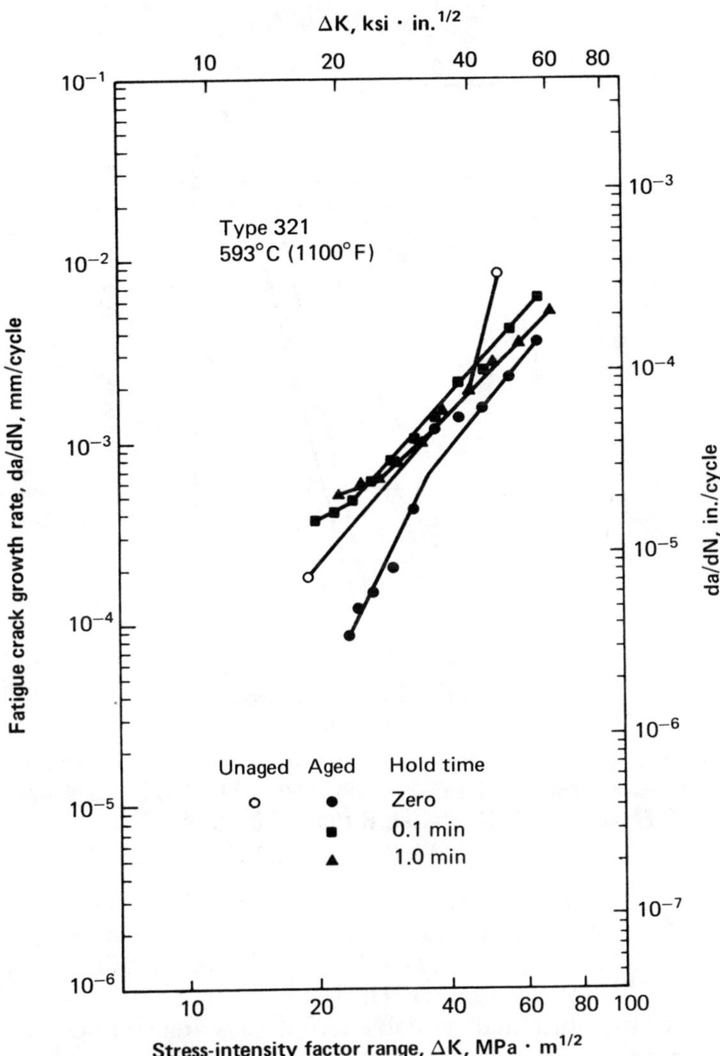

Fig. 5.22. Fatigue crack growth rates for annealed type 321 stainless steel unaged and aged at 593°C (1100°F) for 5000 h and tested in air with continuous sawtooth waveform (0.17 Hz), with 0.1 and 1.0-min hold time and at an R ratio of 0 at 593°C (Ref 5.22)

and Kossowsky (Ref 5.41). The base metal was solution annealed prior to welding. One set of welds was made by the gas tungsten arc welding (GTAW) process with 21-6-9 filler wire, and the other was made by the shielded metal arc welding (SMAW) process with IN 182 covered electrodes. Results of these tests are summarized in Fig. 5.24. Specimens with SMA welds had the same fatigue crack growth rates at room temperature and at −269°C (−452°F). Specimens welded by the GTAW process had higher crack growth rates at −269°C than at room temperature. Examination of the microstructures near the fracture surfaces for the specimens tested at −269°C showed that there was 6 to 7% delta ferrite (produced by welding) in the weld metal along with induced martensite. The SMA weld metal was fully austenitic.

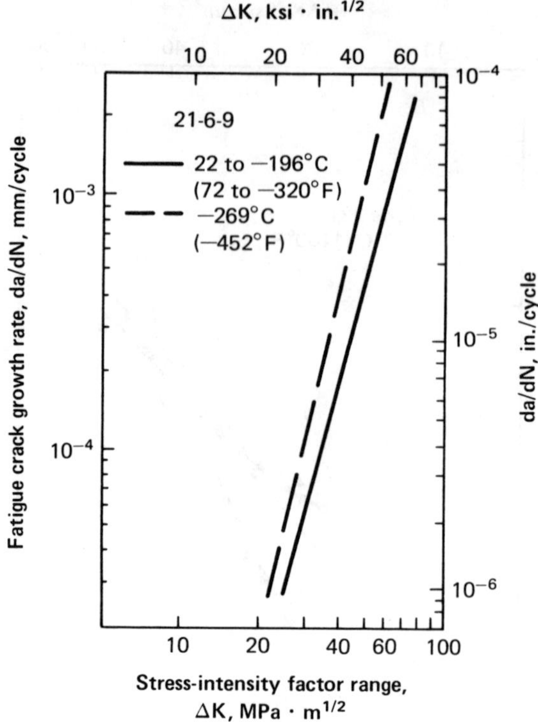

Fig. 5.23. Fatigue crack growth rates in specimens of annealed 21-6-9 stainless steel at 22, −196 and −269°C (72, −320 and −452°F), 20 and 28 Hz, and an R ratio of 0.1 (Ref 5.33)

On the same program, Wells *et al* determined the fracture toughness of welded specimens by the J-integral method at −269°C (−452°F). For the GTAW specimens, K_{Ic}(J) was 82.4 MPa · $m^{1/2}$ (74.4 ksi · $in.^{1/2}$); for the SMAW specimens, it was 176 MPa · $m^{1/2}$ (159 ksi · $in.^{1/2}$). The lower fracture toughness of the GTAW specimens was also attributed to delta ferrite plus transformed martensite in the microstructure.

Fatigue crack growth rate data reported by Shahinian *et al* (Ref 5.21) for tests at room temperature on specimens of 22-13-5 stainless steel showed a growth rate curve substantially the same as that shown for 21-6-9 stainless steel in Fig. 5.23. The 22-13-5 specimens were of the single-edge-notch cantilever type obtained from the L-T orientation and tested at 0.17 Hz in air with an R ratio of zero.

Kromarc 58

Kromarc 58 stainless steel is a nickel-chromium-manganese-molybdenum stainless with small amounts of nitrogen, aluminum, vanadium, zirconium and boron to control grain size and improve strength. Applications for this alloy are primarily structures that are exposed to cryogenic temperatures. In the solution treated condition, the typical yield strength of Kromarc 58 is 371 MPa (54 ksi) at room temperature and 1094 MPa (159 ksi) at −269°C (−452°F) (Wells *et al*; Ref 5.41). The K_{Ic}(J) value for Kromarc 58 plate was 216 MPa · $m^{1/2}$ (195 ksi · $in.^{1/2}$) at −269°C (−452°F) for specimens obtained from the T-L orien-

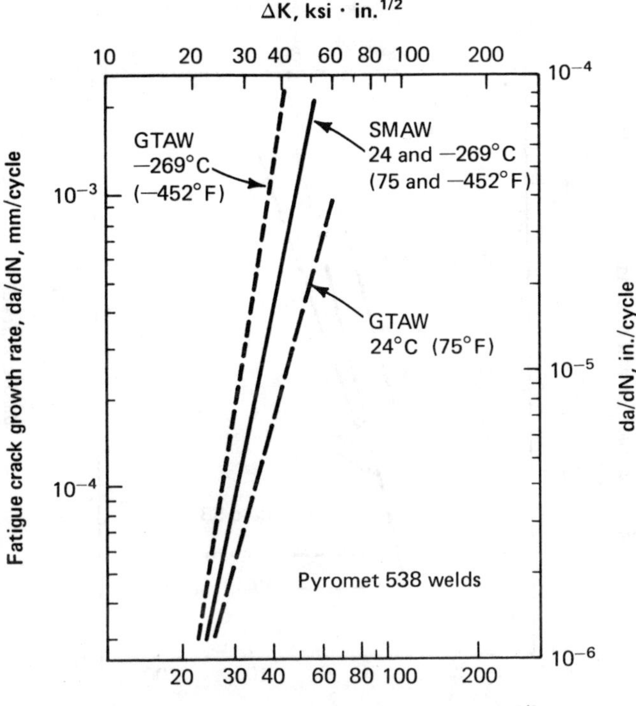

Fig. 5.24. Fatigue crack growth rates in weld metal in Pyromet 538 stainless steel at room temperature and −269°C (−452°F) and at 10 Hz (Ref 5.41)

tation. For the fusion zone of a gas tungsten arc weld made with Kromarc 58 filler metal, the K_{Ic}(J) value was 156 MPa · $m^{1/2}$ (141 ksi · $in.^{1/2}$) at −269°C (−452°F). Fatigue crack growth rate data for the base metal at room temperature and at −269°C and for the weld metal at −269°C are shown in Fig. 5.25 (Ref 5.41). The data were obtained on compact specimens at 10 Hz and at an R ratio of 0.1. Fatigue crack growth rates for tests in liquid helium were lower than at room temperature at the same ΔK values. Therefore, if room temperature crack growth rate data are used to estimate crack growth at cryogenic temperatures, the estimated values will be conservative.

Summary

Effects of variations in testing conditions on fatigue crack growth rates in austenitic stainless steels tend to follow certain trends for all of the alloys. Increasing the exposure temperature, the load ratio R, and the oxidizing and corrosive effects of the environment and decreasing the cyclic frequency tend to increase the fatigue crack growth rates based on fracture mechanics procedures. Thermal aging and neutron irradiation may have no significant effects on the fatigue crack growth rates or may result in increased or decreased growth rates depending on the condition of the material, temperature of exposure, irradiation intensity, and testing conditions. Variations in crack orientation (L-T or T-L),

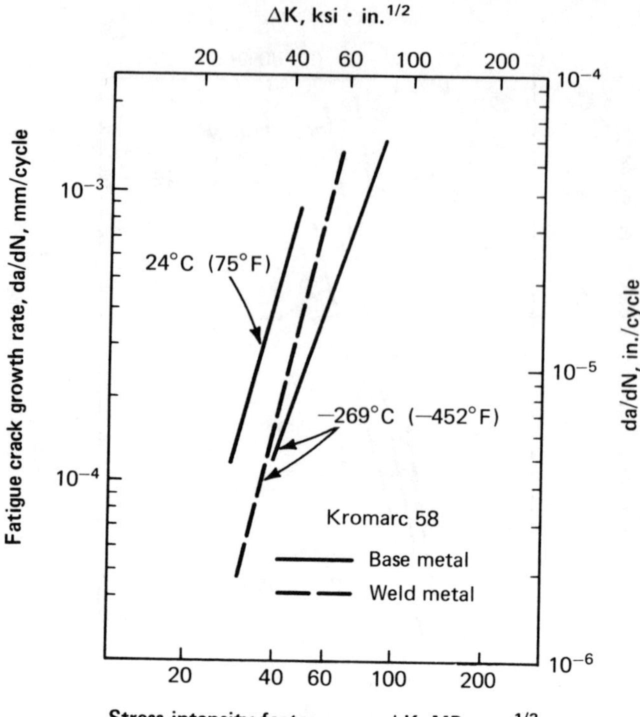

Fig. 5.25. Fatigue crack growth rates for solution treated Kromarc 58 base metal in air at room temperature, and base metal and weld metal at −269°C (−452°F) in liquid helium, at 10 Hz and an R ratio of 0.1 (Ref 5.41)

loading waveforms, grain sizes, melting procedures, and heat-to-heat variations in composition appear to have little influence on fatigue crack growth rates in the austenitic steels. However, long holding times at maximum load during elevated temperature cyclic tests may cause marked increases in crack growth rates in cold worked alloys when compared with test data obtained on continuous cycling. Fatigue crack growth rates in austenitic weld metal usually are no greater than those in the base metal.

5.3.2. Fracture Properties of Martensitic Stainless Steels

Types 403 and 410 stainless steels represent the basic compositions of the martensitic stainless steels. Because these steels harden on air cooling from the hot working processes, the hot worked and air cooled material may be tempered and used in this condition or given a process annealing treatment prior to rehardening. Rehardening is preferred because it improves properties, as discussed in the introductory section.

Available fracture toughness data on martensitic grades are limited to data from tests on specimens of Types 403, 403 modified, 420, 422 and 431.

Type 403 Standard and Modified Versions

Type 403 martensitic stainless steel is used extensively for steam turbine rotor blades and rotors that operate at temperatures up to 480°C (900°F). For this type of application, the components are tempered at 590°C (1100°F) or higher, after which embrittlement at service temperatures is negligible.

Fracture toughness test data have been obtained by Logsdon (Ref 5.53) on three heats of Type 403 modified stainless steel using compact specimens from 2 to 8 in. thick for tests in the temperature range from −196 to +80°C (−320 to +175°F). These specimens were obtained from rotor forgings and were oriented so that crack growth would be in a radial direction. The compositions of these heats constituted a modified Type 403 because they contained nickel and molybdenum, as shown in Table 5.2. Heat treatment of these forgings consisted of preheating to 260°C (500°F), holding at 260°C for 2 or 3 hours, heating slowly to 954°C (1750°F), holding at 954°C for 16 to 20 hours, and oil quenching to 93/149°C (200/300°F). They were tempered by preheating to 260°C (500°F), heating slowly to 593 to 621°C (1100 to 1150°F), holding at temperature for 27 to 32 hours, and cooling in air.

Scatter bands for the fracture toughness (K_{Ic}) data for each heat are shown in Fig 5.26. The thicker specimens were tested at the higher temperatures in order to obtain valid fracture data. These results show that there is some heat-to-heat scatter in the fracture data with small variations in composition and in other heat-to-heat variables. The results also illustrate the requirements for obtaining valid data near room temperature for unusually thick specimens of relatively tough materials.

In a later study by Logsdon (Ref 5.54), 1-in.-thick compact specimens were machined from the broken halves of the larger specimens from Heat 484 from the earlier study. The 1-in.-thick specimens were tested at temperatures from −18 to +150°C (0 to 300°F) using the J-integral method. From −18 to +24°C (0 to 75°F), the $K_{Ic}(J)$ data points were close to the fracture toughness curve defined by the K_{Ic} data points for the larger specimens from the same heat. At higher temperatures, the $K_{Ic}(J)$ data points apparently define a plateau in toughness values, as shown in Fig. 5.27.

Two-inch-thick WOL specimens (modified compact) from each of the three heats of Type 403 modified stainless steel in Logsdon's earlier study (Ref 5.53) were subjected to cyclic loading to obtain fatigue crack growth rate data in room temperature air and in distilled water at 271°C (520°F) under a pressure of 8.3 MPa (1200 psi). The results, which were obtained at a frequency of 10 Hz and

Table 5.2. Compositions of Type 403 modified stainless steels (Ref 5.53)

Heat No.(a)	Composition, %								
	C	Mn	P	S	Si	Ni	Cr	Mo	Cu
637	0.13	0.46	0.013	0.009	0.20	0.32	12.18	0.43	0.12
933	0.15	0.54	0.010	0.008	0.32	0.38	12.37	0.08	0.10
484	0.13	0.57	0.009	0.006	0.33	1.60	12.32	0.55	...

(a) See Fig. 5.26.

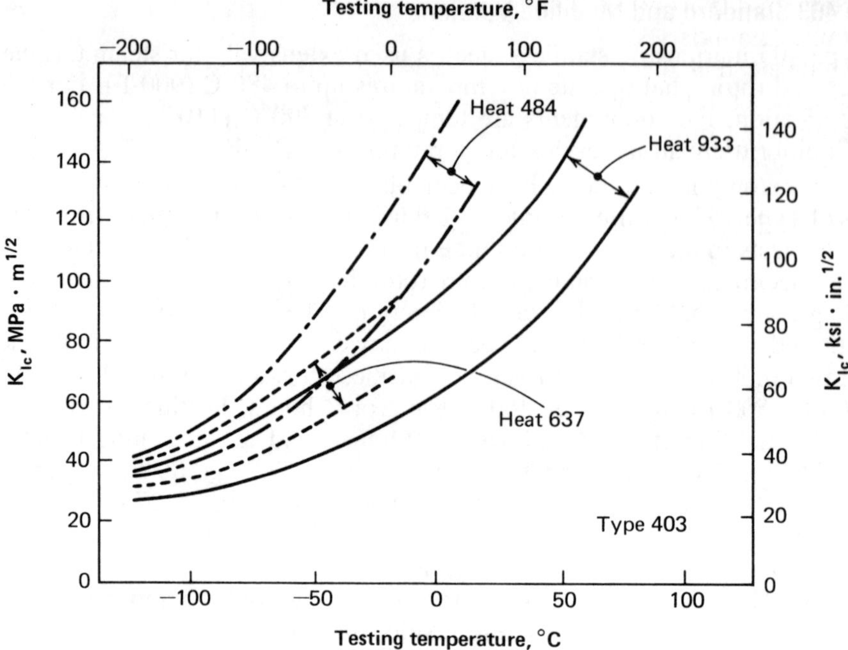

Fig. 5.26. Scatter bands of fracture toughness (K_{Ic}) data for three heats of type 403 stainless steel in the heat treated condition (Ref 5.53)

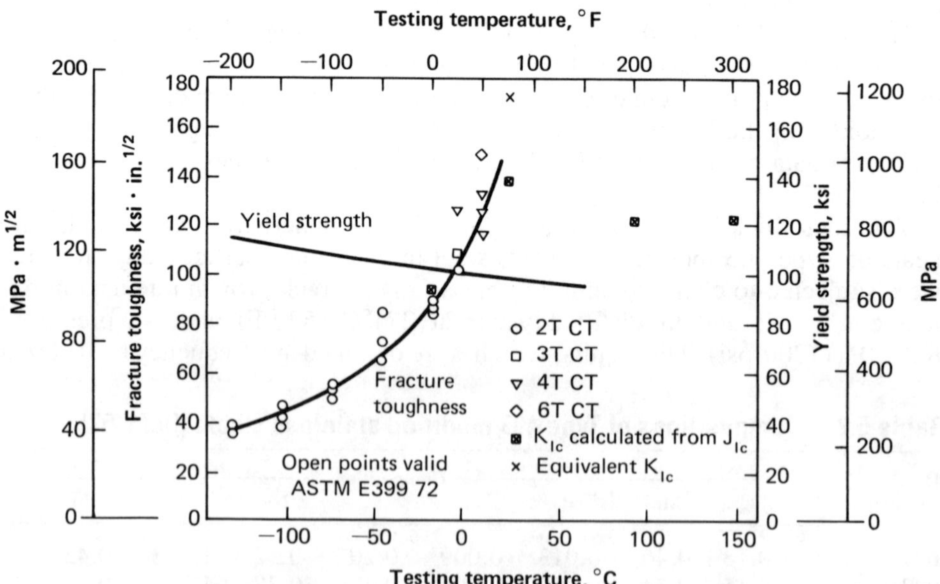

K_{Ic}(J) specimens were 1 in. thick (heat 484; see Fig. 5.26).

Fig. 5.27. Fracture toughness data obtained over ranges of temperature and specimen thickness for type 403 modified stainless steel (Ref 5.54)

an R ratio of 0.083 or 0.067, are summarized in Fig. 5.28. The curves representing the upper boundaries of the scatter bands of the fatigue crack growth rate data indicate that there is some heat-to-heat variation in fatigue crack growth rate properties for these heats. Furthermore, exposure at 271°C (520°F) in distilled water at a pressure of 8.3 MPa (1200 psi) increased the fatigue crack growth rates slightly.

Fatigue crack growth rates in air and in solutions of sodium chloride, sodium sulfate, sodium phosphate and sodium silicate were determined on 1.27-cm-

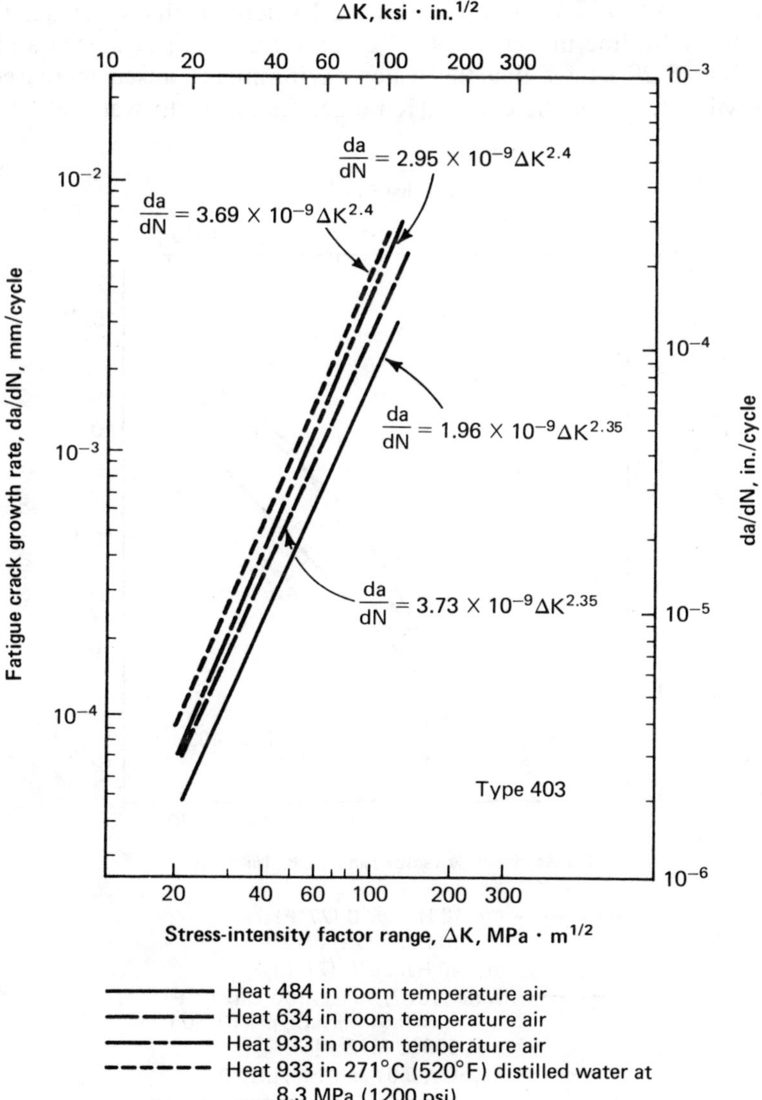

Fig. 5.28. Upper boundaries of fatigue crack growth rate scatter bands for three heats of type 403 modified stainless steel in the heat treated condition, tested at 10 Hz and an R ratio of 0.083 or 0.067 (Ref 5.53)

(0.5-in.-) thick compact specimens of Type 403 stainless steel by Abrego and Begley to show how certain impurities in steam influence fatigue behavior (Ref 5.55). The specimens were obtained, at the L-T orientation, from plate that had been austenitized at 950°C (1750°F), cooled in air, and tempered at 650°C (1200°F) for one hour to obtain a yield strength of 682 MPa (100 ksi). All fatigue crack growth rate tests were conducted at room temperature or at 100°C (212°F), at an R ratio of 0.5 and at frequencies of 0.1 to 40 Hz. The specimens were completely submerged in the solutions during testing. For the slow crack growth rate portions of the da/dN curves, a technique was used to bracket the ΔK_{th} values.

Fatigue crack growth rate data points obtained in air at frequencies of 10 and 40 Hz and at 25°C (77°F) and 100°C (212°F) defined the same da/dN curve, shown as the solid line in Fig. 5.29. The curve for tests in water at a pH of 7 at 100°C in Fig. 5.29 shows that the water environment caused increased fatigue crack growth rates over the entire ΔK range. Exposure to water at 25°C (77°F)

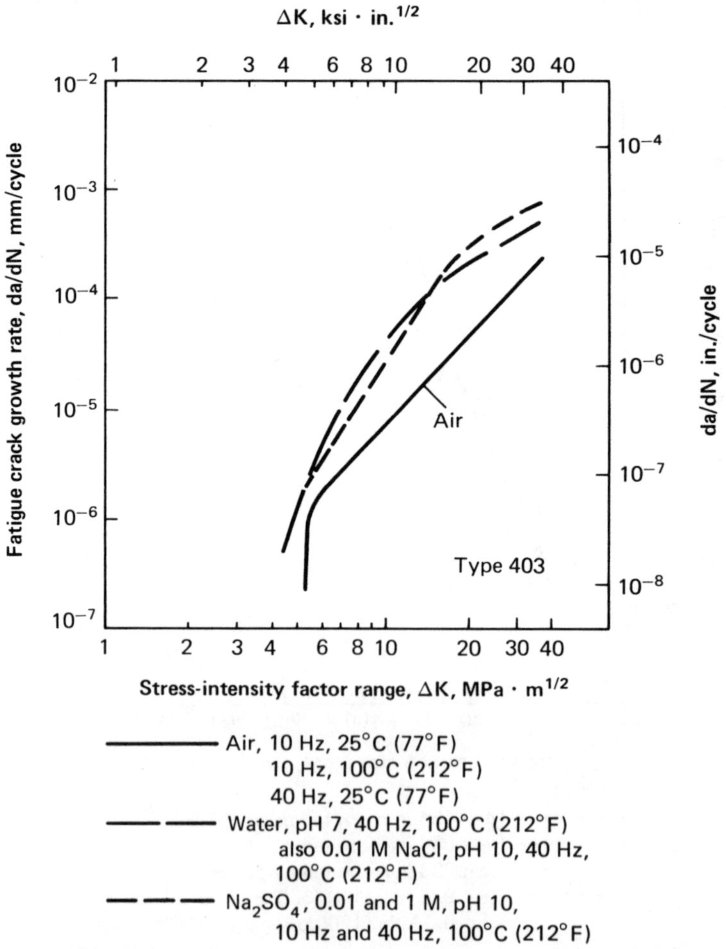

Fig. 5.29. Fatigue crack growth rates in Type 403 stainless steel in air, water, 0.01 M NaCl solution, and 0.01 and 1.0 M Na_2SO_4 solutions (Ref 5.55)

resulted in intermediate crack growth rates between those in air and those in water at 100°C, as shown on a different scale in Fig. 5.30.

Tests in the 0.01 M (molar) and 1.0 M sodium chloride solutions were made with the solutions at pH levels of 2, 7 and 10 and with an open circuit. Fatigue crack growth rates in 0.01 M sodium chloride at pH 10 and 100°C were the same as those in water at 100°C (Fig. 5.29). At lower cyclic frequencies, the fatigue crack growth rates were higher than at 40 Hz at ΔK values above 20 MPa · $m^{1/2}$ (18 ksi · $in.^{1/2}$). For tests in the 1.0 M sodium chloride solution at 100°C (212°F), fatigue crack growth rates were the same as for water at the same temperature (Fig. 5.30). At 100°C (212°F), fatigue crack growth rates in 1.0 M sodium phosphate solution at pH 10 and at 10 and 40 Hz and in 1.0 M sodium silicate at pH 10 and at 10 Hz were practically the same as those in air.

Results of additional studies on fatigue crack growth rates of Type 403 modified stainless steels in various environments have been reported by Clark (Ref 5.5). Compact specimens from two heats of Type 403 modified were tested in air, distilled water, seawater and sulfurous acid at 24 and 93°C (75 and 200°F) and in

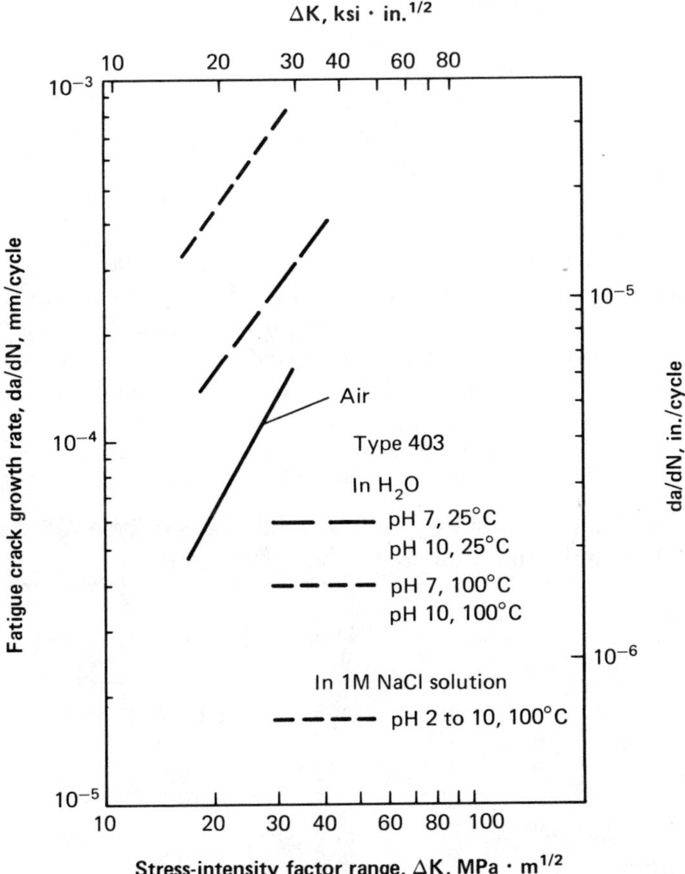

Fig. 5.30. Fatigue crack growth rates in type 403 stainless steel in air, water, and a 1 M NaCl solution at 10 Hz and an R ratio of 0.5 (Ref 5.55)

high-oxygen (40 ppm) and low-oxygen (1 ppm) steam at 100°C (212°F). The specimens had been austenitized at 960°C (1760°F), quenched in oil, and tempered at 663°C (1225°F) for four hours. For these tests, R ratios were from 0 to 0.1 at a testing frequency of 30 Hz—except at low ΔK levels, when a testing frequency of 160 Hz was used. The fatigue crack growth rate properties in air at room temperature and at 93°C (200°F) were the same. At 93°C, the distilled water, seawater and sulfurous acid environments increased the rate of fatigue crack growth by factors of approximately 2.5, 3 and 5, respectively, over that for the air environment. The oxygen content of the steam did not influence the rate of fatigue crack growth.

Type 420

Single-edge-notch cantilever specimens 15 mm (0.60 in.) thick of Type 420 stainless steel were used in a program by Bhangaonkar and Banerjee to determine K_{Ic}, K_{Iscc} threshold values, and sustained-load crack growth rates in a 3.5% sodium chloride solution for specimens representing various heat treated conditions (Ref 5.56). The specimens were obtained from forged blocks which had been cooled in air from the final forging temperature of 900 to 1100°C (1650 to 2000°F), tempered at 700°C (1290°F) for two hours, and cooled in air to improve the machineability. The specimens were reheat treated by preheating at 600°C (1110°F) for 10 minutes and austenitizing at 1040°C (1900°F) for 45 minutes in argon followed by quenching in oil, martempering in oil or martempering in an air blast. They were tempered at 250, 450 or 650°C (480, 840 or 1200°F) for two hours and quenched in oil or cooled in air. Results of the fracture toughness and K_{Iscc} tests are presented in Table 5.3.

The specimens tempered at 450°C (840°F) had the lowest fracture toughness and the lowest K_{Iscc} values because of the 450°C embrittlement. Such embrittlement is caused by formation of fine carbide precipitates and Fe_3C plates, whereas the low corrosion resistance is caused by chromium depletion through precipitation of Cr_7C_3. For the specimens tempered at 250°C (480°F) and 650°C (1200°F), those that were quenched in oil had the highest fracture toughness and the highest K_{Iscc} values. The highest K_{Iscc} value was obtained after quenching in

Table 5.3. Results of K_{Ic} and K_{Iscc} tests on specimens of Type 420 stainless steel given several different heat treatments (Ref 5.56)

Quenching treatment	Tempering temperature, °C	Tempering temperature, °F	Average hardness, VPN	K_{Ic}, MPa·m$^{1/2}$	K_{Ic}, ksi·in.$^{1/2}$	K_{Iscc}(a), MPa·m$^{1/2}$	K_{Iscc}(a), ksi·in.$^{1/2}$
In oil	250	482	465	62.4	56.8	25.8	23.5
Martempered in oil	250	482	425	35.5	32.3	25.3	23.0
Martempered in air	250	482	305	48.1(b)	43.8(b)	23.1	21.0
In oil	450	842	365	49.7	45.2	18.2	16.6
Martempered in oil	450	842	435	45.1	41.0	17.0	15.5
Martempered in air	450	842	402	40.4	36.8	17.0	15.5
In oil	650	1202	267	83.1(b)	75.6(b)	73.4	66.8
Martempered in air	650	1202	251	78.5(b)	71.4(b)	66.5	60.5

(a) K_{Iscc} in 3.5% NaCl solution. (b) Did not meet all requirements of ASTM Method E399.

oil and tempering at 650°C (1200°F). During tempering at 650°C, $M_{23}C_6$ precipitates are formed which reduce the hardness but do not cause chromium depletion.

Crack growth rate data for sustained loading tests in a 3.5% sodium chloride solution showed that the crack growth rates were highest for specimens tempered at 450°C (840°F) and lowest for specimens tempered at 650°C (1200°F).

Type 422

Type 422 stainless steel contains nickel, molybdenum and tungsten, as well as 12% chromium to improve properties. The effects of sodium chloride solutions and elevated temperature exposure on fatigue crack growth rates were determined by Eisenstadt and Rajan in tests of notched round rotating beam specimens in which the numbers of test cycles were marked by minor stress interruptions that produced marking rings (Ref 5.57). Calculations for maximum stress-intensity factors were based on equations for solid round bars subjected to bending loads. The material for these tests apparently had been heat treated to a yield strength of approximately 827 MPa (120 ksi). The specimens were one inch in diameter in the test sections. Each specimen was rotated at 600 cycles per minute (10 Hz) while at constant load with the salt water solution flowing over the notched section. Tests with several concentrations of salt solution indicated that the maximum corrosive effect was obtained with the 4.5% solution. Results of tests with specimens in the 4.5% sodium chloride solution at room temperature, 57°C (135°F) and 71°C (160°F) are shown in Fig. 5.31. Increasing the temperature of the solution substantially increased the fatigue crack growth rates.

Type 431

Type 431 stainless steel contains more chromium and nickel than Type 422, but does not contain additional alloying elements such as molybdenum and tungsten. Type 431 rolled bars 5.7 cm ($2\frac{1}{4}$ in.) thick were obtained for a series of fracture toughness tests by Sprowls *et al* at Alcoa. Specimen blanks were austenitized at 1040°C (1900°F), quenched in oil, cooled to −73°C (−100°F) and held at this temperature for two hours. One set of specimens was tempered at 635°C (1175°F) for two hours (HT125) and the other was tempered at 290°C (550°F) for two hours, cooled in air, and retempered at 290°C for two hours (HT200). Yield strengths were 717 MPa (104 ksi) for the HT125 specimens and 1137 MPa (165 ksi) for the HT200 specimens. Fracture toughness (K_{Ic}) values for 2-in.-thick compact specimens from the T-L orientation were 93 MPa · $m^{1/2}$ (84.6 ksi · $in.^{1/2}$) for the HT125 condition and 75 MPa · $m^{1/2}$ (68.3 ksi · $in.^{1/2}$) for the HT200 condition. Results of K_{Iscc} tests in 20% sodium chloride, for compact specimens (T-L) bolt loaded to 95% of the K_{Ic} value, were 47 MPa · $m^{1/2}$ (43 ksi · $in.^{1/2}$) for the HT125 condition and 13 MPa · $m^{1/2}$ (12 ksi · $in.^{1/2}$) for the HT200 condition.

5.3.3. Fracture Properties of Precipitation Hardening Stainless Steels

The precipitation hardening stainless steels may be divided into three categories: martensitic (maraging), semiaustenitic and austenitic. As noted in Table 5.1, alloys 15-5PH, 17-4PH, PH13-8Mo and Custom 455 are martensitic; 17-7PH, PH15-7Mo and AM355 are semiaustenitic; and A-286 is austenitic. There are other types in each of these categories, but no published fracture mechanics data

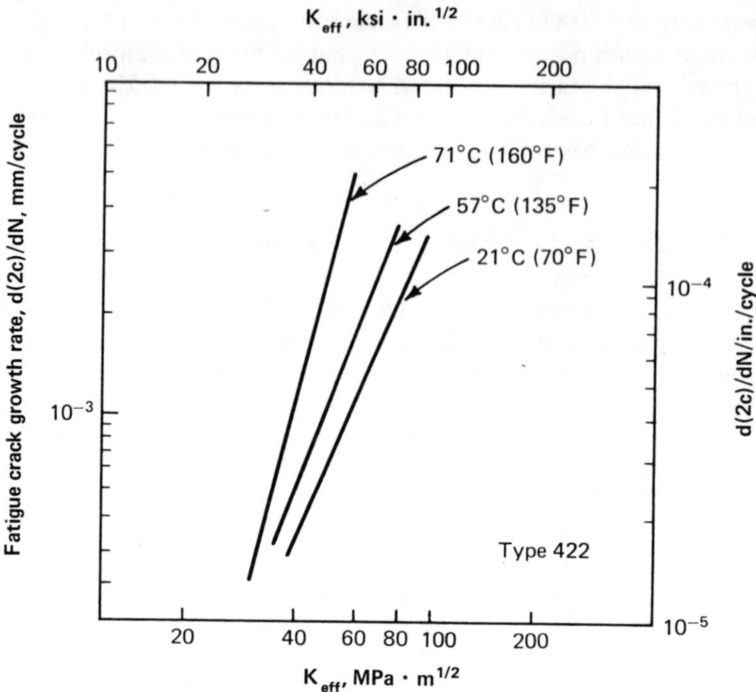

Fig. 5.31. Fatigue crack growth rates in precracked round rotating beam specimens of type 422 stainless steel in 4.5% NaCl solution at room and elevated temperatures, 10 Hz, and an R ratio of −1 (Ref 5.57)

have been located for them. More fracture mechanics data have been obtained on PH13-8Mo than on any of the other precipitation hardening stainless steels, because PH13-8Mo was selected for certain components of the B-1 bomber program (Ref 5.6). All of the precipitation hardening stainless steels require special heat treatments to develop optimum properties for any specific application (Ref 5.58).

Martensitic Precipitation Hardening Stainless Steels

The martensitic grades are usually purchased in the solution annealed condition—i.e., austenitized at some designated temperature from 815 to 1040°C (1500 to 1900°F) and cooled in air or quenched in oil or water. The microstructure after this treatment is predominantly low-carbon martensite. After fabrication, the components are age hardened (maraged) in the temperature range from 480 to 590°C (900 to 1100°F) and cooled in air. Aging develops the precipitation hardened properties in these steels. Heat treatments and corresponding tensile properties are practically the same for Types 15-5PH and 17-4PH. On aging, precipitates of submicroscopic copper compounds form in the martensitic matrix, causing substantial increases in strength and hardness. The composition of 15-5PH is such that no delta ferrite occurs in this alloy, and thus it has good short transverse centerline ductility in wrought products. A small percentage of delta ferrite may be observed in microstructures of 17-4PH stainless steel.

An aluminum-containing intermetallic compound precipitates in the martensitic matrix during aging of PH13-8Mo, while intermetallic compounds of copper and titanium are precipitated on aging of Custom 455.

Plane strain fracture toughness (K_{Ic}) data for 15-5PH, 17-4PH, PH13-8Mo and Custom 455 martensitic precipitation hardening stainless steels are presented in Table 5.4, and data for PH13-8Mo from the B-1 program are given in Table 5.5. For each of these steels, the fracture toughness increases as the aging temperature is increased. The aging temperature, in °F, is indicated by the condition designation in the table. Of these four steels, Custom 455 has the highest strength in the H900 and H950 conditions. For forged and rolled bars in the H1000 condition, however, the strength and toughness of Custom 455 are comparable to those of PH13-8Mo. For many applications, either the H1000 or the RH1000 condition represents the best combination of strength and toughness.

Results of threshold stress corrosion (K_{Iscc}) tests on martensitic precipitation hardening stainless steels are presented in Table 5.6 and in Fig. 5.32. Sump tank residue water and aqueous sodium chloride solutions may influence the threshold stress-intensity factors of specimens obtained in the L-T orientation in these steels depending on the steel condition and the environment, but the effect is more noticeable for specimens of the T-L orientation and probably still more damaging for specimens of the short transverse orientation. However, no K_{Iscc} data were located for specimens of the short transverse orientation.

Fatigue crack growth rates for specimens of 15-5PH (H1050 and H1100) are shown in Fig. 5.33 for various conditions and environments at room temperature. For specimens in the H1050 condition, increasing the R ratio from 0.05 to 0.67 and incorporating a one-minute holding period at maximum load in each cycle substantially increased the crack growth rates at ΔK values over 40 MPa $\cdot$ $m^{1/2}$ (36 ksi $\cdot$ $in.^{1/2}$). For specimens in the H1100 condition, exposure to a salt solution environment during tests with a one-minute holding period at maximum load increased the fatigue crack growth rates over those of specimens tested in air with one-minute holding time or with continuous cycling (Ref 5.67).

Results of fatigue crack growth rate tests on specimens of 17-4PH stainless steel under comparable conditions are presented in Fig. 5.34. Those specimens that were tested in the H1050 condition at a stress ratio of 0.67 with a one-minute holding period at maximum load in each cycle had the highest fatigue crack growth rates (as for 15-5PH) in the upper levels of ΔK values. Specimens in the H1100 condition tested in a salt solution with a one-minute holding period, however, had fatigue crack growth rates only slightly higher than those of comparable specimens tested in air with continuous cycling (Ref 5.67).

Fatigue crack growth rate data collected by Crooker, Hasson and Yoder on specimens of 17-4PH (H1050 and H1150) also show the marked increases in fatigue crack growth rates that accompany increases in load ratio over the range from 0.04 to 0.80 (Ref 5.68). In comparing fatigue crack growth rates for specimens of 17-4PH stainless steel in a dry argon environment and in a 100% humid argon environment, Rack and Kalish (Ref 5.63) have shown that high humidity increases fatigue crack growth rates in the absence of oxygen. The specimens were in the H900, H1000 and H1100 conditions, the frequency was 10 Hz and the R ratio was 0.1.

Fatigue crack growth rate data for room temperature tests on specimens from rolled bar and extrusions of PH13-8Mo (H1000) stainless steel make up the scatter

Table 5.4. Fracture toughness (K_{Ic}) data for 15-5 PH, 17-4 PH, PH 13-8 Mo and Custom 455 martensitic precipitation hardening stainless steels from several sources

Type(a)	Condition(b)	Testing temperature °C	Testing temperature °F	Yield strength MPa	Yield strength ksi	Tensile strength MPa	Tensile strength ksi	Fracture toughness Orientation	Fracture toughness K_{Ic} MPa·m$^{1/2}$	Fracture toughness K_{Ic} ksi·in.$^{1/2}$	Ref
15-5 PH (VAR)	H900	RT		1280	185	1380	200	L-T	96	87	5.59
	H900	RT		1210	175	1320	192	...	81	74	5.60
	H900	RT		1180	171	1330	193	T-L	81	74	5.61
	H1080	22	72	1030	149	1040	151	Random	115-122	104-111	5.62
	H1080	0	32	1044	151	1052	152	Random	96-114	87-104	...
	H1080	−20	−4	1041	151	1054	153	Random	89-101	81-92	...
17-4 PH	H900	RT		1210	176	1380	200	T-L	48	44	5.63
	H975	RT		1160	168	1230	178	L-T	93	85	5.59
	H1100	RT		883	128	972	141	T-L	153(c)	139(c)	5.64
17-4 PH (AM)	H900	RT		1170	170	1310	190	L-T	53	48	5.1
	H900	RT		1210	176	1340	195	...	57	52	5.60
PH 13-8 Mo	H950	RT		1360	197	1550	225	T-L	70	64	5.61
	H1050	RT		1230	178	1320	192	T-L	112	102	5.61
Custom 455 (VAR)	H900	RT		1760	255	...	...	L-T	51	46	5.65
	H950	RT		1700	246	...	...	L-T	79	72	5.65
	H1000	RT		1365(d)	198(d)	...	...	L-T	110	100	5.1

(a) Heat treatments: 15-5 PH and 17-4 PH were austenitized at 1040°C (1900°F), AC; PH 13-8 Mo was austenitized at 1000°C (1825°F), AC; Custom 455 was annealed at 980°C (1800°F), WQ, reheated to 815°C (1500°F), OQ. (b) Aging treatments: H900 at 480°C (900°F), AC; H950 at 510°C (950°F), AC; H975 at 525°C (975°F), AC; H1000 at 540°C (1000°F), AC; H1050 at 565°C (1050°F), AC; H1080 at 580°C (1080°F), AC; H1100 at 595°C (1100°F), AC. (c) $K_{Ic}(J)$ data. (d) Typical.

Table 5.5. Summary of fracture toughness (K_{Ic}) data for compact specimens of PH 13-8 Mo stainless steel from the B-1 Program (Ref 5.6)

Product form	Condition (a)	Yield strength (L), MPa	Yield strength (L), ksi	Tensile strength (L), MPa	Tensile strength (L), ksi	Average fracture toughness, K_{Ic}: L-T orientation, $MPa\cdot m^{1/2}$	L-T orientation, $ksi\cdot in.^{1/2}$	T-L orientation, $MPa\cdot m^{1/2}$	T-L orientation, $ksi\cdot in.^{1/2}$
Forged bar	H950	1410	204	1490	216	66	60	63	57
Rolled bar	RH950	1500	217	1630	236	68	62	...	...
		1510	219	1630	237	64	58	...	...
Rolled bar	RH975	1490	216	1610	233	79	72	...	...
		1510	219	1590	231	72	66	...	...
Forged bar	H1000	1390	201	1460	212	104	95	99	90
		1320	191	1430	208	87	79	89	81
		1460	212	1510	219	113	103	99	90
Rolled bar	H1000	1430	208	1490	216	96	87	82	75
	Welded joint (b)	...	...	...	...	91	83	...	...
	Welded joint (c)	...	...	...	...	97	88	...	...
Rolled bar	RH1000	1480	215	1530	222	122	111	...	...
		1500	218	1560	226	104	95	...	...
Extruded bar	H1000	1480	214	1520	221	74	67	72	66

(a) Heat treatments: H950 and H1000—austenitized at 925°C (1700°F), AC; RH950, RH975 and RH1000—austenitized at 925°C (1700°F), AC, cooled to −73°C (−100°F) for 5 h; H950 and RH950—aged at 510°C (950°F) for 4 h; RH975—aged at 525°C (975°F) for 4 h; H1000 and RH1000—aged at 540°C (1000°F) for 4 h. (b) Weld metal. (c) Heat-affected zone.

Table 5.6. Stress-corrosion threshold (K_{Iscc}) data for martensitic precipitation hardening stainless steels at room temperature

Type	Condition	Environment	K_{Ic} MPa·m$^{1/2}$	K_{Ic} ksi·in.$^{1/2}$	K_{Iscc}(a) MPa·m$^{1/2}$	K_{Iscc}(a) ksi·in.$^{1/2}$	Reference
15-5 PH.......	H900	3.5% NaCl	81	74	62	56	5.60
	H900	20% NaCl	79	72	36 (T-L)	33	5.61
17-4 PH(b)	H900	3.5% NaCl	57	52	57	52	5.60
PH 13-8 Mo ...	H950	3.5% NaCl	81	74	81	74	5.60
Custom 455....	H950	3.5% NaCl	79	72	79	72	5.60

(a) L-T orientation except as noted. (b) Additional data for K_{Iscc} tests on 17-4 PH are presented in Ref 5.66.

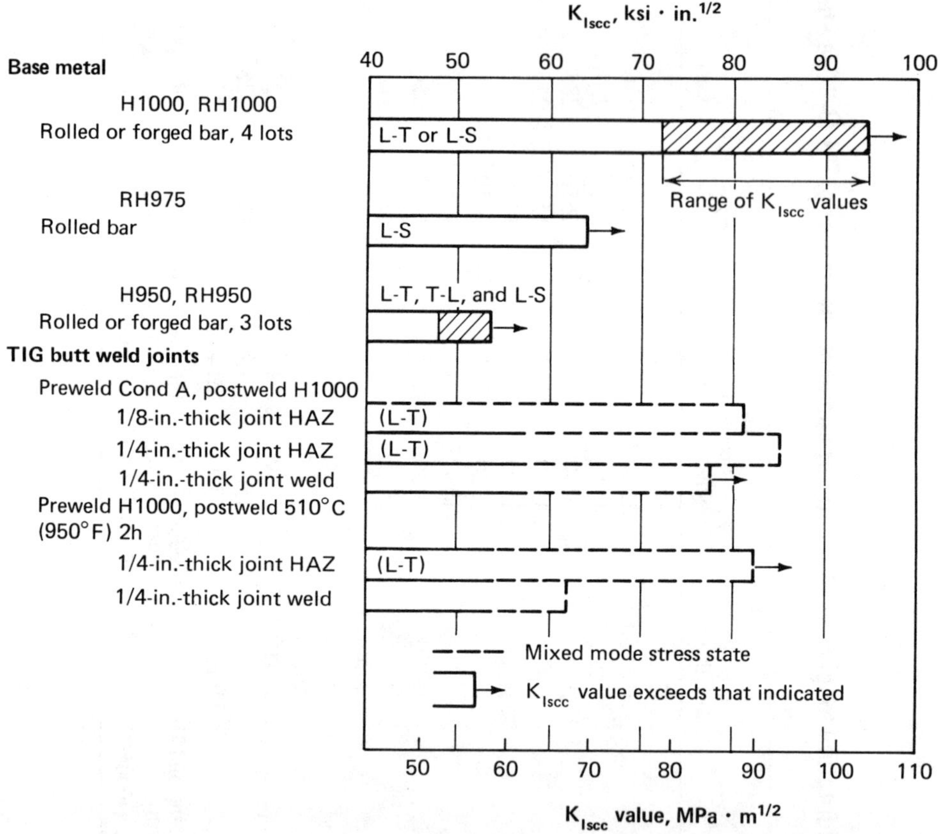

Fig. 5.32. Summary of K_{Iscc} tests on specimens of PH 13-8 Mo stainless steel in sump tank residue water for base metal and weldments (Ref 5.6)

band in Fig. 5.35 (Ref 5.6). Specimens of L-T and T-L orientations were tested in low humidity air and in sump tank residue water at frequencies of 1 and 6 Hz and at an R ratio of 0.08. Under these conditions, variations in frequency and environment had little effect on fatigue crack growth rates. For tests at −54 °C

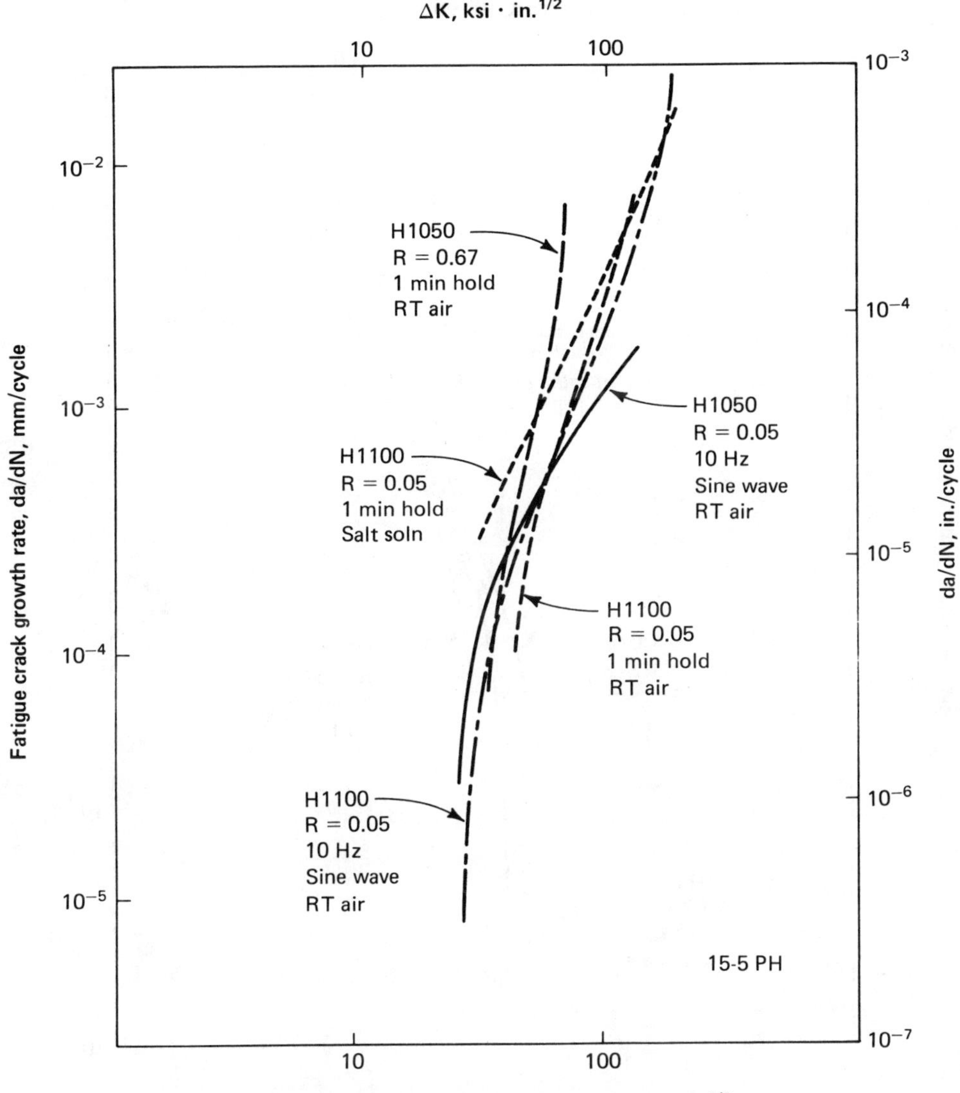

Fig. 5.33. Fatigue crack growth rates in WOL specimens of 15-5 PH stainless steel in the H1050 and H1100 conditions in room temperature air and in a 3.5% NaCl solution (Ref 5.67)

(−65 °F), the rates of fatigue crack growth were lower than those at room temperature over most of the ΔK range.

Effects of increasing the load ratio, R, on fatigue crack growth rates in low humidity air (LHA) and in sump tank residue water (STW) for specimens of PH13-8Mo (H1000) are shown in Fig. 5.36. The highest fatigue crack growth rates in this series were obtained on specimens tested at an R ratio of 0.3 in STW. Increasing the load ratio from 0.08 to 0.3 had a marked effect on the growth rates

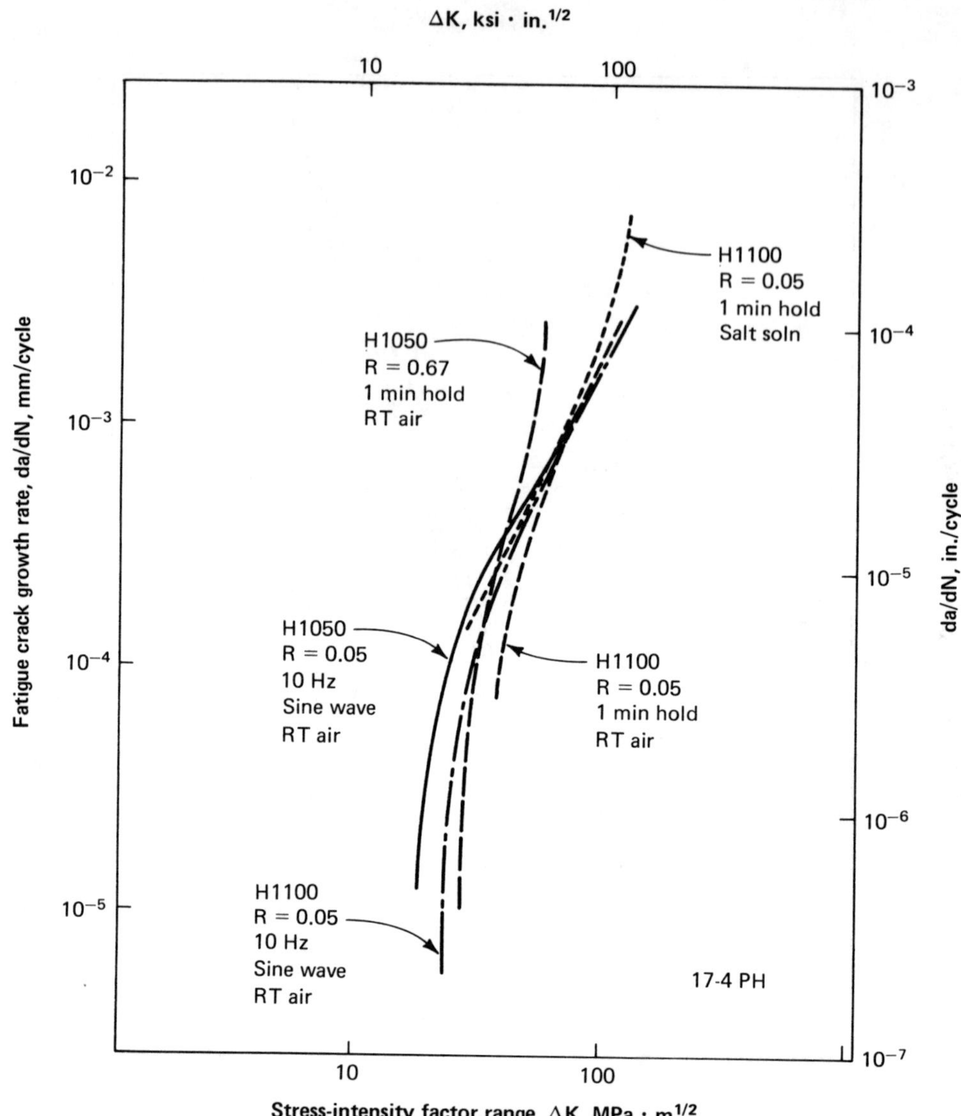

Fig. 5.34. Fatigue crack growth rates in WOL specimens of 17-4 PH stainless steel in the H1050 and H1100 conditions in room temperature air and in a 3.5% NaCl solution (Ref 5.67)

at all ΔK values in the range covered by these tests. Specimen orientation also influenced the results.

Fatigue crack growth rates for the curve shown in Fig. 5.37 for PH13-8Mo (H1100) specimens tested in air at 0.17 Hz and a stress ratio of zero are lower at given ΔK values than the rates for corresponding specimens in the H1000 condition.

Semiaustenitic Precipitation Hardening Stainless Steels

Of the semiaustenitic precipitation hardening stainless steels, most of the available mechanical property data have been obtained on alloys 17-7PH,

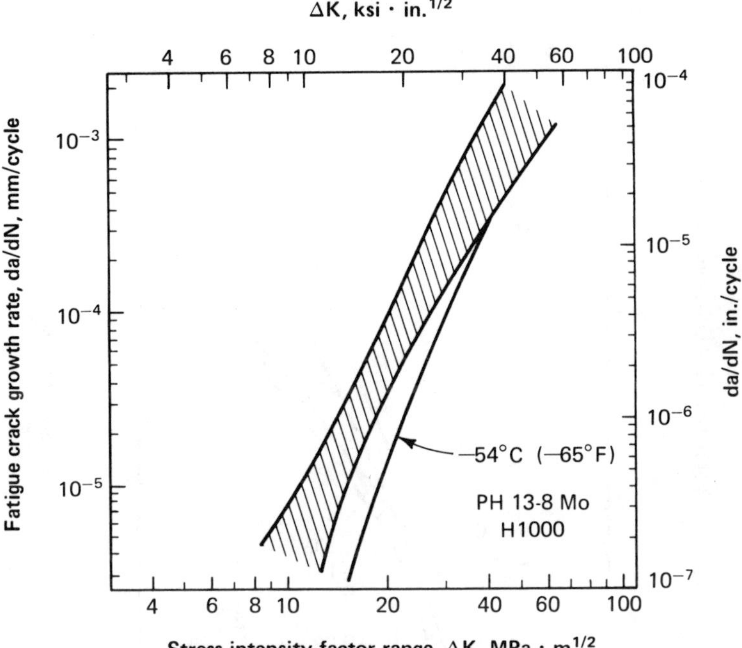

Fig. 5.35. Fatigue crack growth rate scatter band for compact specimens from rolled bar and extrusions of PH 13-8 Mo stainless steel in the H1000 condition for room temperature tests in low-humidity air and in sump tank water at frequencies of 1 and 6 Hz and an R ratio of 0.08 for L-T and T-L orientations (Ref 5.6)

PH15-7Mo and AM355. The compositions of 17-7PH and PH15-7Mo are similar (Table 5.1) except that 2% of the chromium in 17-7PH is replaced by 2% molybdenum in PH15-7Mo. Three or four heat treating steps are required for solution annealing, conditioning and aging these steels. For 17-7PH and PH15-7Mo steels, the solution annealing step at 1065 °C (1950 °F) followed by cooling in air (Condition A) is usually done at the mill. This treatment dissolves the carbides and develops a uniform austenitic microstructure for fabrication processing. After fabricating, the components are heated at 760 °C (1400 °F) for 1½ hours, for the TH treatment, and cooled to 15 °C (60 °F) or lower within one hour (preferably in water) and held at this temperature for at least ½ hour. This treatment destabilizes the austenite by rejecting carbon as carbides and raises the M_s temperature, allowing the austenite to transform to martensite on cooling to 15 °C or lower. For the third step, the components are heated to 565 °C (1050 °F) for 1½ hours and cooled in air. At 565 °C, precipitation of nickel-aluminum compounds occurs, developing the properties associated with the TH1050 condition. If the components are aged at 580 °C (1080 °F), the condition is designated TH1080.

The four-step heat treatment used to develop the RH950 and RH1050 conditions also requires solution annealing at 1065 °C (1950 °F), but step two after fabricating requires heating at 955 °C (1750 °F) for 10 minutes followed by cooling in air to room temperature. Carbides are rejected at 955 °C to destabilize the austenite, but to a lesser degree than at 760 °C (1400 °F). Transformation of the

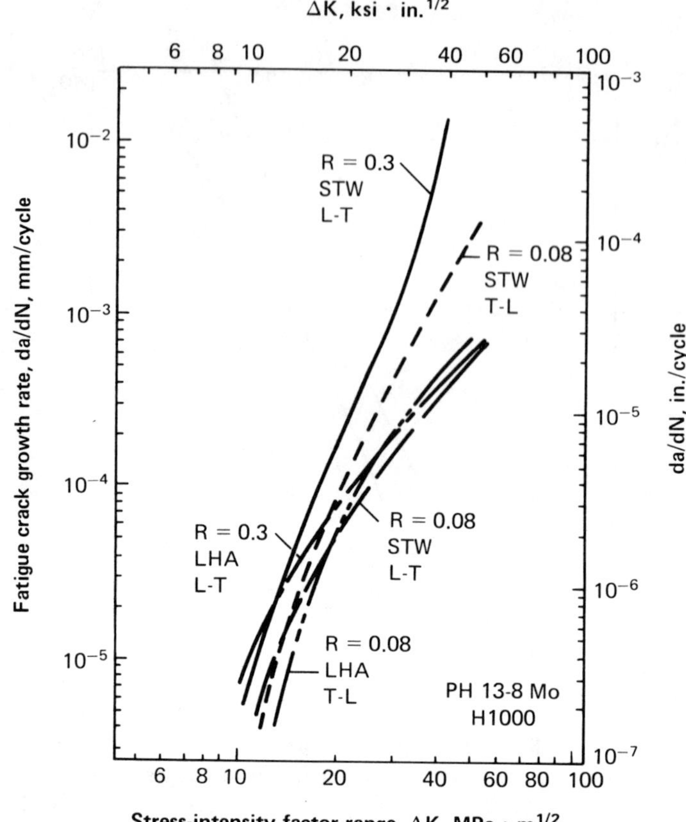

Fig. 5.36. Fatigue crack growth rates in compact specimens of PH 13-8 Mo stainless steel in the H1000 condition for room temperature tests at 1 Hz, R ratios of 0.08 and 0.3, L-T and T-L orientations, in low-humidity air (LHA) or sump tank water (STW) (Ref 5.6)

destabilized austenite to martensite occurs on the third step during cooling to −75 °C (−100 °F) over an eight-hour period. The fourth step of the RH950 treatment is to age the components at 510 °C (950 °F) for one hour. Precipitation of nickel-aluminum compounds on aging further increases the strength of the martensite which formed on the third step. If the components are aged at 565 °C (1050 °F), the condition is designated RH1050.

If these steels are severely cold worked to transform the austenite to martensite, a single aging treatment at 480 °C (900 °F) will suffice to obtain the age hardened condition (CH900) and the highest strength level that can be achieved for 71-7PH or PH15-7Mo.

For AM355 stainless steel, mill products usually are heat treated at the mill to obtain a fabricable condition; then, after fabricating, the components are solution annealed at 1040 °C (1900 °F), quenched in water, subzero cooled to −75 °C (−100 °F), held three hours and warmed to room temperature. The components are then conditioned by reheating to 955 °C (1750 °F), cooled in air and subzero

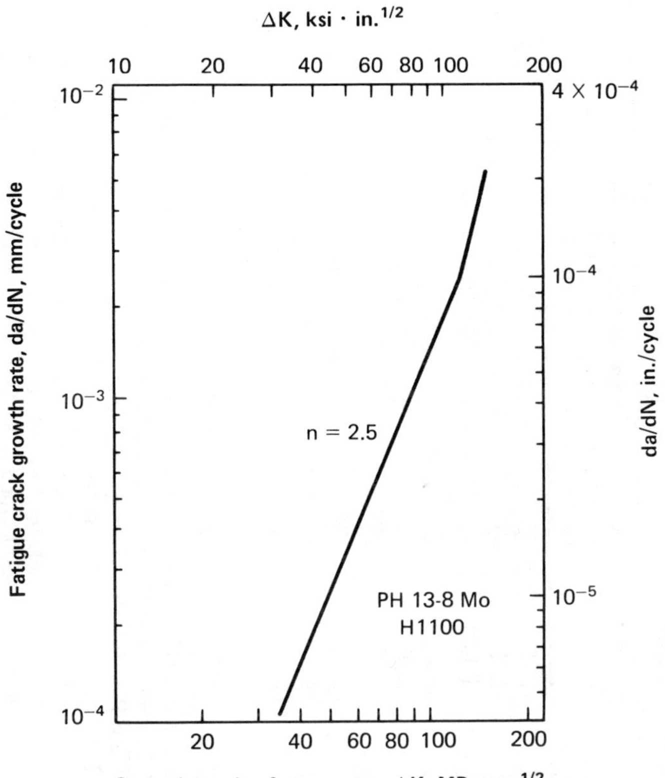

Fig. 5.37. Fatigue crack growth rates in cantilever beam specimens of PH 13-8 Mo (H1100) stainless steel, at L-T orientation, 0.17 Hz, and an R ratio of 0, in room temperature air (Ref 5.21)

cooled to −75 °C (−100 °F), held for three hours, and warmed to room temperature. They are then aged at 455 °C (850 °F) for three hours, for the SCT850 condition, or aged at 540 °C (1000 °F) for three hours, for the SCT1000 condition.

Only limited fracture toughness data have been published on the semiaustenitic precipitation hardening stainless steels. Typical K_{Ic} data are presented in Table 5.7. The highest toughness was obtained for specimens of AM355 (SCT1000) at a yield strength of 1170 MPa (170 ksi). Increasing the aging temperature from 455 °C (850 °F) to 540 °C (1000 °F) reduced the yield strength only slightly but increased the fracture toughness substantially.

Threshold stress-corrosion cracking (K_{Iscc}) data also are presented in Table 5.7 for these steels. Apparently these steels have relatively low resistance to crack growth in aqueous sodium chloride solutions. The highest value for K_{Iscc} in this series was obtained for the AM355 (SCT1000) specimens.

Austenitic Precipitation Hardening Stainless Steel

The austenitic precipitation hardening stainless steel A-286 is the main representative in this category. It contains titanium and small amounts of vanadium and aluminum, which precipitate as intermetallic compounds such as $Ni_3(Al, Ti)$ and

Table 5.7. Fracture toughnes (K_{Ic}) and threshold stress corrosion (K_{Iscc}) data for 17-7 PH, PH 15-7 Mo and AM355 semiaustenitic precipitation hardening stainless steels at room temperature

Type	Condition(a)	Yield strength		Orien-tation	Fracture toughness				Reference
					K_{Ic}		K_{Iscc}		
		MPa	ksi		MPa·m$^{1/2}$	ksi·in.$^{1/2}$	MPa·m$^{1/2}$	ksi·in.$^{1/2}$	
17-7 PH	RH950	1180	171	L-T	35	32	<21(b)	<19(b)	5.60
	TH1050	...	...	L-T	43	39	17.5(b)	16(b)	...
	RH1050	1310	190	T-L	52	47	<20(c)	<18(c)	5.27
PH 15-7 Mo	RH950	1405	204	T-L	34	31	<16(c)	<15(c)	5.27
	RH950	1350	196	L-T	35	32	15(b)	14(b)	5.60
	RH1050	1345	195	T-L	44	40	<22(c)	<20(c)	5.27
	TH1050	1160	168	L-T	37	34	20(b)	18(b)	5.60
	TH1080	...	...	L-T	55	50	...	...	5.1
AM355	SCT850	1240	180	L-T	65	59	35(b)	32(b)	5.60
	SCT850	1240	180	T-L	53	48	9(c)	8(c)	5.27
	SCT1000	1170	170	T-L	115	105	41(c)	37(c)	5.27

(a) **RH heat treatments for 17-7 PH and PH 15-7 Mo:** solution annealed at 1065°C (1950°F) and air cooled; conditioned by heating at 955°C (1750°F) for 10 min, air cooling, subzero cooling to −75°C (−100°F) for 8 h, and warming in air; then aged at 510°C (950°F) for 1 h (RH950) or aged at 565°C (1050°F) for 1 h (RH1050). **TH heat treatments for 17-7 PH and PH 15-7 Mo:** solution annealed at 1065°C (1950°F) and air cooled; conditioned by heating at 760°C (1400°F) for 1½ h, cooling to 16°C (60°F) within 1 h of removal from furnace, and holding for 30 min; then aged at 565°C (1050°F) for 1½ h and air cooled (TH1050) or aged at 580°C (1080°F) for 1½ h and air cooled (TH1080). **SCT heat treatments for AM355:** solution annealed at 1040°C (1900°F), water quenched, subzero cooled to −75°C (−100°F), held for 3 h, reheated to 955°C (1750°F), air cooled or water quenched, subzero cooled to −75°C (−100°F), and held for 3 h; then aged at 455°C (850°F) for 3 h (SCT850) or aged at 540°C (1000°F) for 3 h (SCT1000). (b) In 3.5% NaCl solution. (c) In 20% NaCl solution.

$Ni_4Mo(Fe, Cr)$ Ti on aging. Various mill forms of the alloy are usually supplied in the annealed condition—Condition A (980°C, or 1800°F, for one hour followed by quenching in oil or water). Precipitation hardening occurs on aging in the range from 700 to 760°C (1300 to 1400°F) for 16 hours. Other combinations of heat treatments may be used depending on the application. One variation is to re-solution treat at 900°C (1650°F) for two hours, quench in oil or water, and age at 700°C (1300°F) for 16 hours. This variation results in improved room temperature properties but less desirable stress-rupture properties.

Because of the high toughness of A-286 stainless steel, even at −269°C (−452°F), available fracture toughness data have been obtained only by the J-integral method. Results from several sources are presented in Table 5.8.

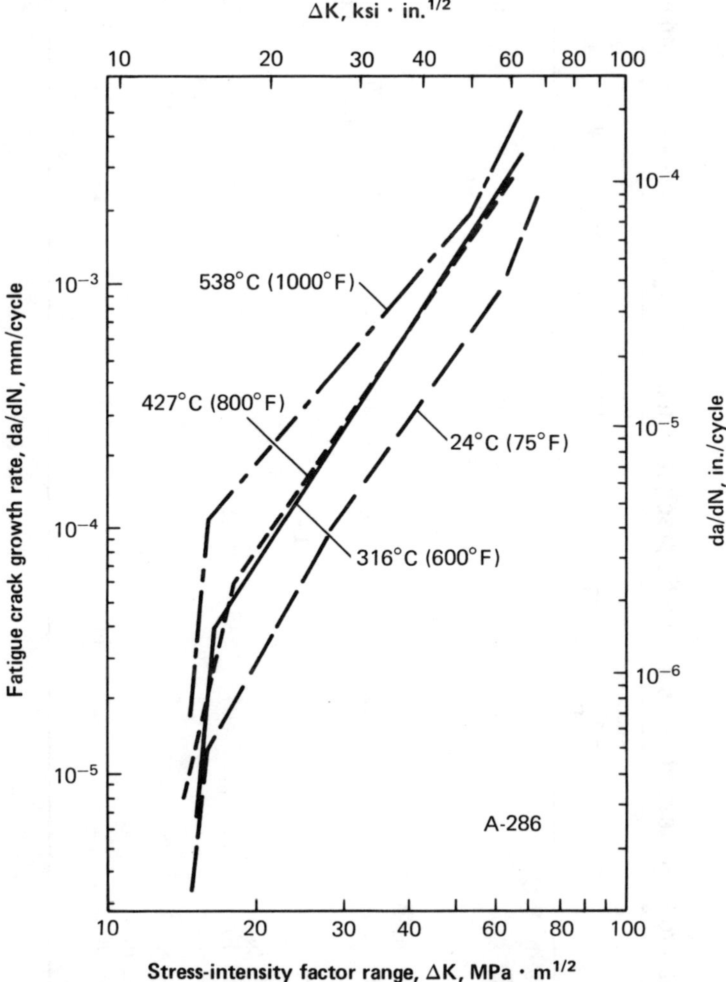

Fig. 5.38. Fatigue crack growth rates for specimens of A-286 stainless steel at room temperature and elevated temperatures for tests in air at 3 Hz (RT) and 0.67 Hz (elevated temperatures), an R ratio of 0.05, and at L-T, T-L, R-L and R-C orientations (Ref 5.72)

Table 5.8. Fracture toughness of A-286 austenitic precipitation hardening stainless steel based on the J-integral method

Heat treatment	Room-temperature yield strength, MPa	ksi	Specimen orientation	Specimen thickness, mm	in.	Testing temperature, °C	°F	J_{Ic}, kJ/m²	in.-lb/in.²	$K_{Ic}(J)$, MPa·m$^{1/2}$	ksi·in.$^{1/2}$	Reference
980°C (1800°F) ½ h, WQ, 720°C (1325°F) 16 h	769	112	T-L	12.6	0.5	25	77	133	758	167	152	5.69
						430	800	92	524	139	126	
						540	1000	81	463	130	119	
980°C (1800°F) ½ h, WQ, 720°C (1325°F) 16 h	722	105	...	3.05	0.12	25	77	120	686	159	144	5.70
						540	1000	99	563	144	131	
STA (solution treated and aged)	...	...	L-T	...	...	25	77	121	692	159	145	5.69
900°C (1650°F) 2 h, OQ, 730°C (1350°F) 16 h	607	88	T-S	38	1.5	25	77	75	426	125	114	5.71
						−196	−320	67	385	123	112	
						−269	−452	61	350	118	107	
900°C (1650°F) 5 h, OQ, 718°C (1325°F) 20 h	822	119	...	12.7	0.5	24	75	121	692	161	146	5.41
						−269	−452	143	815	180	163	

Results shown for the series of tests performed by Reed, Tobler and Mikesell (Ref 5.71) are lower than the others at room temperature. For this series, the heat treatment and the specimen orientation were not the same as for the others. Results of the J_{Ic} tests done by Wells *et al* (Ref 5.41) show that toughness increases as the testing temperature is decreased to −269°C (−452°F).

Fatigue crack growth rate data have been reported by James (Ref 5.72) for compact specimens from A-286 stainless steel flat bar 12.7 mm (0.5 in.) thick and round bar 38 mm (1.5 in.) in diameter which had been annealed at 980°C (1800°F), quenched in water, and aged at 720°C (1325°F) for 16 hours. These specimens were obtained in the L-T, T-L, R-L and R-C orientations and were tested at room temperature and at 316, 427 and 538°C (600, 800 and 1000°F) at an R ratio of 0.05 in air. Testing frequencies were 3 Hz at room temperature and 0.67 Hz at the elevated temperatures. The summary curves in Fig. 5.38 show that fatigue crack growth rates tend to increase as the exposure temperature is increased. Variations in the orientation of the specimens from the two product forms had no effect on the fatigue crack growth rates. For these specimens of A-286 alloy, the trend of the fatigue crack growth rates at each temperature was similar to that for 20% cold worked Type 316 stainless steel tested under similar conditions.

Results of fatigue crack growth rate tests on compact and single-edge-notch specimens of A-286 stainless steel also have been reported by Gamble and Paris (Ref 5.73). Tests were made at room temperature and at 482°C (900°F) in air at R ratios of 0 and −1. At room temperature, variations in load ratio had no effect on fatigue crack growth rates. At 482°C (900°F), however, growth rates at R = −1 were approximately three times greater than for the corresponding R = 0 data. Specimen geometry did not influence the results. Results of the fatigue crack growth rate tests were used in predicting the allowable number of service cycles for gas turbine disks subjected to cyclic thermal stresses.

Fatigue crack growth rates reported by Tobler and Reed (Ref 5.33) for solution treated and aged A-286 stainless steel specimens tested at −196 and −269°C (−320 and −452°F) were lower, at each ΔK value, than rates obtained at room temperature. In this respect, A-286 is similar to the other stable stainless steels. Fatigue crack growth rate data reported by Wells *et al* (Ref 5.41) also showed that crack growth rates at −269°C (−452°F) were lower than those at room temperature for A-286 in the solution treated and aged condition.

5.4 REFERENCES

5.1. *Properties and Selection: Stainless Steels, Tool Materials and Special-Purpose Metals*: Vol 3 of *ASM Metals Handbook* (9th Ed.), American Society for Metals, Metals Park, OH, 1980

5.2. *Toughness of Ferritic Stainless Steels,* edited by R. A. Lula: STP 706, American Society for Testing and Materials, Philadelphia, 1980

5.3. The Super 12% Chromium Steels, by J. Z. Briggs and T. D. Parker: Climax Molybdenum Co., Division of American Metal Climax, Inc., New York, 1965

5.4. Fatigue Crack Propagation in Austenitic Stainless Steels, by L. A. James: *Atomic Energy Review*, Vol 14, No. 1, 1976, p 37-86

5.5. The Fatigue Crack Growth Rate Properties of Type 403 Stainless Steel in Marine Turbine Environments, by W. G. Clark, Jr.: in *Corrosion Problems in Energy Conversion*, Electrochemical Society, 1974, p 368-383

5.6. Fracture Mechanics Evaluation of B-1 Materials, by R. R. Ferguson and R. C. Berryman: Report AFML-TR-76-137 (I and II), Rockwell International Corp., B-1 Div., Los Angeles, 1976

5.7. Influence of Strain Induced Martensitic Transformations on Fatigue Crack Growth Rates in Stainless Steels, by A. G. Pineau and R. M. Pelloux: *Metallurgical Transactions*, Vol 5, No. 5, May 1974, p 1103-1112

5.8. An Effective Strain Concept for Crack Propagation and Fatigue Life with Specific Applications to Biaxial Stress Fatigue, by E. K. Walker: Report AFFDL-TR-70-144, Proceedings of the Air Force Conference on Fatigue and Fracture in Aircraft Structures and Materials, Miami Beach, FL, Dec 1969

5.9. Fatigue Crack Propagation Behavior of Type 304 Stainless Steel at Elevated Temperatures, by L. A. James and E. B. Schwenk: *Metallurgical Transactions*, Vol 2, No. 2, Feb 1971, p 491-496

5.10. Effect of Thermal Aging Upon the Fatigue Crack Propagation of Austenitic Stainless Steels, by L. A. James: *Metallurgical Transactions*, Vol 5, No. 4, Apr 1974, p 831-838

5.11. Fatigue Crack Growth Characteristics of Several Austenitic Stainless Steels at High Temperatures, by P. Shahinian, H. H. Smith and H. E. Watson: STP 520, American Society for Testing and Materials, Philadelphia, 1973, p 387-400

5.12. The Effects of Temperature, Composition, and Carbide Morphology on Crack Growth in Type 304 Stainless Steel, by M. Cahn and J. Shively: Report AI-73-12, Atomics International Div., Rockwell International Corp., Canoga Park, CA, March 12, 1973

5.13. Frequency Effects in the Elevated Temperature Crack Growth Behavior of Austenitic Stainless Steels – A Design Approach, by L. A. James: *Transactions of ASME, Journal of Pressure Vessel Technology*, Vol 101, No. 2, May 1979, p 171-176

5.14. The Effect of Frequency Upon the Fatigue-Crack Growth of Type 304 Stainless Steel at 1000°F, by L. A. James: STP 513, American Society for Testing and Materials, Philadelphia, 1972, p 218-229

5.15. Hold Time Effects on the Elevated Temperature Fatigue-Crack Propagation of Type 304 Stainless Steel, by L. A. James: *Nuclear Technology*, Vol 16, No. 3, Dec 1972, p 521-530

5.16. Creep-Fatigue Crack Propagation in Austenitic Stainless Steel, by P. Shahinian: *Journal of Pressure Vessel Technology*, ASME, Vol 98, 1976, p 166-172

5.17. Influence of Section Thickness on Fatigue Crack Growth in Type 304 Stainless Steel, by P. Shahinian: *Nuclear Technology*, Vol 30, Sept 1976, p 390-397

5.18. Specimen Size Considerations in Fatigue Crack Growth Rate Testing, by L. A. James: Report HEDL-TME-78-99, Hanford Engineering Development Laboratory, Westinghouse Hanford Co., Richland, WA, Dec 1978

5.19. The Effect of Stress Ratio on the Elevated Temperature Fatigue-Crack Propagation of Type 304 Stainless Steel, by L. A. James: *Nuclear Technology*, Vol 14, No. 2, May 1972, p 163-167

5.20. Effect of Heat-to-Heat and Melt Practice Variations Upon Fatigue Crack Growth in Two Austenitic Steels, by L. A. James: STP 679, American Society for Testing and Materials, Philadelphia, 1979, p 3-16

5.21. Fatigue Crack Growth in Selected Alloys for Reactor Applications, by P. Shahinian, H. E. Watson and H. H. Smith: *Journal of Materials*, Vol 7, No. 4, Dec 1972, p 527-535

5.22. Effect of Hold Time and Thermal Aging on Elevated Temperature Fatigue Crack Propagation in Austenitic Stainless Steels, by D. J. Michel and H. H. Smith: Report NRL-MR-3627, Naval Research Laboratory, Washington, Oct 1977

5.23. Effect of Hold Time on Elevated Temperature Fatigue Crack Propagation in Type 304 and 316 Stainless Steel, by D. J. Michel and H. H. Smith: Proceedings of the 1976 ASME-MPC Symposium on Creep-Fatigue Interaction, ASME Winter Meeting, New York, Dec 1976

5.24. Fatigue Crack Growth of Stainless Steel Piping in a Pressurized Water Reactor Environment, by W. H. Bamford: *Transactions of ASME, Journal of Pressure Vessel Technology*, Vol 101, No. 1, Feb 1979, p 73-79

5.25. Fatigue Growth in Piping and RPV Steels in Simulated BWR Water Environment, by D. A. Hale, J. L. Yuen and T. L. Gerber: Report NUREG/CR-0390, General Electric Co., San Jose, CA, March 1979

5.26. Fatigue Crack Propagation Behavior in Liquid Sodium Environment, by L. A. James and R. L. Knecht: *Metallurgical Transactions*, Vol 6A, No. 1, Jan 1975, p 109-116

5.27. Defect Growth Rates in Austenitic Stainless Steels, by L. A. James: *Transactions of the American Nuclear Society*, Vol 26, June 1976, p 381

5.28. Irradiation Effects on Fatigue Crack Propagation in Austenitic Stainless Steels, by D. J. Michel: Report NRL-MR-3610, Naval Research Laboratory, Washington, Sept 1977

5.29. Fatigue Crack Propagation in Neutron Irradiated Type 304 and Type 308 Stainless Steel Plate and Weldments, by D. J. Michel and H. H. Smith: *Journal of Nuclear Materials*, Vol 71, No. 1, Dec 1977, p 173-177

5.30. The Effect of Fast Neutron Irradiation Upon the Fatigue Crack Propagation Behavior of Two Austenitic Stainless Steels, by L. A. James: *Journal of Nuclear Materials*, Vol 59, No. 2, Feb 1976, p 183-191

5.31. Effect of Neutron Irradiation on Fatigue Crack Propagation in Types 304 and 316 Stainless Steels at High Temperatures, by P. Shahinian, H. E. Watson and H. H. Smith: STP 529, American Society for Testing and Materials, Philadelphia, 1973, p 493-508

5.32. Fatigue Crack Propagation in Fast Neutron Irradiated Stainless Steels and Welds, by P. Shahinian: STP 570, American Society for Testing and Materials, Philadelphia, 1976, p 191-204

5.33. Fatigue Crack Growth Resistance of Structural Alloys at Cryogenic Temperatures, by R. L. Tobler and R. P. Reed: in *Advances in Cryogenic Engineering*, edited by K. D. Timmerhaus, Vol 24, Plenum Press, 1978, p 82-90

5.34. Welding Stainless Steels for Structures Operating at Liquid Helium Temperature, by C. E. Witherell: *Welding Journal*, Welding Research Supplement, Vol 59, No. 4, Nov 1980, p 326-s to 342-s

5.35. Fatigue and Creep Crack Propagation in Stainless Steel Weld Metal, by P. Shahinian: *Welding Journal*, Welding Research Supplement, Vol 57. No. 3, March 1978, p 87-s to 92-s

5.36. Fatigue Crack Propagation in Stainless Steel Weldments at High Temperatures, by P. Shahinian, H. H. Smith and J. R. Hawthorne: *Welding Journal*, Welding Research Supplement, Vol 51, No. 11, Nov 1972, p 523-s to 527-s

5.37. Fatigue Crack Growth in Type 304 Stainless Steel Weldments at Elevated Temperature, by L. A. James: *Journal of Testing and Evaluation*, Vol 1, No. 1, Jan 1973, p 52-57

5.38. Crack Propagation Behavior in Type 304 Stainless Steel Weldments, by L. A. James: *Welding Journal*, Welding Research Supplement, Vol 52, No. 4, April 1973, p 173-s to 179-s

5.39. Fatigue Crack Propagation in Types 304 and 308 Stainless Steels at Elevated Temperatures, by D. T. Raske and C. F. Cheng: *Nuclear Technology*, Vol 34, June 1977, p 101-110

5.40. Fatigue Crack Propagation in Austenitic Stainless Steels, by A. W. Thompson: *Engineering Fracture Mechanics*, Vol 7, No. 1, March 1975, p 61-68

5.41. Evaluation of Weldments in Austenitic Stainless Steels for Cryogenic Applications, by J. M. Wells, W. A. Logsdon and R. Kossowsky: in *Advances in Cryogenic Engineering*, edited by K. D. Timmerhaus, Vol 24, Plenum Press, 1978, p 150-160

5.42. Fatigue Crack Growth in Type 316 Stainless Steel at High Temperature, by P. Shahinian, H. H. Smith and H. E. Watson: *Journal of Engineering for Industry, Transactions of ASME*, Series B, Vol 93, No. 4, Nov 1971, p 976-980

5.43. Effect of Thermal Aging and Hold Time on Fatigue Crack Propagation in Type 316 Stainless Steel, by D. J. Michel and H. H. Smith: Proceedings of the Second International Conference on Mechanical Behavior of Materials, Federation of Materials Societies, Boston, Aug 16-20, 1976, p 568-572; published by American Society for Metals, Metals Park, OH, 1976

5.44. Fatigue Crack Propagation in Type 316 Stainless Steel: Effect of Precipitate Formation During Testing at 649°C, by D. J. Michel and H. H. Smith: Report of NRL Progress, Naval Research Laboratory, Washington, Jan 1979, p 11-13

5.45. Accelerated Creep-Fatigue Crack Propagation in Thermally Aged Type 316 Stainless Steel, by D. J. Michel and H. H. Smith: *Acta Metallurgica*, Vol 28, No. 7, July 1980, p 999-1007

5.46. Fatigue Crack Propagation in Thermally Aged Type 316 Stainless Steel, by D. J. Michel and H. H. Smith: Report of NRL Progress, Naval Research Laboratory, Washington, July 1978, p 12-14

5.47. Fatigue Crack Growth in 20% Cold-Worked Type 316 Stainless Steel at Elevated Temperatures, by L. A. James: *Nuclear Technology*, Vol 16, No. 1, Oct 1972, p 316-322

5.48. Fatigue Crack Growth Behavior of Stainless Steel Type 316 Plate and 16-8-2 Weldments in Air and High-Carbon Liquid Sodium, by J. L. Yuen and F. L. Copeland: *Journal for Engineering Materials and Technology, Transactions of ASME*, Vol 101, No. 3, July 1979, p 214-223

5.49. Significance of Welding Variables, Long-Time Aging and Tension Hold Times on Fatigue Crack Growth in Type 316 Stainless Steel Welds, by J. B. Hawthorne and H. E. Watson: Proceedings of the 1976 ASME-MPC Symposium on Creep-Fatigue Interaction, ASME Winter Annual Meeting, New York, Dec 1976

5.50. Fatigue Behavior of Type 316 Stainless Steel Under Simulated Body Conditions, by K. R. Wheeler and L. A. James: *Journal of Biomedical Materials Research*, Vol 5, No. 3, May 1971, p 267-281

5.51. The Influence of Gas Environments on Fatigue Crack Growth Rates in Types 316 and 321 Stainless Steel, by M. W. Mahoney and N. E. Paton: *Nuclear Technology*, Vol 23, No. 3, Sept 1974, p 290-297

5.52. Effect of Neutron Irradiation on Fatigue Crack Propagation in Type 316 Stainless Steel at 649°C, by D. J. Michel and H. H. Smith: NRL Memorandum Report 3936, Naval Research Laboratory, Washington, March 14, 1979

5.53. An Evaluation of the Crack Growth and Fracture Properties of AISI 403 Modified 12 Cr Stainless Steel, by W. A. Logsdon: *Engineering Fracture Mechanics*, Vol 7, No. 1, March 1975, p 23-40

5.54. Elastic Plastic (J_{Ic}) Fracture Toughness Values: Their Experimental Determination and Comparison with Conventional Linear Elastic (K_{Ic}) Fracture Toughness Values for Five Materials, by W. A. Logsdon: STP 590, American Society for Testing and Materials, Philadelphia, 1976, p 43-60

5.55. Fatigue Crack Propagation of 403 Stainless Steel in Aqueous Solutions at 100°C, by L. Abrego and J. A. Begley: Paper No. 235 for the International Corrosion Forum, Palmer House, Chicago, March 3-7, 1980

5.56. Fracture Toughness of 420 Steel in a 3.5 Percent NaCl Solution in Water, by

S. A. Bhangaonkar and S. Banerjee: *Transactions of the Indian Institute of Metals*, Vol 28, No. 1, Feb 1975, p 53-57

5.57. Effect of Salt Water Temperature on Crack Growth Characteristics of 12 Chrome Steel, by R. Eisenstadt and K. M. Rajan: *Journal of Engineering Materials and Technology*, Vol 96, No. 2, 1974, p 81-87

5.58. New Developments in High-Strength Stainless Steels, by A. F. Hoenie and D. B. Roach: DMIC Report 223, Defense Metals Information Center, Battelle Columbus Laboratories, Columbus, OH, Jan 3, 1966

5.59. Correspondence from Armco Steel Corp., Advanced Materials Div., Baltimore, Oct 18, 1972

5.60. Stress Corrosion Properties of High Strength Precipitation Hardening Stainless Steels, by C. S. Carter *et al: Corrosion*, Vol 27, No. 5, May 1971, p 190-197

5.61. Evaluation of Stress-Corrosion Cracking Susceptibility Using Fracture Mechanics Techniques, by D. O. Sprowles *et al*: Aluminum Company of America, Alcoa Center, PA, Contract NAS8-21487, May 31, 1973

5.62. Fracture Mode and Toughness vs. Temperature (−20 to 22 C) for UNS-S15500 (15-5 PH) Stainless Steel, by C. Calabrese: *Engineering Fracture Mechanics*, Vol 11, No. 3, 1979, p 537-546

5.63. The Strength, Fracture Toughness and Low Cycle Fatigue Behavior of 17-4 PH Stainless Steel, by H. J. Rack and D. Kalish: *Metallurgical Transactions*, Vol 5, No. 7, 1974, p 1595-1604

5.64. Elastic-Plastic Fracture Toughness (J_{Ic}) of High Strength Steels and Titanium Alloys, by J. P. Gudas and J. A. Joyce: Report DTNSRDC-78/054, David W. Taylor Naval Ship Research and Development Center, Bethesda, MD, June 1978

5.65. Evaluation of Carpenter Custom 455, by J. M. Uchida: Research Report D6-23928, The Boeing Co., Renton, WA, Nov 18, 1969

5.66. Stress Corrosion Cracking Properties of 17-4 PH Steel, by C. T. Fugii: STP 610, American Society for Testing and Materials, Philadelphia, 1976, p 213-225

5.67. Influence of Microstructural and Load Wave Form Control on Fatigue Crack Growth Behavior of Precipitation Hardening Stainless Steels, by K. R. Kondas: Doctor's Degree Dissertation, University of Missouri, July 1976

5.68. A Fracture Mechanics and Fractographic Study of Fatigue-Crack Propagation Resistance in 17-4 PH Stainless Steel, by T. W. Crooker, D. F. Hasson and G. R. Yoder: NRL Report 7910, Naval Research Laboratory, Washington, July 1975

5.69. The Room Temperature and Elevated Temperature Fracture-Toughness Response of Alloy A-286, by W. J. Mills: *Transactions of ASME, Journal of Engineering Materials and Technology*, Vol 100, No. 2, April 1978, p 195-199

5.70. Techniques Developed for Elevated Temperature Fracture Toughness Testing of Irradiated Materials in Thin Sections, by F. H. Huang and G. L. Wire: *Transactions of ASME, Journal of Engineering Materials and Technology*, Vol 101, No. 4, Oct 1979, p 403-406

5.71. The Fracture Toughness and Fatigue Crack Growth Rate of an Fe-Ni-Cr Superalloy at 298, 76, and 4 K, by R. P. Reed, R. L. Tobler and R. P. Mikesell: in *Advances in Cryogenic Engineering*, edited by K. D. Timmerhaus, Vol 22, Plenum Press, 1977, p 68-79

5.72. The Effect of Temperature on the Fatigue-Crack Propagation Behavior of A-286 Steel, by L. A. James: Report HEDL-TME-75-82, Westinghouse Hanford Co., Richland, WA, Jan 1976

5.73. Cyclic Crack Growth Analysis for Notched Structures at Elevated Temperatures, by R. M. Gamble and P. C. Paris: STP 590, American Society for Testing and Materials, Philadelphia, 1976, p 345-367

Chapter 6

Fracture Properties of Aluminum Alloys

J. G. Kaufman and J. S. Santner

6.1. INTRODUCTION

Early experience in the application of fracture mechanics to aluminum alloys is exemplified by the study in 1967 by Moore *et al* (Ref 6.1) in which natural flaws (fatigue cracks) were produced by cyclic internal pressurizing in a number of test cylinders machined from aluminum die forgings. After the flaws had been developed, the cylinders were subjected to increasing internal pressure until fracture occurred. Ultimate fracture occurred in these cylinders at estimated stress-intensity factors closely paralleling the measured plane strain fracture toughness values of specimens machined from the forgings. The results are shown in the bar graph in Fig. 6.1. Before we discuss the fracture mechanics approach to design for aluminum alloys, however, it is appropriate to begin with an overview of the various types of aluminum alloys, the effects of their major alloying elements, and their hardening mechanisms. This will provide a convenient means of introducing some of the advantages and limitations of fracture mechanics in analysis of structures produced from aluminum alloys.

6.1.1. Major Alloy Systems

The major alloy systems in wrought aluminum alloys are summarized in Table 6.1, which includes the major alloying elements, the principal strengthening mechanisms and the range of yield strength levels that are typical for each alloy system (Ref 6.2). The highest-strength wrought aluminum alloys are those in the 2000 (aluminum-copper) and 7000 (aluminum-zinc-magnesium) series. Yield strengths in the range from 350 to 560 MPa (50 to 80 ksi) are attainable in these alloys. They are strengthened by solution heat treating to place all soluble elements in a supersaturated matrix. The solution treatment is followed by a precipitation hardening treatment called aging, which may be done at room temperature (natural aging) or at mildly elevated temperatures (artificial aging) to achieve the

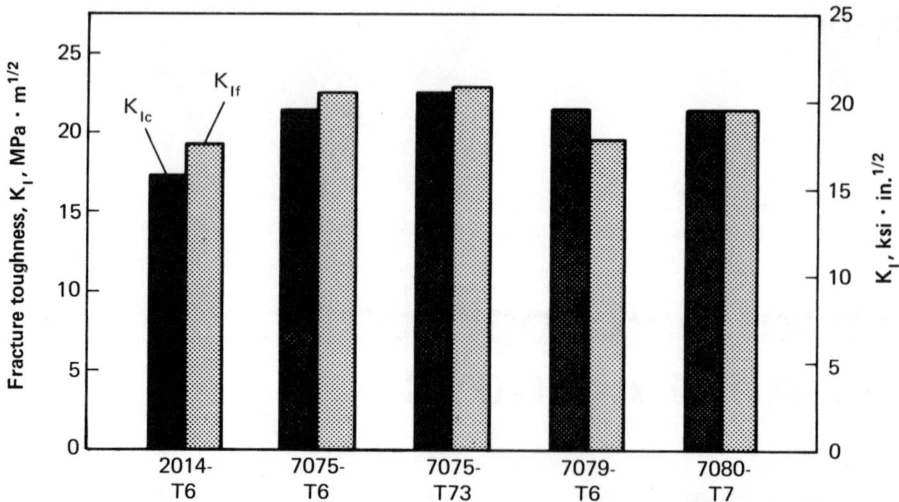

Fig. 6.1. Comparison of conditions in pressurized cylinders (K_{If}) with plane strain fracture toughness measurements (K_{Ic}) for several aluminum alloys (Ref 6.1)

Table 6.1. Wrought aluminum and aluminum alloy designation system (Ref 6.2)

Aluminum Association series	Type of alloy composition	Strengthening method	Range of tensile strength, MPa	Range of tensile strength, ksi
1xxx	Al	Cold working	70-175	10-25
2xxx	Al-Cu-Mg	Heat treating	170-520	25-75
3xxx	Al-Mn	Cold working	140-280	20-40
4xxx	Al-Si	Cold working	105-350	15-50
5xxx	Al-Mg	Cold working	140-380	20-55
6xxx	Al-Mg-Si	Heat treating	150-380	22-55
7xxx	Al-Zn(Mg)	Heat treating	380-520	55-75
	Al-Zn(Mg-Cu)	Heat treating	520-620	75-90

appropriate amount of strengthening by precipitated particles. Peak strengths are obtained through optimized aging of the Guinier-Preston (GP) zones, which are coherent particles composed primarily of the principal alloying elements. Additional aging normally results in larger, partially coherent or noncoherent particles, which are less effective as strengthening agents. Other particles that are present in the microstructures of these alloys contribute to strength and in some cases detract from toughness, as described later. The larger of these particles are referred to as second-phase constituent particles, which are normally composed, at least in part, of impurities such as iron and silicon or of excesses of the principal alloying elements. The smaller particles are composed of residuals from the original ingot solidification process, which are not subsequently redissolved. These particles will be referred to later in respect to their specific effects on fracture toughness.

Alloys of the 5000 (Al-Mg) and 6000 (Al-Mg-Si) series are lower in strength and higher in toughness than those of the 2000 and 7000 series. The 5000 series alloys are strengthened primarily by cold working, although the alloying elements produce small amounts of solid solution strengthening. The alloys of the 6000 series, like the higher-strength alloys, are hardened by solution heat treatment and precipitation aging — in this instance with Mg-Si-bearing particles as the principal strengthening agents. Other types of aluminum alloys, such as the 3000 (Al-Mn) and 4000 (Al-Si) series, are not normally used in situations where high strength and/or high toughness are of interest, and thus they will not be treated further in this discussion.

6.2. FRACTURE TOUGHNESS

Before we deal with data on the fracture toughness of aluminum alloys, a few comments about testing procedures are appropriate. (A general discussion of testing procedures is given in Chapter 3.) Usually, plane strain fracture toughness tests of aluminum alloys are made in accordance with ASTM Method E399, but some additional requirements are needed to ensure that reproducible values have been obtained. These requirements are contained in ASTM Methods B645 and B646. One of the principal additional criteria spelled out by these methods for aluminum alloys is that minimum specimen size requirements are twice those in ASTM Method E399. The systematic programs leading to this observation, which are of particular importance for higher-toughness aluminum alloys, were developed by Kaufman and Nelson (Ref 6.3) and are summarized in Fig. 6.2A and 6.2B. Until both thickness and crack length are approximately five times the square of the ratio of K_{Ic} to yield strength, it is not possible to be certain that size-independent values of plane strain fracture toughness have been obtained. Smaller specimens lead to lower fracture toughness values. The majority of the K_{Ic} data presented hereafter have been obtained for specimens with the above size requirements.

While much of the subsequent discussion will center around plane strain fracture toughness, it is quite obvious that many aluminum alloys — in aerospace applications in particular — are used in thicknesses for which it is not possible to obtain plane strain fracture toughness numbers. Until the advent of crack resistance curve measurements per ASTM Method E561, the usual practice was to measure the critical stress-intensity factor, referred to as K_c, in center-cracked panels of various sizes. For aluminum alloys, many of the data were generated on 406-mm- (16-in.-) wide panels of sheet in thicknesses up to about 3 mm (⅛ in.) (Ref 6.4), and additional data were generated on panels 25 to 38 mm (1 to 1½ in.) thick and 508 mm (20 in.) wide (Ref 6.5). The data obtained on such large panels are useful for comparison purposes as presented in Fig. 6.3A and 6.3B. Note that, for both ranges of panel thickness, the trend line indicates a sharply decreasing critical K value with increasing yield strength.

While these so-called "K_c" data points would now be recognized as representing points on crack resistance (K_R) curves, they provide useful indications of the fact that the crack resistance of even 25-mm- (1-in.-) thick panels of relatively high-strength aluminum alloys is well above the values that would be conservatively

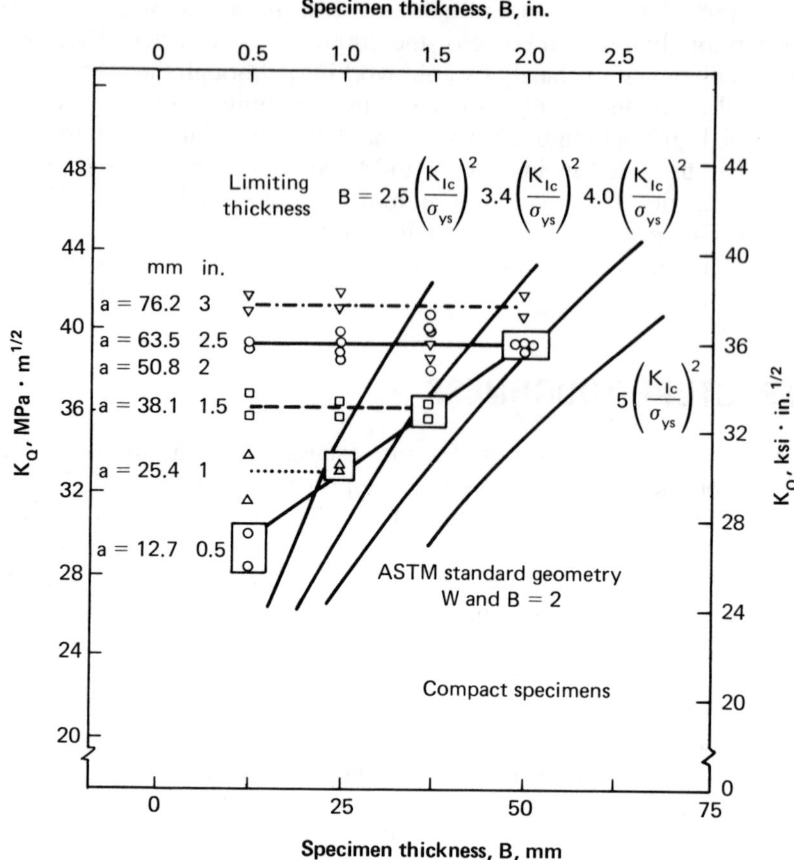

Fig. 6.2A. Influence of specimen thickness in plane strain fracture toughness tests of 3-in. 2219-T851 plate (Ref 6.3)

estimated from plane-strain fracture toughness numbers. It is expected that, in the future, much broader use will be made of crack resistance curve measurements and that they will be produced as whole families of curves for various thicknesses of individual aluminum alloys in order to obtain complete representations of their toughness.

Table 6.2 is a summary of plane strain fracture toughness values for a wide variety of aluminum alloys (Ref 6.6)—including 2024 and 7075, which have been used for components which do not require a high level of fracture toughness. Data for individual lots of these alloys, which will be discussed later, illustrate the influences of product thickness, grain orientation, and temperature on the fracture toughness of aluminum alloys.

It is appropriate to note at this point that much of the early fracture toughness testing of aluminum alloys was carried out using a modification by Kaufman and Holt (Ref 6.7) of a Kahn-type tear test. In this test, the energy required to initiate and propagate a crack is measured in a relatively small specimen containing a sharp machined notch. Although this test has never been adopted as a standard because of its sensitivity to measurement procedure and testing machine stiffness,

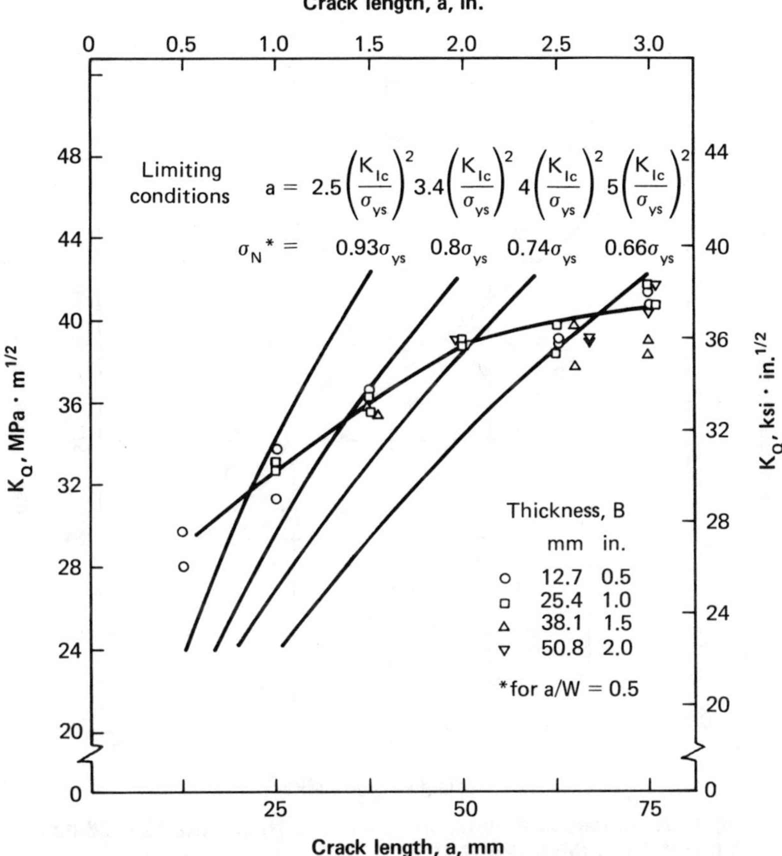

Fig. 6.2B. Influence of crack length independent of thickness in plane strain fracture toughness tests of 3-in. 2219-T851 plate (Ref 6.3)

the unit propagation energy from such tests has been shown to correlate closely with plane strain fracture toughness, as illustrated in Fig. 6.4. In view of the early use of the tear test and its good correlation with fracture toughness, plus its advantages in terms of size and economics, it has been widely employed to study the effects of certain metallurgical variables. This test was first utilized by Kaufman and by Nock and Hunsicker (Ref 6.8) to identify the fact that, of the available aluminum alloys, the alloys of the 7000 series provided the best combination of strength and toughness, as illustrated in Fig. 6.5. Therefore, in the discussion that follows, certain metallurgical effects will be illustrated using the results of tear tests instead of those of the more rigorous fracture toughness tests.

6.2.1. Effects of Microstructure, Composition and Heat Treatment

The role of certain particles in the microstructure as crack initiators has been recognized for some time (Ref 6.9, 6.10 and 6.11), and several excellent descriptions of this role have been developed—notably, by Low and his coworkers

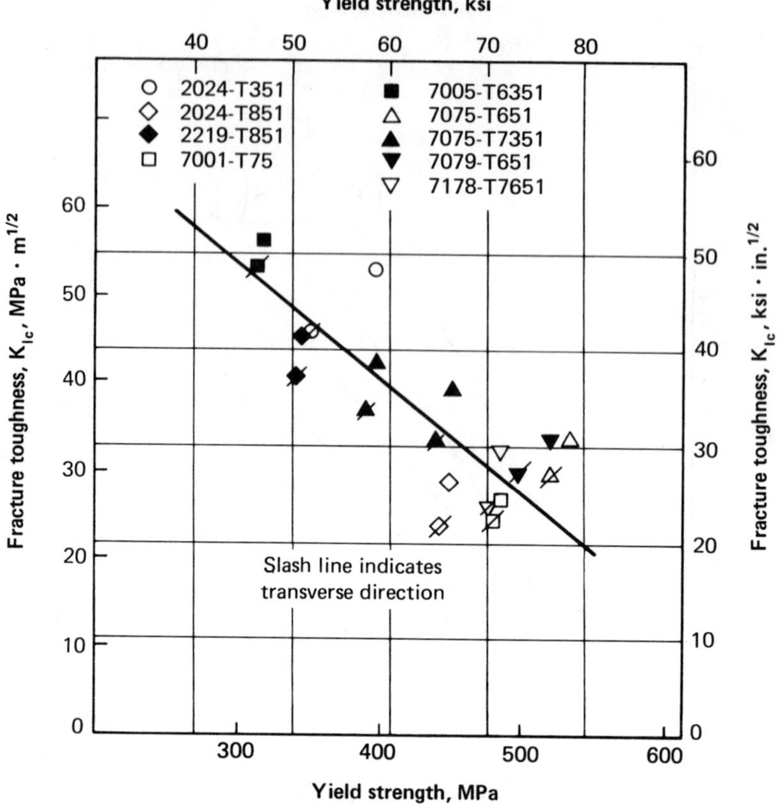

Fig. 6.3A. K_{Ic} measurements on specimens from plate 25 to 38 mm (1.0 to 1.5 in.) thick (Ref 6.5)

(Ref 6.12 and 6.13) in their work on dimpled rupture. The basic fracture mode in all aluminum alloys is dimpled rupture characterized by particle or particle-interface fracture, void growth, and coalescence; therefore, the hardness, size and spacing of particles are all important. Cracks will nucleate in regions where nonuniform deformations occur on stressing and where cross-slip dislocation glide is restricted. Precipitate particles not only restrict dislocation glide but also provide sites for initial microcracks, because of either their relatively low fracture toughness or the incoherency of the particle-matrix interfaces. The larger the particles, the lower the stresses at which they fracture, which is consistent with linear elastic fracture mechanics concepts. Once microcracking occurs, void growth begins, followed by complete fracture resulting from void coalescence or impingement by void sheet formation. The distance between particles, as well as particle size, controls the void growth stage. It requires more deformation for the voids to coalesce if they are far apart than if they are close together. Also, with particles relatively far apart there is less opportunity for numerous smaller particles to form void sheets. Dimple size is recognized as an index to toughness, which is consistent with the fact that the large dimples are associated with the more widely spaced particles. Hunsicker (Ref 6.14) and Staley (Ref 6.15) have isolated

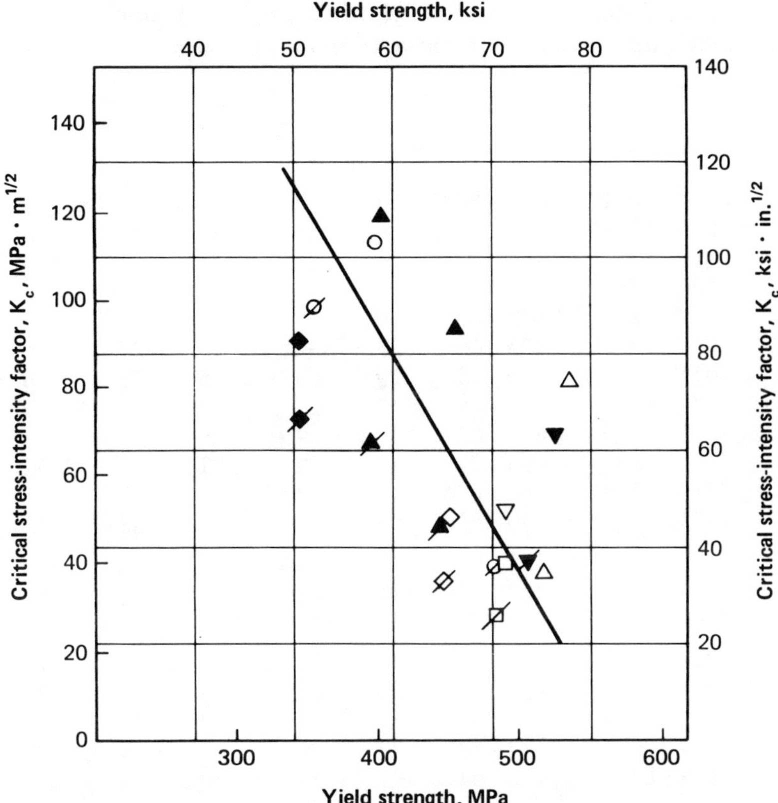

Fig. 6.3B. K_c measurements on specimens from sheet up to 3 mm (1/8 in.) thick (Ref 6.5)

the role of specific particles in aluminum alloys most clearly and have shown that three levels of particle interaction are involved.

The largest constituent particles (>1 μm), which form by eutectic decomposition during ingot solidification and which cannot be taken into solid solution during processing, are insolubles such as Al_7Cu_2Fe, Mg_2Si and $(Fe,Mn)Al_6$ or are relatively soluble $CuAl_2$ to $CuAl_2Mg$. These particles fracture first on stressing, and void growth follows in the manner described by Van Stone *et al* (Ref 6.13) and by Staley (Ref 6.15), as illustrated in Fig. 6.6 and 6.7. These cracked particles were also considered to be the initial sources of fatigue cracks (Ref 6.16), although this now appears to be an oversimplification, as will be discussed later. The size and number of insoluble constituents can be controlled by limiting the iron and silicon contents, and this improves toughness as illustrated in Fig. 6.8 (Ref 6.17). This has been one of the important factors in developing the high-toughness alloys now available, which utilize the highest purities justifiable with regard to quantity and cost. The effectiveness of this is illustrated in part by the comparison of alloys 2024/7075 with alloys 2124/7475 presented in Fig. 6.9 (Ref 6.6).

The second grouping of particles (0.03 to 0.5 μm) consists of dispersoids such as $Al_{12}Mg_2Cr$ or $Al_{20}Cu_2Mn$, which form by solid state precipitation and, once

Table 6.2. Typical room temperature yield strength and plane-strain fracture toughness values for several high-strength aluminum alloys (Ref 6.6)

Product	Alloy	Temper	Yield strength(a) MPa	Yield strength(a) ksi	Plane-strain fracture toughness, K_{Ic} L-T MPa·m$^{1/2}$	L-T ksi·in.$^{1/2}$	T-L MPa·m$^{1/2}$	T-L ksi·in.$^{1/2}$	S-L MPa·m$^{1/2}$	S-L ksi·in.$^{1/2}$
Plate	2014	T651	440	64	24	22	22	20	19	17
	2024	T351	325	47	36	33	33	30	26	24
	2024	T851	455	66	24	22	23	21	18	16
	2124	T851	440	64	32	29	25	23	24	22
	2219	T851	435	63	39	35	36	33	...	...
	7050	T73651	455	66	35	32	30	27	29	26
	7075	T651	505	73	29	26	25	23	20	18
	7075	T7651	470	68	30	27	24	22	20	18
	7075	T7351	435	63	32	29	29	26	21	19
	7475	T651	495	72	43	39	37	34	32	29
	7475	T7651	460	67	47	43	39	35	31	28
	7475	T7351	430	62	53	48	42	38	35	32
Die forgings	7050	T736	455	66	36(b)	33(b)	25(c)	23(c)	25(c)	23(c)
	7149	T73	460	67	34(b)	31(b)	24(c)	22(c)	24(c)	22(c)
	7175	T736	490	71	33(b)	30(b)	29(c)	26(c)	29(c)	26(c)
Hand forgings	2024	T852	430	62	29	26	21	19	18	16
	7050	T73652	455	66	36	33	23	21	22	20
	7075	T7352	365	53	37	34	29	26	23	21
	7079	T652	440	64	29	26	25	23	20	18
	7175	T736	470	68	37	34	30	27	26	24
Extrusions	7050	T7651x	495	72	31	28	26	24	21	19
	7050	T7351x	459	65	45	41	32	29	26	24
	7075	T651x	490	71	31	28	26	24	21	19
	7075	T7351x	435	63	35	32	29	26	22	20

(a) At 0.2% offset (longitudinal). (b) Parallel to grain flow. (c) Nonparallel to grain flow.

formed, cannot be dissolved. The dispersoids seem to have dual, but contradictory, roles; they suppress recrystallization or limit the growth of grains, and they promote the transgranular mode of fracture associated with the highest energy absorption. They also nucleate microvoids by decohesion at the matrix interface, leading to the formation of void sheets between larger voids. As a result, careful control of these dispersoids (volume fraction, size and spacing) is imperative. One means of reducing volume fraction of dispersoids is by elimination of chromium and substitution of zirconium, which forms a finer dispersion; see Fig. 6.10 (Ref 6.18).

The third and finest classification of particles (<0.01 μm) which seem important in the fracture process is the age-hardening precipitates of major alloying elements (GP zones), which impede dislocation movement and lead to optimum combinations of strength and toughness. Forsyth (Ref 6.16) also credits them with contributing to high fatigue limits. The morphology of these precipitates is largely dependent on quenching rate and degree of aging. Slow quenching (as in thick sections) and extended aging tend to produce coarser grains and grain boundary

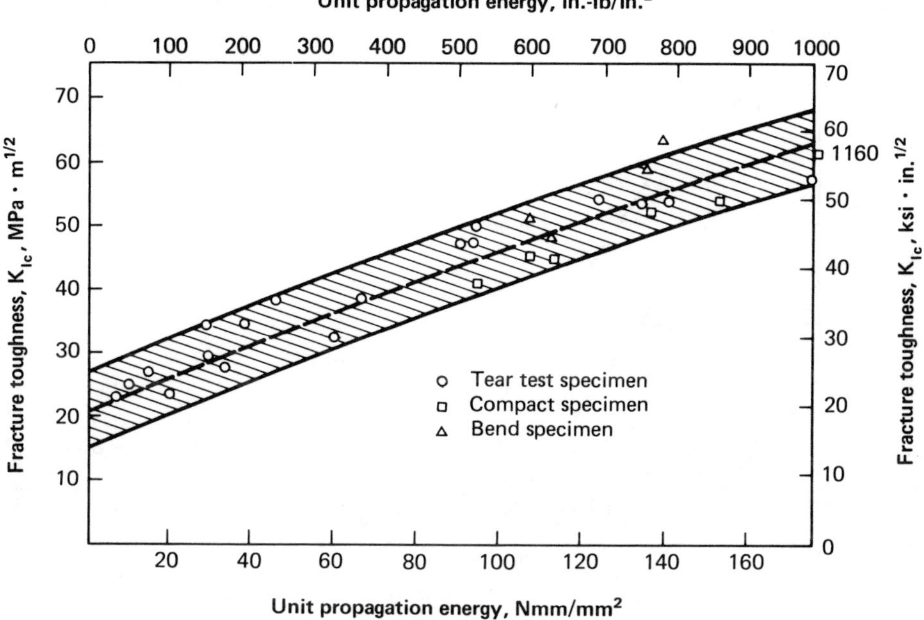

Fig. 6.4. Relationship between plane strain fracture toughness (K_{Ic}) and unit propagation energy from tear tests (Ref 6.7)

particles, which, particularly for alloys of the 7000 series, result in intergranular fracture and low toughness. Faster quenching and stepped precipitation treatments will help in minimizing these problems. For optimum combinations of stress-corrosion resistance and/or high-temperature properties together with toughness, the alloys are often over-aged. If over-aging is not properly controlled, the precipitates may grow to sizes at which they can be significant contributors to void sheet formation.

While the volume fraction, size and spacing of precipitate particles are critical features with regard to toughness, grain structure is also important. Hunsicker (Ref 6.14) and Staley (Ref 6.15) observed that toughness increased as recrystallized grain size decreased, and developed process controls to consistently provide finer grain sizes. Recent evidence (Ref 6.19) suggests that even greater toughness can be obtained with a lamellar recrystallized microstructure. This, like control of quenching rate, tends to minimize the potential for intergranular fracture, which, as indicated above, reduces toughness significantly.

With this background, and using illustrations from recent alloy design programs specifically aimed at improving fracture toughness, the following specific observations can be made with respect to the compositions of 2000 and 7000 series aluminum alloys:

Fe and Si: Reducing iron and silicon contents generally increases toughness by reducing the size and number of insoluble constituent particles (Ref 6.19). This effect has been used in the development of alloys 2124, 7050, 7175 and 7475. Data comparing K_{Ic} values for 2024, 2124, 7075 and 7475 plate (Fig. 6.9;

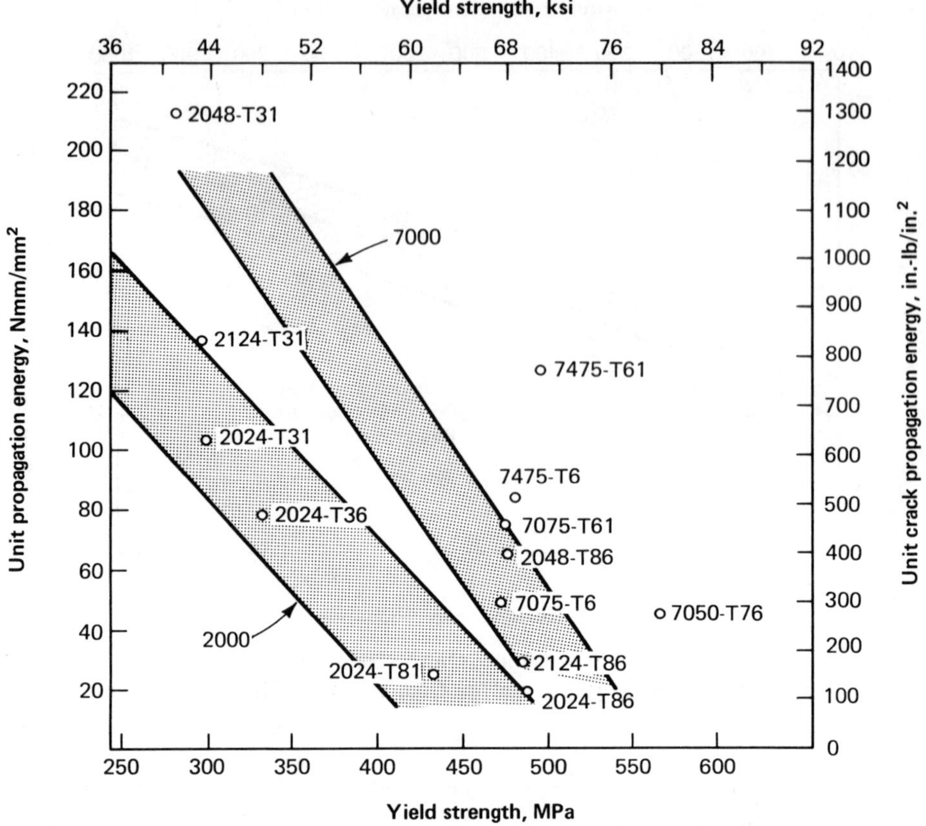

Toughness measured as unit crack propagation energy on specimens from sheet 1.6 to 3.2 mm (0.063 to 0.125 in.) thick.

Fig. 6.5. Variation in toughness of 2000 and 7000 series aluminum alloys with yield strength (Ref 6.8)

Ref 6.6) and 7075 and 7175 die forgings (Fig. 6.11; Ref 6.18) illustrate this effect.

Zr, Cr and Mn: Among the dispersoid-forming elements—zirconium, chromium and manganese—zirconium produces the finest particles and therefore the highest toughness (Fig. 6.10; Ref 6.18). Manganese is least effective in this regard.

Mg: Holding down magnesium levels to the minimum required to provide needed strength reduces the insoluble constituents present, and provides some improvement in fracture resistance. This effect, along with optimization of zirconium and copper contents, was utilized in the development of alloys 7050 and 7475.

With respect to thermal treatments, it is now recognized that over-aging beyond the peak strength (commonly done to improve resistance to stress corrosion) results in some improvement in fracture toughness. This improvement in tough-

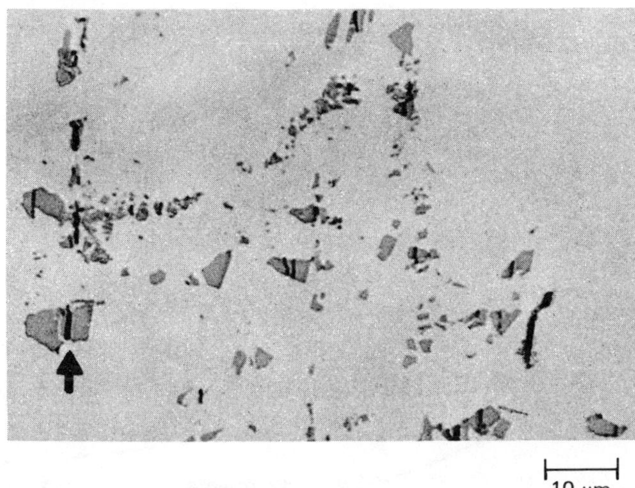

Fig. 6.6. Cracked particles in alloy 2024-T851 strained 6.1%, illustrating start of void formation (Ref 6.15)

Fig. 6.7. Fractograph of alloy 2024-T851, showing regions of large dimples (A), regions of small dimples (B), and a cracked particle (C) (Ref 6.15)

ness is accomplished at some sacrifice in strength/toughness combinations. However, the optimum strength/toughness combination occurs for peak-aged alloys (Fig. 6.12; Ref 6.15), and the best stable combination of properties at any given

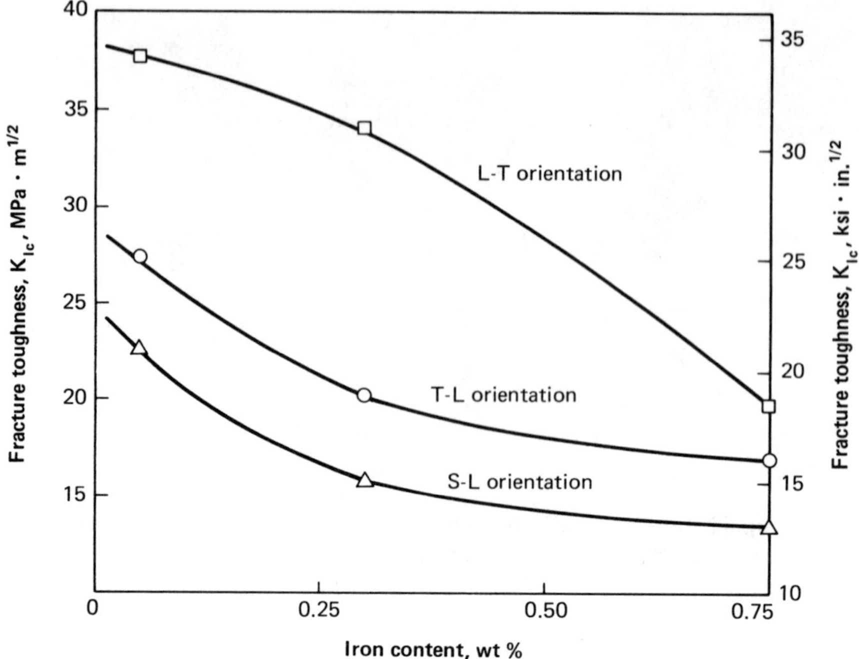

Fig. 6.8. Effect of iron content on fracture toughness of an Al-5.9Zn-2.5Mg-0.7Cu-0.3Mn-0.03Ti-0.02Si alloy in the T6 temper (Ref 6.17)

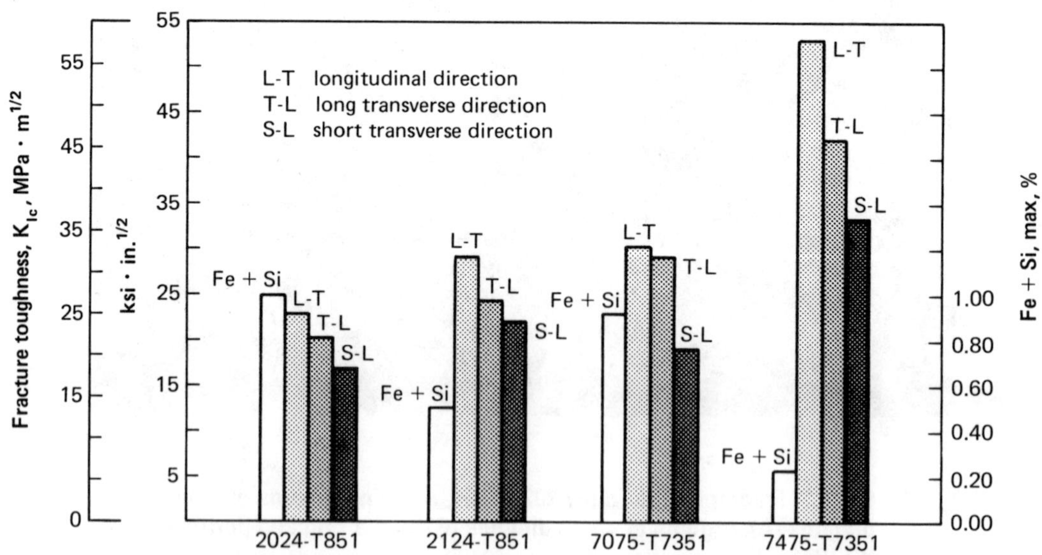

Fig. 6.9. Fracture toughness comparisons of aluminum alloys 2024, 2124, 7075 and 7475 (Ref 6.6)

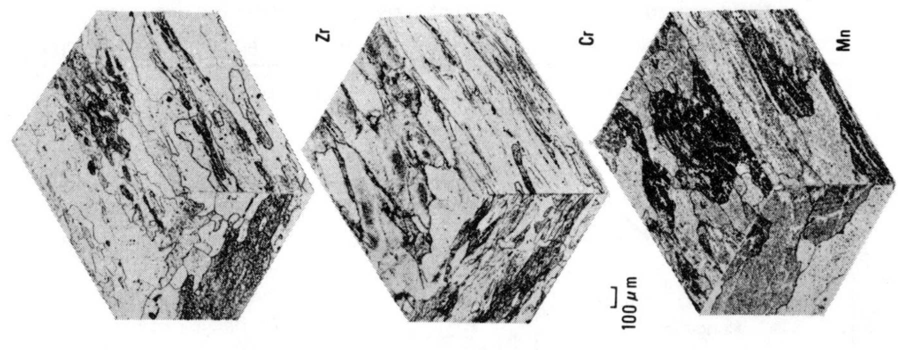

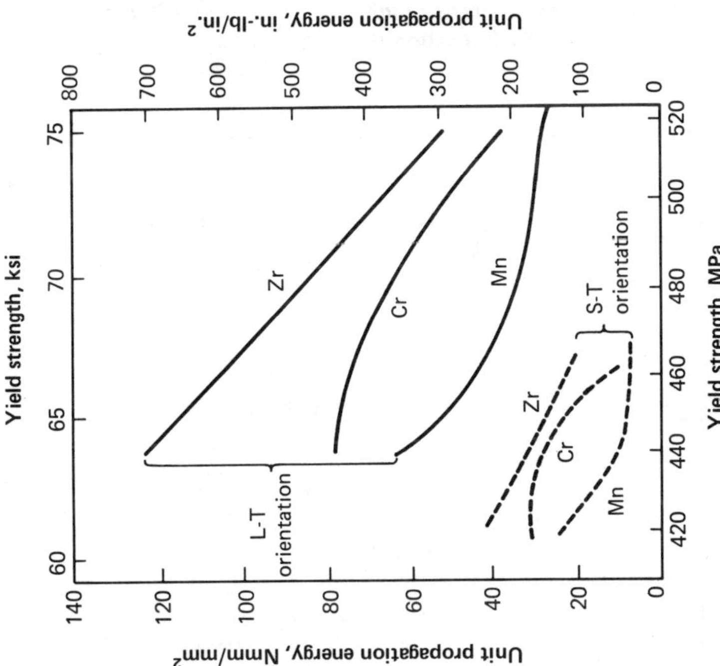

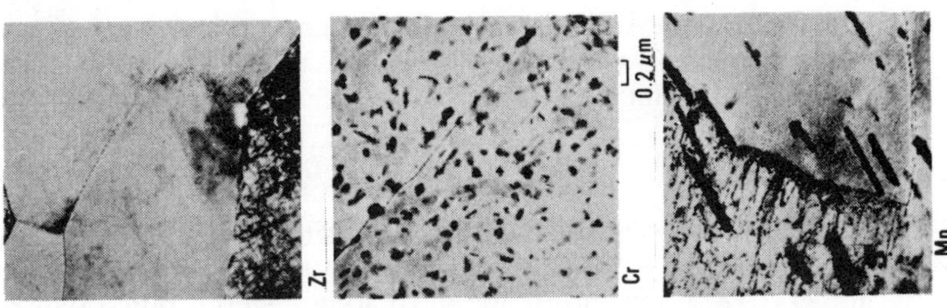

Fig. 6.10. Effects of ancillary element additions on the energy to propagate a crack in high-purity 7075 plate (Ref 6.18)
Micro- and macrodistributions of Zn, Cr and Mn are also shown.

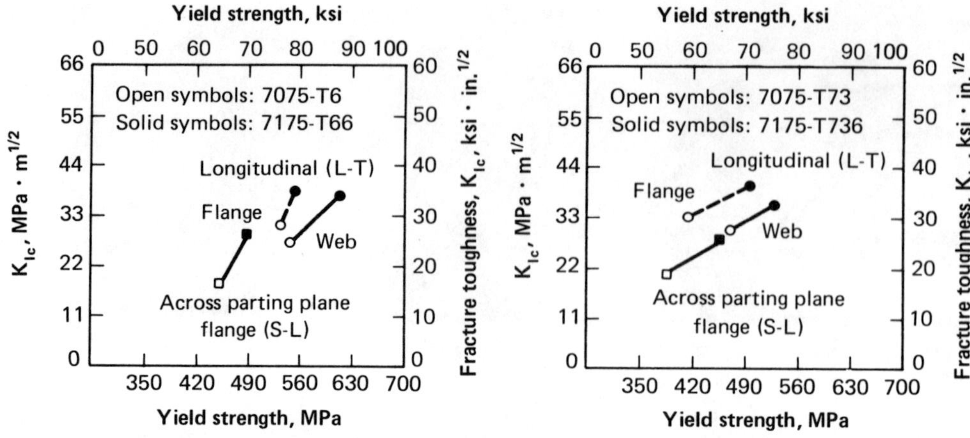

Fig. 6.11. Plane strain fracture toughness of 7075 and 7175 die forgings of the same configuration (Ref 6.18)

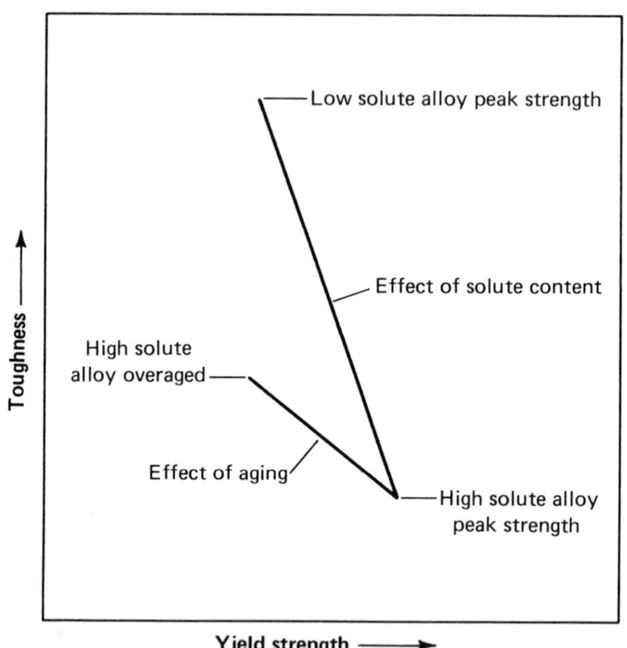

Fig. 6.12. Effects of decreasing solute content and over-aging on the toughness–yield strength relationship for 7000 series aluminum alloys (Ref 6.15)

strength level is achieved by modifying the composition to achieve the desired strength level with peak aging for that composition.

Quenching rate from solution heat treatment is also recognized as a key factor (Fig. 6.13; Ref 6.18) in providing the most advantageous precipitate morphology. Grain size is another key factor; finer grain size favors transgranular fracture with high-energy absorption, as indicated by the data in Fig. 6.14 for the 7475 alloy.

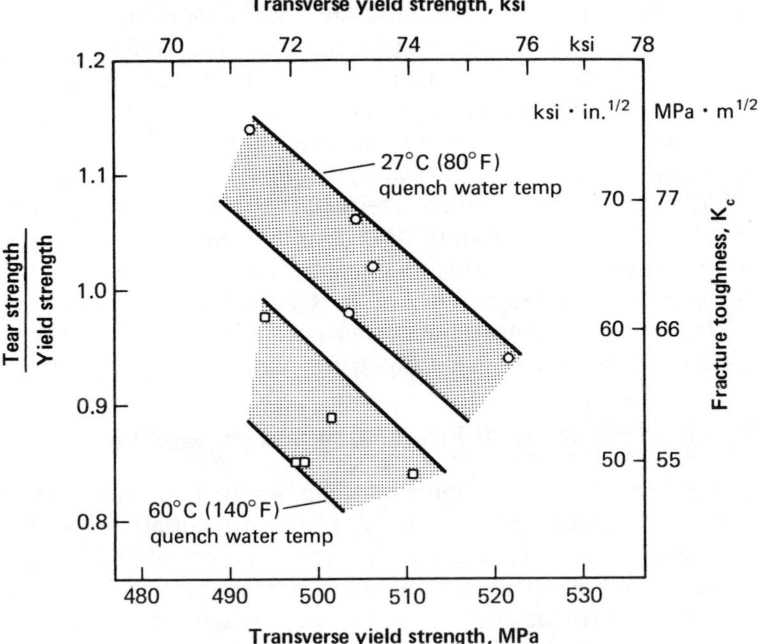

Fig. 6.13. Effect of quench water temperature on the tear strength–yield strength ratio for 7075 sheet (Ref 6.18)

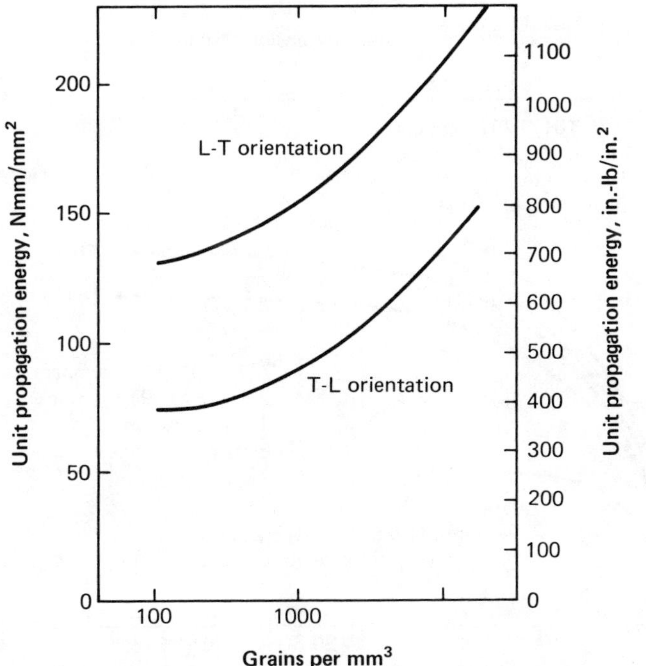

Fig. 6.14. Effect of recrystallized grain size on unit crack propagation energy values for 7475-T6 sheet (Ref 6.18)

Alloy 7475 represents one of the most successful applications of alloy design techniques. Its composition and properties are modified from those of alloy 7075 by (*a*) reducing iron and silicon contents, (*b*) optimizing dispersoids, (*c*) altering precipitates, (*d*) controlling quenching rate and (*e*) controlling grain size. These modifications result in the toughest aluminum alloy available commercially at high strength levels. Crack resistance curves from several investigators (Fig. 6.15; Ref 6.18) further emphasize the improvement that may be made in toughness by reducing the impurity levels of alloy 7075. For designers, this influence is shown most clearly by information on crack lengths for unstable crack growth at specific design stresses such as that shown in Fig. 6.16 (Ref 6.18). The crack tolerance of the 7475-T761 alloy is almost three times greater than that of conventional 7075-T6. Similar effects have been noted for plate.

6.2.2. Effects of Mechanical Processing and Orientation

Mechanical processing contributes to the microstructural control used to improve toughness, as indicated above, but there are certain fabrication formats for aluminum alloys which lead to structures and degrees of grain orientation that are beneficial, as discussed in the following paragraphs. The data presented in Table 6.2 (Ref 6.6) provide some evidence of how toughness varies with orientation and product type.

Orientation

Data for plate of four different alloy/temper combinations (Ref. 6.20) in Fig. 6.17 illustrate how toughness differs substantially depending on whether

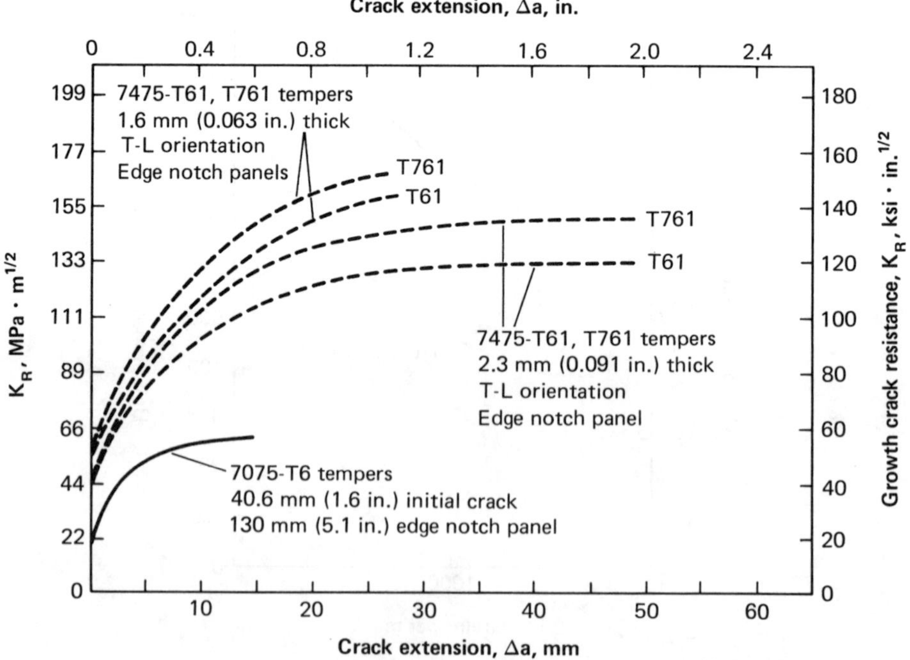

Fig. 6.15. Crack growth resistance curves for 7075 and 7475 sheet (Ref 6.18)

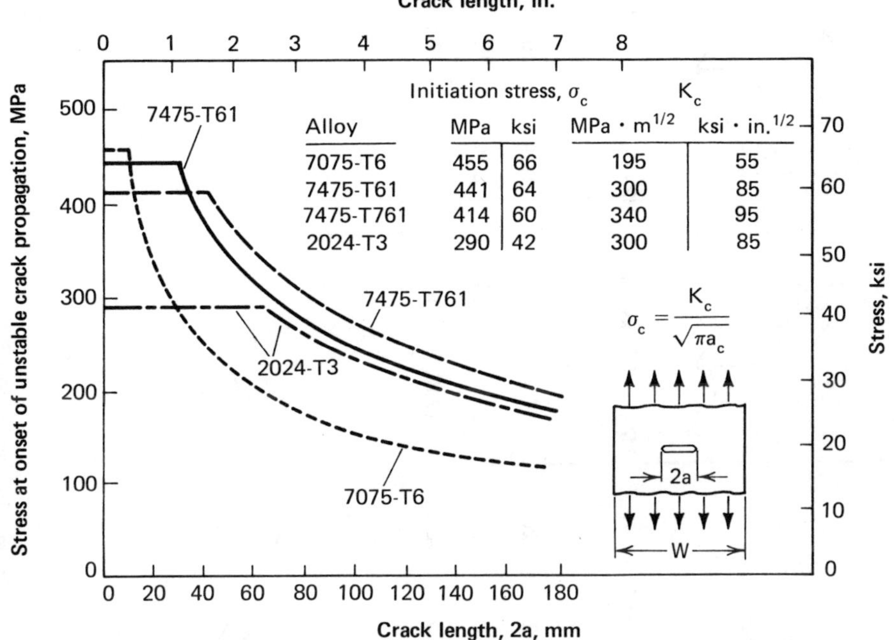

Fig. 6.16. Gross section stress at initiation of unstable crack propagation vs crack length for wide sheet panels of four aluminum alloy/temper combinations (Ref 6.18)

principal stresses are parallel or perpendicular to the major plane of the product. With stress perpendicular to the plane of the product, and crack growth on the natural planes which result from rolling of the plate product (i.e., the S-L and S-T orientations), K_{Ic} values may be only two-thirds as great as those for other orientations. These differences are minimized in high-toughness alloys like 2124 and 7475, but significant variation may still exist. Therefore, it is important to identify toughness numbers with crack-plane orientations as defined in ASTM Method E399 and in Fig. 3.2.

Product Form

For a given alloy and temper, values of K_{Ic}, like other toughness indicators, may be fairly consistent from product to product, but greater orientation effects are commonly noted for extrusions than for plate and forgings because of the more extreme grain flow patterns in extrusions. For die forgings, it is frequently difficult if not impossible to obtain valid K_{Ic} values because of the restrictions in product dimensions. In interpretation of data, strict adherence to ASTM requirements is needed to avoid improper judgements of product-to-product variations.

Product Size

Because of the effects of such factors as grain size, grain orientation and quenching rate on toughness, relatively thick plate, forgings and extrusions frequently are lower in strength and toughness than thinner items of the same alloy and temper (Ref 6.21). It is useful to consider those aluminum alloys especially

	L-T		L-S		T-L		T-S		S-L		S-T	
	MPa · m$^{1/2}$	ksi · in.$^{1/2}$	MPa · m$^{1/2}$	ksi · in.$^{1/2}$	MPa · m$^{1/2}$	ksi · in.$^{1/2}$	MPa · m$^{1/2}$	ksi · in.$^{1/2}$	MPa · m$^{1/2}$	ksi · in.$^{1/2}$	MPa · m$^{1/2}$	ksi · in.$^{1/2}$
2014-T651	21.1	19.2	26.4	24.0	21.2	19.3	24.3	22.1	17.8	16.2	18.4	16.7
2024-T851	23.1	21.0	24.6	22.4	20.5	18.7	22.2	20.2	17.9	16.3	17.4	15.8
7075-T651	30.7	28.0	30.2	27.5	26.6	24.2	28.8	26.2	21.1	19.2	20.4	18.6
7079-T651	29.7	27.0	30.6	27.9	26.3	23.9	27.0	24.6	17.8	16.2	18.8	17.1

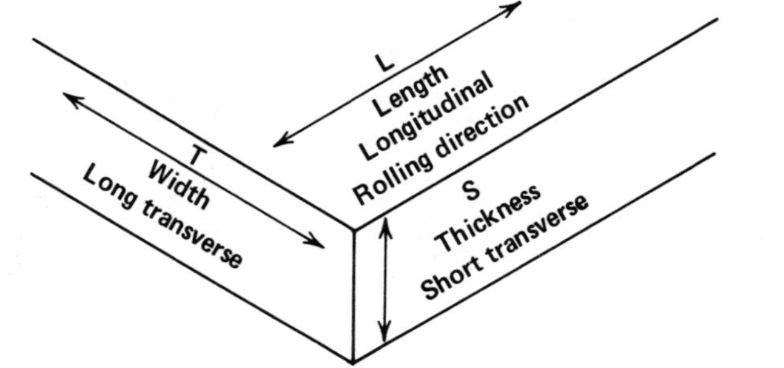

Fig. 6.17. Influence of specimen orientation on K_{Ic} for aluminum alloy plate (Ref 6.21)

designed for thick-section use, such as 7050 (Ref 6.22), when both strength and toughness are critical in thick structural components. Even for a given sheet thickness, however, variations in toughness can be observed as a result of the variations in microstructure which may be obtained, as illustrated in Fig. 6.18 (Ref 6.23) for a range of sheet thicknesses.

6.2.3. Effect of Temperature

The relationship between fracture toughness and temperature for aluminum alloys normally follows that derived from smaller-scale notch tensile and tear tests. As illustrated by the data in Fig. 6.19 (Ref 6.24), K_{Ic} values for 2000, 6000 and 7000 series alloys are at least as high at subzero temperatures as at room

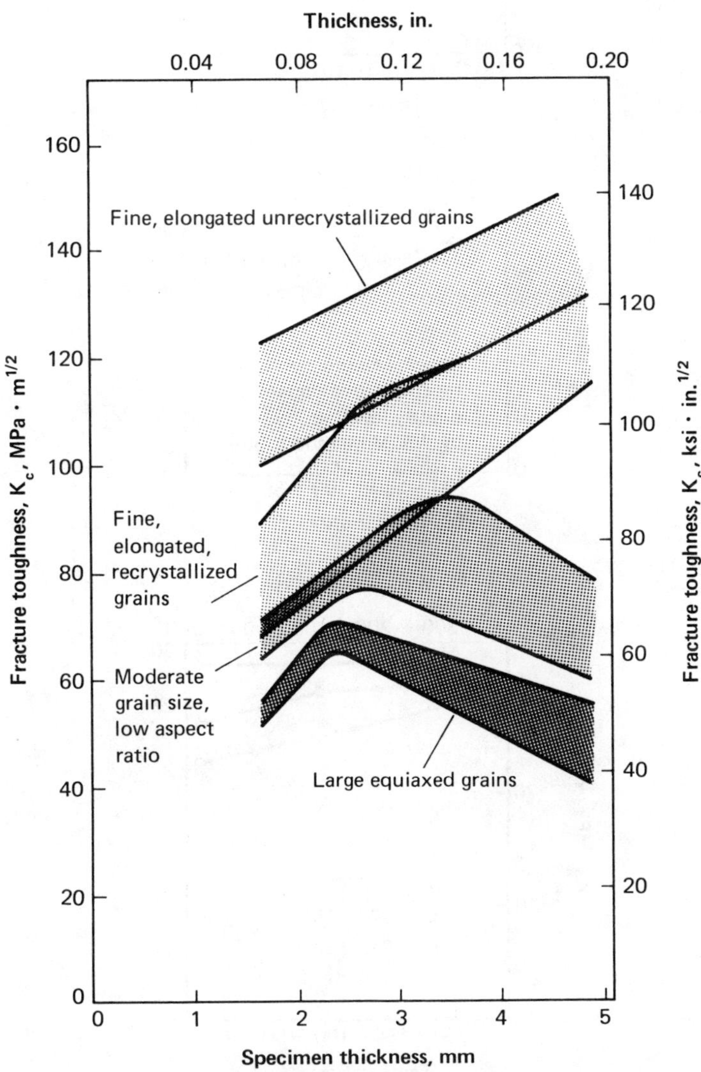

Fig. 6.18. Effect of sheet thickness and microstructure on fracture toughness of over-aged 7000 series aluminum alloys (Ref 6.23)

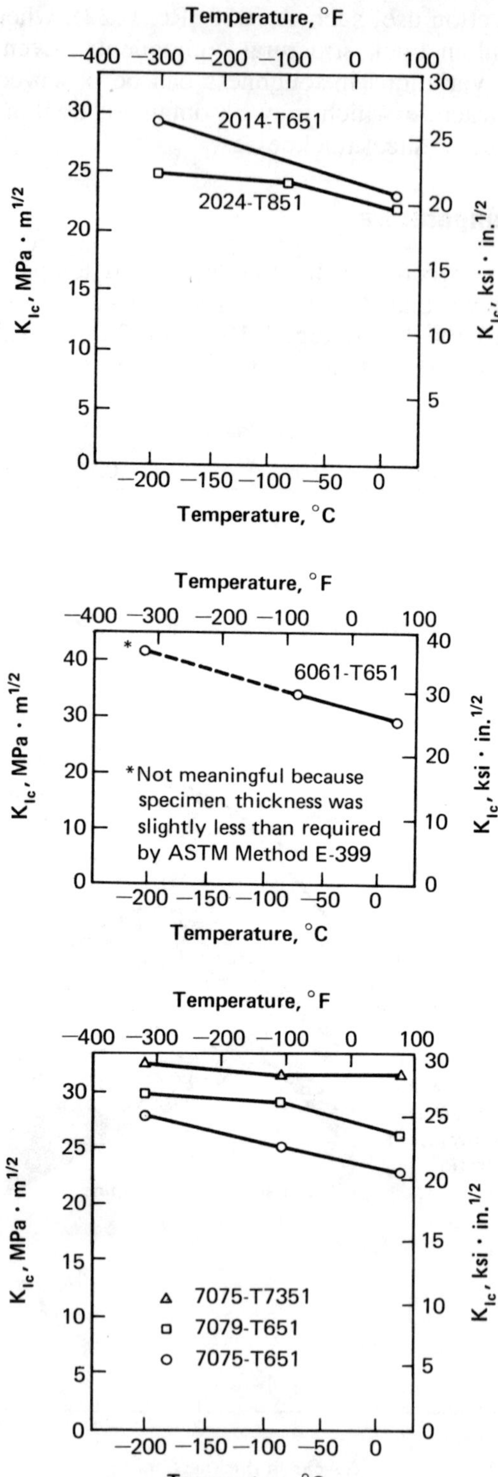

Fig. 6.19. K_{Ic} vs temperature for 2000, 6000 and 7000 series aluminum alloys (Ref 6.24)

temperature. Valid tests have not been made for alloys of the 5000 series, but indications are that the same trend is present to an even greater extent, although the toughness may level off at temperatures below liquid nitrogen temperature (Ref 6.25). Data from notch tensile and tear tests at liquid helium temperature (−269°C, or −452°F) suggest that some reduction in toughness may occur at temperatures below −196°C (−320°F); but even there, toughness is likely to remain well above room temperature values.

Few K_{Ic} data are available for temperatures above room temperature, but limited testing of alloy 2219-T851 (Ref 6.26) suggest that trends present in tear test data are indicative of those in K_{Ic} data. With increasing temperature, toughness gradually increases for most alloys, the only exceptions being those which are not aged to peak strength prior to high-temperature exposure.

6.3. FATIGUE CRACK PROPAGATION

Considerable use has been made of the fracture mechanics approach in measurement of fatigue crack growth rates in aluminum alloys (Ref 6.20 and 6.27 to 6.29). These data have been generated by methods comparable to those of ASTM Method E647 for measuring fatigue crack growth rates. In general, fatigue crack growth rates are found to fall within a relatively narrow scatter band, with only small systematic effects of composition, fabricating practice or strength level, as illustrated by the data in Fig. 6.20 (Ref 6.29).

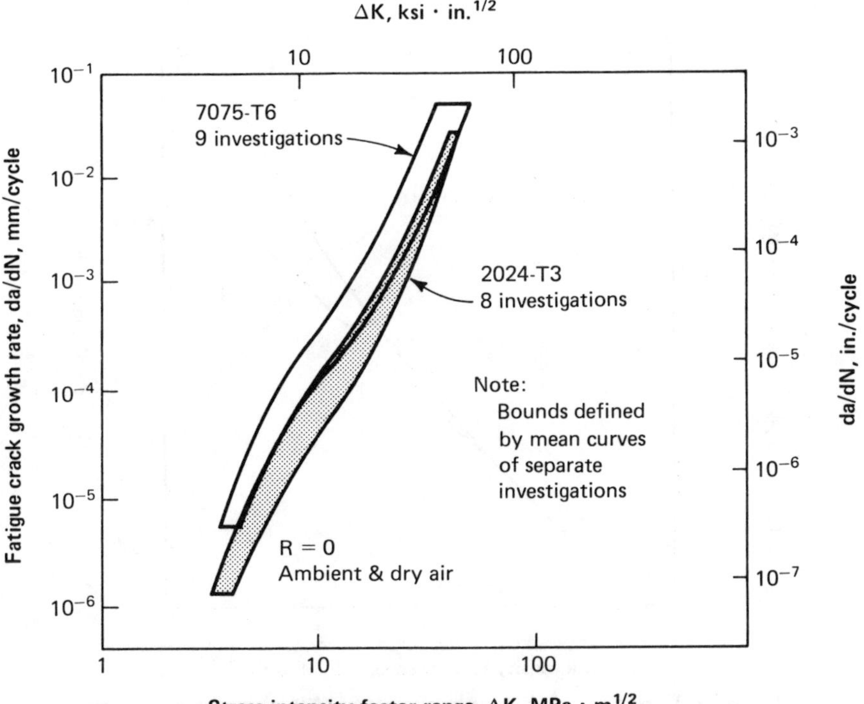

Fig. 6.20. Summary of fatigue crack growth rate data for aluminum alloys 7075-T6 and 2024-T3 (Ref 6.29)

6.3.1. Effects of Composition, Microstructure and Thermal Treatments

There are many sources of fatigue crack growth rate data which show the effects of various physical and microstructural variables on fatigue life of aluminum alloys (Ref 6.20 and 6.27 to 6.40), but there is little agreement on the key variables and there are few significant approaches for improving the fatigue crack growth resistance of these alloys. However, some generalizations can be made (Ref 6.41 and 6.42). Metallurgical microstructures which distribute plastic strain and avoid strain concentration can be produced. In this regard, the presence of a minimum number of initiation sites and fine dispersoids which will pin dislocations should be helpful. The effects of various types of particles on fatigue resistance have been broadly explored, but the specific effects of grain structure, shape, morphology, and degree of recrystallization have not been well defined. However, those metallurgical factors which contribute to increased fracture toughness (see Section 6.2) also generally contribute to increased resistance to fatigue crack propagation at relatively high ΔK levels. For example, as illustrated in Fig. 6.21 (Ref 6.6), at low stress intensities the fatigue crack growth rates for 7475 are about the same as those for 7075. However, the factors that contribute to the higher fracture toughness of 7475 also contribute to the retardation of fatigue crack growth, resulting in two or more times slower growth for 7475 than for 7075 at ΔK levels equal to or greater than about 16 MPa · $m^{1/2}$ (15 ksi · $in.^{1/2}$). A similar trend has been observed for 2124-T851, which exhibits slower growth than

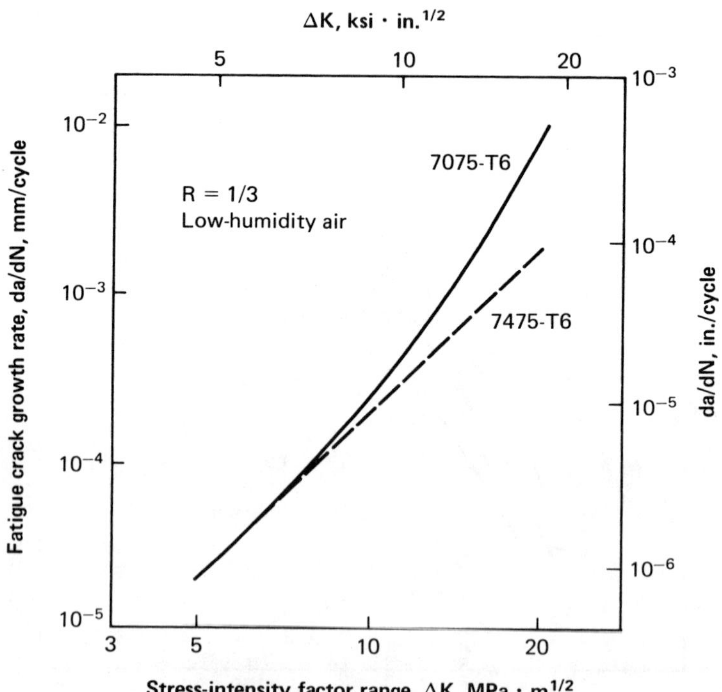

Fig. 6.21. Benefit of high-toughness alloy 7475 at intermediate and high stress intensity (Ref 6.6)

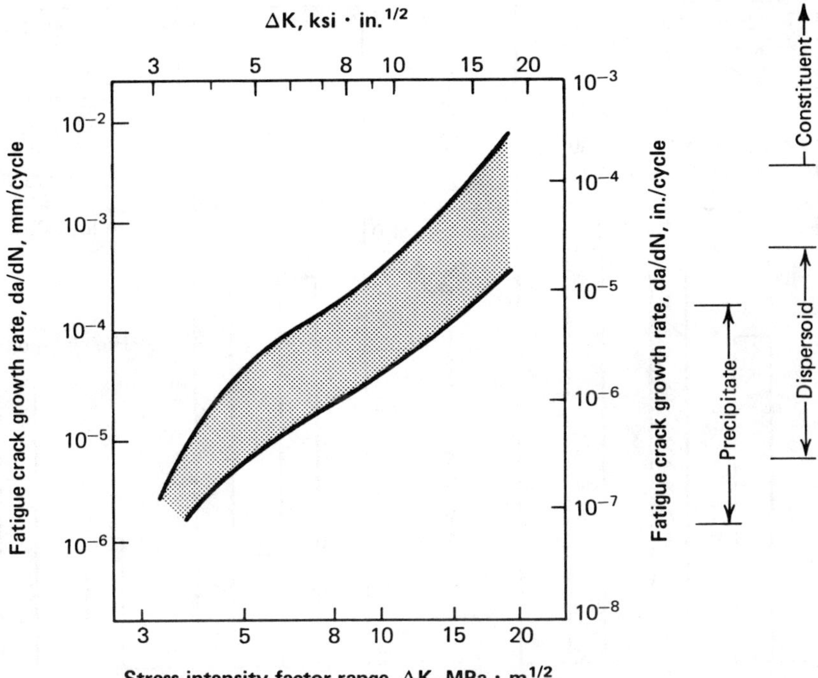

Fig. 6.22. Comparison of typical particle sizes in aluminum alloys with crack advance per cycle on fatigue loading (Ref 6.43)

2024-T851. Smooth specimens of alloys 2024 and 2124 exhibit quite similar fatigue behavior. Because fatigue in smooth specimens is dominated by initiation, this suggests that the large insoluble particles may not be significant contributors to fatigue crack initiation. However, once the crack is initiated, crack propagation is slower in material with relatively few large particles (2124) than in material with a greater number of large particles (2024).

Staley (Ref 6.43) has summarized the role of particle size in influencing fatigue crack growth in aluminum alloys, as shown in Fig. 6.22. The influence of alloy composition on dispersoid effect is shown in Fig. 6.23. The general trend in Fig. 6.23 is that for more finely dispersed particles the fatigue crack propagation life is increased. Whereas dispersoid type appears to have a relatively small effect on mean calculated life, the smaller precipitates provided by aging produce a much larger effect. There is some evidence that new processing practices may provide the fine microstructures needed to enhance fatigue resistance. The potential of intermediate working (commonly referred to as ITMT treatments) remains attractive but has not been proven for notched specimens (Ref 6.44, 6.45 and 6.46).

More promising results have been obtained with products produced from atomized aluminum powder, such as improvements in smooth specimen fatigue life (Ref 6.47, 6.48 and 6.49) and in notched specimen fatigue life. These alloys also offer attractive new strength levels along with excellent stress-corrosion resistance and toughness.

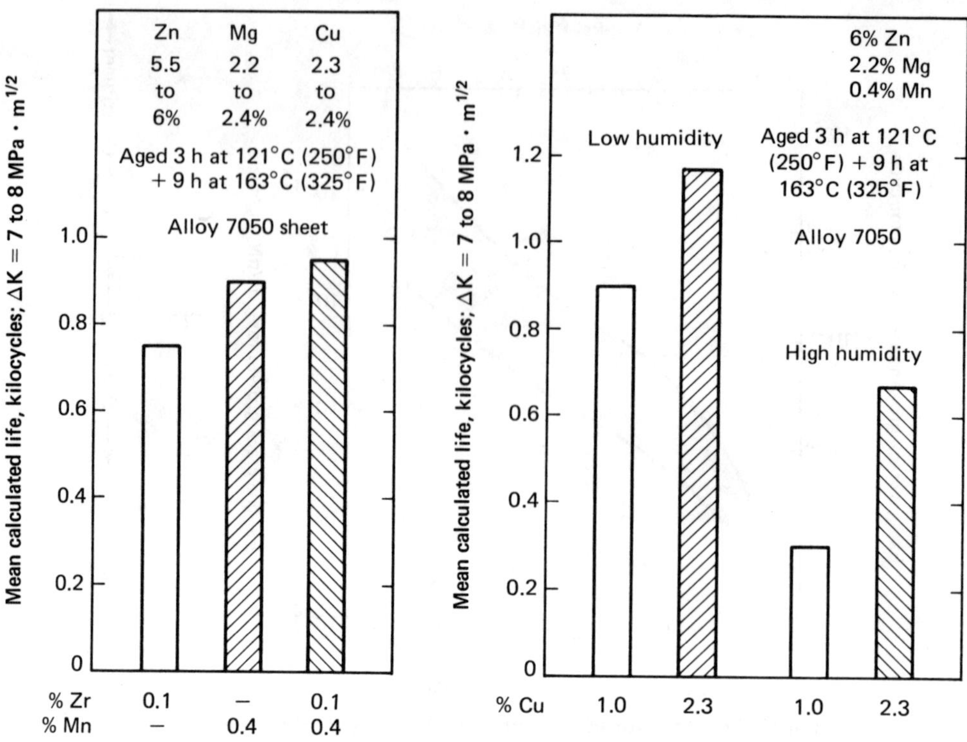

Fig. 6.23. Effect of dispersoid type (based on composition) on fatigue crack propagation life of 7050 alloy sheet (Ref 6.43)

6.3.2. Effects of Product Form and Orientation

The rate of fatigue crack propagation in aluminum alloys is relatively insensitive to product form and orientation. This is illustrated in Fig. 6.24 for thick 5083-O plate that was evaluated for use in tankage for liquified natural gas; growth rates in specimens from four orientations were well within the range for replicate tests in any one orientation (Ref 6.50). It would be expected that in the high ΔK range, differences in growth rate would reflect differences in toughness, and therefore would indicate somewhat higher growth rates for stressing normal to the plane of the product—that is, in the S-L and S-T orientations. Product thickness also seems to have a small effect, as illustrated by the data for 5083-O in Fig. 6.25 (Ref 6.50).

6.3.3. Effects of Exposure Temperature and Humidity

Although no significant amount of fatigue testing has been done at temperatures above room temperature, there has been a great amount of testing at subzero temperatures—particularly at −196°C (−320°F). In general, fatigue crack growth rates below room temperature are about the same as or lower than those at room temperature (Fig. 6.26). The role of humidity seems to be much more

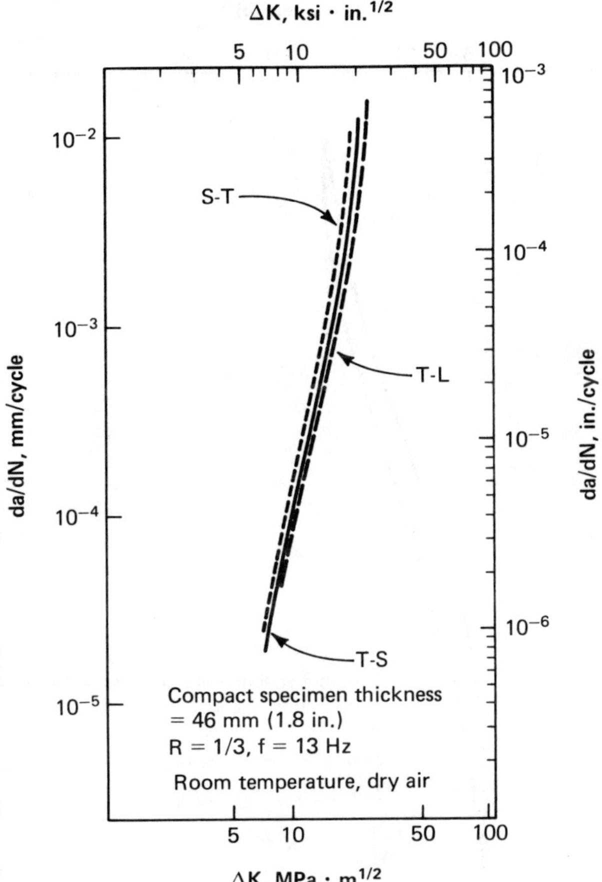

Fig. 6.24. Effect of orientation on fatigue crack growth rates in 180- and 196-mm (7.0- and 7.7-in.) 5083-O plate (Ref 6.50)

important than that of temperature. As shown in Fig. 6.26, growth rates for alloy 5083-O are appreciably higher in moist air than in dry air (Ref 6.50). Growth rates in water solutions of sodium chloride are similar to those in moist air.

Data for Alclad 7075-T6 in Fig. 6.27 (Ref 6.51) illustrate that even relatively low levels of moisture can accelerate crack growth rates. Unless relative humidity is below 3 to 5%, it seems best to consider that accelerated fatigue crack growth rates are likely in service.

6.3.4. Effect of Load-Time History

Selection of the proper type of load cycle to be used in evaluating fatigue crack propagation rates of aluminum alloys has been found to be particularly critical. Staley (Ref 6.43) and Bucci *et al* (Ref 6.52) noted that the variable amplitude crack propagation test offers the following advantages: (*a*) increased sensitivity to microstructural effects; (*b*) relevancy to design through consideration of important

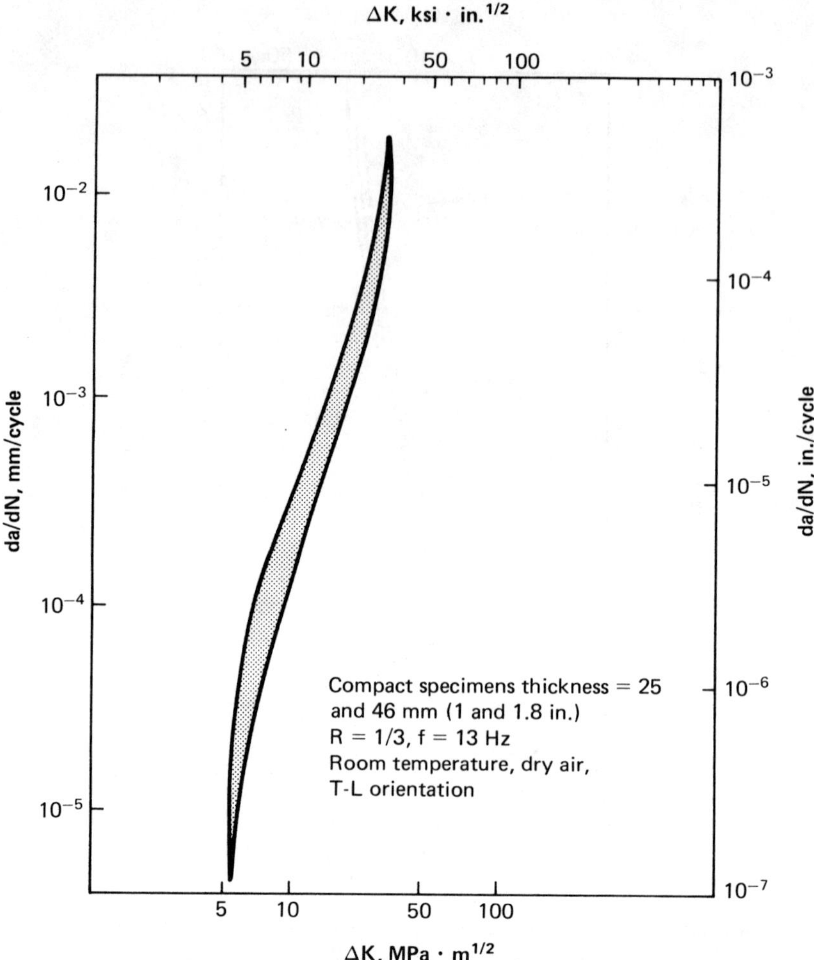

Fig. 6.25. Fatigue crack growth rates in 5083-O plate in thicknesses of 25, 70 and 178 mm (1, 1.8 and 7 in.) (Ref 6.50)

effects of load history, especially overload/plastic zone interactions; (*c*) practical interpretation when results are expressed in terms of crack size versus fatigue life; and (*d*) suitability for automated testing and analysis. Bucci further indicated that there are important interactions between microstructure and crack growth under variable amplitude cyclic loading that are not accounted for in constant amplitude testing. Ratings of alloys based on constant amplitude testing are not very likely to provide realistic indications of fatigue performance under the usual service-type variable amplitude loading. The primary cause of this difference is that variable amplitude loading includes the effects of crack growth retardation on fatigue crack propagation—effects that cannot be demonstrated in constant amplitude testing. Crack growth retardation is caused by tension overloading during fatigue testing. The variable amplitude test is believed to be more sensitive to alloy difference, and it clearly provides more useful information for alloy development investigations.

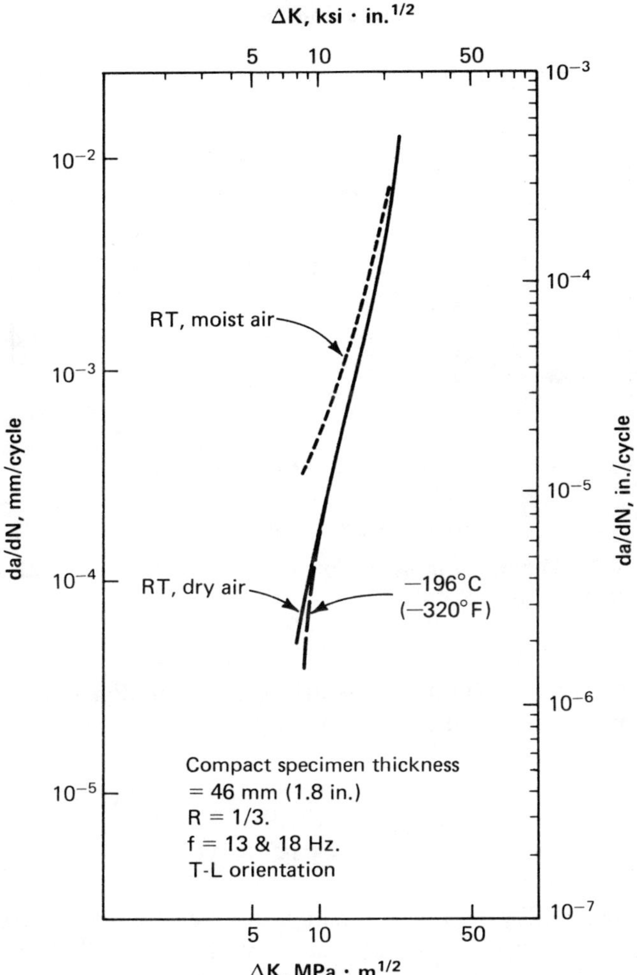

Fig. 6.26. Effect of temperature and humidity on fatigue crack growth in 180-mm (7.0-in.) 5083-O plate (Ref 6.50)

For example, as illustrated by the data for alloys 7075 and 7050 in Fig. 6.28 (Ref 6.53), quite different results are obtained in constant amplitude tests than in tests with single overloads every 4000 or 8000 cycles. Thus, information on the variation in load level during fatigue cycling is required for correct characterization of the fatigue behavior of aluminum alloys.

6.3.5. Effect of Load Ratio, R

The effect of load ratio on fatigue crack growth rate (load ratio, R, is the ratio of minimum to maximum load in the fatigue cycle) is well known. This effect is shown for alloy 7075-T6 sheet about 2.5 mm (0.1 in.) in Fig. 6.29 (Ref 6.54). In general, and in all the data presented here, an increase in R at the same ΔK level

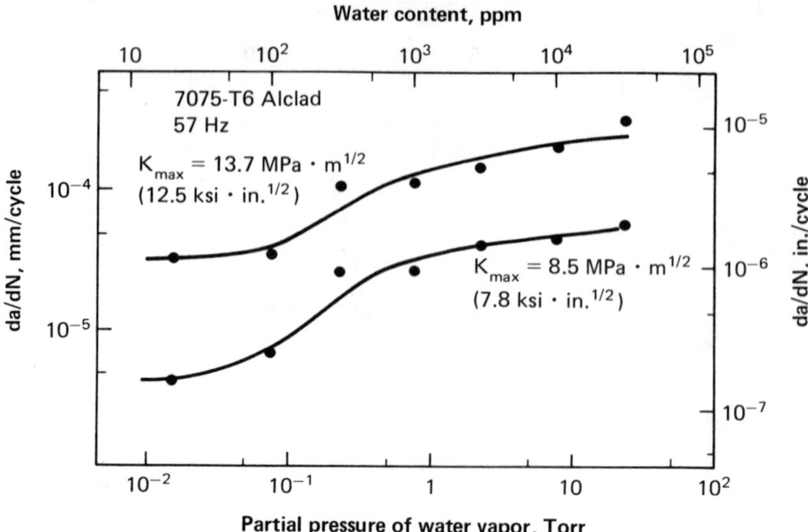

Fig. 6.27. Effects of moisture on fatigue crack growth rates in aluminum alloys (Ref 6.51)

causes an increase in growth rate. A modification of the Paris equation (Eq 2.16) that accounts for the effects of load ratio is the Forman equation (Eq 3.5), which bears repeating here:

$$\frac{da}{dN} = \frac{C(\Delta K)^n}{(1 - R)K_c - \Delta K} \quad \text{(Eq 6.1)}$$

For the data in Fig. 6.29, the Forman equation gives a good representation of the variation of da/dN with R for a wide range of crack growth rates.

6.4. SUSTAINED-LOAD CRACK GROWTH IN AGGRESSIVE ENVIRONMENTS

Fracture mechanics has been used in studies of stress-corrosion crack growth in aluminum alloys to determine crack growth rates as functions of time in various environments (see Fig. 6.30; Ref 6.55) and to compare various alloys in a given environment (see Fig. 6.31; Ref 6.55). The data in Fig. 6.32 (Ref 6.56) provide a summary of crack growth rates obtained for a series of aluminum alloys in a NaCl-dichromate-acetate solution designed to approximate the behavior of these alloys in various high-humidity environments, such as seacoast environments.

Before specific roles of composition and microstructure in determining crack growth behavior are considered, some general features of stress-corrosion crack growth will be discussed. First, note in Fig. 6.30 to 6.32 that stress-corrosion crack growth in aluminum alloys can be characterized by a two-part curve. One part is the plateau of stress-intensity-independent crack growth, which might be

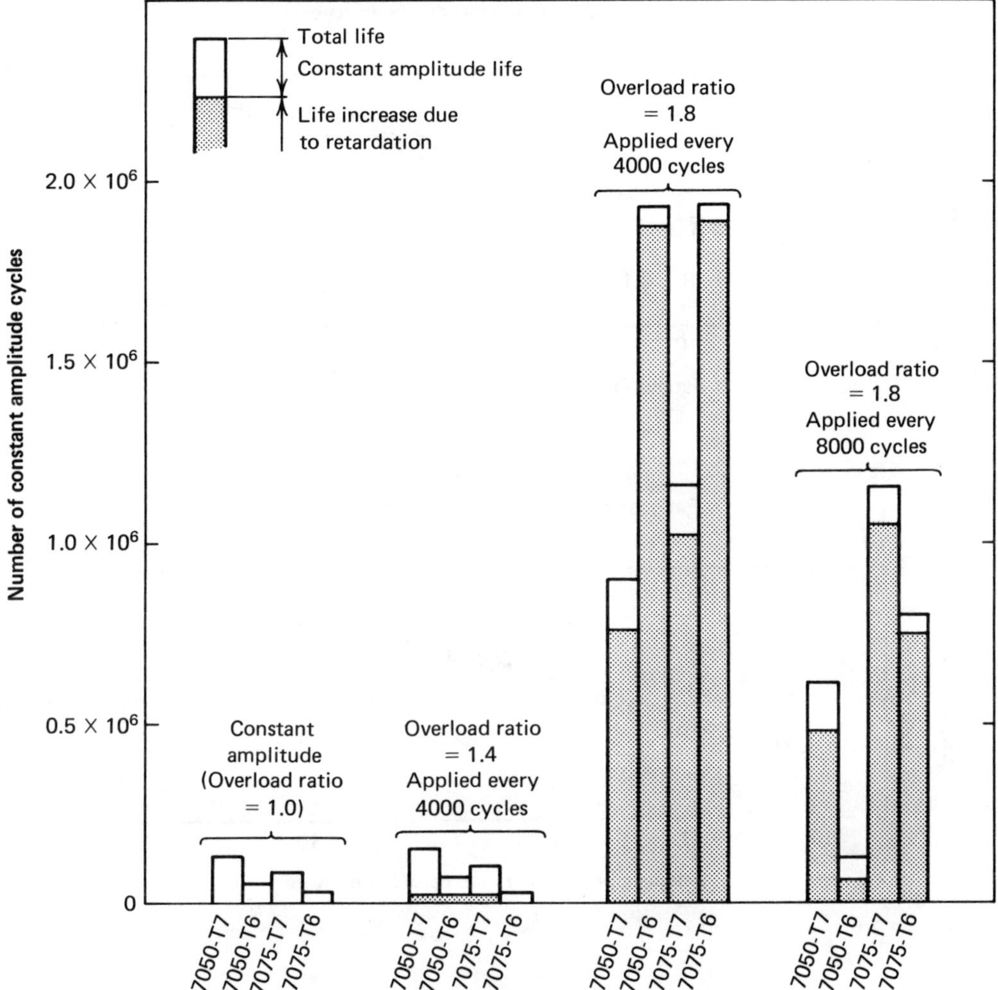

Fig. 6.28. Relative ranking of fatigue life of 7075 and 7050 aluminum alloys under constant amplitude and periodic single overload conditions (Ref 6.53)

useful in life predictions where some stress-corrosion cracking is expected. The other part is the vertical portion indicative of the threshold level below which the alloys may presumably be safely stressed without fear of subcritical crack growth. For aluminum alloys, considerable caution is called for in identifying this "threshold" value of K_{Iscc}, because development of crack growth has been observed after hundreds of hours of incubation under stress. As a guideline, observed crack growth rates should be less than 10^{-10} m/s (4×10^{-9} in./s) before it is assumed that threshold behavior is in effect.

The emphasis of this discussion is on the fracture mechanics approach to stress-corrosion behavior. However, experience to date has indicated that it is

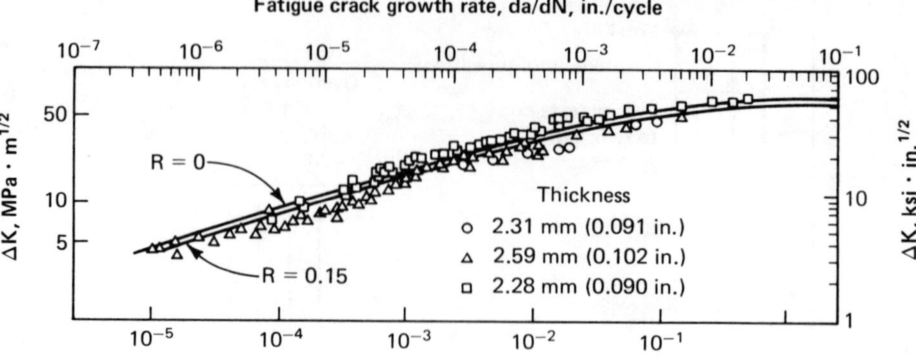

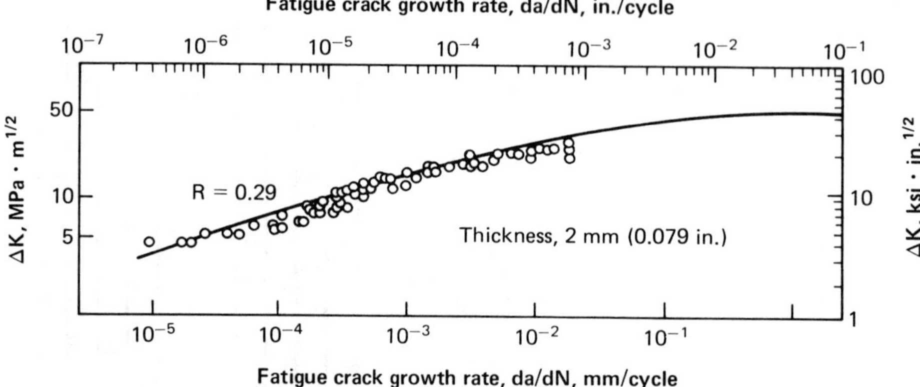

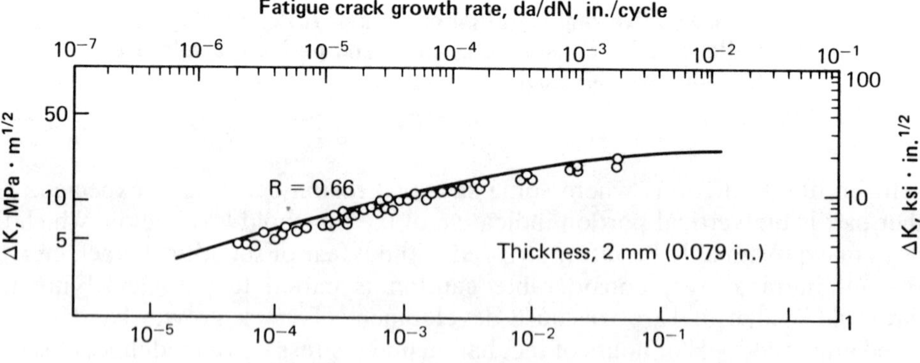

Fig. 6.29. Effect of load ratio, R, on fatigue crack growth rates in aluminum alloy 7075-T6 (Ref 6.54)

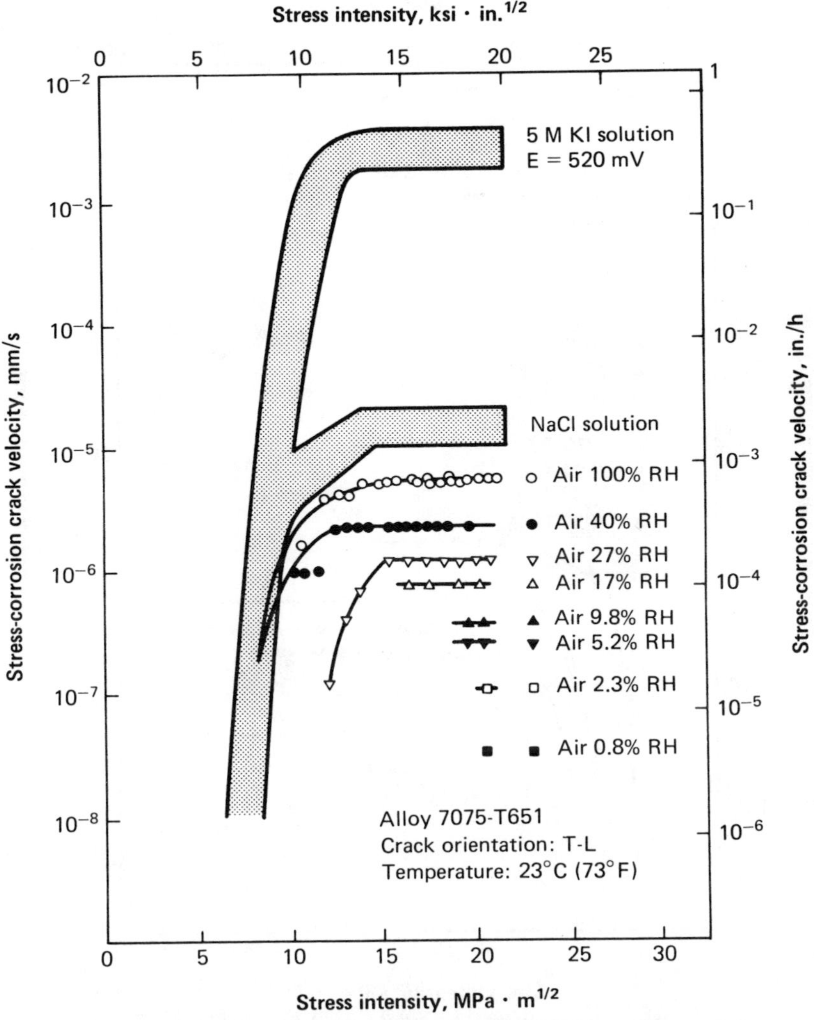

Fig. 6.30. Effect of humidity and stress intensity on stress corrosion crack velocity in aluminum alloy 7075-T651 (Ref 6.55)

unsatisfactory to assume that the fracture mechanics approach will always provide a conservative assessment of the stress-corrosion behavior of aluminum alloys (Ref 6.57 and 6.58). When data from smooth and precracked specimens are shown together, as in Fig. 6.33 (Ref 6.19), two things are apparent. First, cracks will initiate and grow rapidly to complete fracture at stresses well below those that would be predicted from data for precracked specimens alone. Smooth specimen data correctly demonstrate that cracks will initiate and propagate to complete failure in a relatively short period of time. On the other hand, the smooth specimen data would be overly conservative for predicting the relationship between stress and crack growth rate as crack length increases. Consideration of both types of

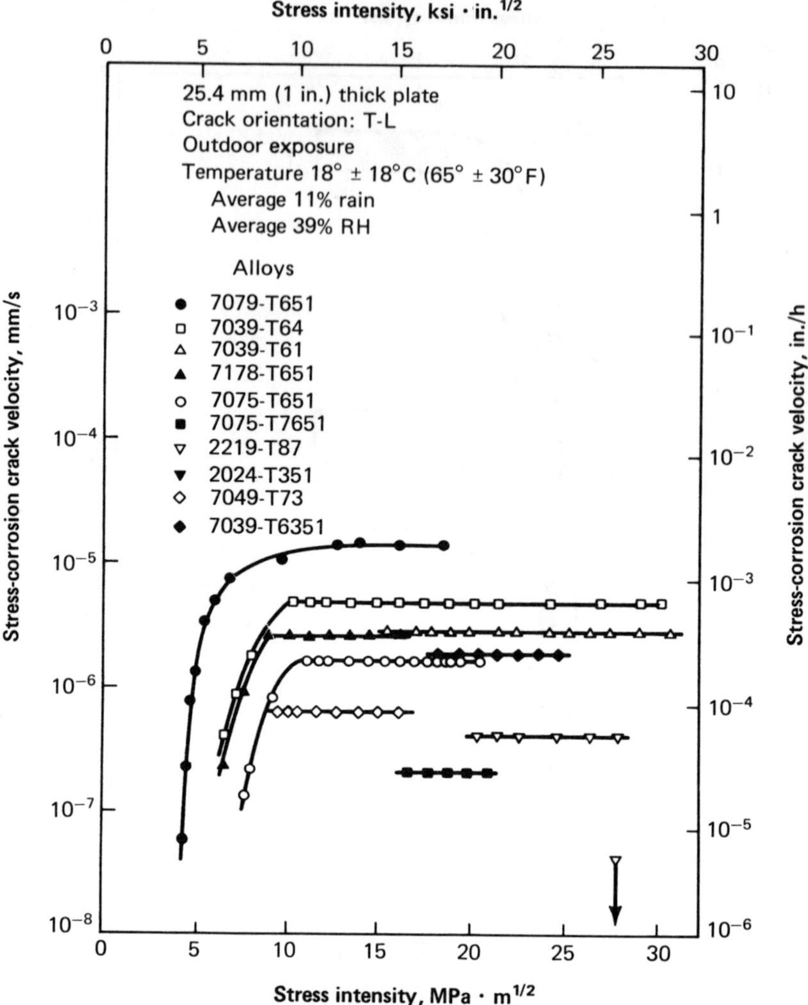

Fig. 6.31. Effect of outdoor exposure and stress intensity on stress corrosion crack velocity in several high-strength aluminum alloys (Ref 6.55)

data using "safe zone analysis" plots (see Fig. 6.33) is a critical part of any stress-corrosion susceptibility evaluation.

6.4.1. Effects of Composition, Microstructure and Thermal Processing

Subcritical crack growth in the form of stress-corrosion cracking (SCC) has been observed in aluminum alloys in the Al-Au, Al-Cu, Al-Mg, Al-Zn, Al-Cu-Mg, Al-Mg-Si, Al-Mg-Zn and Al-Mg-Zn-Cu series; it is generally limited to instances where the stresses are normal to the plane of the product—that is, in the short transverse orientations, S-L and S-T. It is not possible to make many

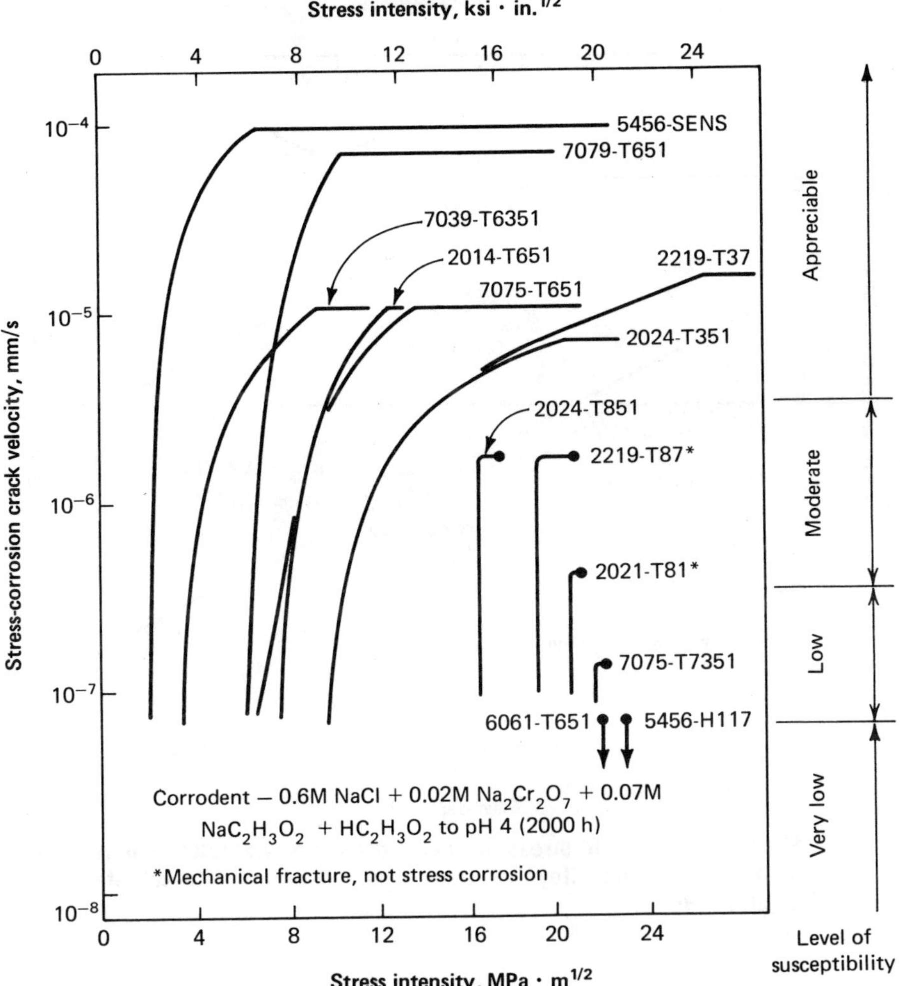

Fig. 6.32. Stress corrosion crack velocity curves for several aluminum alloys, showing various plateau levels and various apparent K_{Iscc} levels (Ref 6.56)

systematic generalizations about the effects of individual alloying elements, but some comments about the effects of individual elements can be made regarding the type of subcritical crack growth observed in aqueous solutions:

1. Susceptibility to SCC growth increases with the amounts of alloying additions, although in many instances the degree of susceptibility is affected as much or more by the relative amounts of the elements present as by their absolute amounts.
2. Additions of dispersoid-forming elements such as chromium, manganese, zirconium, titanium and vanadium tend to reduce the amount of subcritical crack growth that takes place, probably because of the finer structures which result.

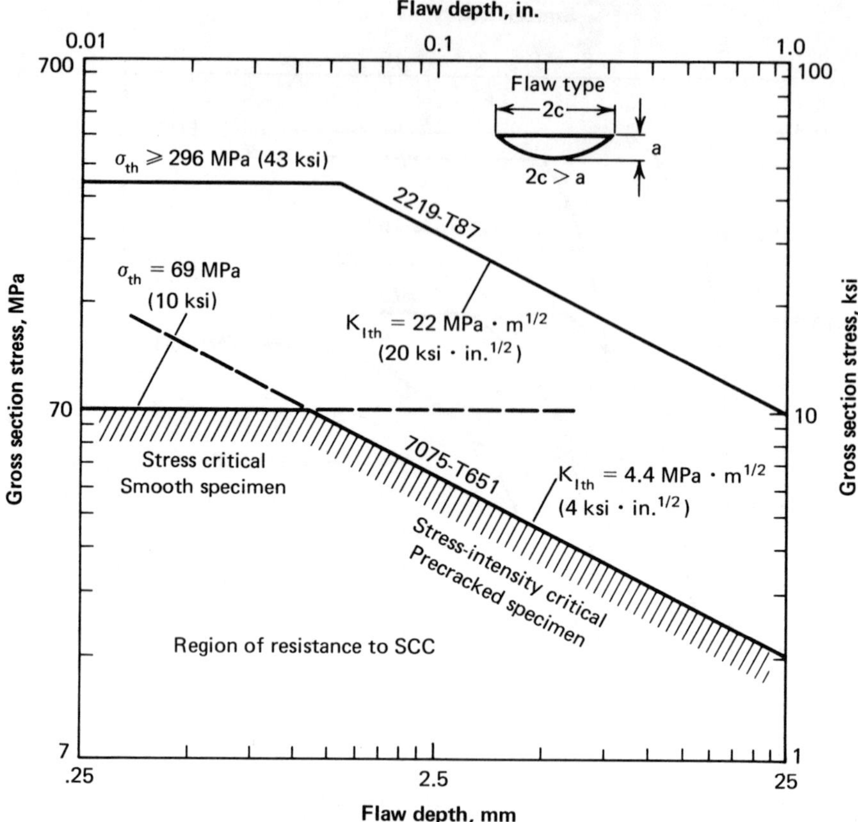

Fig. 6.33. Stress and stress-intensity threshold characterizations for two aluminum alloys exposed in a salt-dichromate-acetate solution (Ref 6.18)

Regarding thermal practices, precipitation hardening treatments of aluminum alloys are very effective in controlling resistance to crack growth. Those alloys which exhibit SCC are least resistant to crack growth in the freshly quenched condition, but aging can significantly increase their resistance. The greatest resistance is observed in the Al-Zn-Mg and Al-Zn-Mg-Cu alloys: when they are aged beyond peak hardness (example: 7075-T7351 vs 7075-T651 in Fig. 6.32), essentially SCC-free behavior can be obtained. Speidel (Ref 6.59) has illustrated an inverse interaction between yield strength and SCC resistance (Fig. 6.34). However, there is no general correlation for aluminum alloys between strength and SCC resistance (Fig. 6.35; Ref 6.60) or between toughness and SCC resistance.

Despite the ability to categorize alloy behavior by alloy and thermal treatment, it is still difficult to define the specific metallurgical mechanisms which control crack growth. The nature of SCC in aluminum alloys is basically that of corrosion occurring in grain boundaries as a result of differences in electrochemical potential between the grain boundary precipitates and the grain interior, with stress appar-

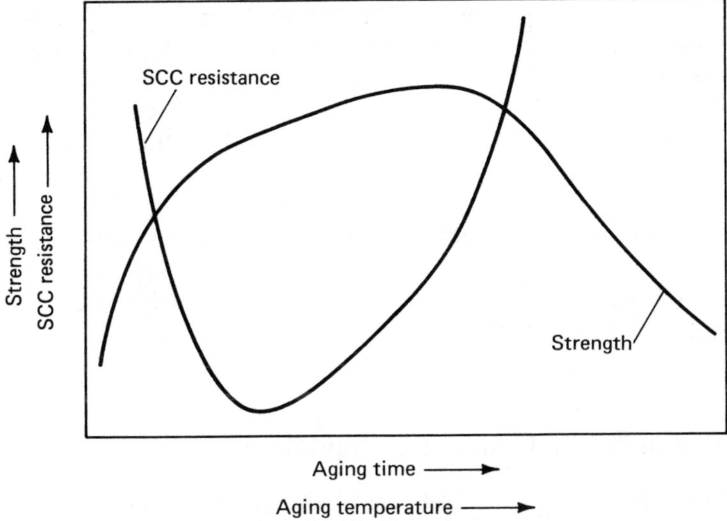

Fig. 6.34. Variations in strength and stress corrosion resistance during aging of precipitation hardening aluminum alloys (Ref 6.59)

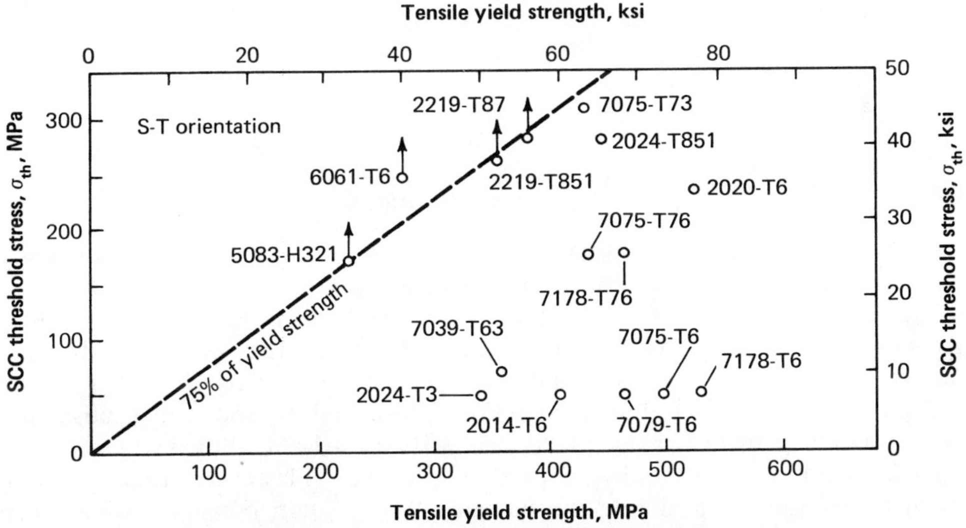

Fig. 6.35. Plot of SCC threshold stress vs tensile yield strength for a wide variety of aluminum alloys and tempers (3.5% NaCl alternate immersion) (Ref 6.60)

ently playing a role in determining the degree to which the grain boundary fissures are opened. Recent electrochemical and transmission electron microscopy studies strongly indicate that hydrogen embrittlement at grain boundaries may control SCC in high-strength aluminum alloys such as those of the Al-Mg-Zn series

(Ref 6.61 and 6.62). The precipitate structure in the grain itself is thought to be a factor by virtue of the degree to which it forces bands of high dislocation density to pile up against the grain boundaries, with the greatest resistance occurring in situations where precipitates cannot be readily sheared. There is no consensus on the role of the precipitate-free zone, and individual studies could be cited that would tend to predict that either enlarging or reducing the width of the zone would be detrimental.

There are questions about some of the basic metallurgical mechanisms of subcritical crack growth in aqueous solutions. Nevertheless, rather complete characterizations of most commercial alloys are available. Such characterizations are presented in Table 6.3 (Ref 6.6).

6.4.2. Effects of Product Form and Orientation

As indicated earlier, subcritical crack growth may occur at room temperature by stress corrosion cracking in certain aluminum alloys provided that stresses are applied in the short transverse direction. With that limitation, the major differences among the various product forms are differences in how each form affects the degree of "short transverseness" of the microstructure. For plate, for relatively wide planar extrusions and for forgings, subcritical crack growth by SCC is likely to be limited to the short transverse direction. In nearly round or square extrusions or rolled shapes of susceptible alloy/temper combinations, stress corrosion cracks may grow in directions parallel to the principal grain flow.

6.4.3. Effects of Temperature and Environment

Aqueous chloride-bearing solutions in combination with certain aluminum alloy compositions and grain structures are likely to produce substantial amounts of subcritical crack growth, by a stress corrosion mechanism, at room temperature. At temperatures above room temperature, subcritical crack growth under sustained loading occurs to a substantial degree only in the Al-Cu alloys. Work carried out on alloy 2219, illustrated in Fig. 6.36(a) and (b), shows that substantial "creep crack growth" takes place primarily in Al-Cu alloys where grain-boundary-locking dispersoids are present to prevent plastic deformation along grain boundaries by a sliding mechanism. Where grain boundary sliding can take place to accommodate reasonable levels of strain, creep crack growth is not found. In alloy 2219, for example, zirconium and vanadium dispersoids are present in the grain boundaries to provide higher strength, but, while prohibiting grain boundary sliding, they contribute to relatively low-strain cracking along those boundaries after exposure for only 100 hours.

Temperatures below room temperature seem to suppress all types of environmentally controlled subcritical crack growth observed in aluminum alloys. At subzero temperatures, fatigue crack propagation is far more likely to control most fracture events in aluminum alloys.

Table 6.3. Relative stress corrosion (SCC) ratings(a) for high-strength aluminum alloy products (Ref 6.6)

Alloy and temper(b)	Test direction(c)	Rolled plate(d)	Rod & bar(e)	Extruded shapes	Forgings
2011-T3, T4	L	···	B	···	···
	LT	···	D	···	···
	ST	···	D	···	···
2011-T8	L	···	A	···	···
	LT	···	A	···	···
	ST	···	A	···	···
2014-T6	L	A	A	A	B
	LT	B*(f)	D	B*(f)	B*(f)
	ST	D	D	D	D
2024-T3, T4	L	A	A	A	···
	LT	B*(f)	D	B*(f)	···
	ST	D	D	D	···
2024-T6	L	···	A	···	A
	LT	···	B	···	A*(f)
	ST	···	B	···	D
2024-T85xx	L	A	A	A	A
	LT	A	A	A	A
	ST	B	A	B	C
2048 and 2124-T851	L	A	···	···	···
	LT	A	···	···	···
	ST	B	···	···	···
2219-T3, T37	L	A	···	A	···
	LT	B	···	B	···
	ST	D	···	D	···
2219-T6, T85xx	L	A	A	A	A
	LT	A	A	A	A
	ST	A	A	A	A
6061-T6	L	A	A	A	A
	LT	A	A	A	A
	ST	A	A	A	A
7005-T53, T63	L	···	···	A	A
	LT	···	···	A*(f)	A*(f)
	ST	···	···	D	D
7039-T63, T64	L	A	···	A	···
	LT	A*(f)	···	A*(f)	···
	ST	D	···	D	···
7049-T735xx	L	A	···	A	A
	LT	A	···	A	A
	ST	A	···	B	B
7149-T735xx	L	···	···	A	A
	LT	···	···	A	A
	ST	···	···	B	B

(continued)

Table 6.3 (continued).

Alloy and temper(b)	Test direction(c)	Rolled plate(d)	Rod & bar(e)	Extruded shapes	Forgings
7049-T765xx	L	...	...	A	...
	LT	...	...	A	...
	ST	...	...	C	...
7050-T7365xx	L	A	...	A	A
	LT	A	...	A	A
	ST	B	...	B	B
7050-T765xx	L	A	...	A	...
	LT	A	...	A	...
	ST	C	...	C	...
7x75-T6	L	A	A	A	A
	LT	B*(f)	D	B*(f)	B*(f)
	ST	D	D	D	D
7x75-T735xx	L	A	A	A	A
	LT	A	A	A	A
	ST	A	A	A	A
7x75-T736	L	...	...	...	A
	LT	...	...	...	A
	ST	...	...	...	B
7x75-T765xx	L	A	...	A	...
	LT	A	...	A	...
	ST	C	...	C	...
7178-T765xx	L	A	...	A	...
	LT	A	...	A	...
	ST	C	...	C	...
7079-T6	L	A	...	A	A
	LT	B*(f)	...	B*(f)	B*(f)
	ST	D	...	D	D

(a) Ratings shall be based on service experience, if available, or on standard SCC tests (ASTM G47) as required by ASTM Materials Specifications. To rate a new material and test direction, tests shall be performed on at least 10 random lots, and the test results shall have 90% compliance at a 95% level of confidence for one of the following stress levels: A – OK at stresses up to and including 75% specified tensile yield strength; B – OK at stresses up to and including 50% specified tensile yield strength; C – OK at stresses up to and including 25% specified tensile yield strength; D – no specified test stress. (b) Ratings also apply to the TX5x and TX5xx (stress relieved) tempers where not listed. For 7x75 alloys the x denotes either 7075, 7175 or 7475 modifications.

(c) Test direction refers to orientation of the stressing direction *relative to the directional grain structure* typical of wrought materials, which in the case of extrusions and forgings may not be predictable from the geometrical cross section of the product. Refer to ASTM G47 for sampling procedures. L – longitudinal: parallel to direction of principal metal extension during manufacture of the product. LT – long transverse: perpendicular to direction of principal metal extension. In products whose grain structures clearly show directionality (width-to-thickness ratio greater than two), it is that perpendicular direction parallel to the major grain dimension. ST – short transverse: perpendicular to direction of principal metal extension and parallel to minor dimension of grains in products with significant grain directionality.

(d) Ratings for rolled sheet are the same as for rolled plate less than 40 mm (1.5 in.) thick. (e) Sections with width-to-thickness ratios equal to or less than two, for which there is no distinction between LT and ST. (f) Ratings marked with asterisks (*) are one class lower for thicker sections: extrusions, 25 mm (1 in.) and over; plate and forgings, 40 mm (1.5 in.) and over.

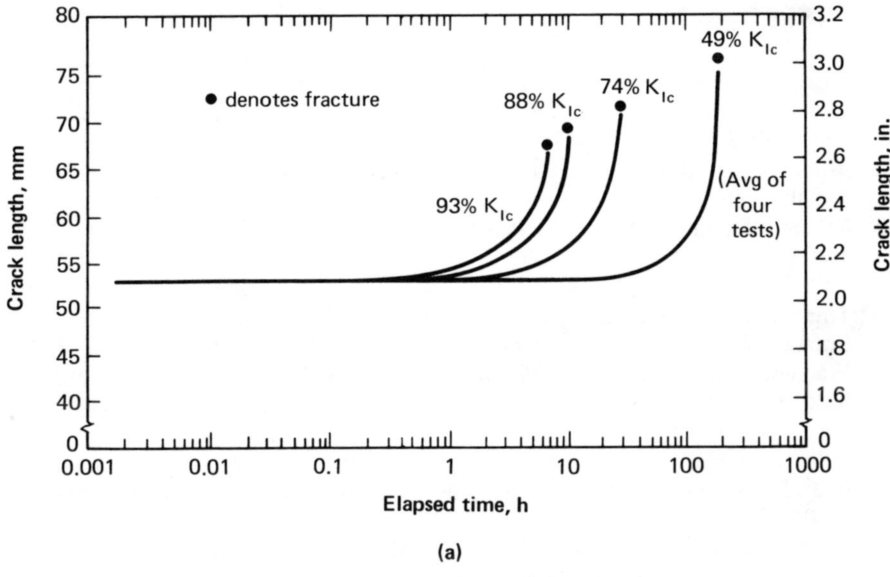

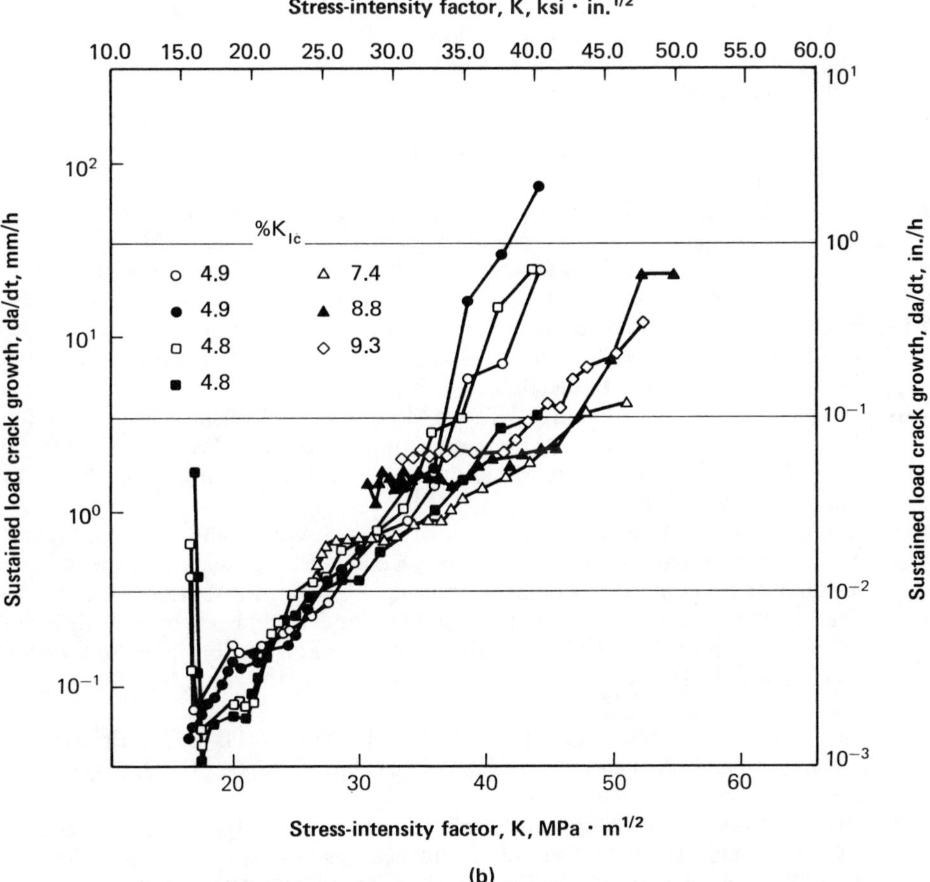

Fig. 6.36. Creep crack growth for 76-mm (3-in.) alloy 2219-T851 plate at 150°C (300°F) (Ref 6.26)

6.5. REFERENCES

6.1. Fatigue and Fracture Characteristics of Aluminum Alloys, by R. L. Moore, G. E. Nordmark and J. G. Kaufman: *Engineering Fracture Mechanics*, Vol 4, Pergamon Press, 1972, p 51-63

6.2. *Aluminum* (in three volumes), edited by Kent R. Van Horn: American Society for Metals, 1967

6.3. More on Specimen Size Effects in Fracture Toughness Testing, by J. G. Kaufman and F. G. Nelson: STP 559, American Society for Testing and Materials, Philadelphia, 1974, p 74-85

6.4. The Aluminum Association Position on Fracture Toughness Requirements and Fracture Toughness Testing: Document T5, The Aluminum Association, Sept 1974

6.5. Fracture Toughness of Aluminum Alloy Plate – Tension Tests of Large Center Slotted Panels, by J. G. Kaufman: *Journal of Materials*, ASTM, Vol 2, No. 4, 1967, p 889-914

6.6. Selecting Aluminum Alloys to Resist Failure by Fracture Mechanisms, by R. J. Bucci: *Engineering Fracture Mechanics*, Vol 12, Pergamon Press, 1979, p 407-441

6.7. Fracture Characteristics of Aluminum Alloys, by J. G. Kaufman and M. Holt: Technical Paper No. 18, Alcoa Laboratories, Alcoa Center, PA, 1965

6.8. High Strength Aluminum Alloys, by J. A. Nock, Jr., and H. Y. Hunsicker: *Journal of Metals*, Vol 15, No. 3, March 1963, p 216-224

6.9. The Fracture of Metals, by J. R. Low, Jr.: *Progress in Material Science,* Vol 12, Pergamon Press, 1963

6.10. The Mechanisms of Ductile Rupture of Metals Containing Inclusions, by J. Gurland and J. Plateau: *Transactions of the American Society for Metals*, Vol 56, 1963, p 442-454

6.11. On the Mechanics of Fracture from Inclusions, by F. A. McClintock: in *Ductility*, American Society for Metals, Metals Park, OH, 1968, p 524-527

6.12. Effects of Microstructure in Fracture Toughness of High Strength Alloys, by J. R. Low, Jr.: *Engineering Fracture Mechanics*, Vol 1, No. 1, June 1968, p 47-53

6.13. Investigation of the Plastic Fracture of High Strength Aluminum Alloys, by R. H. Van Stone, J. R. Low, Jr., and R. H. Merchant: STP 556, American Society for Testing and Materials, Philadelphia, 1974, p 93-124

6.14. Metallurgical Structure Control – Key to More Fracture Resistant Strong Aluminum Alloys, by H. Y. Hunsicker: Aluminum Company of America, Alcoa Center, PA, presented at the controlled Metallurgical Structures Conference, New York University, Jan 13-14, 1975

6.15. Fracture Toughness of Microstructure of High Strength Alloys, by J. T. Staley: Aluminum Company of America, Alcoa Center, PA, presented at the American Institute of Metallurgical Engineers' Spring Meeting, Pittsburgh, May 23, 1974

6.16. The Metallurgical Aspects of Fatigue and Fracture Toughness, by P. J. E. Forsyth: AGARD Report No. 610, Metallurgical Aspects of Fatigue and Fracture Toughness, NATO Advisory Group for Aerospace Research and Development, Dec 1973, p 1-22

6.17. The Fracture Toughness of Al-Zn-Mg-Cu-Mn Alloy OTD 5024, Effect of Iron Content, by C. J. Peel and P. J. E. Forsyth: Technical Report 70162, Royal Aircraft Establishment, Sept 1970

6.18. Design of Aluminum Alloys for High Toughness and High Fatigue Strength, by J. G. Kaufman: AGARD Conference Proceedings No. 185, Specialist Meeting on Alloy Design for Fatigue and Fracture Resistance, Brussels, April 1975

6.19. The Effect of Composition on the Fracture Properties of 7178-T6 Aluminum Alloy Sheet, by D. E. Piper, W. E. Quist and W. E. Anderson: Metallurgical Society Conference, Vol 31, 1966, p 227-280

6.20. Program to Improve the Fracture Toughness and Fatigue Resistance of Aluminum Sheet and Plate for Airframe Applications, by M. V. Hyatt: Report AFML-TR-73-224, Wright-Patterson Air Force Base, OH, 1973

6.21. Fracture Toughness of Aluminum Alloys, by J. G. Kaufman, P. E. Schilling and F. G. Nelson: *Metals Engineering Quarterly*, Vol 9, No. 3, Aug 1969, p 39-47

6.22. New Aluminum Alloy X7050, by J. T. Staley, H. Y. Hunsicker and R. Schmidt: Aluminum Company of America, Alcoa Center, PA, and Naval Air Systems Command, Washington, presented at the American Society for Metals' Metal Congress and Fall Meeting of TMS, AIME, Detroit, Oct 1971

6.23. Research on Synthesis of High Strength Aluminum Alloys, Part 1: Relation Between Precipitate, Microstructure, and Strength in Aluminum Alloys, by A. R. Rosenfeld, C. W. Price, C. J. Martin, D. S. Thompson and R. E. Zinkham: Report AFML-TR-74-129, Dec 1974

6.24. Plane Strain Fracture Toughness of Aluminum Alloys at Room and Subzero Temperatures, by F. G. Nelson and J. G. Kaufman: STP 496, American Society for Testing and Materials, Philadelphia, 1971, p 27-39

6.25. Tensile Properties and Notch Toughness of Aluminum Alloys at −425 F in Liquid Helium, by J. G. Kaufman, K. O. Bogardus and E. T. Wanderer: *Advances in Cryogenic Engineering*, Vol 13, Plenum Press, 1968, p 294-308

6.26. Creep Cracking of 2219-T851 Plate at Elevated Temperatures, by J. G. Kaufman, K. O. Bogardus, D. A. Mauney and R. C. Malcolm: STP 590, American Society for Testing and Materials, Philadelphia, March 1976, p 149-168

6.27. Effects of Constituent Particles on the Notch-Sensitivity and Fatigue-Crack-Propagation Characteristics of Aluminum-Zinc-Magnesium Alloys, by L. H. Glassman and A. J. McEvily, Jr.: NASA Technical Note, April 1962

6.28. The Effect of Intermetallic Particles on Fatigue Crack Propagation in Aluminum Alloys, by D. Broek: in *Fracture 1969*, Proceedings of the 2nd International Conference on Fracture, Chapman and Hall Ltd., London, 1969, p 754

6.29. A Review of Fatigue Crack Growth in High Strength Aluminum Alloys and the Relevant Metallurgical Factors, by C. T. Hahn and R. Simon: *Engineering Fracture Mechanics*, Vol 5, No. 3, Sept 1973, p 523-540

6.30. A New Thermomechanical Procedure for Improving the Ductility and Toughness of Al-Zn-Mg-Cu Alloys in the Transverse Directions, by E. DiRusso, M. Conserva, M. Buratti and F. Gatto: *Materials Science and Engineering*, Vol 14, No. 1, April 1974, p 23-26

6.31. The Effect of Ingot Processing Treatments on the Grain Size and Properties of Aluminum Alloy 7075, by J. Waldman, H. Sulinski and H. Markus: *Metallurgical Transactions,* Vol 5, 1974, p 573-584

6.32. Some Recent Developments in Fatigue and Fracture, by A. R. Rosenfeld and A. J. McEvily: AGARD Report No. 610, Metallurgical Aspects of Fatigue and Fracture Toughness, NATO Advisory Group for Aerospace Research and Development, Dec 1973, p 23-55

6.33. Strengthening Mechanisms in Fatigue, by C. E. Feltner and P. Beardmore: STP 467, American Society for Testing and Materials, Philadelphia, 1969, p 77-112

6.34. Strengthening and Fracture in Fatigue (Approaches for Achieving High Fatigue Strength), by J. C. Grosskreutz: *Metallurgical Transactions*, Vol 3, No. 5, May 1972, p 1255-1262

6.35. Dependence of Fatigue Life and Flow Stress on the Microstructure of Precipitation-

Hardened Al-Cu Alloys, by B. K. Park, V. Greenhut, G. Lütjering and S. Weissman: Report AFML-TR-70-195, Wright-Patterson Air Force Base, OH, Aug 1970

6.36. Mechanism of Fatigue Enhancement in Selected High Strength Aluminum Alloys: Progress Report NADC-MA-7171, Naval Air Development Center, Washington, Dec 10, 1971

6.37. Influence of Inclusion Content on Fatigue Crack Propagation in Aluminum Alloys, by S. M. El-Sondoni and R. M. Pelloux: *Metallurgical Transactions*, Vol 14, No. 2, Feb 1973, p 519-531

6.38. Microstructure and Fatigue Behavior of Al-Alloys, by G. Lütjering, H. Döker and D. Munz: in *The Microstructure and Design of Alloys*, Proceedings of the 3rd International Conference on Strength of Metals and Alloys, Vol 1, Cambridge, England, Aug 1973, p 427-431

6.39. The Effect of Grain Size on the Fatigue of an Al-Mg Alloy, by L. P. Karjalainen: *Scripta Metallurgica*, Vol 7, No. 1, Jan 1973, p 43-48

6.40. The Effect of Precipitate Size on Crack Propagation and Fracture of an Al-Cu-Mg Alloy, by D. Broek and C. Q. Bowles: *Journal of the Institute of Metals*, Vol 99, Aug 1971, p 255-257

6.41. Review of Fatigue and Fracture Research on High Strength Aluminum Alloys, by T. H. Sanders, Jr., and J. T. Staley: presented at the 1978 ASM Materials Science Seminar, St. Louis, Oct 14-15, 1978

6.42. Improving Fatigue Resistance of Aluminum Aircraft Alloys, by J. T. Staley, W. G. Truckner, R. J. Bucci and A. B. Thakker: WESTEC 77 Paper No. 3, American Society for Metals, Metals Park, OH, 1977

6.43. How Microstructure Affects Fatigue and Fracture of Aluminum Alloys, by J. T. Staley: in *Fracture Mechanics*, edited by N. Perrone *et al*, University Press of Virginia, 1978, p 671

6.44. Improved Fracture Resistance of 7075 Through Thermomechanical Processing, by W. H. Reimann and A. W. Brisbane: *Engineering Fracture Mechanics*, Vol 5, No. 1, Feb 1973, p 67-78

6.45. Thermomechanical Processing and Fatigue of Aluminum Alloys, by F. G. Ostermann and W. H. Reimann: STP 467, American Society for Testing and Materials, Philadelphia, 1969, p 169-187

6.46. Improved Fatigue Resistance of Al-Zn-Mg-Cu (7075) Alloys Through Thermomechanical Processing, by F. G. Ostermann: *Metallurgical Transactions*, Vol 2, No. 10, Oct 1971, p 2897-2902

6.47. High Strength Aluminum Powder Metallurgy Mill Products, by W. S. Cebulak, E. W. Johnson and H. Markus: Aluminum Company of America, Alcoa Center, PA, presented at the 1974 Western Metal and Tool Exposition, Los Angeles, March 13, 1974

6.48. Properties of High Strength Aluminum P/M Products, by J. P. Lyle, Jr., and W. S. Cebulak: *Metals Engineering Quarterly*, Vol 14, No. 1, Feb 1974, p 52-63

6.49. Powder Metallurgy Approach for Control of Microstructure and Properties, by J. P. Lyle, Jr., and W. S. Cebulak: Aluminum Company of America, Alcoa Center, PA, presented at the TMS-AIME Spring Meeting, Advances in Physical Metallurgy of Aluminum Alloys, Philadelphia, May 29 to June 1, 1973

6.50. Fracture Toughness and Fatigue Properties of 5083-O Plate and 5183 Welds for Liquified Natural Gas Application, by J. G. Kaufman and R. A. Kelsey: STP 579, American Society for Testing and Materials, Philadelphia, 1975, p 138-158

6.51. Some Aspects of Environmentally Enhanced Fatigue Crack Growth, by R. P. Wei: *Engineering Fracture Mechanics*, Vol 1, No. 4, April 1970, p 633-651

6.52. Ranking 7XXX Aluminum Alloy Fatigue Crack Growth Resistance Under Constant Amplitude and Spectrum Loading, by R. J. Bucci, A. B. Thakker, T. H. Sanders,

R. R. Sawtell and J. T. Staley: STP 714, American Society for Testing and Materials, Philadelphia, Oct 1980, p 41-78

6.53. Effect of Microstructure on Fatigue Crack Growth of 7XXX Aluminum Alloys Under Constant Amplitude and Spectrum Loading, by T. H. Sanders, R. R. Sawtell, J. T. Staley, R. J. Bucci and A. B. Thakker: Final Report, Contract No. N00019-76-C-0482, Naval Air Systems Command, 1978

6.54. Numerical Analysis of Crack Propagation in Cyclic Loaded Structures, by R. G. Forman, R. E. Kearney and R. M. Engle: *Transactions of ASME, Journal of Basic Engineering*, Vol 89, No. 3, Sept 1967, p 459-464

6.55. Stress Corrosion Cracking of High Strength Aluminum Alloys, by M. O. Speidel and M. V. Hyatt: *Prior Advances in Corrosion Science and Technology*, Vol 2, Plenum Press, 1972

6.56. Stress Corrosion Cracking Control Measures, by B. F. Brown: NBS Monograph 156, U.S. Government Printing Office, June 1977

6.57. Stress Corrosion and Corrosion Fatigue Susceptibility of High Strength Alloys, by G. E. Nordmark, B. W. Lifka, M. S. Hunter and J. G. Kaufman: Technical Report AFML-TR-70-259, Wright-Patterson Air Force Base, OH, Nov 1970

6.58. Stress Corrosion Cracking of Aluminum Alloys, by M. O. Speidel: in *ACPA Handbook on Stress Corrosion Cracking* (in preparation)

6.59. Stress Corrosion Cracking in Aluminum Alloys, by M. O. Speidel: *Metallurgical Transactions A*, Vol 6A, No. 4, April 1975, p 631-651

6.60. High Strength Aluminum Alloys With Improved Resistance to Corrosion and Stress Corrosion Cracking, by D. O. Sprowls: presented at the Tri-Service Corrosion Conference, Philadelphia, Oct 26-28, 1976

6.61. Pre-exposure Embrittlement and Stress Corrosion Failure in Al-Zn-Mg Alloys, by C. M. Scamans, R. Alani and P. R. Swann: *Corrosion Science*, Vol 16, No. 7, July 1976, p 443-459

6.62. Reduced Ductility of High-Strength Aluminum Alloy During or After Exposure to Water, by D. Hardie and N. J. Holroyd: *Metal Science*, Vol 13, No. 11, Nov 1979, p 603-610 [illegible] T. [illegible] American Society for Testing and Materials, Philadelphia, [illegible]

[illegible] Microfractography of Fatigue Crack Growth in 2XXX Aluminum Alloys [illegible] Constant Amplitude and Spectrum Loading [illegible]

[illegible] Final Report, Contract No. [illegible]

[illegible] Numerical Analysis of Crack [illegible]

[illegible] Fracture Toughness [illegible]

[illegible] Aluminum Alloys [illegible]

Chapter 7

Fracture Properties of Titanium Alloys

H. W. Rosenberg, J. C. Chesnutt and H. Margolin

7.1. INTRODUCTION

Titanium is used for two primary reasons: (*a*) structural efficiency, which derives from its combination of high strength and low density; and (*b*) resistance to corrosion by chlorides and oxidizing media, which derives from its strong passivation tendencies. This chapter describes how the structural efficiency of titanium alloys may be influenced by fracture toughness and crack growth parameters.

Titanium, like most structural materials, is supplied in all mill product forms, and several titanium alloys are available to meet specific needs. Most important among titanium alloys is Ti-6Al-4V. Offering a strength-to-density ratio of 25×10^6 mm (1×10^6 in.), Ti-6Al-4V has found application for a wide variety of aerospace hardware. Jet aircraft manufacturers are the principal consumers of titanium in this market. The Ti-6Al-4V alloy is often specified for critical parts the failure of which could result in loss of an entire system. In these situations, the higher acquisition cost of titanium can be more than offset by its reduced costs of ownership.

Because of its great popularity, Ti-6Al-4V has become the best understood of all titanium alloys, and much of the property data obtained on this alloy has been stored in computer data banks and is available for statistical analysis. The first part of this chapter, therefore, will deal primarily with the properties of Ti-6Al-4V.

Several metallurgical and environmental variables have been identified that influence the fracture behavior of titanium alloys in general and of alloy Ti-6Al-4V in particular; the effects of these variables will be discussed in detail in this chapter. It is beyond the scope of this chapter, however, to provide design-type data to the user requiring the ultimate in performance. That type of information must either be generated by the user or be obtained from other sources.

The references listed at the end of this chapter contain additional mechanical property data. The purpose of this chapter is to provide the reader with general guidelines that indicate what variables have what effects on toughness and fatigue crack propagation. Among the metallurgical variables of importance are composition, microstructure (as it depends on processing and heat treatment), and crystallographic texture. Environmental factors are discussed also.

In the following sections, the metallurgy of alloy Ti-6Al-4V is treated first, followed by a brief discussion indicating how certain other titanium alloys differ metallurgically from Ti-6Al-4V. The remainder of this chapter reviews, in highlight fashion, what is known about fracture toughness and fatigue crack propagation, including the effects of testing environments, for the most common titanium alloys. The references that are cited refer primarily to alloys for which standard specifications exist.

7.2. METALLURGY OF TITANIUM ALLOYS

7.2.1. Phase Relationships and Chemistry

Titanium exists in two crystalline states. In pure titanium, the low-temperature α phase is stable at temperatures below about 883°C (1621°F) and crystallizes in the hexagonal close packed structure with a c/a ratio of 1.58, which is slightly less than the ideal ratio for packing of rigid spheres. The high-temperature β form is a body centered cubic phase that is stable from about 883°C (1621°F) to the melting point.

The transformation temperature and phase compositions of titanium can be altered by alloying additions. Elements that increase the transformation temperature are known as alpha stabilizers, and those that decrease it are called beta stabilizers. Other, sparingly soluble elements, when present in excess of their solubility limits, may form compounds or second phases of essentially pure solute. Of the elements commonly present in Ti-6Al-4V: carbon is a compound former; vanadium, iron and hydrogen are beta stabilizers; and aluminum, oxygen and nitrogen are alpha stabilizers.

7.2.2. Metallurgy of Ti-6Al-4V

Standard grade Ti-6Al-4V becomes 100% beta phase at temperatures above about 1000°C (1832°F). Below this temperature, alpha and beta phases coexist. Ti-6Al-4V is thus a two-phase alloy with the beta phase present even at cryogenic temperatures. This comes about because 4 wt % vanadium exceeds the alpha solubility limit. When additional phases occur, it is usually because the alloy has been contaminated with an impurity (such as boron) or because an element such as yttrium has been added for grain refining purposes. To avoid problems caused by impurities, maximum impurity levels are limited by specifications that cover composition limits. At room temperature, Ti-6Al-4V is about 90 vol % alpha phase. Thus, the alpha phase dominates the physical, chemical and mechanical properties of this alloy.

Alloy Ti-6Al-4V may be obtained in two basic ranges of composition: the standard grade and the ELI grade. The acronym ELI, as in Ti-6Al-4V ELI, denotes the "extra low interstitial" grade. In the ELI grade, oxygen is held to less than 0.13 wt %, whereas the maximum oxygen content of the standard grade is commonly 0.20 wt %. The two grades have the following typical composition ranges:

Element	Composition, wt % Standard	ELI
Aluminum	5.75-6.75	5.50-6.50
Vanadium	3.5 -4.5	3.5 -4.5
Iron	0.25 max	0.25 max
Oxygen	0.20 max	0.13 max
Nitrogen	0.05 max	0.05 max
Hydrogen	0.015 max	0.015 max
Carbon	0.08 max	0.08 max

Oxygen, nitrogen, hydrogen and carbon are the interstitial elements. Except for carbon, they are all readily soluble in titanium. In general, they increase strength and decrease ductility, and in this sense have effects quite similar to those of metallic alloying additions. Carbon has limited solubility and is a strong compound former, but carbon levels are so low in commercial products that carbides are virtually nonexistent. Hydrogen, aside from being a beta stabilizer, has other unique features. It is soluble as well as highly mobile in titanium. Hydrogen can, therefore, be picked up during processing operations such as forging, heat treating and pickling. By the same token, hydrogen can be removed from titanium by vacuum annealing operations at temperatures on the order of 700 to 900°C (1292 to 1652°F). In vacuum annealing operations, both the metal and the furnace surfaces must be clean to ensure effective outgassing. Depending in part on the amount of beta phase present, hydrogen at sufficiently high levels is manifested by hydride precipitation and embrittlement. Embrittlement may occur as a delayed reaction. Residual stress gradients lead to hydrogen gradients which may localize the hydrides (Ref 7.1 and 7.2). Aluminum tends to increase the apparent solubility of hydrogen in alpha titanium (Ref 7.3). Hydrogen also is highly soluble in beta titanium and in the beta phase of alpha + beta alloys. For these reasons, welding of unalloyed titanium to alloys such as Ti-6Al-4V is not recommended lest the hydrogen normally present in Ti-6Al-4V migrate to the unalloyed titanium and cause embrittlement (Ref 7.1). Hydrogen in titanium is readily controlled, and specifications limit the maximum allowable content.

7.2.2.1. Microstructure and Transformation Behavior

Control of microstructure is the primary key to successful application of alloy Ti-6Al-4V. It depends on both processing history and heat treatment. The microstructure that combines highest strength and ductility is not the microstructure that provides optimum fracture toughness or resistance to crack growth. The over-all effects of processing history and heat treatment on microstructure are very complex. However, the present discussion will illustrate those features most likely to be found in the alloy by the user.

Figure 7.1 illustrates the effect of solution temperature on the microstructure obtained at a cooling rate equivalent to that observed in parts of moderate thickness. In Fig. 7.1(a), the Widmanstätten-like transformed microstructure arises

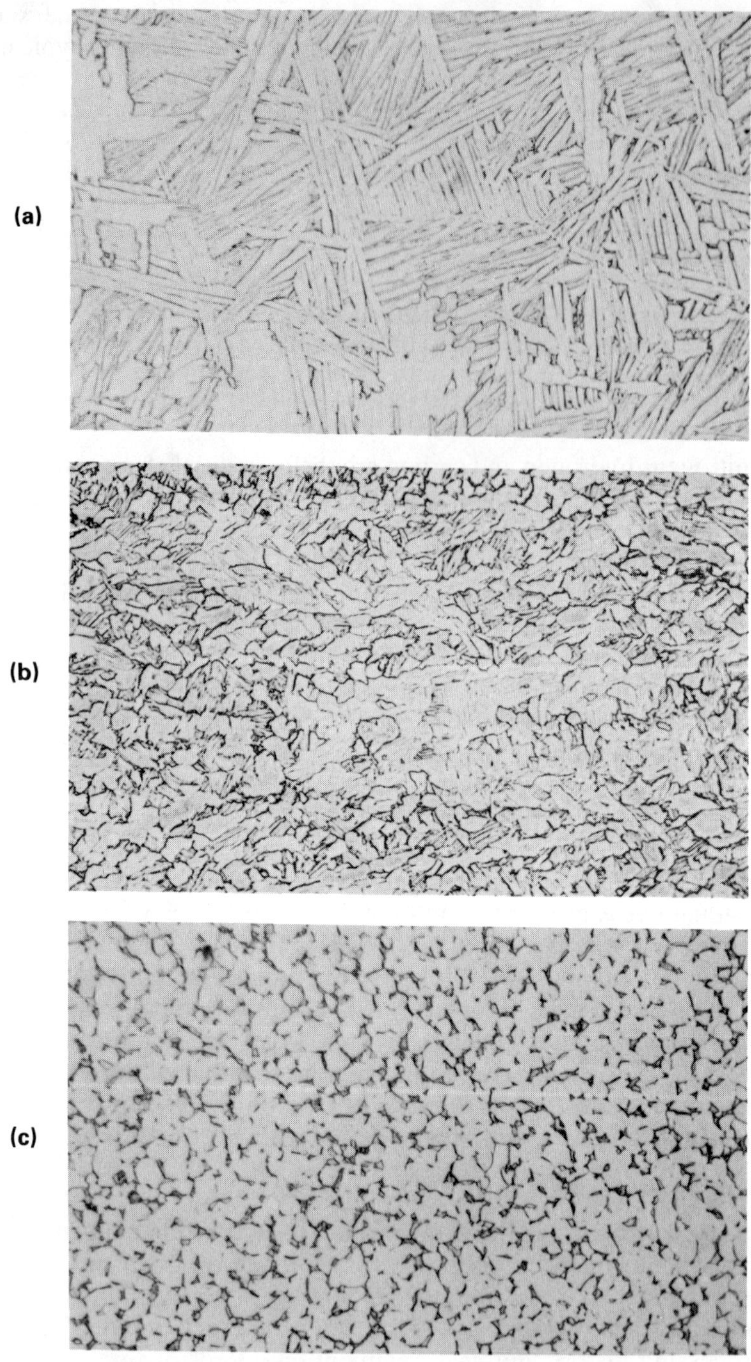

(a) 1010°C (1850°F), 1 h, encapsulated cool; 500×. (b) 982°C (1800°F), 1 h, encapsulated cool; 500×. (c) 927°C (1700°F), 1 h, encapsulated cool; 500X.

Fig. 7.1. Typical microstructures of alloy Ti-6Al-4V, showing effect of solution temperature

from $\beta \rightarrow \alpha + \beta$ transformation during cooling from the beta-phase field. The light, platelike and patchy areas are composed of alpha phase that nucleated and grew during cooling, whereas the retained beta phase is restricted primarily to the dark-etching outlines between the plates. Platelet colonies (that is, groups of platelets) in various orientations are apparent in the microstructure. The prior beta grain size exceeded the entire field of view in Fig. 7.1(a). The alloy in Fig. 7.1(b) was cooled to room temperature from about 20°C (36°F) below the $\beta \rightleftharpoons \alpha + \beta$ transformation temperature (the beta transus). In this photomicrograph, the more or less equiaxed light areas are primary alpha phase, which is the alpha phase that was present at the solution temperature. The background is a transformed structure similar to that in Fig. 7.1(a) except that the platelet colonies are smaller. In Fig. 7.1(c), after cooling to room temperature from about 70°C (126°F) below the transformation temperature, the predominant feature is equiaxed primary alpha. The transformed background is a minor constituent.

Another microstructure of special interest in obtaining maximum fracture toughness is one produced by so-called recrystallization annealing. In this process, the alloy is heated to a temperature about 70°C (126°F) below the transformation temperature, held for a time and then very slowly cooled. The resulting microstructure is shown in Fig. 7.2. Most of this structure is continuous primary alpha phase and regrowth alpha (that is, alpha which has formed on the pre-existing primary alpha phase). The primary alpha existed at the upper temperature and nucleated the regrowth alpha during cooling. Stable beta phase decorates the alpha grain boundary triple points. This particular microstructure provides an excellent combination of fracture toughness and ductility. Another significant feature of the recrystallization annealed microstructure in Fig. 7.2 is that the alpha phase is fully recrystallized and has a very low dislocation density.

The effects of prior processing on microstructure are quite varied. Extensive hot working in the alpha + beta field is required to produce the microstructure typified by Fig. 7.1(c). If hot working of the alpha + beta phases has been limited,

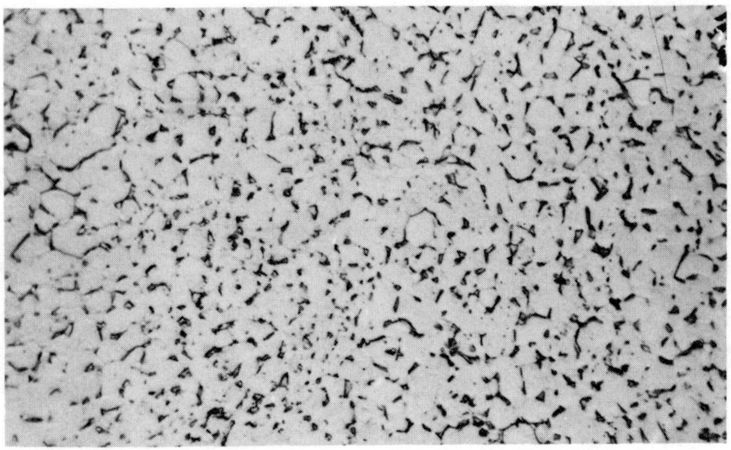

927°C (1700°F), 1 h, very slow cool; 500×.

Fig. 7.2. Microstructure of alloy Ti-6Al-4V after recrystallization annealing

microstructures such as that shown in Fig. 7.3 will occur. As may be observed, the Widmanstätten platelets are quite distorted but are not yet broken up. This condition is not particularly detrimental to fracture toughness, although it may affect fatigue crack propagation.

When alloy Ti-6Al-4V is processed improperly after heating into the beta field, alpha phase can form preferentially along the prior beta grains. Extensive hot work is required to break up such structures. An example of grain boundary alpha not completely broken up is shown in Fig. 7.4. Because cracks tend to propagate in or near interfaces (Ref 7.4, 7.5 and 7.6), this type of structure can provide loci for crack initiation and propagation and thereby lead to premature failure.

Microstructural control is effected by using proper combinations of hot work and heat treatment. Heat treatment alone does not suffice to convert the Widmanstätten structure to an equiaxed form; therefore, heat treatment alone is not used unless a transformed structure is desired. Grain refinement cannot be obtained by heat treatment, and, after one $\beta \rightarrow \alpha + \beta$ sequence has been accomplished, additional $\alpha + \beta \rightarrow \beta \rightarrow \alpha + \beta$ cycles have no effect on the basic crystallographic texture although the texture may be coarsened as a consequence of beta grain growth.

Alpha arising from the $\beta \rightarrow \alpha + \beta$ transformation has a crystallographic (Burgers) relationship with the parent beta. Generally, two or more variants are present (Ref 7).

At the highest cooling rates, Ti-6Al-4V can form hexagonal close packed martensite which contains vanadium in supersaturation. Crystallographically, the martensite constituent and the alpha phase are nearly identical and are difficult to distinguish by optical means. Electron or x-ray metallography is required to distinguish quantitatively between martensite and fine Widmanstätten alpha. Figure 7.5 illustrates the effect that cooling rate has on the microstructure of Ti-6Al-

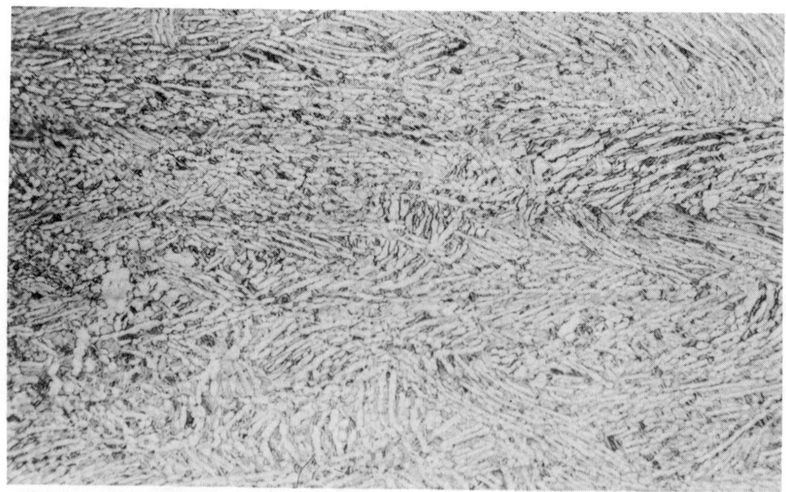

Rolled at 955°C (1750°F); 100×.

Fig. 7.3. Distorted Widmanstätten alpha remaining as a result of limited working in the $\alpha + \beta$ field

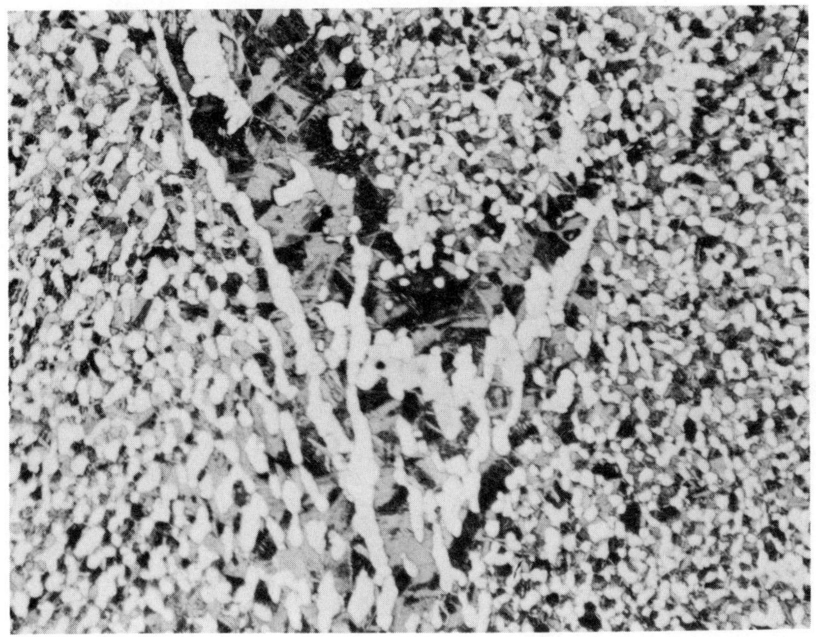

Rolled at 940°C (1725°F); 250×.

Fig. 7.4. Grain boundary alpha remnants which were not broken up after forging because of improper cooling from the β field

4V when quenching is effected from the beta and high alpha + beta fields. No martensite is present in Fig. 7.5(b) and (d). Martensite *may* be present in Fig. 7.5(a) and (c).

At certain cooling rates, the $\beta \rightarrow \alpha + \beta$ transformation involves an interface phase (Ref 7.8) which is generally not observable by optical microscopy but which can be indexed crystographically as face centered cubic. The FCC form of interface phase usually is transient, in which case it transforms to hexagonal close packed alpha. Whether or not the interface phase significantly affects toughness or fatigue crack propagation has not been resolved, although there is evidence that toughness is insensitive to the interface phase whereas a broad interface phase leads to improved resistance to fatigue crack growth (Ref 7.9).

7.2.2.2. Deformation Modes

Crucial to the toughness question is the size of the plastic zone that can form ahead of a propagating crack. That size, simplistically, depends on the yield strength, and this in turn depends on the factors discussed in the upcoming discussion on sources of strength. Having a high yield strength and a relatively low elastic modulus, Ti-6Al-4V can store more energy elastically than most metals before plastic deformation begins. The elastic modulus depends on the direction in which it is measured in a single crystal of titanium. This feature is illustrated in Fig. 7.6.

The flow stress also depends on the direction of measurement. In directions normal to the hexagonal axis of an alpha titanium crystal, $\bar{a}$ slip, on the prism,

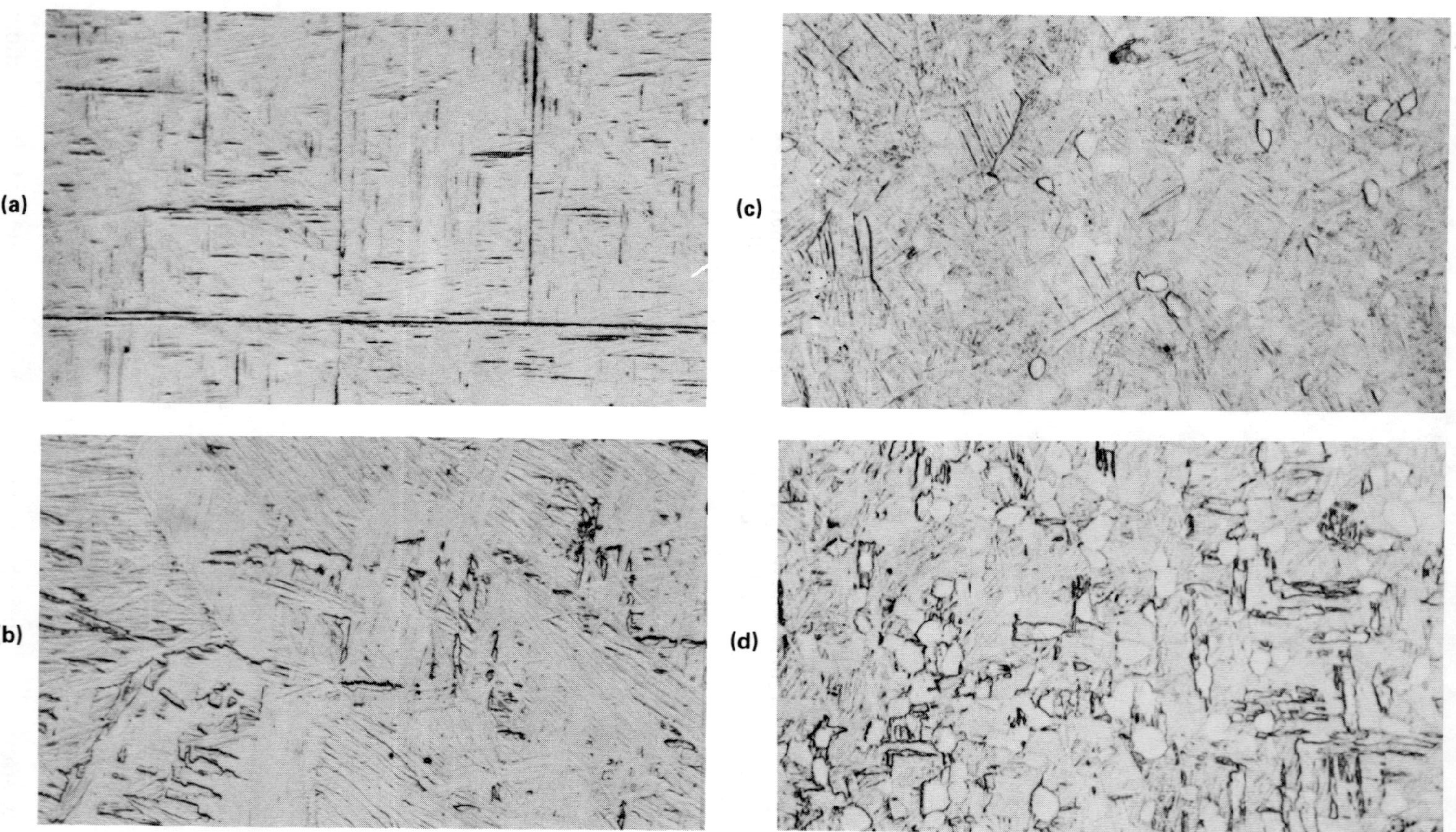

(a) 1010°C (1850°F), 1 h, WQ; 500×. (b) 1010°C (1850°F), 1 h, AC; 500×. (c) 980°C (1800°F), 1 h, WQ; 500×. (d) 980°C (1800°F), 1 h, AC; 500×. (a) and (b) are from the β field; (c) and (d) are from the high $\alpha + \beta$ field.

Fig. 7.5. Illustration of quenching rate effect on microstructures of alloy Ti-6Al-4V

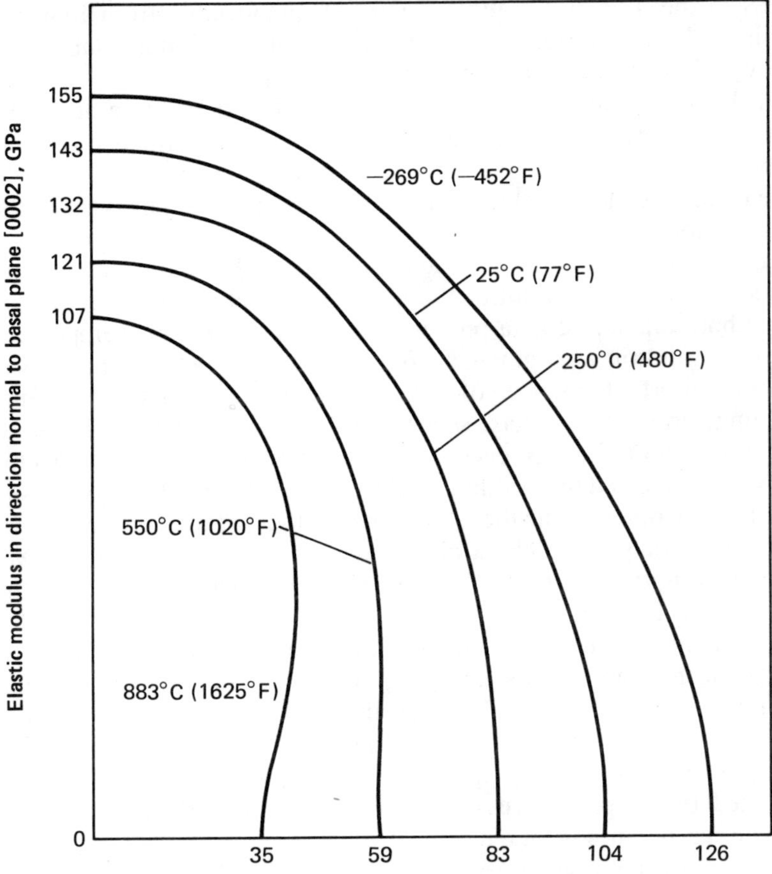

Distance from origin to isotherm in direction of interest is equal to modulus.

Fig. 7.6. Plot of elastic modulus vs direction in single crystal of titanium for various temperatures (Ref 7.10)

pyramidal and basal planes, is the primary deformation mode. In directions parallel to the hexagonal axis, twinning and $\bar{c} + \bar{a}$ slip act to accommodate the plastic strain. Each slip and twinning mode has its own unique flow stress and amount of strain that be accommodated.

The plastic zone ahead of an advancing crack is, therefore, not uniform in cross section. It varies in a macroscopic sense in response to the microstructure (phase, shape) discussed in the previous section. It also varies from grain to grain in accordance with crystal type, whether alpha or beta, and with crystal orientation. To complicate matters further, Poisson's ratio and its plastic counterparts necessarily depend on direction on both the macro and micro scales. Furthermore, the true strain at ultimate load is not a linear function of alloy content even in the Ti-Al binary system (Ref 7.11). Events occurring in and around a crack tip and its associated plastic zone in Ti-6Al-4V are, for these reasons, complex indeed.

Toughness in the Ti-6Al-4V alloy is not yet quantifiable from first principles. Nevertheless, there is a great deal of empirical information available for Ti-6Al-4V from which some general rules can be developed.

The mechanical properties of titanium alloys thus depend on alloy chemistry, microstructure, and metallographic texture through its influence on elastic and plastic anisotropy. The influence of these factors on strength, toughness and resistance to environmental effects on crack propagation is discussed further in the following sections.

The Ti-6Al-4V alloy derives its annealed strength from several sources, the principal source being substitutional and interstitial alloying of elements in solid solution in both alpha and beta phases. Aluminum is the most important substitutional solid solution strengthener. Its effect on strength is linear (Ref 7.12). Other, less important sources of strengthening are interstitial solid solution strengthening, grain size effects, second phase (beta) effects, ordering in alpha, age hardening, and effects of crystallographic texture (Ref 7.13). Aluminum in Ti-6Al-4V, as suspected by Williams and Blackburn (Ref 7.14), gives rise to some tendency toward ordering in the alpha phase, the ordered product being Ti_3Al (Ref 7.15). Ordering in the alpha phase contributes perhaps 15 to 35 MPa (2 to 5 ksi) to the strength of standard Ti-6Al-4V, and contributes less than this to the strength of the ELI grade. Ordering also appears to degrade toughness. The effect of crystallographic texture is to introduce directionality into the strength equation. Relative to the hexagonal axis in alpha, strength (and modulus) is high in the parallel direction and low normal to that direction. Because metalworking operations tend to produce preferred crystallographic orientations in alpha grains, strength becomes an anisotropic quantity in most product forms. This feature can be minimized by proper processing and is rarely of direct concern. In some instances, it can be an advantage.

Because the beta phase present in alloy Ti-6Al-4V can be manipulated in amount and composition by heat treatment, the alloy is responsive to heat treatment. The $\beta \rightarrow \alpha+\beta$ reaction at low temperature leads to increased strength. The key is to quench from high in the $\alpha+\beta$ field and then age at a lower temperature. A typical strengthening heat treatment consists of heating for 1 h at 955°C (1750°F) and water quenching, followed by heating for 4 h at 540°C (1000°F) and air cooling. Response is limited in a practical sense, however, by two factors: (*a*) the small amount of beta in Ti-6Al-4V and (*b*) section size. The first factor puts an intrinsic ceiling on the increased strengthening response available—about 280 MPa (40 ksi) in thin-gage material. The second factor relates to depth of hardening, because Ti-6Al-4V is not effectively hardenable in sections greater than 25 mm (1 in.) in thickness. The Ti-6Al-4V alloy is, therefore, most commonly used in the annealed condition.

7.2.3. Metallurgy of High-Strength Alpha-Beta Alloys

Two alloys that fall in the high-strength alpha-beta class are Ti-6Al-6V-2Sn, which is used in airframes, and Ti-6Al-2Sn-4Zr-6Mo, which is used in jet engines. Alloy Ti-6Al-2Sn-4Zr-6Mo is also often classified as a super alpha alloy. Both of these alloys are stronger and more readily heat treated than Ti-6Al-4V. These features arise from the increased solid solution strengthening afforded by tin and zirconium, which have relatively small effects on the transformation temperature, and from the increased amounts of beta phase that result from the larger vanadium and molybdenum additions. Both vanadium and molybdenum are beta stabilizers. The Ti-6Al-6V-2Sn alloy contains the beta stabilizers copper and iron in combined amounts up to 1.4 wt % for enhanced strength and response to aging. Alloy Ti-6Al-2Sn-4Zr-6Mo is also useful at the moderately elevated temperatures from 425 to 480°C (800 to 900°F). This alloy combines high tensile strength with good creep resistance. The alpha phase tends to order more readily in these alloys than in alloy Ti-6Al-4V. Moreover, the transformed alpha platelets in Ti-6Al-2Sn-4Zr-6Mo tend to be narrower than those in Ti-6Al-4V, and formation of packets of parallel platelets is less likely. For both Ti-6Al-6V-2Sn and Ti-6Al-2Sn-4Zr-6Mo, the nose of the C curve defining the $\beta \rightarrow \alpha+\beta$ transformation as it depends on time and temperature is shifted to lower temperatures and longer times than for Ti-6Al-4V. Martensite does not form in ordinary situations.

Alpha is the dominant phase in these alloys but to a lesser extent than in Ti-6Al-4V. The physical metallurgy of these alloys is otherwise very similar to that of Ti-6Al-4V.

7.2.4. Metallurgy of "Super Alpha" Alloys

Alloys Ti-6Al-2Sn-4Zr-2Mo and Ti-8Al-1Mo-1V are in the "super alpha" class. They are used primarily in jet engine applications and are useful at temperatures above the normal range for Ti-6Al-4V. Alloy Ti-6Al-2Sn-4Zr-2Mo may be modified with silicon additions of up to 0.1%, and, when beta annealed (i.e., annealed by heating above the transformation temperature), the modified alloy provides the highest creep strength and temperature capability of all commercial titanium alloys currently (1980) produced in the United States. The Ti-8Al-1Mo-1V alloy has the highest modulus and lowest density of any commercial titanium alloy. Each of these alloys tends to order in the alpha phase more readily than does Ti-6Al-4V. Also, the nose of the C curve defining the $\beta \rightarrow \alpha+\beta$ transformation is shifted upward and to the left, or to higher temperatures and shorter times, in comparison with Ti-6Al-4V. Martensite forms more readily in either of these alloys than in Ti-6Al-4V.

Generally speaking, these alloys contain less beta phase than Ti-6Al-4V. Age hardening treatments are thus not very effective and are, moreover, deleterious to creep resistance. These alloys therefore are usually employed as solution annealed and stabilized. Solution annealing may be done at a temperature some 35°C (63°F) below the transformation temperature, and stabilization is commonly produced by heating for 8 h at about 590°C (1100°F).

At high temperatures, dynamic strain aging arising from aluminum, silicon and tin, and perhaps oxygen and zirconium, is thought to contribute to the creep resistance of these materials.

The alpha phase dominates the properties of these alloys to a greater extent than it does in Ti-6Al-4V. The metallurgy of the super alpha alloys is otherwise similar to that of Ti-6Al-4V.

7.3. TENSILE PROPERTIES

Table 7.1 lists typical minimum property guarantees for titanium alloy mill products. In Table 7.2, the effects of temperature on strength are shown for the same alloys, and data for unalloyed titanium is included to illustrate that the alloys not only have higher room temperature strengths but also retain much larger fractions of that strength at elevated temperatures.

Table 7.3 lists typical specifications for the alloys discussed here. These alloys are covered also by numerous commercial specifications, and design information is readily available. More extensive listings of specifications are given in the general references at end of this chapter.

In terms of the principal heat treatments used for titanium, beta annealing decreases strength by 35 to 100 MPa (5 to 15 ksi) depending on prior grain size, average crystallographic texture, and testing direction. Solution treating and aging can be used to enhance strength at the expense of fracture toughness in alloys containing sufficient beta stabilizer (that is, 4 wt % V or more).

7.4. FRACTURE TOUGHNESS

Fracture toughness can be varied within a nominal titanium alloy by as much as a multiple of two or three by manipulating alloy chemistry, microstructure and texture. Some trade-offs of other desired properties may be necessary to achieve

Table 7.1. Typical mill-guaranteed room temperature tensile properties for selected titanium alloys

Alloy	Ultimate strength MPa	Ultimate strength ksi	Yield strength MPa	Yield strength ksi	Ductility: Elongation, %	Ductility: Reduction in area, %
Ti-6Al-4V	895	130	825	120	10	20
Ti-6Al-6V-2Sn	1065	155	995	145	10	20
Ti-6Al-2Sn-4Zr-6Mo	1030	150	965	140	10	20
Ti-6Al-2Sn-4Zr-2Mo	895	130	825	120	10	25
Ti-8Al-1Mo-1V	895	130	825	120	10	20

Table 7.2. Fraction of room-temperature strength retained at elevated temperature for several titanium alloys(a)

Temperature °C	Temperature °F	Unalloyed Ti TS	Unalloyed Ti YS	Ti-6Al-4V TS	Ti-6Al-4V YS	Ti-6Al-6V-2Sn TS	Ti-6Al-6V-2Sn YS	Ti-6Al-2Sn-4Zr-6Mo TS	Ti-6Al-2Sn-4Zr-6Mo YS	Ti-6Al-2Sn-4Zr-2Mo TS	Ti-6Al-2Sn-4Zr-2Mo YS	Ti-8Al-1Mo-1V TS	Ti-8Al-1Mo-1V YS
93	200	0.80	0.75	0.90	0.87	0.91	0.89	0.90	0.89	0.93	0.90	0.93	0.92
204	400	0.57	0.45	0.78	0.70	0.81	0.74	0.80	0.80	0.83	0.76	0.84	0.79
316	600	0.45	0.31	0.71	0.62	0.76	0.69	0.74	0.75	0.77	0.70	0.76	0.67
427	800	0.36	0.25	0.66	0.58	0.70	0.63	0.69	0.71	0.72	0.65	0.67	0.60
482	900	0.33	0.22	0.60	0.53	...	...	0.66	0.69	0.69	0.62	0.61	0.55
538	1000	0.30	0.20	0.51	0.44	...	...	0.61	0.66	0.66	0.60	0.53	0.46

(a) Short time tensile test with less than 1 h at temperature prior to test.

Table 7.3. Typical specifications for titanium and titanium alloys

Material	AMS	Military MIL-I
Unalloyed Ti	4900	9046, 9047
Ti-6Al-4V	4911	9046, 9047
Ti-6Al-4V ELI	4907	9046, 9047
Ti-6Al-6V-2Sn	4918	9046, 9047
Ti-6Al-2Sn-4Zr-6Mo	4981	9047
Ti-6Al-2Sn-4Zr-2Mo	4929	9046, 9047
Ti-8Al-1Mo-1V	4915, 4916	9046, 9047

high fracture toughness. Plane-strain fracture toughness, K_{Ic}, is of special interest because the critical crack size at which unstable growth can occur is proportional to $(K_{Ic})^2$ and strength is often achieved in titanium alloys at the expense of K_{Ic}.

The main purpose of this section is to indicate the scope of possibilities, as well as some of the property trade-offs required, for obtaining high levels of fracture toughness in titanium alloys. A further purpose is to review some of the specific variables that are known to affect fracture toughness.

7.4.1. Effects of Alloy Chemistry

There are significant differences among titanium alloys (Ref 7.16), but there is also appreciable overlap in their properties. Table 7.4 gives examples of typical plane-strain fracture toughness ranges for alpha-beta titanium alloys. From these data it is apparent that the basic alloy chemistry affects the relationship between strength and toughness. From Table 7.4 it is also evident that transformed microstructures may greatly enhance toughness while only slightly reducing strength.

Within the permissible range of chemistry for a specific titanium alloy and grade, oxygen is the most important variable insofar as its effect on toughness is concerned. This is readily shown by the data of Ferguson and Berryman (Ref 7.17), who reported strength and K_{Ic} values for specimens of alpha-beta

Table 7.4. Typical fracture toughness of high-strength titanium alloys (Ref 7.16)

Alloy	Alpha morphology	Yield strength		Fracture toughness (K_{Ic})	
		MPa	ksi	$MPa{\cdot}m^{1/2}$	$ksi{\cdot}in.^{1/2}$
Ti-6Al-4V	Equiaxed	910	130	44-66	40-60
	Transformed	875	125	88-110	80-100
Ti-6Al-6V-2Sn	Equiaxed	1085	155	33-55	30-50
	Transformed	980	140	55-77	50-70
Ti-6Al-2Sn-4Zr-6Mo	Equiaxed	1155	165	22-23	20-30
	Transformed	1120	160	33-55	30-50

processed and recrystallization annealed Ti-6Al-4V. Regression analysis of their data shows that for each 0.01% increase in oxygen, toughness is reduced by about 3.7 MPa · $m^{1/2}$ (3.4 ksi · $in.^{1/2}$). Whether this is a direct effect or an indirect effect, in the sense that oxygen increases strength and the strength increase reduces K_{Ic}, remains to be determined. Multiple regression analysis of the Ferguson and Berryman data, where both oxygen content and tensile strength are assumed to be independent variables, shows that tensile strength is the dominant variable (the residual effect of oxygen does not reach statistical significance). This implies that, if oxygen affects K_{Ic}, it does so through its strengthening effect. The solid solution strengthening effect of oxygen is further complicated by the fact that oxygen tends to promote the formation of Ti_3Al (Ref 7.15). Finally, the precision and accuracy to which oxygen can be analyzed in titanium do not compare with the relative precision and accuracy common to strength measurements.

The work of Rosenberg and Parris (Ref 7.18) on the Ti-*x*Al-2Mo system (*x* varies from 4 ro 6%) gives some evidence that both aluminum and oxygen can exert influences on toughness independent of strength. These authors, however, did not employ recrystallization annealing, which should minimize formation of Ti_3Al in the continuous alpha which dominates the toughness properties. For the recrystallization annealed condition, the continuous alpha is regrowth alpha, which is low in both aluminum and oxygen; the solute-rich alpha in which Ti_3Al can form in thus isolated at the core of each primary grain.

Whatever that situation, the data of Ferguson and Berryman (Ref 7.17) on the oxygen effect are given in Fig. 7.7. Tensile strengths for each oxygen level are

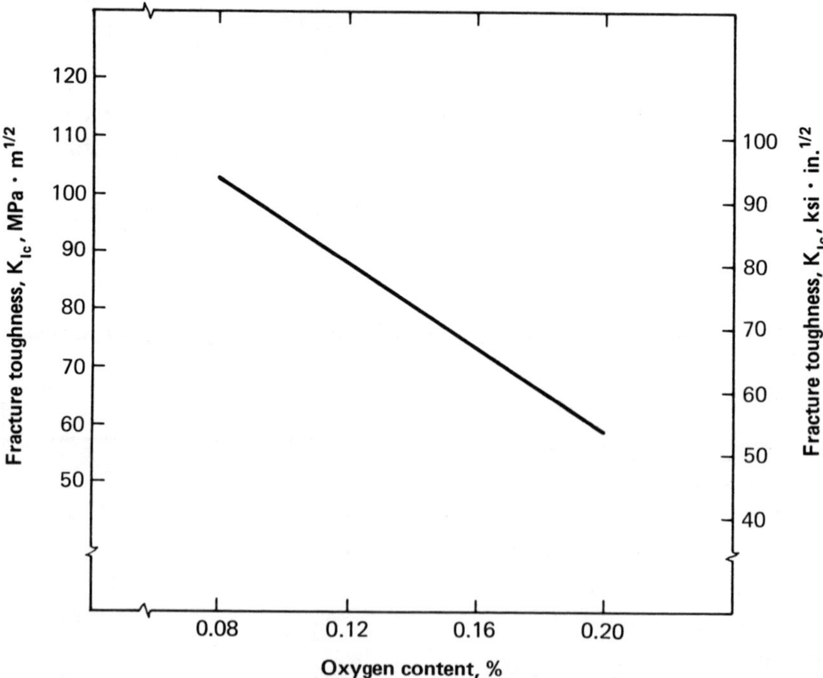

Fig. 7.7. Influence of oxygen content on fracture toughness of recrystallization annealed alloy Ti-6Al-4V (Ref 7.17)

also shown. In essence, if high fracture toughness is required, oxygen must be kept low, other things being equal. Reducing aluminum, as in Ti-6Al-4V ELI, is also indicated, but the effect is not as strong as it is for oxygen (Ref 7.18).

The effect of chemistry on K_{Ic} for Ti-6Al-4V was shown in another way by Cooper (Ref 7.19), who summed the expected changes in the $\beta \rightleftharpoons \alpha+\beta$ transformation temperature, ΔT, with alloy additions, and correlated that with K_{Ic}. He found a negative effect of ΔT on K_{Ic}. Since oxygen and aluminum additions have the effect of increasing ΔT, the data are consistent.

The effect of oxygen on K_{Ic} is not limited to alpha-beta alloys at ambient temperatures. Van Stone and his coworkers (Ref 7.20 and 7.21) reported a much higher K_{Ic} value for Ti-5Al-2.5Sn ELI (which has a low oxygen content) than for the standard grade in the temperature range from −253 to +22°C (−423 to +72°F). Slow cooling of Ti-5Al-2.5Sn ELI from the solution temperature was found to decrease K_{Ic}, whereas this effect of cooling rate was absent for the standard grade. Van Stone attributed the change in toughness to a change in deformation mode morphology. The combination of ordering and normal interstitial content did not significantly change the slip character, whereas ordering in ELI plate resulted in coarser slip bands.

As might be expected, hydrogen also has an effect on toughness. The work of Meyn (Ref 7.22) shows that very low hydrogen contents (less than about 40 ppm) enhance toughness. This effect is particularly dramatic with hydrogen contents below 10 ppm. Table 7.5 illustrates the essential results for Ti-6Al-4V at two different oxygen levels. Meyn used a high loading rate. However, the effect of hydrogen content on K_{Ic} has been confirmed by Chen (Ref 7.23). The work of Chesnutt *et al* (Ref 7.24) on hydrogen effects with nonvalid K_{Ic} tests shows the same trend; in this study, however, the hydrogen effect may have depended on microstructure.

7.4.2. Effects of Microstructure

Improvements in K_{Ic} can be obtained by providing either of two basic types of microstructures: (*a*) transformed structures, or structures transformed as much as possible, because such structures provide tortuous crack paths; and (*b*) equiaxed structures composed mainly of regrowth alpha that have both low dislocation densities and low concentrations of aluminum and oxygen (the so-called "re-

Table 7.5. Effect of hydrogen content on room-temperature K_{Ic} in alloy Ti-6Al-4V after furnace cooling from 927°C (1700°F) (Ref 7.22)

Hydrogen content, ppm	K_{Ic} at room temperature(a) MPa·$m^{1/2}$	ksi·$in.^{1/2}$	Hydrogen content, ppm	K_{Ic} at room temperature(a) MPa·$m^{1/2}$	ksi·$in.^{1/2}$
At 0.16 wt % oxygen			**At 0.05 wt % oxygen**		
8	145	132	9	133	121
36	118	107	36	125	114
53	104	95	50	96	87
122	100	91	125	101	92

(a) Specimens were tested in accord with ASTM E399, but were loaded rapidly (total testing time = 10 s).

crystallization annealed" structures). It is not yet known (in 1980) whether or not combinations of these two types of structures would further enhance K_{Ic} values.

Transformed structures appear to be tough primarily because fractures in such structures must proceed along tortuous, many-faceted crack paths. According to the work of Hall and Hammond (Ref 7.25), K_{Ic} is proportional to the fraction of transformed microstructure in alloy Ti-6Al-4V (see Table 7.6). These authors, however, propose that it is strain induced transformation of the retained laths of beta phase that leads to enhanced fracture toughness. Evidently, their idea is that this "TRIP" mechanism enhances "ductility" in front of each crack tip. However, in comparing beta alloys deformed by either slip or "TRIP" mechanisms, Wardlaw *et al* (Ref 7.26) could find no advantage in ductility for the "TRIP" alloys. Curtis and Spurr (Ref 7.27) suggested that it is primarily the alpha platelet size and efficient dispersion of the beta phase that enhance toughness. In any event, the work of Chesnutt and Spurling (Ref 7.6) provides direct evidence that crack tortuosity is an important factor in determining the form of fracture topography–microstructure correlations for the same sample. Hall and Pierce (Ref 7.28) made similar observations concerning microstructures for alloy Ti-6Al-6V-2Sn. Hall *et al* (Ref 7.29) recommended that, for the best combination of fracture toughness and tensile ductility in Ti-6Al-2Sn-4Zr-6Mo, a microstructure containing 10% primary alpha be employed. There is strong evidence that crack tortuosity is an important variable affecting K_{Ic}.

Rogers (Ref 7.4) looked at the toughness relationship in another way. The alloy he studied was IMI550 (Ti-4Al-2Sn-4Mo-0.5Si) that had been heated for 1 h at 900°C (1625°F) and air cooled, then heated for 24 h at 500°C (932°F) and air cooled. His data show that an increase in the mean phase boundary intercept distance between primary alpha and transformed microstructure diminishes fracture toughness. Rogers' data are presented in Table 7.7. Whether this relationship holds true for all alpha-beta titanium alloys is not known. Rogers' rationale is that, because crack loci and void formation tend to occur at the interfaces between alpha and transformed beta, then a decrease in the distance between phase boundaries causes an increase in the spatial frequency of microvoid formation so as to make blunting and arresting of cracks more likely.

In a similar vein, Gerberich and Baker (Ref 7.30) proposed that a proper balance between platelet thickness and spacing in the transformed microstructure

Table 7.6. Relationship between K_{Ic} and fraction of transformed structure in alloy Ti-6Al-4V (Ref 7.25)

Heat treating temperature(a) °C	°F	Fraction of transformed structure, %	K_{Ic} MPa·m$^{1/2}$(b)	K_{Ic} ksi·in.$^{1/2}$
1050	1922	100	69.0 (69.9)	64
950	1742	70	61.5 (60.4)	55
850	1562	20	46.5 (44.6)	40
750	1382	10	39.5 (41.5)	38

(a) Heated for 1 h at indicated temperature and then air cooled. (b) Values in parentheses calculated from linear least-squares expression relating % transformation to K_{Ic}

Table 7.7. Effect of primary alpha dispersion on K_{Ic} for alloy Ti-4Al-2Sn-4Mo-0.5Si (IMI550) plate(a) (Ref 7.4)

d(b), μm	K_{Ic} MPa·$m^{1/2}$(c)	K_{Ic} ksi·$in.^{1/2}$	d(b), μm	K_{Ic} MPa·$m^{1/2}$(c)	K_{Ic} ksi·$in.^{1/2}$
7.58	59.3 (59.2)	54	3.71	67.5 (66.0)	60
5.83	62.9 (62.3)	57	2.26	67.8 (68.6)	62
3.78	63.1 (65.9)	60	1.96	71.1 (69.1)	63
2.50	67.4 (68.1)	62			

(a) 30-mm plate heated for 1 h at 900°C (1652°F) and air cooled, then heated for 24 h at 500°C (932°F) and air cooled. (b) d is mean phase boundary intercept distance. (c) Values in parentheses were calculated from the linear least-squares expression relating K_{Ic} to d.

is required to achieve highest toughness. Platelets need to be thick enough to turn a crack while being spaced such that turns are frequent.

Greenfield and Margolin (Ref 7.31) studied a complex experimental alpha-beta alloy for which strength was held constant in both equiaxed alpha and transformed microstructural conditions. For the equiaxed alpha data, toughness increased with beta grain boundary area per unit volume. For the transformed condition, the data showed that toughness increased with percent primary alpha, which is always elongated whether transformed or originating in the grain boundary.

In any case, beta forging can be substituted for beta heat treating. See Chesnutt *et al* (Ref 7.24) for data on alloy Ti-6Al-2Sn-4Zr-6Mo, Petrak (Ref 7.32) for data on alloy Ti-6Al-4V, Ulitchny *et al* (Ref 7.33) for data on alloy Ti-6Al-6V-2Sn, and Chesnutt *et al* (Ref 7.34) for data on all three of these alloys. The results of Chesnutt *et al* (Ref 7.24) are presented in Table 7.8. Curtis and Spurr (Ref 7.27) reported a similar effect of rolling temperature on Ti-6Al-4V; beta rolling enhances K_{Ic}. Chesnutt *et al* (Ref 7.35) and Berryman *et al* (Ref 7.36) presented similar results for the experimental alpha-beta alloy Corona 5 (Ti-4.5Al-5Mo-1.5Cr). Bohanek (Ref 7.37) demonstrated the same effect of transformed structure enhancement of toughness in Ti-6Al-4V billet. He also showed that this effect does not necessarily carry through to a forged part.

Because welds in alloy Ti-6Al-4V will contain transformed products, one would expect such welds to be relatively high in toughness. This is, in fact, the case, as the data of Ferguson and Berryman (Ref 7.17) show. Their data are summarized in Table 7.9.

Table 7.8. Effect of forging procedure on fracture toughness of alloy Ti-6Al-2Sn-4Zr-6Mo(a) (Ref 7.24)

Forging temperature	K_{Ic} MPa·$m^{1/2}$	K_{Ic} ksi·$in.^{1/2}$
55°C (100°F) below beta transus	41, 40	37, 36
40°C (70°F) above beta transus	71, 72	65, 66

(a) Heated for 1 h at 885°C (1625°F) and air cooled, then heated for 8 h at 595°C (1100°F) and air cooled.

Table 7.9. Fracture toughness of alloy Ti-6Al-4V (0.11 wt % O_2) in welds and heat affected zones (Ref 7.17)

	K_{Ic}					
	Weld		Heat-affected zone		Base metal(a)	
Postweld stress relief	MPa·m$^{1/2}$	ksi·in.$^{1/2}$	MPa·m$^{1/2}$	ksi·in.$^{1/2}$	MPa·m$^{1/2}$	ksi·in.$^{1/2}$
2 h at 590°C (1100°F), AC	87(b)	79(b)	81(c)	74(c)	92	84
1 h at 650°C (1200°F), AC	85(d)	77(d)	77(d)(e)	70(d)(e)	92	84
1 h at 760°C (1400°F), AC	...	...	76(d)	69(d)	92	84

(a) Recrystallization anneal from Fig. 7.7 at 0.11 wt % O_2. (b) Based on data from 2 samples. (c) Based on data from 20 samples. (d) Based on data from 1 sample. (e) Annealed for 2 h at 650°C (1200°F), AC.

Grain size apparently is not always a definite variable. Margolin *et al* (Ref 7.38) showed that toughness first decreased, and then increased, as beta grain size decreased in equiaxed Ti-5.25Al-5.5V-0.9Fe-0.5Cu at a yield strength of 1240 MPa (180 ksi), whereas at 1140 MPa (165 ksi) the toughness of the same equiaxed alloy decreased continuously with decreasing grain size (Ref 7.31). Mahajan and Margolin (Ref 7.39) also showed that cracks tend to develop at interfaces between primary alpha and transformed beta and along slip bands in alpha on surfaces of Ti-6Al-2Sn-4Zr-6Mo.

Recrystallization annealing is a very efficient method of obtaining high fracture toughness. Also, the relatively precise temperature control necessary to limit primary alpha to a low volume fraction is not needed. As discussed above, a recrystallization anneal is effected simply by heating to a temperature about 70°C (126°F) below the transformation temperature, holding for a time sufficient to reach equilibrium, and then cooling slowly to about 700°C (1300°F). A transformed structure is thereby completely avoided; the resulting microstructure is nearly free of dislocations, and it is this characteristic that justifies the term "recrystallization anneal". This procedure is well established for the Ti-6Al-4V ELI alloy, and typical data for the recrystallization annealed alloy are shown in Fig. 7.7. The slow cooling should be terminated at about 700°C (1300°F) to avoid ordering and Ti_3Al formation. This is less important for the ELI grade than for the standard grade.

A real virtue of the recrystallization annealed microstructure is that substantial toughness is achieved while maintaining ductility and fatigue resistance at high strength levels. Such a microstructure also tends to reduce scatter in the data, thus permitting higher design criteria.

7.4.3. Effects of Texture

Effects of texture arise from preferred crystallographic orientation in the material of interest. With all $\bar{c}$ axes of the alpha grains tending to lie along one (or a few) direction(s) in the product, the physical and mechanical property values necessarily will depend on the direction in which they are measured. Toughness is no exception. The effects that crystallographic texture can have on properties of Ti-6Al-2Sn-4Zr-6Mo (Ref 7.40) are shown in Table 7.10. Table 7.10 is not typical of mill or forged products; the data are given only for illustration. The

Table 7.10. Effect of test direction on mechanical properties of textured Ti-6Al-2Sn-4Zr-6Mo plate (Ref 7.40)

Test direction(a)	Tensile strength, MPa	Yield strength, MPa	Elongation, %	Reduction in area, %	Elastic modulus, GPa	K_{Ic} MPa·m$^{1/2}$	K_{Ic} ksi·in.$^{1/2}$	K_{Ic} specimen orientation
L	1027	952	11.5	18.0	107	75	68	L-T
T	1358	1200	11.3	13.5	134	91	83	L-T
S	938	924	6.5	26.0	104	49	45	S-T

(a) High basal pole intensities reported in the transverse direction, 90° from normal, and also intensity nodes in positions 45° from the longitudinal (rolling) direction and about 40° from the plate normal.

Ti-6Al-4V alloy shows similar behavior (Ref 7.41 to 7.45). Harmsworth (Ref 7.46) gives similar data for Ti-6Al-6V-2Sn.

Tchorzewski and Hutchinson (Ref 7.45) showed that effects of alpha phase crystallographic texture can override effects of microstructure and grain size in Ti-6Al-4V. They found that texture affects both the onset of crack extension and the maximum load sustainable in the presence of a fatigue crack. Moreover, texture was found to influence the shape of the plastic zone ahead of the crack. Finally, these authors concluded that the conditions for plane strain fracture may be more stringent in materials having certain textures than in isotropic materials; this remains to be verified.

Toughness directionality is not restricted to alpha or to alpha-beta alloys. Williams *et al* (Ref 7.47) demonstrated that toughness in near-beta alloys is a directional property. In this context, near-beta alloys are those alloys in which primary alpha is the minor phase. Such alloys commonly contain beta-stabilizing additions of 8 wt % or more.

Because the basic crystallographic texture arises during hot working operations and cannot be entirely eliminated by heat treatment, there are no known methods of producing wrought titanium parts having completely random texture and thus zero directionality. However, by balancing the hot working operations as much as possible along all three reference axes, it is possible to reduce directionality to acceptable levels. But every part is different, and it has become usual practice in the aerospace industry for critical parts to be qualified by the fabricator.On the other hand, by tailoring the alpha phase crystallographic texture to a specific need—say, modulus along a fiber axis—texturing offers a significant potential for enhancing properties. In practice to date (1980), attempts to develop specific textures have been somewhat inconsistent. Paton *et al* (Ref 7.48) suspect that textured material may have increased susceptibility to environmental effects in specific directions.

7.4.4. Effects of Environment

Effects of temperature on toughness are usually less abrupt for titanium than for common low-alloy steel. For example, Tobler (Ref 7.49) reported a gradual K_{Ic} transition temperature between −196 and −143°C (−320 and −215°F) for recrystallization annealed Ti-6Al-4V ELI. For temperatures at and above −143°C (−215°F), his K_{Ic} values were typically about 90 MPa · $m^{1/2}$ (82 ksi · in.$^{1/2}$). At −196°C (−320°F), his values were typically 60 to 65 MPa · $m^{1/2}$ (55 to 60 ksi · in.$^{1/2}$). The loss is about 30%. The early conclusion by Christian and Hurlich (Ref 7.50) that Ti-6Al-4V ELI may be used to cryogenic temperatures thus has some justification. The same may not be true of standard grade Ti-6Al-4V. Tobler's product contained 62 ppm hydrogen, whereas that of Christian and Hurlich contained 70 to 100 ppm. Other authors who have provided temperature-dependent toughness results for alloy Ti-6Al-4V are Cervay (Ref 7.51) and Hall *et al* (Ref 7.52).

Van Stone *et al* (Ref 7.20) found no significant K_{Ic} transition temperature for Ti-5Al-2.5Sn ELI at temperatures down to −263°C (−442°F). Values for K_{Ic} did decrease with temperature, however.

Chait and Lum (Ref 7.53) determined the toughness trend (Charpy energy per unit area) with temperature for Ti-6Al-6V-2Sn. Because of the rich beta content, their low-temperature toughness values were quite low. There was no sharp transition temperature. The toughness vs temperature curve had a simple "S" shape over the temperature range from −196 to +377°C (−320 to +700°F). Harmsworth (Ref 7.46) reported K_{Ic} data for alloy Ti-6Al-6V-2Sn over the temperature range from −54 to +93°C (−65 to +200°F) the trends of which were in agreement with those of Chait and Lum.

In aerospace applications, there is a natural concern that chemical environmental factors such as water, salt water or jet fuel will alter toughness in critical components. Data obtained at McDonnell Aircraft (Ref 7.54) on annealed standard grade Ti-6Al-4V indicate the following environmental effects on the apparent value of fracture toughness:

Laboratory air	56 MPa · $m^{1/2}$ (51 ksi · in.$^{1/2}$)
JP4 fuel	47 MPa · $m^{1/2}$ (43 ksi · in.$^{1/2}$)
3.5% salt solution....	34 MPa · $m^{1/2}$ (31 ksi · in.$^{1/2}$)

The results obtained by Ferguson and Berryman (Ref 7.17) and by Hall *et al* (Ref 7.52) are similar in that salt water degrades K_{Ic}. The early results of Hatch *et al* (Ref 7.55) also showed that salt water can degrade crack growth resistance even in thin sheet where plane strain conditions do not exist. These authors also found significant alloy effects on crack growth in salt water.

Curtis and Spurr (Ref 7.27) charted the effects of quenching temperature on K_{Ic} and K_{Iscc} (3.5% salt solution) for alloys Ti-6Al-4V and Ti-4Al-3Mo-1V. The K_{Ic} and K_{Iscc} curves parallel each other for each alloy. The K_{Ic} and K_{Iscc} curves are also parallel for Ti-6Al-4V where rolling temperature is the independent variable and for Ti-4Al-3Mo-1V where either aging temperature or aging time is the independent variable. These authors also showed beta annealing to benefit both K_{Ic} and K_{Iscc} in analogous ways. For both alloys, however, $K_{Iscc} < K_{Ic}$. Although these experiences are limited, one might expect that most factors which enhance K_{Ic} also enhance K_{Iscc}.

7.5. FATIGUE CRACK PROPAGATION

Just as K_{Ic} is important in calculating loads that a structural member can carry in the presence of a flaw, so also it is important in many cases to know what the remaining fatigue life is in the presence of a fatigue crack or other sharp crack. In a very general way, fatigue crack propagation (FCP) behavior in titanium parallels fracture toughness—that is, for a given alloy those conditions giving highest toughness tend also to give lowest cyclic growth rates under fatigue loading. Amateau *et al* (Ref 7.56) have collected some crack propagation models, a few of which recognize an inverse K_{Ic}-FCP relationship. The correlation, however, is certainly not so good that one variable can be calculated from the other. Furthermore, there may be an additional variable, such as strength, that affects (or even controls) both K_{Ic} and FCP in similar ways. This possibility is illustrated in Fig. 7.8. It is also quite possible that fatigue crack propagation and toughness do not arise from the same mechanism, especially when the stress-intensity factor range $\Delta K \ll K_{Ic}$.

In FCP measurements, there can be a significant amount of scatter in the data. The example shown in Fig. 7.9 is one of the more extreme cases encountered and arises for the mill annealed condition where uncertainties of microstructure, tex-

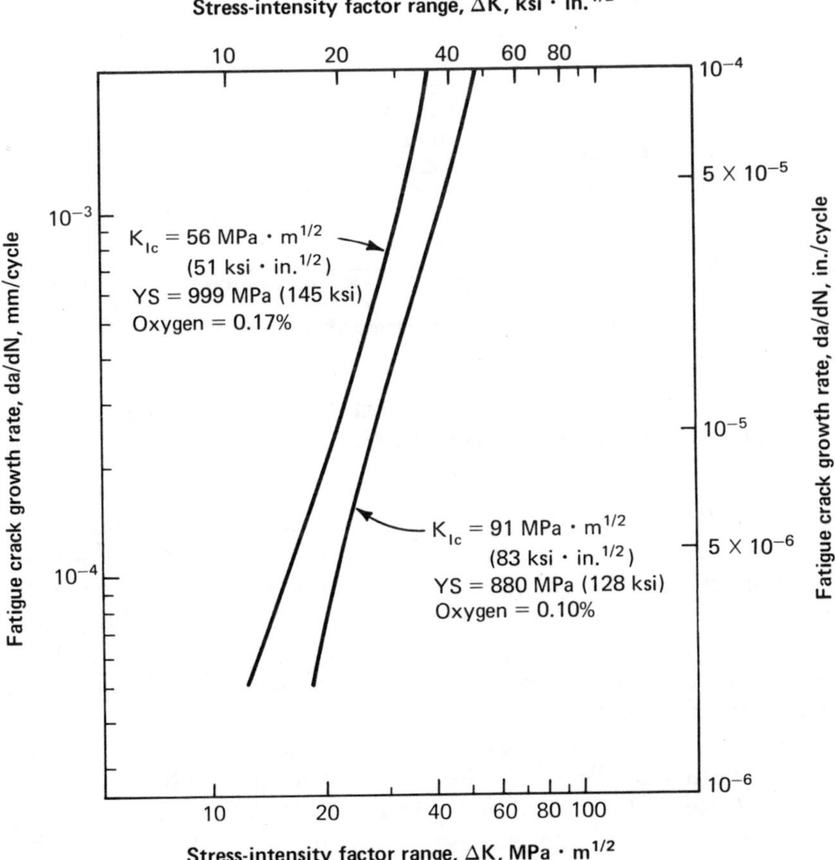

R = 0.02; 10 Hz; air, argon and JP4 environments pooled; average of six tests per trend line.

Fig. 7.8. Variation of fatigue crack propagation rate with yield strength, K_{Ic} and other variables for annealed Ti-6Al-4V forgings (Ref 7.54)

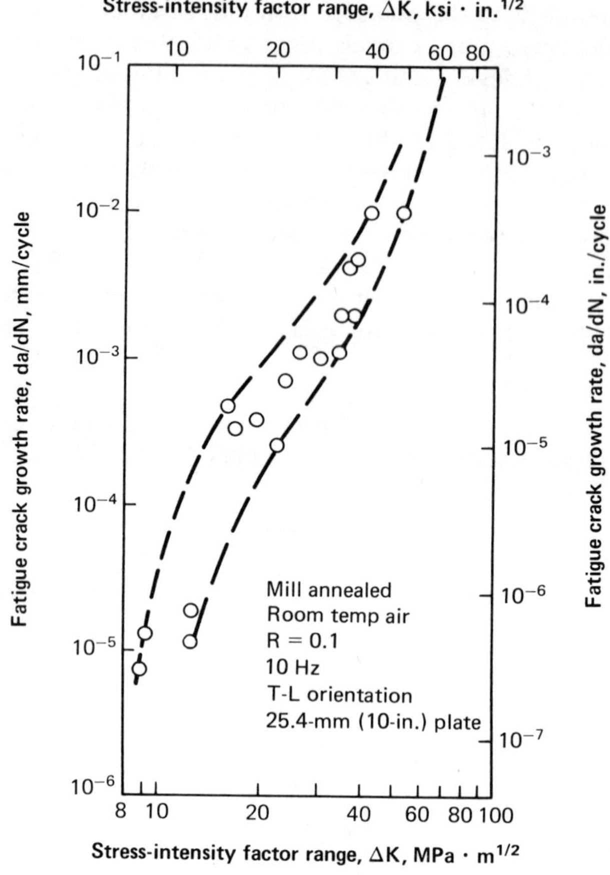

The data are for six heats of mill annealed alloy Ti-6Al-4V.

Fig. 7.9. Illustration of the scatter that can occur in fatigue crack propagation measurements (Ref 7.41)

ture and strength may exist. Part of the scatter is due to test reproducibility. There may also be a point-to-point material variability. The latter is due to minor processing and material inhomogeneities. Variations in chemistry, microstructure, and texture effects within a given lot may, in some cases, be additive even under controlled conditions. There are, of course, lot-to-lot differences. For design purposes, users are well advised to use statistical data derived from information in digital form from several lots.

Much less scatter than that illustrated in Fig. 7.9 was observed for most other conditions reported by Bjeletich (Ref 7.41). See also Tobler (Ref 7.49) and Chesnutt *et al* (Ref 7.24) for additional data. Like Bjeletich, these authors present data in scatter-plot form. One purpose of showing Fig. 7.9 is to give the reader a feel for what is required before an effect can be confirmed. For example, two separate curves within the scatter band do not necessarily indicate a trend, and more data would be required.

The basic Paris equation, $da/dN = C(\Delta K)^n$, is only approximately obeyed for titanium. Deviations from the Paris equation, while often fairly reproducible in very specific material forms under very specific conditions, are not systematic. Generally, log-log plots of da/dN and ΔK yield three regions. A Stage I threshold region usually lies somewhere below about ΔK = 12 MPa · $m^{1/2}$ (11 ksi · $in^{1/2}$) and is characterized by steep slopes where da/dN increases very rapidly with ΔK. The Stage II Paris law region of more moderate slope extends upward from ΔK values of about 12 MPa · $m^{1/2}$. At still higher ΔK values, a Stage III portion of the curve slopes more and more steeply as ΔK approaches K_{Ic}.

Studies of fracture surfaces have shown that there is a transition from cyclic cleavage to striations as ΔK increases beyond the threshold stress. For the microstructures studied by Yuen *et al* (Ref 7.57), the transition occurs at about 13 MPa · $m^{1/2}$ (12 ksi · $in.^{1/2}$). Paton *et al* (Ref 7.7), however, found the transition to occur over a range of ΔK. Yuen *et al* (Ref 7.57) proposed that the transition occurs when the cyclic plastic zone size is on the order of the alpha grain size and corresponds to a transition from single to multiple slip. During a study of Widmanstätten structures, Yoder *et al* (Ref 7.58) found that the transition occurs when the mean packet size equals the cyclic plastic zone size.

Wells (Ref 7.59) reported that there is a material thickness effect on FCP, but, because his data lie within a scatter band similar to that illustrated in Fig. 7.9, his results require confirmation.

The following section addresses the variables that influence fatigue crack propagation, FCP, in titanium. As in the previous section, the effects of chemistry, microstructure, texture and environment are discussed in order.

7.5.1. Effects of Alloy Chemistry

Titanium alloys have different FCP characteristics just as they have different K_{Ic} characteristics. Selected data from Beyer *et al* (Ref 7.60), summarized in Table 7.11, indicate that fatigue cracks propagate more rapidly in Ti-6Al-2Sn-4Zr-6Mo than in Ti-8Al-1Mo-1V or Ti-6Al-2Sn-4Zr-2Mo under the same test conditions. This may be a simple effect of strength. However, the relative amounts of beta phase may lead to intrinsically different fatigue crack propagation characteristics. The Ti-6Al-2Sn-4Zr-6Mo alloy is also more easily textured. In addition, different phases, such as orthorhombic martensite, may exist and could effect FCP problems after aging.

Table 7.11. Selected data on effect of alloy type on fatigue crack propagation resistance in room temperature air at 0.6 Hz (Ref 7.60)

Fatigue crack propagation, da/dN			Stress-intensity factor range, ΔK: Ti-8Al-1Mo-1V		Ti-6Al-2Sn-4Zr-2Mo		Ti-6Al-2Sn-4Zr-6Mo	
m/cycle	in./cycle	R(a)	MPa·$m^{1/2}$	ksi·$in.^{1/2}$	MPa·$m^{1/2}$	ksi·$in.^{1/2}$	MPa·$m^{1/2}$	ksi·$in.^{1/2}$
2.54×10^{-7}	1×10^{-5}	0.1	22	20	23	21	17	15
		0.5	17	15	18	16	15	14
2.54×10^{-8}	1×10^{-6}	0.1	13	12	11(b)	10(b)	...	...
		0.5	10	9	10	9	7	6

(a) Load ratio. (b) Extrapolated.

Just as for K_{Ic}, it appears that oxygen has an effect on FCP. Results of one investigation are presented in Fig. 7.8. In addition, the results obtained by Ferguson and Berryman (Ref 7.17), which are summarized in Table 7.12, show that a lower oxygen content improves FCP resistance. The effect of oxygen in these data is significant only in a statistical sense for ΔK values above about 20 MPa · $m^{1/2}$ (18 ksi · $in.^{1/2}$). In that region, increases in oxygen content cause increases in crack propagation rate. Such a trend might be anticipated on the premise that K_{Ic} approximates a limiting value for ΔK in FCP and that K_{Ic} is influenced negatively by oxygen.

Yoder *et al* (Ref 7.61) proposed that an apparent effect of oxygen content on FCP in beta annealed Ti-6Al-4V can be ascribed to an influence of oxygen on the Widmanstätten α packet or colony size and that it is the latter that controls FCP. Lower da/dN values were apparently associated with higher oxygen contents and larger Widmanstätten α colonies. This idea needs confirmation. However, Yoder and coworkers (Ref 7.58, 7.62 and 7.63) have adequately established the packet size effect.

Chesnutt *et al* (Ref 7.24) reported similar trends and, moreover, showed that the effect of oxygen in Ti-6Al-4V extends to at least 317°C (600°F) for the beta annealed condition. Variations in oxygen content between 0.082 and 0.185% had essentially no effect on FCP for the recrystallization annealed condition.

Chesnutt *et al* (Ref 7.24) also reported effects of hydrogen content on FCP. Hydrogen in the 5 to 325 ppm range appears to accelerate FCP in beta annealed Ti-6Al-4V but has no discernible effect on FCP in recrystallization annealed Ti-6Al-4V. At 317°C (600°F), hydrogen has no effect on FCP in either microstructure. Additional FCP studies by Mahoney and Paton (Ref 7.64) on a Si containing alloy at ambient temperature and with a hold at K_{max} during fatigue cycling have shown that significant accelerations can occur when hydrogen content is in the 150 to 350 ppm range. Further studies by Chesnutt and Paton (Ref 7.65) have shown that a combination of subambient temperature with a hold at K_{max} during fatigue cycling can cause significant accelerations in FCP in Ti–6 wt % Al and Ti-6Al-4V alloys in the recrystallization annealed condition. Kennedy *et al* (Ref 7.66) showed the same effect in Ti-6Al-4V and also demonstrated that hydrogen migrates into the region of the crack tip under the influence of the stress gradient.

Table 7.12. Effect of oxygen on fatigue crack propagation in room temperature air for Ti-6Al-4V, RA or RA/DB(a), plate at R = 0.3 and frequencies of 1 to 6 Hz(b) (Ref 7.17)

Fatigue crack propagation, da/dN		Stress-intensity factor range, ΔK			
		0.20% O_2		0.08-0.10% O_2	
m/cycle	in./cycle	MPa·$m^{1/2}$	ksi·$in.^{1/2}$	MPa·$m^{1/2}$	ksi·$in.^{1/2}$
2.54×10^{-5}	10^{-3}	56,53,46	51,48,42	60,64	55,58
2.54×10^{-6}	10^{-4}	39,33,33,33	35,30,30,30	43,43	39,39
2.54×10^{-7}	10^{-5}	19,19,18	17,17,16	21,21	19,19
2.54×10^{-8}	10^{-6}	11,11,12	10,10,11	12,15	11,14

(a) RA = recrystallization annealed; DB = diffusion bonded or DB thermal cycle. (b) No frequency effect over range used.

7.5.2. Effects of Microstructure

Here also there are general parallels between K_{Ic} and FCP. Like K_{Ic}, FCP is favorably influenced by transformation microstructure and also by application of recrystallization-anneal-type thermal cycles. Microstructure in a given lot of Ti-6Al-4V can affect FCP by a factor of more than ten, and can affect ΔK by 5 to 30 MPa · $m^{1/2}$ (4 to 27 ksi · $in.^{1/2}$), depending on where these parameters are measured on the da/dN curve; see, for example, Paton *et al* (Ref 7.7), who reported the work of Williams and Rauscher. Generally speaking, beta annealed microstructures have the lowest fatigue crack growth rates, whereas mill annealed microstructures yield the highest growth rates. A typical example of such behavior is shown in Fig. 7.10, after Ferguson and Berryman (Ref 7.17). Wells (Ref 7.59) also reported lower FCP rates in Ti-6Al-4V after beta annealing. Bjeletich (Ref 7.41) and Tobler (Ref 7.49) reported results for weldments in Ti-6Al-4V. The welds were equivalent in FCP to the base metal. Yoder *et al* (Ref 7.67)

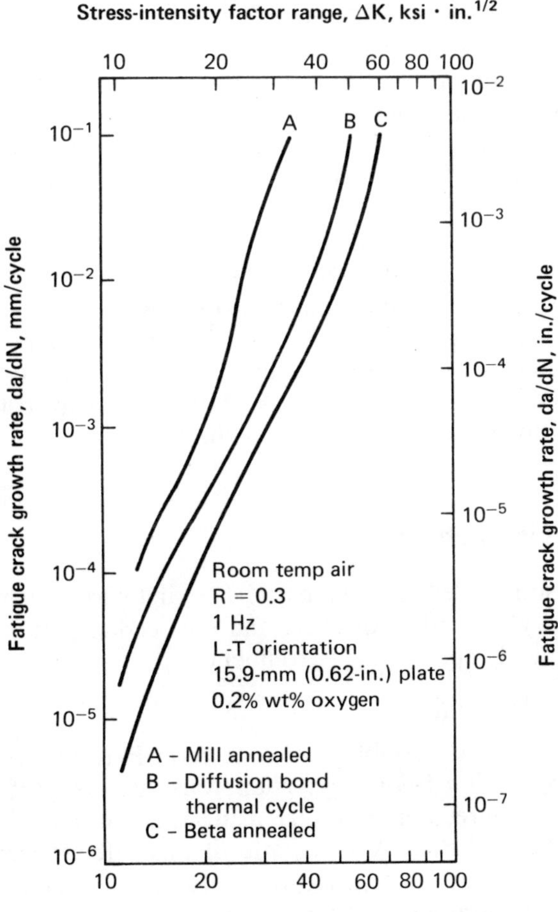

Fig. 7.10. Effect of heat treatment on fatigue crack growth rate of alloy Ti-6Al-4V (Ref 7.17)

reported a 50-fold difference in Region II FCP, but both microstructure and oxygen level varied in the specimens they studied.

Amateau *et al* (Ref 7.56) showed that the effects of Widmanstätten α microstructure on FCP in Ti-6Al-6V-2Sn is entirely analogous to that in Ti-6Al-4V. However, Yoder *et al* (Ref 7.68) found crack branching in Ti-6Al-4V but not in Ti-6Al-6V-2Sn. Chesnutt *et al* (Ref 7.24) found that transformed microstructures provide generally lower rates of FCP in Ti-6Al-2Sn-4Zr-6Mo.

7.5.3. Effects of Crystallographic Texture

Ferguson and Berryman (Ref 7.17) found effects of FCP test direction to be minimal in Ti-6Al-4V plate and forgings for a range of oxygen levels and heat treatment conditions. Because section sizes were generally above 14 mm (1/2 in.) in this work, it can be surmised that crystallographic texture was not strongly developed. Hall *et al* (Ref 7.52) reported a similar result for beta annealed Ti-6Al-4V ELI involving a single comparison at 60 Hz.

Stubbington (Ref 7.5) described his own work and the results of Bowen (Ref 7.69) which showed a significant effect of test direction in a strongly basal textured Ti-6Al-4V bar at room temperature. He concluded that, whereas a five-minute dwell at maximum load had no effect when the maximum stress was parallel to the basal plane, this was not the case when the maximum stress was normal to the basal plane. In the latter instance, a five-minute hold caused a seven-fold increase in FCP rate. Moreover, he found that this increase depended on the volume fraction of primary alpha phase.

Stubbington (Ref 7.5) also found the factor n in the Paris equation to vary with test direction. For the L-S, S-L and T-S orientations, n was about 2.5 and the fatigue striations were normal to the direction of growth. For the L-T, S-T and T-L orientations, n values ranged from 3.1 to 4.1. For these orientations, the fracture topography was complex and no single crack growth model could be proposed.

In summary, the effects of texture exist, but, at least for certain mill product forms, they may not be discernible unless the crystallographic texture is fairly well developed.

7.5.4. Environmental Effects

This section is subdivided into two parts. The first part deals with the mechanical factors of load ratio and frequency, and the second part addresses the temperature and chemical aspects of environment.

7.5.4.1. Mechanical Factors

The most often studied variables are load ratio R (minimum load/maximum load) and load application frequency. Numerous authors (Ref 7.5, 7.17, 7.52, 7.60 and 7.64) are in agreement that increasing R at the same ΔK level generally leads to increased FCP rates. A schematic of this effect (after Chesnutt *et al*; Ref 7.24) appears in Fig. 7.11. Load ratio effects are usually strongest in the Stage I and Stage III portions of the FCP curves.

Yuen *et al* (Ref 7.57) reported that this effect extends into the region of negative R, at least to $R = -5.0$ for Ti-6Al-4V. The results of Beyer *et al* (Ref 7.60) on

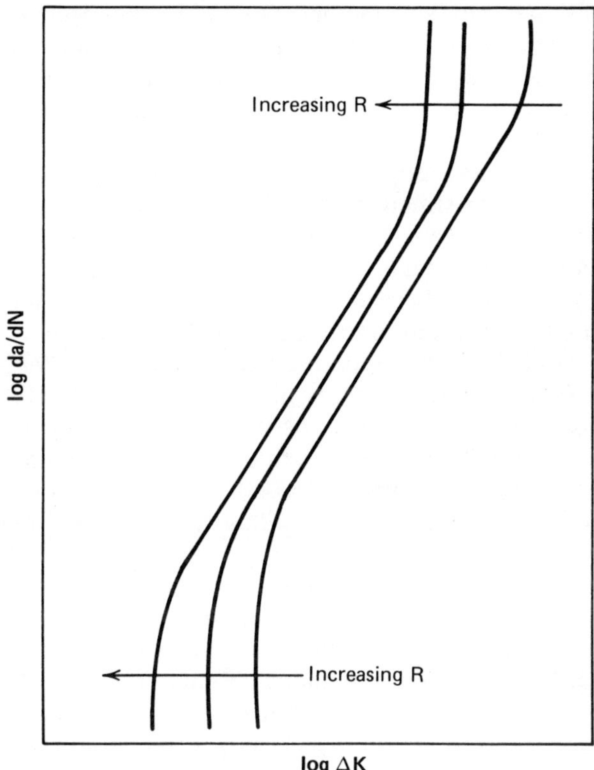

Note that the effect of R on da/dN is more pronounced in Stage I and Stage III.

Fig. 7.11. Schematic illustration of da/dN-ΔK curve behavior as a function of increasing R in titanium alloys (Ref 7.24)

Ti-8Al-1Mo-1V, Ti-6Al-2Sn-4Zr-2Mo and Ti-6Al-2Sn-4Zr-6Mo are not in agreement, however; here, increasing negative R ratios gave lower FCP rates.

The data of Ferguson and Berryman (Ref 7.17) on Ti-6Al-4V support the results of Chesnutt *et al* (Ref 7.24). In these cases, increasing positive R ratios increased the FCP rate.

Hall *et al* (Ref 7.52) agree with both of the above references and show that the extent to which Stage III diverges as R changes depends on test conditions and environment.

Frequency effects are complex. Lower frequencies lead to higher FCP rates only in a general, over-all sense. Many comparisons do not show frequency effects, whereas others do. There appear to be some reasons for this.

The work of Beyer *et al* (Ref 7.60) addresses one important aspect. In their work on Ti-8Al-1Mo-1V at 482°C (900°F), frequency had little or no effect at R = 0.1 but had rather large effects at R = 0.5 and 0.7. Similar but not identical trends were noted for Ti-6Al-2Sn-4Zr-2Mo and Ti-6Al-2Sn-4Zr-6Mo alloys at high temperatures by the same authors.

Bjeletich (Ref 7.41) studied another aspect. In the heat affected zones of electron beam weldments in Ti-6Al-4V, he found no frequency effect over the range from 0.1 to 10 Hz at R = 0.1 in laboratory air at 79°C (175°F). For these test conditions, however, the introduction of a 3.5% salt solution produced a significantly higher FCP rate for the lower frequency.

Chesnutt *et al* (Ref 7.24) reported a frequency effect (1 vs 20 Hz) on FCP in Ti-6Al-4V in several microstructures in a 3.5% salt solution at R = 0.1 at room temperature, whereas Ferguson and Berryman (Ref 7.17) reported little or no effect of frequency over the range from 1 to 30 Hz for testing in air. The latter authors also reported that neither frequency (1 to 6 Hz range) nor R ratio (0.08 to 0.5) affects the FCP rate in mill annealed Ti-6Al-4V sheet.

Chesnutt *et al* (Ref 7.24) also reported results of tests at 20 Hz on Ti-6Al-4V (in six microstructural conditions) in dry air that were interrupted periodically for five-minute holds at maximum load. For three of the six microstructures, hold time had no effect. Two other microstructures were well within the scatter band. Only for the recrystallization annealed condition were the results conceivably indicative of an effect, with hold times leading to higher FCP rates. Even in that case, however, four of seven data points were within the scatter band.

Frequency effects, however, may depend on factors such as chemistry and environment. See the works of Chesnutt and Paton (Ref 7.65) and Mahoney and Paton (Ref 7.64) for hydrogen and temperature interactions in this regard. Bania and Eylon (Ref 7.70) reported a possible beneficial effect of a five-minute dwell at maximum load for Ti-6Al-4V that, if real, depends on heat treatment. This effect is suppressed for the solution treated and aged condition, and is present for three different annealing treatments.

7.5.4.2. Thermal and Chemical Environments

The data of Bjeletich (Ref 7.41) show insignificant differences in FCP rates over the temperature range from −53 to +77°C (−64 to +170°F) for electron beam weldments in Ti-6Al-4V plate. The results of Ferguson and Berryman (Ref 7.17) are similar except that, in the threshold region, FCP is faster at room temperature than at −53°C (−64°F).

Chesnutt *et al* (Ref 7.24), however, studied a wider range of temperatures (−83 to +317°C; −118 to +600°F) and found that increasing temperature accelerates FCP at 20 Hz and R = 0.7 in Ti-6Al-4V pancake forgings given recrystallization anneals or alpha-beta solution treatments followed by aging. For a beta-finish forging given a beta solution treatment and an over-aging treatment, however, these authors found little or no temperature effect. For R = 0.3, the converse of the foregoing microstructure effects appears to apply.

Tobler (Ref 7.71) examined the FCP behavior of Ti-6Al-4V ELI over the temperature range from −269 to +22°C (−452 to +72°F) and found that all data fell within the scatter band.

In summary, the effect of test temperature is not always present, but when it is it seems to depend on both load ratio and microstructure. At least in the literature cited, the effect of increasing temperature, when it exists, is to increase the FCP rate. In theory, however, the opposite effect could be observed if hydrogen were present and its effect diminished with increasing temperature.

Being a reactive metal, titanium will react with all nonmetallic elements excluding the inert gases. For titanium oxides, nitrides and carbides, the free energies of compound formation are higher than for most other metals. Titanium will also readily absorb any nascent hydrogen that happens to be reduced on a fresh surface, such as at the root of an advancing fatigue crack. Such surfaces also absorb oxygen, and perhaps nitrogen, if environmental oxygen becomes depleted.

Because of the reactive nature of titanium, therefore, it should come as no surprise that environment affects FCP rates in titanium just as it affects fracture toughness. The only surprise in the available data is that the chemical environment does not have a larger effect than it does. In general, only the more severe environments (such as a 3.5% NaCl solution) affect FCP rates by an order of magnitude or more, and these effects may in part depend on mechanical factors as discussed in the previous section. Considerably larger effects of environment on crack growth rates may be found near the fatigue threshold.

Chesnutt *et al* (Ref 7.24) tested specimens of Ti-6Al-4V in several environments. The effects of argon, dry air and wet air environments seem to depend at least on microstructure, as shown in Fig. 7.12 and 7.13. The effect of a mildly reactive environment is not always negative.

Ferguson and Berryman (Ref 7.17) studied FCP rates in recrystallization annealed Ti-6Al-4V plate as affected by sump tank water and JP4 fuel environments using low-humidity air as a basis for comparison. Although the effect of sump tank water was a slight increase in FCP rate in most cases, this effect was less than that of an increase in R from 0.3 to 0.5. Where material chemistry was reported, the effect of sump tank water was independent of alloy oxygen content beyond the 0.08 to 0.15% range. The work of Ferguson and Berryman on FCP rates in JP4 fuel is less complete, but the available data indicate that JP4 fuel is a mild environment.

Hall *et al* (Ref 7.52) found FCP rates for beta annealed Ti-6Al-4V ELI plate in JP4 fuel, water, and sump tank water environments to lie between those in air and those in salt water. Salt water is a relatively harsh environment, and its effect is to enhance FCP rates. Increasing R or decreasing frequency enhances the effect of salt water on FCP. Figures 7.14 and 7.15 illustrate the effect of frequency rather dramatically. These authors also found that dye penetrant type ZL-2A has no influence on FCP rate relative to the rate in dry air for both beta annealed Ti-6Al-4V ELI.

Data obtained at McDonnell Aircraft (Ref 7.54) also show JP4 fuel to affect FCP in Ti-6Al-4V to a lesser degree than does salt water (3.5% NaCl).

Harmsworth (Ref 7.46) reported data showing that water and JP4 fuel are relatively mild environments for Ti-6Al-6V-2Sn. As for Ti-6Al-4V, the testing frequency and a salt water environment have a combined effect on FCP for Ti-6Al-6V-2Sn.

7.6. SUSTAINED-LOAD CRACK PROPAGATION

Although rising load and dynamic load spectra are important considerations in vehicle design, the ability of a material to sustain a load without failure in the

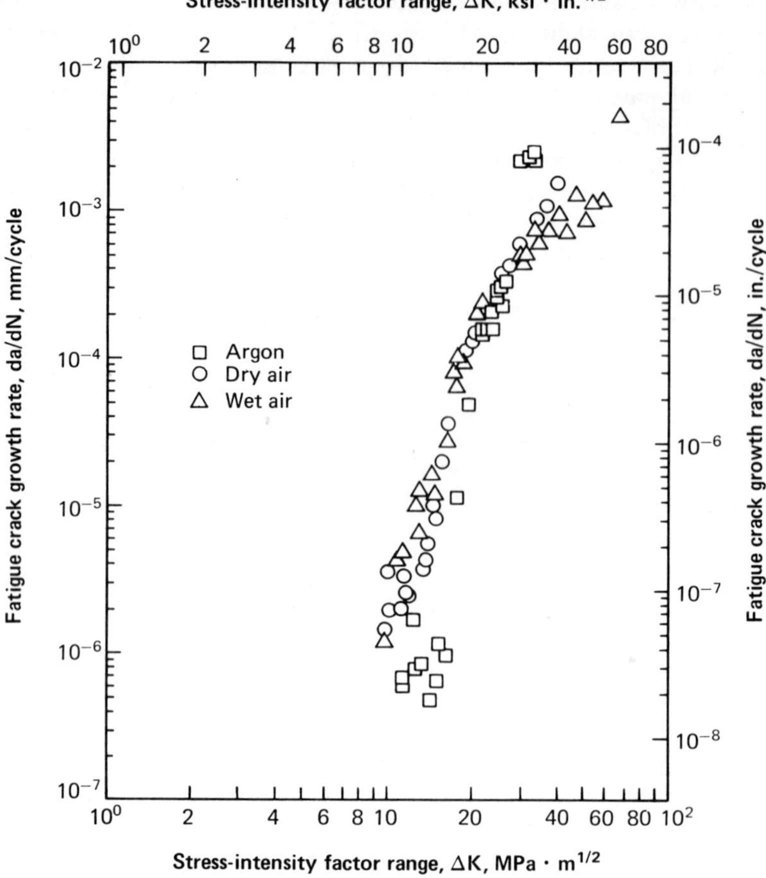

Testing at 20 Hz and R = 0.1. Pancake alpha-beta forged and then heated at 973°C (1780°F) for 4 h, cooled at 50°C (90°F) per hour to 760°C (1400°F), and air cooled.

Fig. 7.12. Dependence of FCP in $\alpha-\beta$ forged Ti-6Al-4V pancake forging on gaseous environment (Ref 7.24)

presence of a fatigue crack or other sharp crack in a specific environment also may be important.

Only a few studies have been reported on sustained-load crack propagation for titanium alloys, and these studies have been concerned primarily with static load crack growth in the presence of JP4 fuel or a 3.5% NaCl solution. Results obtained at McDonnell Aircraft (Ref 7.54) are presented in Fig. 7.16. The rank order of severity of effect of JP4 fuel and 3.5% NaCl on sustained-load crack propagation rate in Ti-6Al-4V is the same as for fatigue crack propagation rate: 3.5% NaCl has the more severe effect.

Bjeletich (Ref 7.41) reported similar results for Ti-6Al-4V weldments compared with mill annealed base metal. He also found a microstructure and temperature effect, which is shown in Fig. 7.17. Just as is the case for K_{Ic} and FCP,

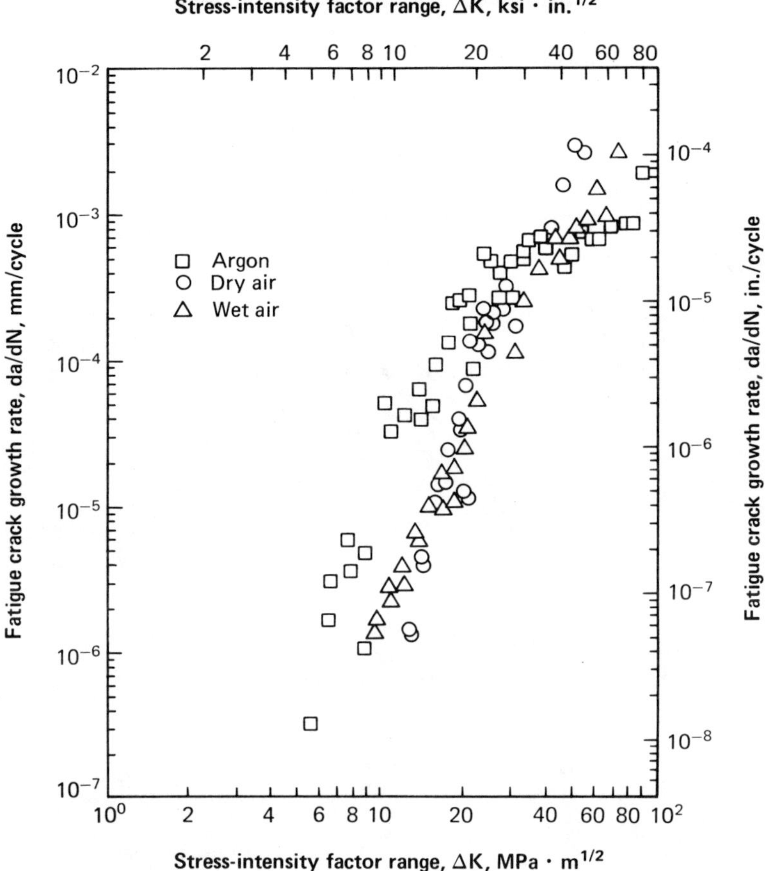

Tested at 20 Hz and R = 0.1. Pancake beta forged and then heated at 1037°C (1900°F) for 1/2 h, water quenched, reheated to 704°C (1300°F), and air cooled.

Fig. 7.13. Dependence of FCP in *β* forged Ti-6Al-4V pancake forging on gaseous environment (Ref 7.24)

transformed microstructures (in the welds) lead to lower propagation rates. The negative effect of temperature is probably due to crack blunting via enhanced corrosion at the crack tip.

For the Ti-4Al-3Mo-1V alloy, Williams (Ref 7.72) found no difference between vacuum and moist air environments in time to failure under the static load. In limited work, he also found little effect of salt water. This lack of effect is probably related to the low aluminum content of this alpha-beta alloy.

In early work, Hatch *et al* (Ref 7.55) pointed out that sustained-load crack propagation in 3.5% NaCl solution is not limited to thick materials. These authors showed that a number of alloys are susceptible in sheet form and moreover that gage effects can exist within a given alloy.

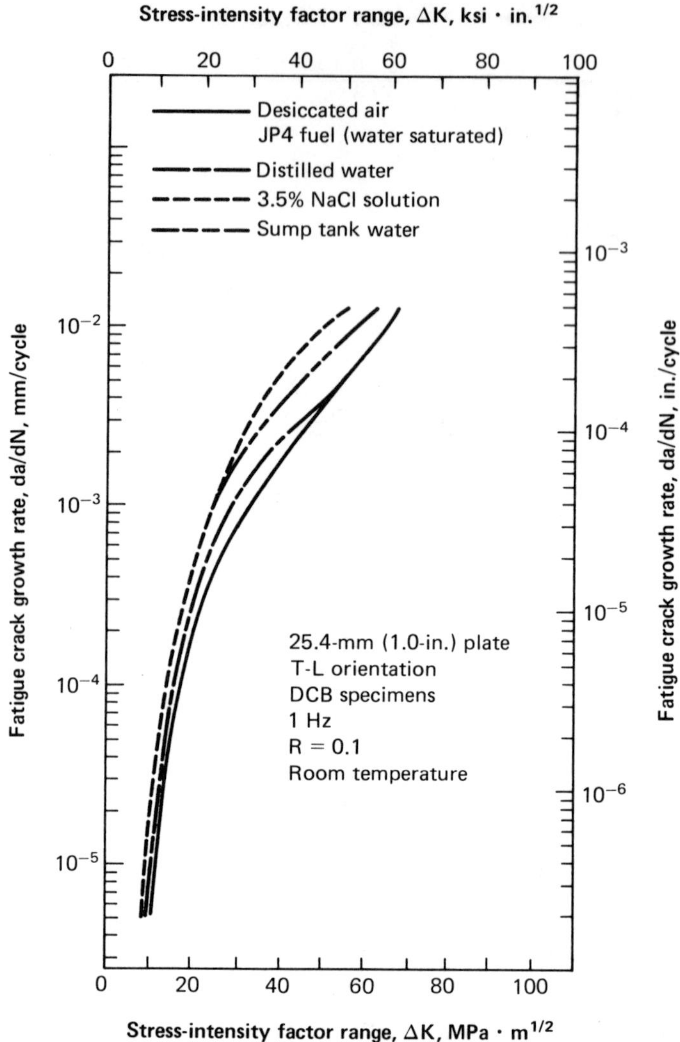

Fig. 7.14. Effect of environment on fatigue crack growth rates at R = 0.1 in recrystallization annealed alloy Ti-6Al-4V (1 Hz) (Ref 7.52)

It is also evident that some care should be taken during sustained-load testing. Petrak (Ref 7.73) found that exposure to room air between the termination of fatigue precracking and static-load testing in salt water may affect K_{Iscc} results in Ti-5Al-2.5Sn. Overnight (or longer) exposures in room air appeared to have the effect of blunting the crack and increasing the K_{Iscc} value. The data are too sparse to indicate the magnitude of the effect.

7.7. CONCLUDING REMARKS

Titanium and titanium alloys have a well-earned reputation for reliability in service. In no small measure, this is a result of the double and triple vacuum arc

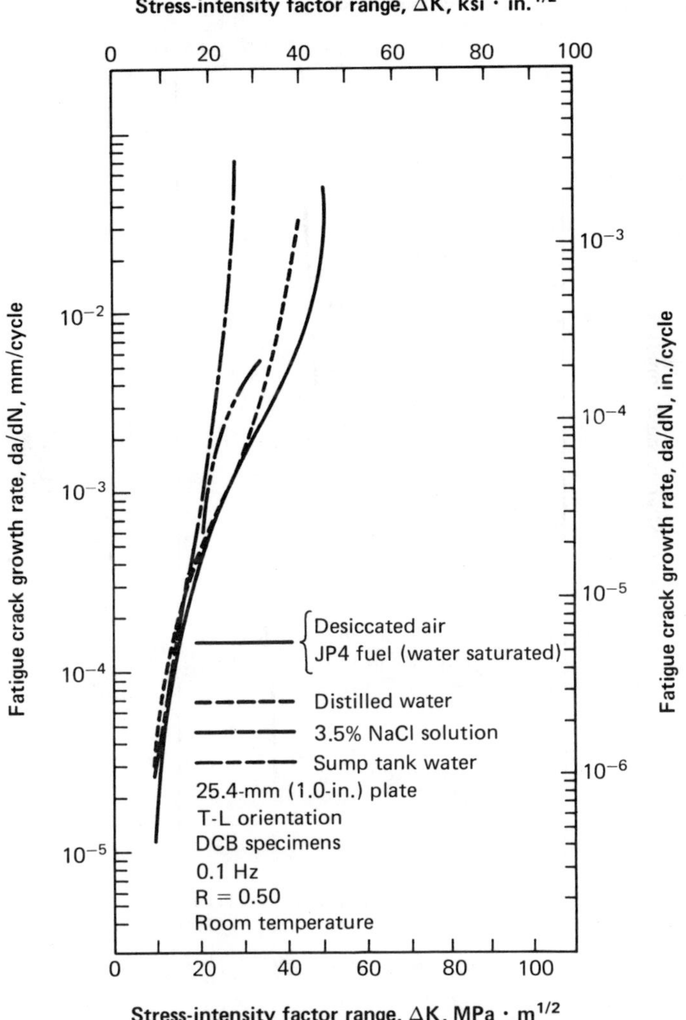

Fig. 7.15. Effect of environment on fatigue crack growth rates at R = 0.50 in recrystallization annealed alloy Ti-6Al-4V (0.1 Hz) (Ref. 7.52)

melting procedures employed throughout the industry in producing the alloys. That reputation is protected also by the excellent corrosion resistance exhibited by titanium. Titanium does not corrode in salt water. Crack initiation in titanium is almost always mechanically induced; only under very special circumstances will cracks initiate due to a combination of an environment and static stress (Ref 7.74).

Because of the many possible effects of chemistry, microstructure, texture, environment and loading, it is not possible to quantify the crack growth behavior of titanium alloys unless these factors are closely controlled.

There are several themes evident in this chapter. Alloys within a given class, such as alpha-beta alloys, show parallel trends in their fracture toughness and crack propagation behaviors. To the extent that they have been studied, the trends

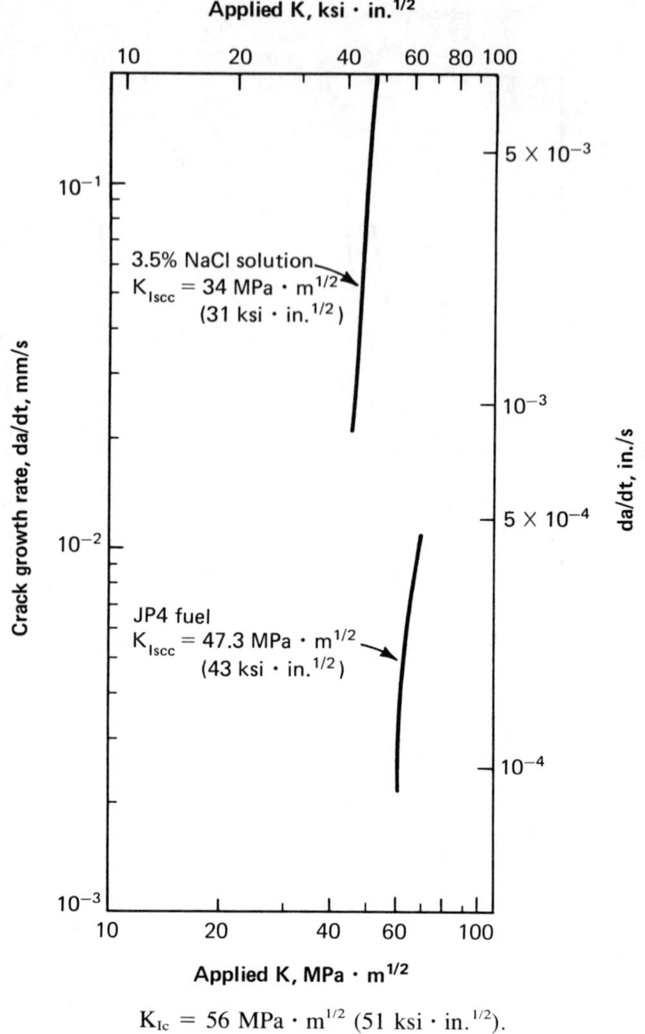

K_{Ic} = 56 MPa · m$^{1/2}$ (51 ksi · in.$^{1/2}$).

Fig. 7.16. Sustained load crack growth behavior of alloy Ti-6Al-4V in two environments (Ref 7.54)

for interstitial effects are similar for all alloys, the higher levels of interstitials leading to faster FCP and lower K_{Ic}. A similar trend is observed for variations in microstructure. Those microstructures (Widmanstätten or recrystallization annealed) that give the highest K_{Ic} values generally yield the lowest crack growth rates whether under fatigue or sustained loads. Moreover, the environmental media studied tend to exhibit similar rank orders of severity among K_{Ic}, FCP, and sustained load crack propagation. Salt water appears to be the most severe of the media studied. Readers interested in additional quantitative comparisons may consult the original references from which this chapter was drawn or may perform their own tests. Finally, those readers who are interested in the available design information or the underlying metallurgy may consult the general references appended.

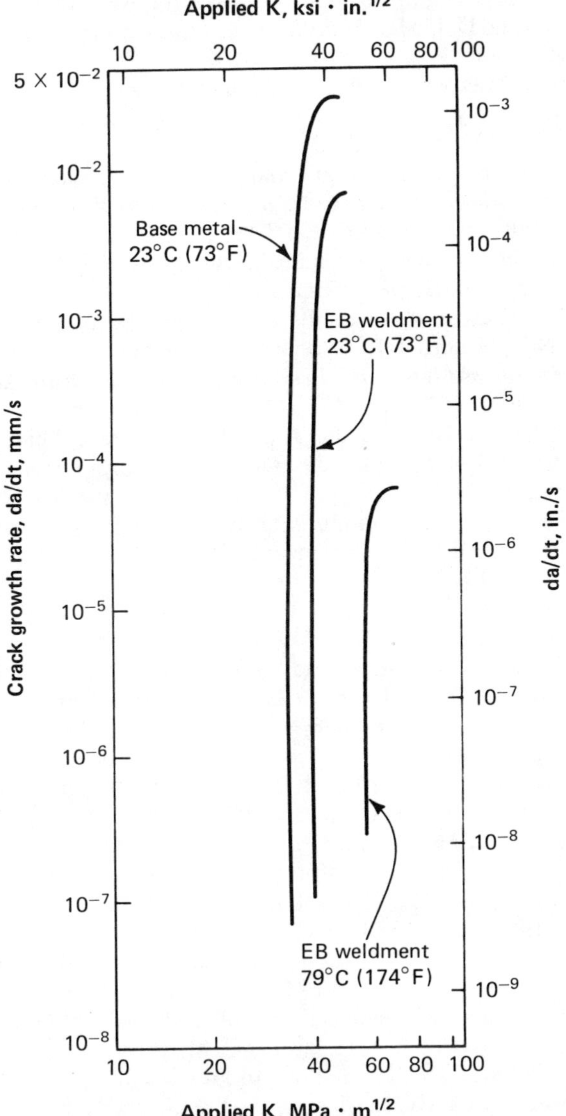

Fig. 7.17. Effect of microstructure and temperature on sustained load crack propagation in Ti-6Al-4V in 3.5% NaCl solution (Ref 7.41)

7.8. REFERENCES CITED IN TEXT

7.1. Uphill Diffusion and Progressive Embrittlement–Hydrogen in Titanium, by J. L. Waisman, R. Toosky and G. Sines: *Metallurgical Transactions A,* Vol 8, No. 8, 1977, p 1249-1256

7.2. Stress-Induced Hydrogen Migration in Beta-Phase Titanium, by P. N. Adler and R. L. Schulte: *Scripta Metallurgica,* Vol 12, No. 7, 1978, p 669-672

7.3. Behavior of Hydrogen in Alpha-Phase Ti-Al Alloys, by N. E. Paton, B. S. Hickman and D. Leslie: *Metallurgical Transactions,* Vol 2, No. 10, Oct 1971, p 2791-2796

7.4. The Effect of Microstructure and Composition on the Fracture Toughness of Titanium Alloys, by D. H. Rogers: in *Titanium Science and Technology,* Vol 3, Plenum Press, 1973, p 1719-1730

7.5. Metallurgical Aspects of Fatigue and Fracture in Titanium Alloys, by C. A. Stubbington: in *Alloy Design for Fatigue and Fracture Resistance,* AGARD-CP-185, NATO, Brussels, Belgium, 1976, p 3-1

7.6. Fracture Topography – Microstructure Correlations in SEM, by J. C. Chesnutt and R. A. Spurling: *Metallurgical Transactions A,* Vol 8, No. 1, 1977, p 216-218

7.7. The Effects of Microstructure on the Fatigue and Fracture of Commercial Titanium Alloys, by N. E. Paton, J. C. Williams, J. C. Chesnutt and A. W. Thompson: in *Alloy Design for Fatigue and Fracture Resistance,* AGARD-CP-185, NATO, Brussels, Belgium, 1976, p 4-1

7.8. Formation Characteristics of the Alpha-Beta-Interface Phase in Ti-6Al-4V, by C. G. Rhodes and N. E. Paton: *Metallurgical Transactions A,* Vol 10, No. 2, 1979, p 209-216

7.9. Mechanical Behavior of Titanium Alloys, by C. G. Rhodes and N. E. Paton: ONR Final Report, SC5056.4FR, Science Center, Rockwell International, Thousand Oaks, CA, April 1979

7.10. Deformation Mechanics of Alpha Titanium Alloys, by H. W. Rosenberg: Ph. D. Thesis, Stanford University, May 1971

7.11. Room Temperature Flow Stress and Strain, by H. W. Rosenberg: in *Scientific and Technological Aspects of Titanium and Titanium Alloys,* Vol 1, edited by J. C. Williams and G. F. Belov, Plenum Press, New York, 1982, p 739-745

7.12. Solid Solution Strengthening in Ti-Al Alloys, by H. W. Rosenberg and W. D. Nix: *Metallurgical Transactions,* Vol 4, No. 5, May 1973, p 1333-1338

7.13. Heat Treating of Titanium Alloys, by H. W. Rosenberg: Paper EM77-16, Society for Manufacturing Engineers, 1976

7.14. A Comparison of Phase Transformations in 3 Commercial Titanium Alloys, by J. C. Williams and M. J. Blackburn: *ASM Transactions Quarterly,* Vol 60, No. 3, Sept 1967, p 373-383

7.15. Deformation Characteristics of Age Hardened Ti-6Al-4V, by G. Welch *et al: Metallurgical Transactions A,* Vol 8, No. 1, 1977, p 169-177

7.16. The Effect of Microstructure on the Control of Mechanical Properties in Alpha-Beta Titanium Alloys, by M. A. Greenfield, C. M. Pierce and J. A. Hall: in *Titanium Science and Technology,* Vol 3, Plenum Press, 1973, p 1731

7.17. Fracture Mechanics Evaluation of B-1 Materials, by R. R. Ferguson and R. G. Berryman: Report AFML-TR-76-137, Vol 1, Rockwell International, Los Angeles, 1976

7.18. Alloy, Texture and Microstructural Effects on Yield Stress and Mixed Mode Fracture Toughness of Titanium, by H. W. Rosenberg and W. M. Parris: in *Fatigue and Fracture Toughness – Cryogenic Behavior,* STP 556, American Society for Testing and Materials, Philadelphia, 1974, p 26-43

7.19. Correlation Study of Fracture Toughness of Airframe Forgings, by D. E. Cooper: TIMET Internal Report, Toronto Quality Control Dept., Jan 31, 1974

7.20. Influence of Composition, Annealing Treatment, and Texture on Fracture Toughness of Ti-5Al-2.5Sn Plate at Cryogenic Temperatures, by R. H. Van Stone *et al*: in *Toughness and Fracture Behavior of Titanium,* STP 651, American Society for Testing and Materials, Philadelphia, 1978, p 154-179

7.21. Investigation of Fracture Mechanism of Ti-5Al-2.5Sn at Cryogenic Temperatures, by R. H. Van Stone, J. R. Low and J. L. Shannon: *Metallurgical Transactions A,* Vol 9, No. 4, 1978, p 539-552

7.22. Effect of Hydrogen on Fracture and Inert-Environment Sustained Load Cracking Resistance of Alpha-Beta Titanium Alloys, by D. A. Meyn: *Metallurgical Transactions A,* Vol 5, No. 11, Nov 1974, p 2405-2414

7.23. C. Chen (private communication): Wyman Gordon Co. data

7.24. Influence of Metallurgical Factors on the Fatigue Crack Growth Rate in Alpha-Beta Titanium Alloys, by J. C. Chesnutt, A. W. Thompson and J. C. Williams: Report AFML-TR-78-68, Rockwell International, Thousand Oaks, CA, 1978

7.25. The Relation Between Crack Propagation Characteristics and Fracture Toughness in Alpha + Beta Titanium Alloys, by T. W. Hall and C. Hammond: in *Titanium Science and Technology,* Vol 2, Plenum Press, 1973, p 1365-1376

7.26. Development of Economical Sheet Titanium Alloy, by T. L. Wardlaw, H. W. Rosenberg and W. M. Parris: Report AFML-TR-73-296

7.27. Effect of Microstructure on Fracture Properties of Titanium Alloys in Air and Salt Solution, by R. E. Curtis and W. F. Spurr: *ASM Transactions Quarterly,* Vol 61, No. 1, March 1968, p 115-127

7.28. Improved Properties of Ti-6Al-6V-2Sn Through Microstructure Modification, by J. A. Hall and C. M. Pierce: Report AFML-TR-70-312, Air Force Flight Dynamics Laboratory, Wright-Patterson AFB, OH, Feb 1971

7.29. Property-Microstructure Relationships in Titanium Alloy Ti-6Al-2Sn-4Zr-6Mo, by J. A. Hall *et al*: Report AFML-TR-71-206, Air Force Materials Laboratory, Wright-Patterson AFB, OH, Nov 1971

7.30. On the Toughness of Two-Phase 6Al-4V Titanium Microstructures, by W. W. Gerberich and G. S. Baker: in *Applications Related Phenomena in Titanium Alloys,* STP 432, American Society for Testing and Materials, Philadelphia, 1967, p 80-99

7.31. Interrelationship of Fracture Toughness and Microstructure in a Ti-5.25Al-5.5 V-0.9Fe-0.5Cu Alloy, by M. A. Greenfield and H. Margolin: *Metallurgical Transactions,* Vol 2, No. 3, March 1971, p 841-847

7.32. Mechanical Property Evaluation of Beta Forged Ti-6Al-4V, by G. J. Petrak: Report AFML-TR-70-291, Dayton University Research Institute, OH, Jan 1971

7.33. Mechanical and Microstructural Properties Characterization of Heat-Treated, Beta-Extruded Ti-6Al-6V-2Sn, by M. G. Ulitchny, H. J. Rack and D. B. Dawson: in *Toughness and Fracture Behavior of Titanium,* STP 651, American Society for Testing and Materials, Philadelphia, 1978, p 17-42

7.34. The Relationship Between Mechanical Properties, Microstructure and Fracture Topography in Alpha-Beta Titanium Alloys, by J. C. Chesnutt, C. G. Rhodes and J. C. Williams: in *Fractography-Microscopic Cracking Processes,* STP 600, American Society for Testing and Materials, Philadelphia, 1976, p 99-138

7.35. The Effect of Microstructure on Fracture of a New High Toughness Titanium Alloy, by J. C. Chesnutt *et al*: in *Fracture 1977,* Vol 2, University of Waterloo Press, Ontario, Canada, 1977, p 195-202

7.36. A New Titanium Alloy With Improved Fracture Toughness: Ti-4.5Al-5Mo-1.5Cr, by R. G. Berryman *et al: Journal of Aircraft,* Vol 14, No. 12, 1977, p 1182-1185

7.37. Relationship of Fracture Toughness to Ti-6Al-4V Billet Microstructure and Processing, by E. Bohanek: TIMET Technical Report No. 49, Project 99-5, TIMET, Pittsburgh, Aug 2, 1971

7.38. Fracture Toughness, Aging Behavior, Grain Growth, and Hardness of Alpha-Beta Titanium Alloys, by H. Margolin *et al*: Report AFML-TR-73-172, Sept 1973

7.39. Surface Cracking in Alpha-Beta Titanium Alloys Under Unidirectional Loading, by Y. Mahajan and H. Margolin: *Metallurgical Transactions A,* Vol 9, No. 3, 1978, p 427-431

7.40. The Effect of Rolling Texture on the Fracture Mechanics of Ti-6Al-2Sn-4Zr-6Mo Alloy, by M. J. Harrigan *et al*: in *Titanium Science and Technology,* Vol 2, Plenum Press, 1973, p 1297-1317

7.41. Development of Engineering Data on Thick-Section Electron Beam Welded Titanium, by J. G. Bjeletich: Report AFML-TR-73-197, Aug 1973

7.42. Effects of Heat Treatment and Microstructure on Tensile and Fracture Toughness Properties of Titanium 6Al-4V Alloy Plate, by P. L. Hendricks: AIAA Paper 74-374, Las Vegas, NV, April 1974

7.43. Effect of Specimen Width on Fracture Toughness of Ti-6Al-4V Plate, by G. S. Hall, S. R. Seagle and H. R. Bomberger: in *Toughness and Fracture Behavior of Titanium,* STP 651, American Society for Testing and Materials, Philadelphia, 1978, p 227-245

7.44. Influence of Crystallographic Orientation on Fracture Toughness of Strongly Textured Ti-6Al-4V, by A. W. Bowen: *Acta Metallurgica,* Vol 26, No. 9, 1978, p 1423-1433

7.45. Anisotropy of Fracture Toughness in Textured Titanium-6Al-4V Alloy, by R. M. Tchorzewski and W. B. Hutchinson: *Metallurgical Transactions A,* Vol 9, No. 8, Aug 1978, p 1113-1123

7.46. Fracture Toughness and Fatigue Properties of Ti-6Al-6V-2Sn Annealed for Possible F-15 Applications, by C. L. Harmsworth: Report No. LA71-1, Wright-Patterson AFB, OH, Feb 2, 1971

7.47. Development of High Fracture Toughness Titanium Alloys, by J. C. Williams *et al*: in *Toughness and Fracture Behavior of Titanium,* STP 651, American Society for Testing and Materials, Philadelphia, 1978, p 64-114

7.48. N. E. Paton, J. C. Williams, J. C. Chesnutt and A. W. Thompson: Paper SC-PP-75-60, AGARD Conference on Alloy Design for Fatigue and Fracture Resistance, Belgium, April 1975

7.49. Low-Temperature Fracture Behavior of a Ti-6Al-4V Alloy and the Electron-Beam Welds, by R. L. Tobler: in *Toughness and Fracture Behavior of Titanium,* STP 651, American Society for Testing and Materials, Philadelphia, 1978, p 267-294

7.50. Physical and Mechanical Properties of Pressure Vessel Materials for Applications in a Cryogenic Environment, by J. L. Christian and A. Hurlich: Report ASD-TDR-628, Part 2, April 1963

7.51. Mechanical Properties of Ti-6Al-4V Annealed Forgings, by R. R. Cervay: Report AFML-TR-74-4, March 1974

7.52. Corrosion Fatigue Crack Growth in Aircraft Structural Materials, by L. R. Hall, R. W. Finger and W. F. Spurr: Report AFML-TR-73-204, Sept 1973

7.53. Influence of Test Temperatures on Fracture Toughness and Tensile Properties of Ti-8Mo-8V-2Fe-3Al and Ti-6Al-6V-2Sn Alloys Heat-Treated to High-Strength Levels, by R. Chait and P. T. Lum: in *Toughness and Fracture Behavior of Titanium,* STP 651, American Society for Testing and Materials, Philadelphia, 1978, p 180-199

7.54. Report MDC-A0913, Phase B Test Program: McDonnell Aircraft Co., McDonnell Douglas Corp., St. Louis, May 18, 1971

7.55. Effects of Environment on Cracking in Titanium Alloys, by A. J. Hatch, H. W. Rosenberg and E. F. Erbin: in *Stress Corrosion Cracking of Titanium,* STP 397, American Society for Testing and Materials, Philadelphia, 1966, p 122-136

7.56. The Effect of Microstructure on Fatigue Crack Propagation in Ti-6Al-6V-2Sn Alloy, by M. F. Amateau, W. D. Hanna and E. G. Kendall: International Conference on Mechanical Behavior of Materials, Kyoto, Japan, Aug 17, 1971, p 77-89

7.57. Correlations Between Fracture Surface Appearance and Fracture Mechanics Parameters for Stage-II Fatigue Crack-Propagation in Ti-6Al-4V, by A. Yuen *et al: Metallurgical Transactions,* Vol 5, No. 8, 1974, p 1833-1842

7.58. Quantitative Analysis of Microstructural Effects of Fatigue Crack Growth in Widmanstätten Ti-6Al-4V and Ti-8Al-1Mo-1V, by G. R. Yoder, L. A. Cooley and T. W. Crooker: *Engineering Fracture Mechanics,* Vol 11, No. 4, 1979, p 805-816

7.59. New Alloys for Advanced Metallics Fighter-Wing Structures, by R. R. Wells: AIAA Paper 74-372, Las Vegas, NV, April 1974

7.60. Titanium Damage Tolerant Design Data for Propulsion Systems, by J. R. Beyer, D. L. Sims and R. M. Wallace: Report AFML-TR-77-101, Air Force Materials Laboratory, Wright-Patterson AFB, OH, June 1977

7.61. Fatigue Crack Propagation Resistance of Beta-Annealed Ti-6Al-4V Alloys of Differing Interstitial Oxygen Contents, by G. R. Yoder, L. A. Cooley and T. W. Crooker: *Metallurgical Transactions A,* Vol 9, No. 10, 1978, p 1413-1420

7.62. Observations on Microstructurally Sensitive Fatigue Crack Growth in a Widmanstätten Ti-6Al-4V Alloy, by G. R. Yoder, L. A. Cooley and T. W. Crooker: *Metallurgical Transactions A,* Vol 8A, No. 11, Nov 1977, p 1737-1743

7.63. On the Effect of Colony Size on Fatigue Crack Growth in Widmanstätten Structure $\alpha+\beta$ Titanium Alloys, by G. R. Yoder and D. Eylon: *Metallurgical Transactions A,* Vol 10A, No. 11, Nov 1979, p 1808-1810

7.64. Fatigue and Fracture Characteristics of Silicon-Bearing Titanium Alloys, by M. W. Mahoney and N. E. Paton: *Metallurgical Transactions A,* Vol 9, No. 10, 1978, p 1497-1501

7.65. Fatigue Crack Propagation in Ti-6Al and Ti-6Al-4V Containing 100-300 ppm Hydrogen at Ambient and Sub-ambient Temperatures, by J. C. Chesnutt and N. E. Paton: (in preparation)

7.66. Hydrogen Related Fatigue Fracture of Ti-6Al-4V, by J. R. Kennedy, P. N. Adler and R. L. Schulte: *Scripta Metallurgica,* Vol 14, No. 3, March 1980, p 299-301

7.67. 50-Fold Difference in Region-II Fatigue Crack Propagation Resistance in Titanium Alloys: A Grain Size Effect, by G. R. Yoder, L. A. Cooley and T. W. Crooker: *Journal of Engineering Materials and Technology, Transactions of ASME,* Vol 101, No. 1, Jan 1979, p 86-91

7.68. Enhancement of Fatigue Crack Growth and Fracture Resistance in Ti-6Al-4V and Ti-6Al-6V-2Sn Through Microstructural Modification, by G. R. Yoder, L. A. Cooley and T. W. Crooker: *Journal of Engineering Materials and Technology, Transactions of ASME,* Series H, Vol 199, No. 4, Oct 1977, p 313-318

7.69. Some Relationships Between Crystallography and Stage II Fatigue Crack Growth in a Ti-6Al-4V Alloy, by A. W. Bowen: Cambridge U. K. Conference on Microstructure and Design of Alloys, Aug 1973, Institute of Metals, 1973, p 446-450

7.70. Fatigue Crack-Propagation of Titanium-Alloys Under Dwell-Time Conditions, by P. J. Bania and D. Eylon: *Metallurgical Transactions A,* Vol 9, No. 6, 1978, p 847-855

7.71. Fatigue Crack Growth and J-Integral Fracture Parameters of Ti-6Al-4V at Ambient and Cryogenic Temperatures, by R. L. Tobler: in *Cracks and Fracture,* STP 601, American Society for Testing and Materials, Philadelphia, 1976, p 346-370

7.72. Subcritical Crack Growth Under Sustained Load, by D. N. Williams: *Metallurgical Transactions,* Vol 5, No. 11, Nov 1974, p 2351-2358

7.73. Crack Arrest and Crack Initiation in a Titanium Alloy, by G. J. Petrak: *Engineering Fracture Mechanics,* Vol 4, 1972, p 347-355

7.74. Properties and Processing of Ti-6Al-4V, by H. W. Rosenberg: TIMET Brochure, TIMET, Pittsburgh, 1979

7.9. GENERAL REFERENCES

Engineering Data

E-1 *Titanium Alloys Handbook,* MCIC-HB-02, Metals and Ceramics Information Center, Battelle-Columbus, Ohio, R. A. Wood and R. J. Favor, Eds.

E-2 *Damage Tolerant Design Handbook,* MCIC-HB-01, Metals and Ceramics Information Center, Battelle-Columbus, Ohio. Part 1 deals with fracture toughness; Part 2 deals with fatigue crack growth.

E-3 *Aerospace Structural Metals Handbook,* AFML-TR-68-115, Mechanical Properties Data Center, Belfour Stulen, Traverse City, Michigan. Vol. 3 deals with titanium.

E-4 *Metals Handbook,* 8th Edition, American Society for Metals, Metals Park, Ohio.

E-5 *Military Standardization Handbook–Metallic Materials and Elements for Aerospace Vehicle Structures,* MIL-HDBK-5, Department of Defense, Washington, D. C. 20025. Obtainable from Naval Publications and Forms Center, 5801 Tabor Avenue, Philadelphia, Pa.

E-6 AMS (Aerospace Material Specification), Society of Automotive Engineers, 400 Commonwealth Drive, Warrendale, Pa. Specifications cover specific alloy products.

E-7 Annual Book of ASTM Standards, American Society for Testing and Materials, 1916 Race Street, Philadelphia, Pa. Current Edition. Parts 8, 9, 10, 11 and 41 are useful in various ways.

E-8 ASME Boiler and Pressure Vessel Code, American Society of Mechanical Engineers, United Engineering Center, 345 East Forty-Seventh Street, New York, N. Y. 10017. Section VIII provides rules for construction of pressure vessels.

E-9 Trade brochures, available from metal producers.

Metallurgy of Titanium

M-1 Open Literature: *Acta Metallurgica, Scripta Metallurgica* and *Metallurgical Transactions* are among the more useful journals.

M-2 A. D. McQuillan and M. K. McQuillan, *Titanium,* Academic Press, New York, N. Y., 1956. Now out of print but still a useful source of information.

M-3 *The Science, Technology, and Application of Titanium,* R. I. Jaffee and N. E. Promisel, Eds., Pergamon Press, New York, N. Y., 1970. Reports the First International Conference on Titanium.

M-4 *Titanium Science and Technology,* R. I. Jaffee and H. M. Burte, Eds., Plenum Press, New York, N. Y., 1973. Reports the Second International Conference on Titanium.

M-5 *Titanium Alloys for Modern Technology,* Sazhin *et al,* Eds., NASA TT F-596, Clearinghouse for Federal Scientific and Technical Information, Springfield, Virginia. Translated from Russian.

M-6 *Physical Metallurgy of Titanium,* Kornilov *et al,* Eds., NASA TT F-338, Clearinghouse for Federal Scientific and Technical Information, Springfield, Virginia. Translated from Russian.

M-7 *Applications Related Phenomena in Titanium Alloys,* STP 432, American Society for Testing and Materials, Philadelphia, 1968. Environmental effects dealt with include salt water.

M-8 *Stress Corrosion Cracking of Titanium,* STP 397, American Society for Testing and Materials, Philadelphia, 1966. Deals mainly with hot salt stress corrosion.

M-9 Metals and Ceramics Information Center, Battelle-Columbus, Ohio. Center maintains extensive library on titanium and for a fee will perform literature search on specific subjects.

Fabrication of Titanium

F-1 See Reference E-4.

F-2 See Reference E-1.

F-3 Trade brochures, available from metal producers.

F-4 See Reference M-9.

Chapter 8

Fracture Properties of Superalloys

Stephen D. Antolovich and J. E. Campbell

8.1. INTRODUCTION

Superalloys have been developed for use in elevated temperature applications in which the structural materials must have high strength, excellent corrosion resistance, and good creep and fatigue resistance. To date, the primary application of superalloys has been in air-breathing jet engines for critical components such as turbine discs, turbine blades, vanes, and burner cans. The operating conditions for these components are very severe. For example, in discs, the temperatures vary from around 150°C (300°F) at the hub to about 550°C (1025°F) at the rim, and, during certain periods in the operation of modern aircraft, the discs may be loaded in excess of 80% of yield strength. In the outer portions of the engines, gas inlet temperatures may exceed 1480°C (2700°F). To prevent re-solutioning or localized melting, turbine blades are often designed with small cooling holes. The turbine blades can experience temperatures ranging from about 550°C (1025°F) at the base to about 1050°C (1925°F) at the tips. In addition to the high temperatures and corrosive environments, temperature gradients and load variations arise as operating conditions change, giving rise to thermomechanical fatigue problems. Because an engine operates for considerable periods of time during which the conditions are constant, the effects of hold time are also of prime importance, giving rise to what have become known as creep/fatigue interactions. However, limited experimental evidence indicates that environmental interactions may, in some cases, be more damaging than creep.

Other elevated temperature applications of superalloys are found in the nuclear industry, in steam power generation, in coal gasification, in pollution control, and in hydrogen production equipment as well as in space vehicles for both engine components and airframe skins that may be aerodynamically heated. For all of these applications, the properties that make these alloys desirable for aircraft engine applications are of prime importance.

Most of the superalloys used in these applications are nickel-base alloys, but iron-base and cobalt-base alloys also have been developed for severe elevated temperature service. However, available fracture mechanics data relating to elevated temperature service are generally limited to fatigue crack propagation rates and sustained load crack growth rates in nickel-base alloys.

Recent studies of materials for superconducting machinery, in which the materials may be exposed to cryogenic temperatures as low as −269°C (−452°F), have shown that several nickel-base superalloys have high strength and toughness at the lowest service temperatures. These nickel-base superalloys have been used in prototype models of superconducting motors and generators and will be used more extensively in advanced designs of superconducting devices. At cryogenic temperatures, the chemical environment has little influence on the properties of these alloys and the strengths are higher than at room temperature. For the selected alloys, the plane-strain fracture toughness is relatively high even at the lowest exposure temperatures. Valid K_{Ic} data have been obtained on one of the nickel-base alloys at room temperature and at subzero temperatures. Otherwise, the available fracture mechanics data at room temperature and at subzero temperatures for these alloys are primarily limited to fatigue crack propagation.

The initial section of this chapter gives a brief review of the physical metallurgy of nickel-base superalloys. Subsequent sections are devoted to a critical survey of fatigue crack propagation (FCP), constant-load crack growth, and fracture behavior at cryogenic temperatures. The effect of microstructure, and the possibility of making improvements by varying the microstructure through heat treatment, are considered.

8.2. PHYSICAL METALLURGY OF NICKEL-BASE SUPERALLOYS

The nickel-base superalloys are compositionally complex but microstructurally simple when compared with steels or with titanium alloys. The microstructure consists of an austenitic matrix which is solid solution strengthened, precipitates that are coherent with the matrix and various types of carbides and other phases which are distributed throughout the matrix and along grain boundaries. Appropriate compositional and morphological control results in alloys that are extremely corrosion resistant and that have excellent high temperature strength, creep resistance and ductility. These alloys may be produced in the cast, wrought and powder forms, with each process either improving key properties or resulting in economic advantages for certain applications.

8.2.1. Alloys Strengthened by Gamma Prime Phase

8.2.1.1. General Comments

Most nickel-base alloys in use today are strengthened by the so-called gamma prime (γ') precipitate, which is based on the ordered Ni_3Al structure and which is coherent with the parent FCC γ lattice (Ref 8.1). The alloys also contain carbides, borides and other phases which may be undesirable. These phases may be distributed throughout the matrix and/or located at grain boundaries, depending on the exact type (Ref. 8.1 and 8.2). Certain elements are added to favor formation of desirable phases, to promote oxidation resistance and to control the nature and properties of the grain boundaries. These features are discussed in subsequent sections.

8.2.1.2. Composition

The compositions of γ'-strengthened nickel-base superalloys are extremely complex, as these alloys may contain up to 15 elements. Typical compositions for commercial alloys are given in Table 8.1. Each element or class of elements has one or more functions which have been discussed in detail elsewhere (Ref 8.3) and which are listed below:

a. **Solid solution strengtheners:** primarily V, Cr, Mo, W, Fe and Co. It should be noted that most elements will enter into a solid solution in the matrix to some extent even though they are added for other purposes. For example, Al is added to form γ', but in solid solution is a very potent strengthener.
b. **γ' formers:** primarily Al, but Ti, Nb and Ta can be substituted for the Al.
c. **Carbide formers:** V, Ti, W, Mo, Cr, Nb and Ta, in decreasing order of effectiveness.
d. **Oxide formers:** Al and Cr. Adherent oxides are formed on the surfaces of superalloys containing Al and Cr and greatly enhance the high temperature oxidation resistance.
e. **Grain boundary modifiers:** elements that do not have any of the above effects and that segregate to grain boundaries because of large differences in atomic size. Such elements are Mg, B, C, Zr and Hf.

It is evident from the above brief classification that a given element may be added for more than one reason and can occur in more than one phase. For example, aluminum occurs in both the matrix and γ' phases. In the matrix, it is a very potent solid solution strengthener (Ref 8.3) and confers oxidation resistance. The γ' phase may contain numerous other elements in addition to the γ' formers (Ref. 8.1). The various phases are discussed in more detail below.

8.2.1.3. Phases

Austenite Matrix. As mentioned previously, the matrix phase is FCC and contains solid solution strengtheners whose strengthening effects are proportional to the difference in atom size between the matrix (nickel) and the solute atom. The efficacy of various solid solution elements has been estimated elsewhere (Ref 8.3) and, as expected, chromium, molybdenum and tungsten are the most effective strengtheners.

Gamma Prime Precipitates. The γ' phase is based on the ordered Ni_3Al structure with nickel atoms at face centers and aluminum atoms (or atoms of other elements) at cube corners. The γ' precipitates can strengthen the alloys in two ways. First, coherency strains make it difficult for dislocations to penetrate the precipitates and second, when dislocations do penetrate the γ', antiphase boundary (APB) energy must be created because of the ordered structure. Copley and Kear (Ref 8.4) have shown that dislocation penetration of the γ' particles is a key step controlling deformation of these alloys. Based on these ideas, they have demonstrated that the strength of these alloys obeys an equation of the form:

Table 8.1. Nominal compositions of nickel-base superalloys

Alloy designation	UNS No.	Nominal composition, wt %: C	Mn	Si	Cr	Co	Mo	W	Nb	Ti	Al	B	Zr	Fe	Others
Astroloy	...	0.06	...	...	15.0	15.0	5.25	...	...	3.5	4.4	0.03	...	...	...
IN-100(a)	...	0.18	...	...	10.0	15.0	3.0	...	...	4.7	5.5	0.014	0.06	...	1.0 V
IN-792(a)	...	0.21	...	...	12.7	9.0	2.0	3.9	...	4.2	3.2	0.02	0.10	...	...
Inconel 625	N06625	0.05	0.2	0.2	21.5	...	9.0	...	3.6	0.2	0.2	...	...	2.5	...
Inconel 706(b)	...	0.03	0.2	0.2	16.0	...	...	...	2.9	1.8	0.2	...	...	40	0.2 Cu
Inconel 718	N07718	0.04	0.2	0.2	18.5	...	3.0	...	5.1	0.9	0.5	...	...	18.5	0.2 Cu
Inconel X-750	N07750	0.04	0.5	0.2	15.5	...	...	...	1.0	2.5	0.7	...	...	7.0	0.2 Cu
René 41	N07041	0.09	...	...	19.0	11.0	10.0	...	...	3.1	1.5	0.010(c)	...	1.0(c)	...
René 77	...	0.15(c)	...	...	15.0	18.5	5.2	...	...	3.5	4.25	0.05(c)	...	1.0(c)	...
René 80	...	0.17	...	...	14.0	9.5	4.0	4.0	...	5.0	3.0	0.015	0.03	...	...
René 95	...	0.15	...	...	14.0	8.0	3.5	3.5	3.5	2.5	3.5	0.01	0.05	...	...
Udimet 630	...	0.04	0.2(c)	0.2(c)	17.0	1.0(c)	3.1	3.0	6.0	1.1	0.6	0.005	...	17.5	...
Udimet 700	...	0.07	...	...	15.0	18.5	5.0	...	...	3.5	4.4	0.025	...	0.5(c)	...
Waspaloy A	N07001	0.07	0.5(c)	0.5(c)	19.5	13.5	4.3	...	...	3.0	1.4	0.006	0.09	2.0(c)	0.03 S(c), 0.10 Cu(c)
Waspaloy B	N07001	0.07	0.75(c)	0.75(c)	19.5	13.5	4.3	...	...	3.0	1.4	0.006	0.07	2.0(c)	0.02 S(c), 0.10 Cu(c)

(a) Cast alloy. (b) Iron-nickel superalloy. (c) Maximum.

$$\tau_c = \frac{\gamma_o}{2b} - \frac{T}{br_o} + \frac{1}{2}\left(\tau_o + \tau_p\right) \qquad \text{(Eq 8.1)}$$

where τ_c is the critical resolved shear stress, γ_o is the APB energy, r_o is the precipitate radius, b is the Burger's vector, T is the dislocation line tension, τ_o is the lattice friction stress of the matrix and τ_p is the lattice friction stress of the particle.

The γ' phase is quite stable with respect to temperature. It has a virtually constant yield strength at temperatures as high as 900°C (1650°F), with the amount of work hardening increasing with temperature to a peak that occurs in the range 700°C to 900°C (1300 to 1650°F) depending on the composition (Ref 8.5). This temperature dependence of work hardening has been explained in terms of interactions between ⟨110⟩ and ⟨100⟩ dislocations.

The shape and structural stability of the γ' depend on the misfit parameter, δ, which is defined as:

$$\delta = \frac{a_p - a_m}{\bar{a}} = 2\,\frac{a_p - a_m}{a_p + a_m} \qquad \text{(Eq 8.2)}$$

where a_p is the lattice parameter of precipitate, a_m is the lattice parameter of matrix and $\bar{a}$ is the average lattice parameter.

When either the particle size or the misfit parameter is small, the γ' particles tend to assume spherical shape. This effect is shown in Fig. 8.1 for René 77, which has a relatively high misfit, and for Udimet 500, in which the misfit is low. For alloys having large misfit parameters, continued aging results in cuboidal precipitates whose faces are parallel to {100} matrix planes. This effect can be seen quite clearly in Fig. 8.1(a) and can be explained simply by noting that *surface* energy effects dominate when particles are small, whereas *strain* energy effects dominate when particles are large. Because {100} planes are "soft", elastic deformation is most easily accommodated when these planes are the matching planes. The misfit parameter depends on composition and has been studied in detail elsewhere (Ref 8.6). These studies showed that elements such as niobium and titanium tend to increase the misfit parameter whereas iron and molybdenum tend to decrease it. The misfit parameter is important because it plays a large role in determining deformation characteristics. When the misfit is large, deformation tends to occur by looping of precipitates by dislocations; when the misfit is small, shearing occurs. The deformation mode plays an important role in determining fracture toughness and fatigue properties and will be discussed later.

The *morphological stability* of γ' has been studied as a function of applied stress, and it has been shown that the equilibrium microstructure depends on the sense of the applied stress and the misfit parameter (Ref 8.7 and 8.8). For example, in alloys for which the misfit parameter is positive, a stress along [001] will convert cubes into plates lying on (001). An example is shown in Fig. 8.2 for René 80.

Carbides. Most nickel-base superalloys contain carbides, both in the boundaries and dispersed throughout the matrix. The most frequently observed types are MC, M_6C and $M_{23}C_6$, which are discussed below.

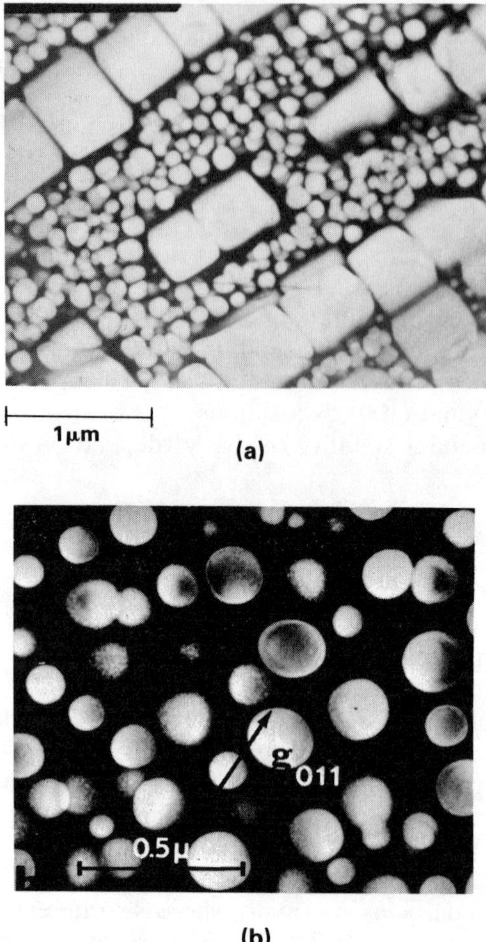

(a) René 77 (*Ref 8.42*). (b) Udimet 500 after aging at 980°C (1800°F) for 4 h (*Ref 8.1*). In (a), the misfit is large, and the small γ' particles are spherical whereas the large ones are cuboidal. In (b), the misfit is near zero and all γ' particles tend to be spherical. Both micrographs were taken using dark field transmission microscopy.

Fig. 8.1. Gamma prime (γ') precipitate structures in René 77 and Udimet 500

(i) MC carbides have a blocky morphology and an FCC structure, and are thought to form below the freezing temperature. They have a tendency to decompose with increasing temperature unless the alloys are high in niobium and tantalum, in which case they are stabilized. The prototype carbides that form are TaC, NbC, TiC and VC, in order of decreasing stability. It has been pointed out (Ref 8.3) that this sequence is not predicted by thermodynamics and may be due to the presence of molybdenum or tungsten, which weakens the bonding forces. Depending on the composition of the alloy in question, not only may the "M"

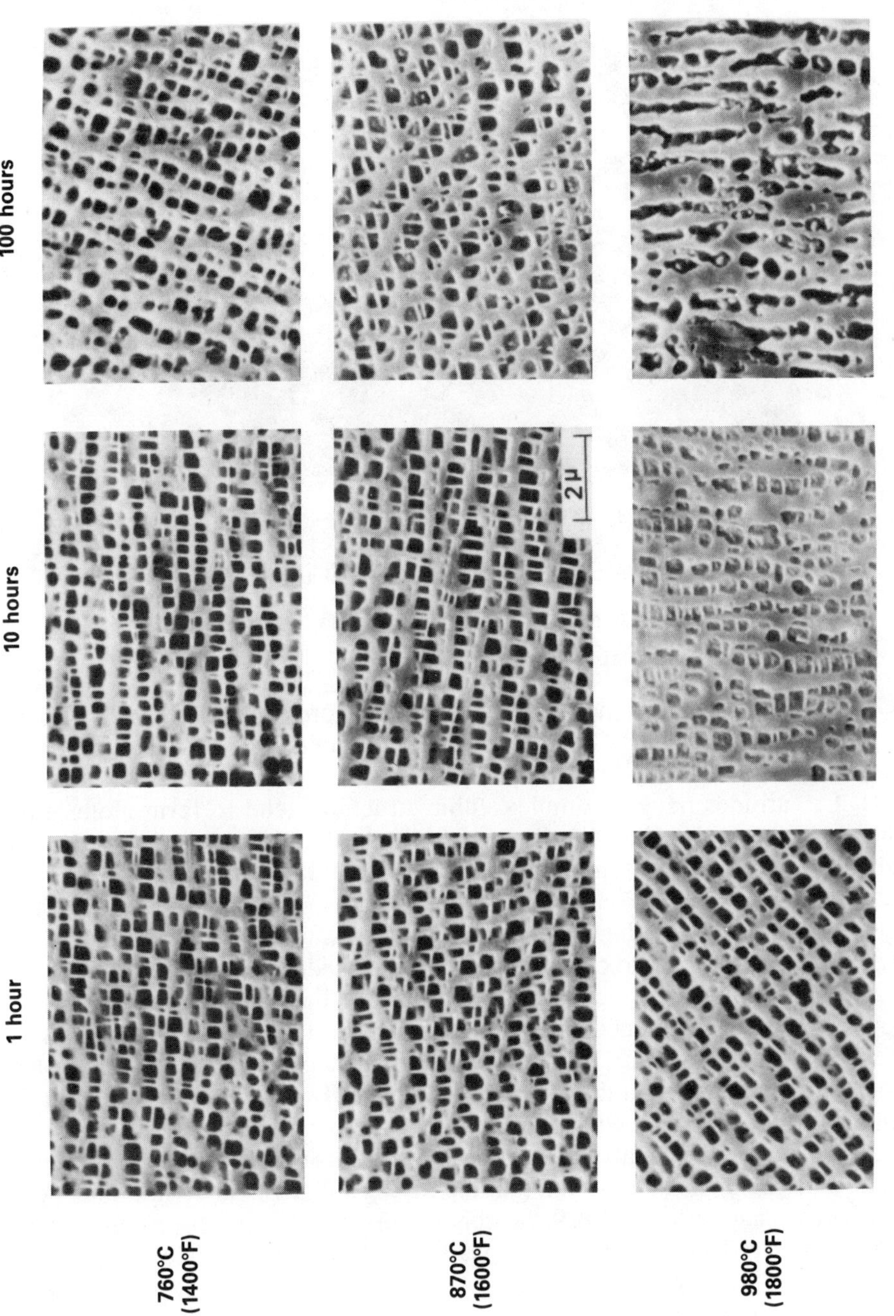

Each specimen was held at ⅓ the yield strength at the indicated temperature and for the indicated time. The stress axis is horizontal. The linear scale at lower right in center micrograph applies to all nine micrographs.

Fig. 8.2. Effects of time, temperature and stress on the morphology in René 80 (Ref 8.42)

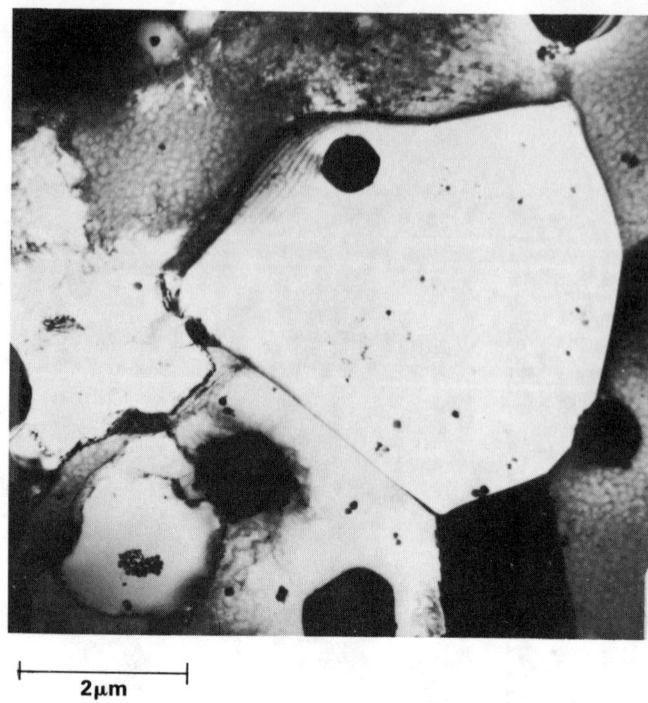

The large carbides have been identified as MC.

Fig. 8.3A. Carbide structure in René 95 (Ref 8.28)

atoms substitute for each other, but other less reactive elements will also be incorporated in the carbides. In René 95, for example, MC carbides occur, as shown in Fig. 8.3A.

(ii) $M_{23}C_6$ carbides have a complex cubic structure, tend to form along grain boundaries, and are abundant in alloys of high chromium content. They tend to be stable at intermediate temperatures — 870 to 980°C (1600 to 1800°F) — depending on the composition. As was the case for MC carbides, there can be considerable variation in the "M" content. For example, when tungsten and molybdenum are present, the composition of the carbide may be $Cr_{21}(Mo,W)_2C_6$. It has also been found that nickel can substitute for the chromium.

Because these carbides occur at grain boundaries, they tend to promote creep resistance by inhibiting grain boundary sliding. However, the deformation that must be transferred from grain to grain is eventually accommodated either by cracking of these carbides or by interfacial decohension leading to intergranular fracture. Carbides are usually the initiation sites for high temperature creep cracking. It has also been suggested that the $M_{23}C_6$ and γ' form an interconnected grain boundary network (Ref 8.9) which is very susceptible to environmental attack and which reduces ductility. A typical grain boundary morphology is shown in Fig. 8.3B for René 77.

(iii) M_6C carbides have a complex cubic crystal structure and generally occur as a grain boundary precipitate. The composition varies widely, and formulas such

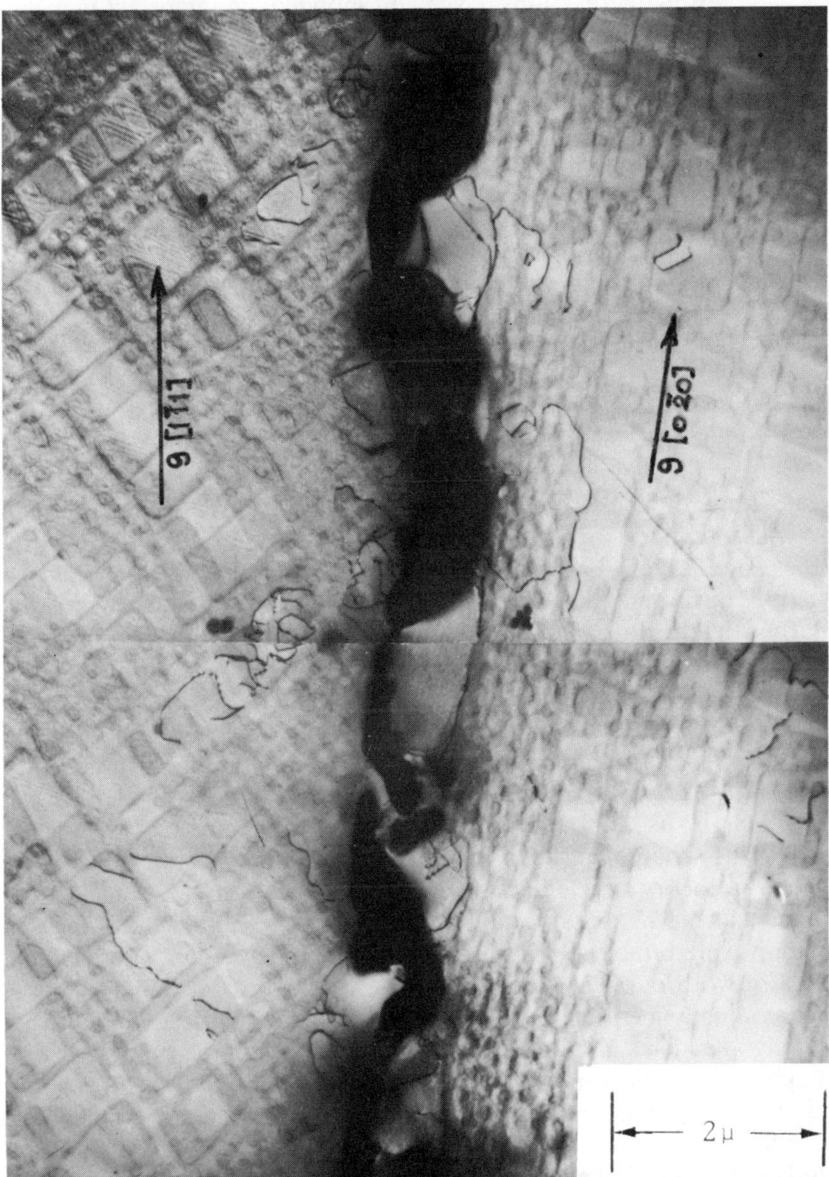

Both $M_{23}C_6$ and large γ' particles are present in the boundary.

Fig. 8.3B. Grain boundary structure of René 77 (Ref 8.42)

as $(Ni,Co)_2Mo_3C$ and $(Ni,Co)_2W_4C$ have been suggested. The high temperature stability of these carbides has been used to control grain size during high temperature heat treatments.

These carbides may react with other phases to form other carbides and phases. These reactions occur relatively slowly, in some cases continuing over the life of the alloy. The following reactions have been shown to be important in nickel-base alloys (Ref 8.3):

$$MC + \gamma \rightarrow M_{23}C_6 + \gamma'$$

$$MC + \gamma \rightarrow M_6C + \gamma'$$

$$M_6C + M' \rightarrow M_{23}C_6 + M''$$

where M′ and M″ represent metals in the matrix.

Other Phases. In addition to the phases already discussed, η, σ, μ and Laves phases, as well as borides, can form in superalloys; except for the borides, which improve creep rupture properties, they are generally deleterious.

The η phase has a Ni_3X composition and an HCP structure, and forms from the γ' when the titanium, niobium or tantalum content is sufficiently high. It can precipitate at grain boundaries in a cellular mode which reduces notch stress-rupture strength, or it can precipitate intergranularly in a Widmanstätten morphology which reduces strength but not ductility (Ref 8.3).

The σ, μ and Laves phases are classified as topologically close packed (TCP) and generally have platelike structures which seem to promote crack formation. Particularly deleterious in nickel-base superalloys is formation of σ phase, which occurs in the temperature range from 650 to 925°C (1200 to 1700°F), its rate of formation being accelerated by an applied stress (Ref 8.2). It has also been determined that formation of σ is sensitive to the average electron vacancy number of the γ' ($\overline{N}_{vz}^{\gamma'}$) and tends to form when $\overline{N}_{vz}^{\gamma'}$ is in the range from 2.26 to 2.41 and above (Ref 8.2). Avoidance of σ is incorporated in the so-called PHACOMP scheme which is used in the development of new nickel-base superalloys (Ref 8.10). This scheme has been computerized and incorporates the major ideas of the electron theory of metals as well as empirical observations. It has been pointed out that the $M_{23}C_6$ structure would closely resemble the σ phase if the carbon atoms were removed and that there is a high degree of coherency between $M_{23}C_6$ and σ, with σ plates frequently nucleating from $M_{23}C_6$ particles.

Boron, as already mentioned, segregates to grain boundaries, where, in addition to occupying vacancies and decreasing diffusion rates in the boundaries, it leads to formation of borides, generally of the composition M_3B_2. The boride particles are hard, refractory precipitates that delay the onset of grain boundary tearing during creep (Ref 8.3).

Grain Boundaries. When it is recognized that high temperature creep and low cycle fatigue (LCF) failures frequently initiate in grain boundaries, then the importance of grain size, grain boundary chemistry and grain boundary morphology becomes clear.

In general, the grain size must not be too fine, because creep resistance is reduced by fine grains. On the other hand, large grains, especially in relation to the section size, can cause sliding and lead to premature failure. Grain size is controlled by the presence of γ', carbides and other particles that may serve to pin grains by a Zener mechanism.

The grain boundary composition deviates from the nominal composition and is rich in elements such as boron, zirconium and magnesium, small additions of which can increase creep life by an order of magnitude by mechanisms that are not

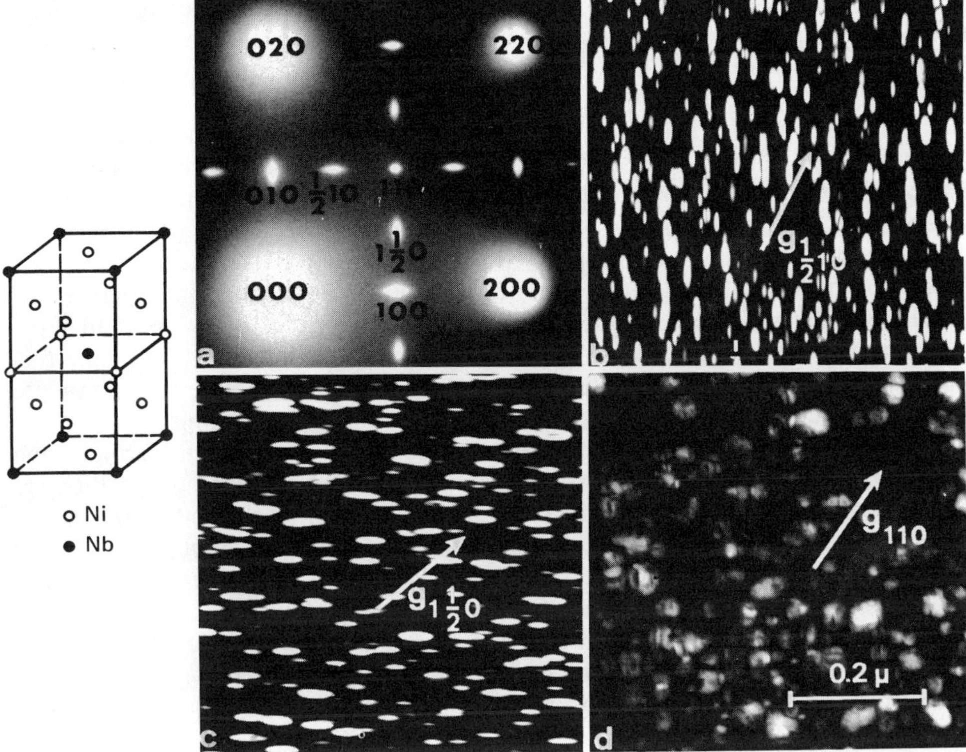

(a) Diffraction pattern showing superlattice reflections indexed with respect to the FCC matrix, as shown in unit cell above. (b) The (100)γ'' variant is preferentially imaged with g = [½ 1 O]. (c) The (010) variant with g = [1 ½ O]. (d) The (001) variant with g = [110].

(g is the reciprocal lattice vector.)

Fig. 8.4A. DO_{22} γ'' precipitate in Inconel 718 (Ref 8.1)

entirely clear. The large size difference between these elements and others in solid solution (20 to 30%) would account for their positions at grain boundaries where they fill vacancies and reduce the rate of grain boundary diffusion. A reduced grain boundary diffusion rate is consistent with an increased creep life (Ref 8.2).

8.2.2. Alloys Strengthened by Gamma Double-Prime

An important group of superalloys—alloys that contain considerable amounts of iron and about 2 to 6% niobium—are strengthened primarily by the so-called γ'' precipitate, which has an ordered BCT (body centered tetragonal, $D0_{22}$) crystal structure based on the compound Ni_3Nb (Fig. 8.4A). The most well-known and most widely used member of this group is Inconel 718, the composition of which is given in Table 8.1*. The γ'' precipitates usually are considerably finer than the

*Other alloys which are believed to be strengthened by γ'' precipitates are Inconel 706 and Udimet 630.

γ' precipitates in other superalloys and, when the alloy is heat treated for maximum strength, deformation occurs as a result of dislocations interacting with the coherency strain fields and eventually shearing the precipitates (Ref 8.11 and 8.12). The orientation relationships are $\{100\}_{\gamma''} \parallel \{100\}_{\gamma}$; $[001]_{\gamma''} \parallel \langle 001 \rangle_{\gamma}$. A typical micrograph of γ'' is shown in Fig. 8.4A. If the alloy is used above about 650°C (1200°F), some of the γ'' may convert to δ phase, which is an ordered Ni_3Nb phase having an orthorhombic crystal structure and a generally platelike morphology, as can be seen in Fig. 8.4B.

Formation of σ phase is generally felt to be damaging to strength, ductility and fatigue properties, although the effect on fatigue is not clear from experimental evidence. This will be discussed in a later section. It is noteworthy that the conventional heat treatment for Inconel 718, discussed later, results in the formation of a globular δ phase in the grain boundaries which is believed to promote good stress-rupture properties by inhibiting grain boundary sliding.

In addition to γ', γ'', δ and σ phases, these alloys also contain carbides and μ, χ and Laves phases. The most common carbides are MC carbides, which are quite stable during both processing and service. They can occur both in the matrix and at grain boundaries. These alloys also are prone to formation of $M_{23}C_6$, which may form as a grain boundary film during service, leading to degraded ductility. The other phases have effects that have not been completely documented. A more detailed review of the physical metallurgy of nickel-base alloys containing considerable amounts of iron is beyond the scope of this chapter. The interested reader is referred to the review by D. R. Muzyka (Ref 8.13).

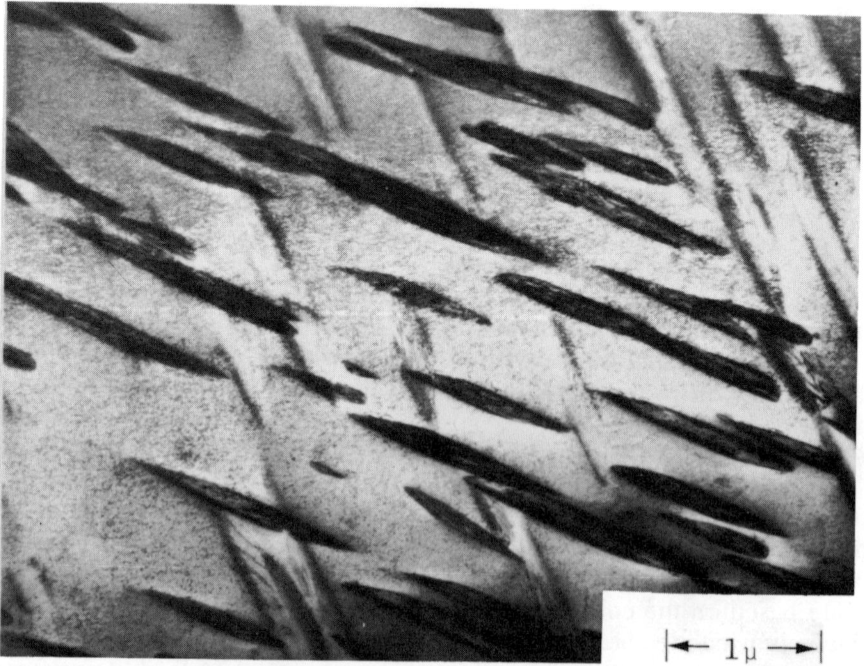

Fig. 8.4B. Platelike δ-Ni_3Nb in Inconel 718 (Ref 8.42)

8.3. FATIGUE CRACK PROPAGATION

8.3.1. Empirical Models for Data Representation

Fatigue crack propagation (FCP) behavior of nickel-base superalloys can be correlated with the stress intensity parameter at room temperature and at elevated temperatures (Ref 8.14). An empirical model which was not discussed in Chapter 3 but which has been proposed as a correlating tool (Ref 8.15) has the form:

$$\log\frac{da}{dN} = C_1 \sinh \{C_2 \log \Delta K + C_3\} + C_4 \qquad \text{(Eq 8.3)}$$

where C_1 is a material constant and the other constants are functions of load ratio, temperature and frequency. These constants can be determined by a statistical regression analysis. It has been pointed out that the above model is symmetric about the inflection point and that this limitation is not justifiable on physical or empirical grounds. Instead, a model that allows for a more general fit has been proposed (Ref 8.16):

$$\frac{1}{da/dN} = \frac{A_1}{(\Delta K)^{n_1}} + \frac{A_2}{(\Delta K)^{n_2}} + \frac{A_2}{(K_c(1-R))^{n_2}} \qquad \text{(Eq 8.4)}$$

where A_1, A_2, n_1, n_2, K_c are fitting parameters and R is the load ratio (i.e., P_{min}/P_{max}). The parameters A_1 and A_2 are functions of load ratio which can be determined by a regression analysis. This formulation is capable of representing a great variety of engineering materials over a wide range of stress intensities, and, in practice, the constants can be characterized as functions of load ratio, temperature and frequency. When the FCP rate exceeds about 2.5×10^{-5} mm/cycle (10^{-6} in./cycle), the well-known Paris equation (Ref 8.17) represents the data very well:

$$\frac{da}{dN} = C(\Delta K)^n \qquad \text{(Eq 8.5)}$$

where C and n are material constants for given values of temperature, frequency and load ratio (as discussed in Chapter 3). Equations 8.3 and 8.4 also represent the data reasonably well in this range.

8.3.2. Effects of Microstructure

The effects of microstructure on FCP have been considered by several investigators at both low and high temperatures (Ref 8.18 to 8.20). The effects of γ' particle size in Waspaloy was studied at 650°C (1200°F), and relatively little effect was reported at test frequencies of 1.0 Hz (Ref 8.18). The heat treatments employed produced uniform γ' sizes of 0.016 to 0.018 μm in one case and 0.020 to 0.023 μm in another case. In a third heat treatment, a bimodal γ' structure was produced in which the size of the fine γ' precipitate was 0.01 μm and the size of the coarse γ' precipitate was 0.08 μm. The results of FCP tests are shown in Fig. 8.5.

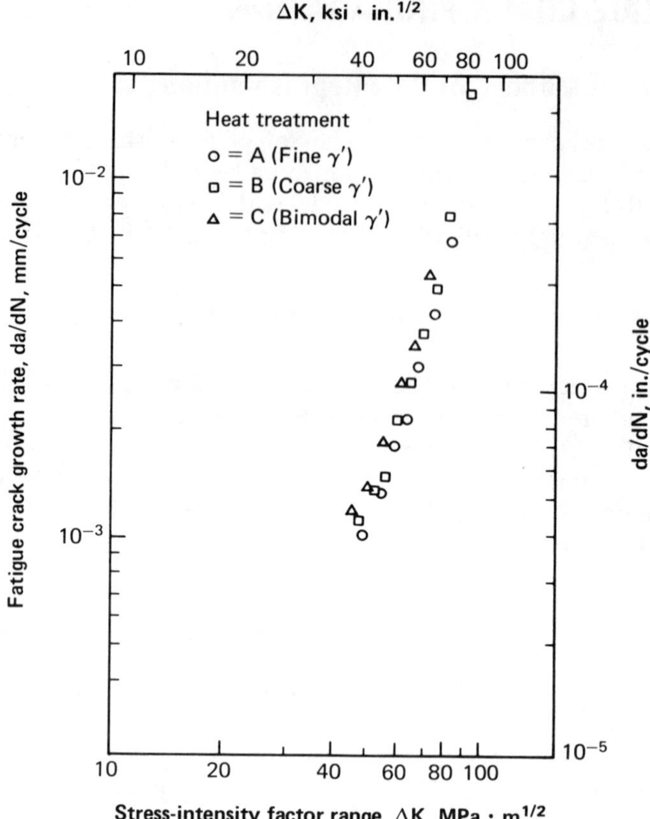

Frequency, 1.0 Hz. Temperature, 650°C (1200°F).

Fig. 8.5. Effect of γ' size on cyclic crack growth rate in Waspaloy (Ref 8.18)

The effect of grain size was also studied using specimens of Astroloy. Heat treatments were carried out in which the γ' particle size was not well controlled but in which the grain size varied from ASTM 8 (fine) to ASTM 5 (coarse). The fine grained specimens had a bimodal γ' distribution of 0.05 and 0.30 μm. Finally, a coarse grained specimen was produced by heat treatment in which the grain boundaries were serrated and there was a bimodal γ' distribution with coarse γ' both in the grain boundaries and in the matrix along with fine γ' (0.065 μm) produced by aging. Subsequent FCP testing showed that the FCP rates tended to be grouped solely in terms of grain size, independent of γ' and grain boundary morphology. The results are shown in Fig. 8.6.

These results are to some extent contradicted by two other studies in which both the γ' particles and the grain size were well controlled (Ref 8.19 and 8.20). In one study (Ref 8.19) using specimens of René 95, a disc material, γ' particle sizes and grain sizes were systematically varied and testing was carried out at 540°C (1000°F). It was found that the lowest FCP rates were associated with fine γ' (0.08 μm) and coarse grains (~50 μm in diameter) while the highest FCP rates occurred when the γ' size was 0.5 μm and the grain size was 10 μm. The

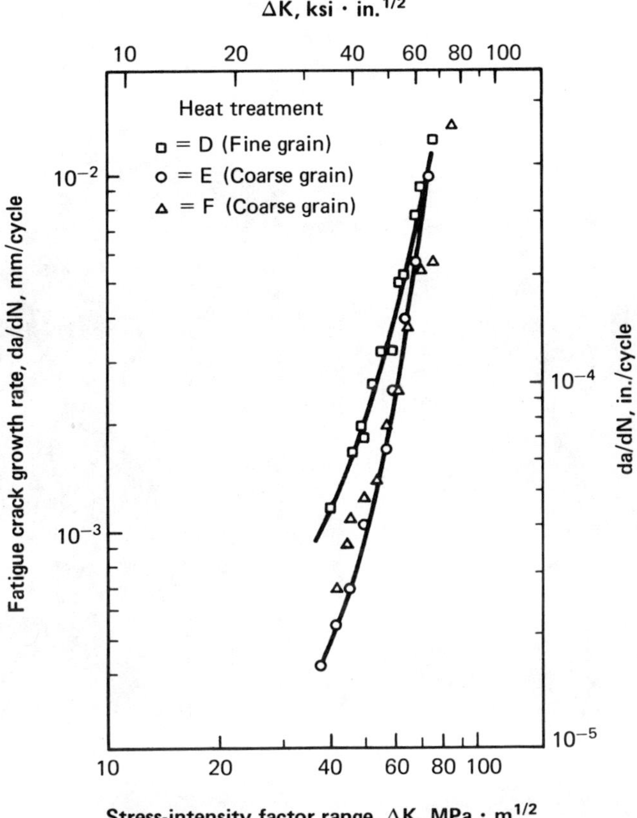

Frequency, 1.0 Hz. Temperature, 650°C (1200°F).

Fig. 8.6. Effect of different grain sizes on the cyclic crack growth rate of P/M Astroloy (Ref 8.18)

microstructures corresponding to these treatments are shown in Fig. 8.7. The results are shown in terms of residual fatigue life in Fig. 8.8. It is clear that, above about 0.2 μm, the γ' particle size has little or no effect on residual life, whereas below this size residual life increases rapidly, especially for material of coarse grain size. It is interesting to note that the most fatigue resistant materials have microstructures that increase slip planarity, and the beneficial effect of planar slip has been pointed out for other materials (Ref 8.21). The crack paths are shown for materials with low and high FCP rates in Fig. 8.9. Note the strong crystalline features and longer crack path associated with the material having the lowest FCP rate. Results similar to those quoted for René 95 at 540°C (1000°F) were recently obtained for Waspaloy at room temperature (Ref 8.20). In this case the γ' particle sizes and grain sizes were systematically varied between two grain sizes (ASTM #8 and #3) and two γ' sizes (0.008 and 0.08 μm). The FCP rates as functions of ΔK are shown in Fig. 8.10. Again, the material having coarse grains and fine γ' exhibited vastly superior FCP resistance. The combination of large grains and fine γ' promotes planar slip. Other investigators (Ref 8.21) have shown that for a given grain size the strengths of alloys containing 0.008 μm and

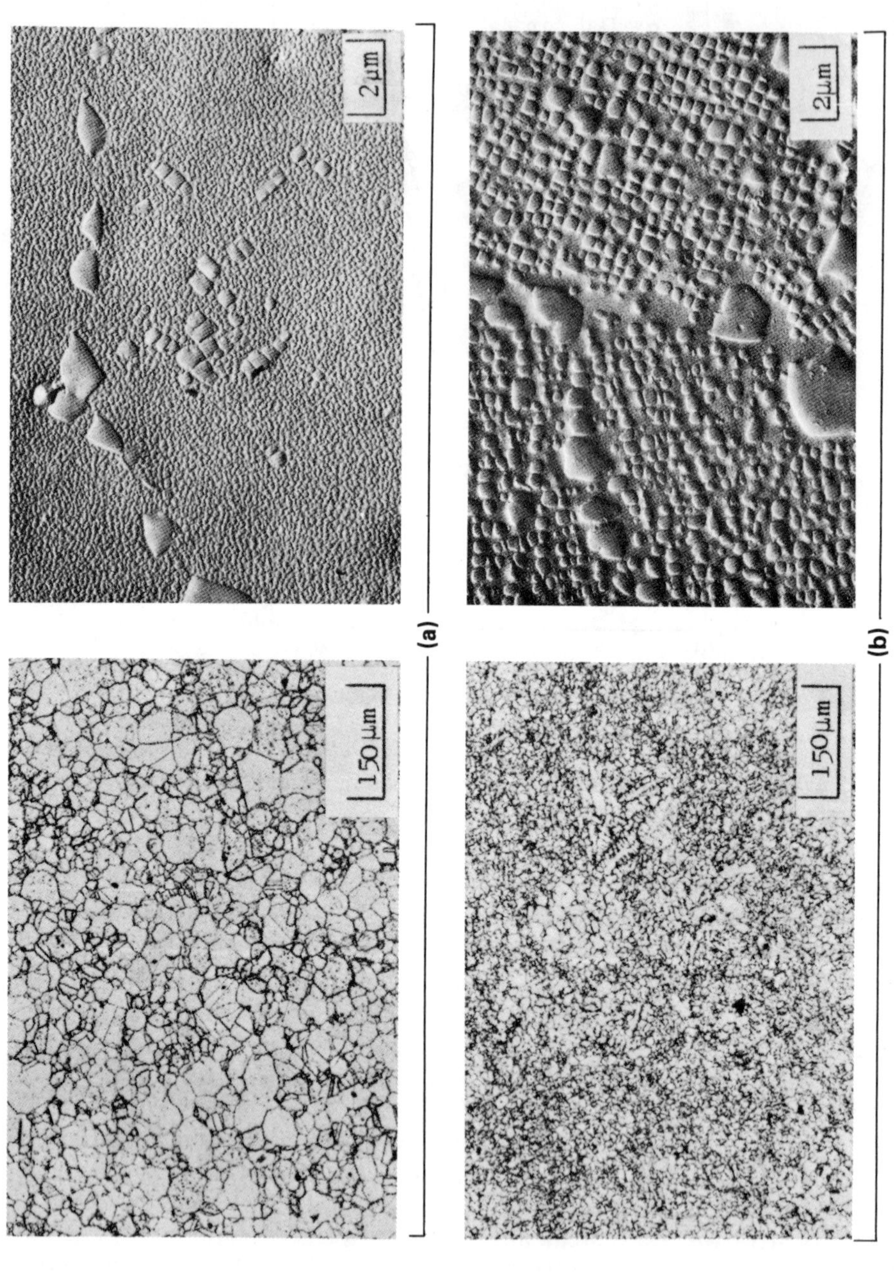

Fig. 8.7. Light optical and replica micrographs of René 95 (Ref 8.43)

The coarse grain/fine γ' structures in (a) exhibited superior FCP resistance in comparison with the fine grain/coarse γ' structures in (b).

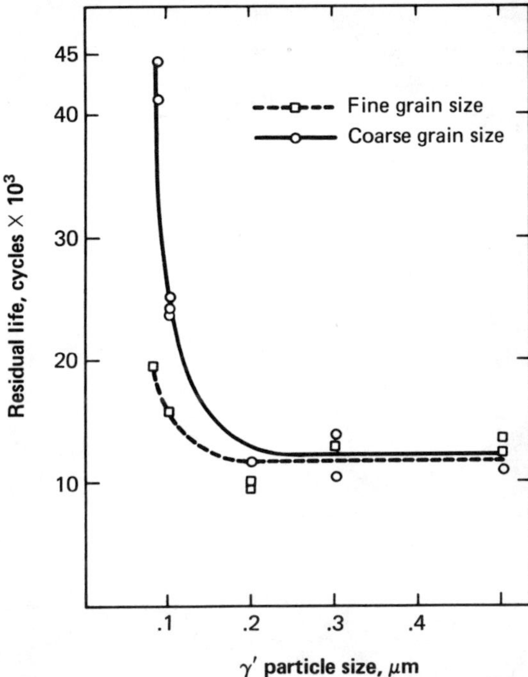

ΔK = 30 to 66 MPa·m$^{1/2}$ (27 to 60 ksi·in.$^{1/2}$). Temperature, 540°C (1000°F). Frequency, 0.3 Hz.

Fig. 8.8. Effect of γ' particle size and grain size on residual cycle life of René 95 (Ref 8.19)

0.08 μm precipitates are equivalent. Consequently, differences in FCP behavior are very likely to be slip-mode related* and not related primarily to a strength effect. Differences in the fracture surface appearances are shown in Fig. 8.11, where the coarse grained, fine γ' material has much more distinctive crystalline features. The results quoted here are in agreement with theoretical predictions made elsewhere in which the prospect of microstructural enhancement of FCP properties was clearly pointed out (Ref 8.22). The divergence of the above results from those of Merrick and Floreen (Ref 8.18) can be explained at least in part by the lack of systematic variation in grain size and by noting that the γ' size in the Merrick and Floreen study was probably too large to produce any pronounced sensitivity such as shown in Fig. 8.7 and 8.10. It is likely that each alloy will have a different γ' size at which improvement will be noticed. Granting for the moment that planar slip is desirable in promoting resistance to FCP, as appears probable from the variety of systems and test conditions quoted above, then the transition γ' particle size should be a decreasing function of the misfit parameter, because, for a large misfit, planar slip occurs only for very small γ' particle sizes.

A word about the mechanism of planar slip in promoting FCP resistance is in order. It has been suggested (Ref 8.22) that the plastic strain levels ahead of an

*Another possibility is microbranching of the crack, which in many cases is a mechanism that depends on the slip mode.

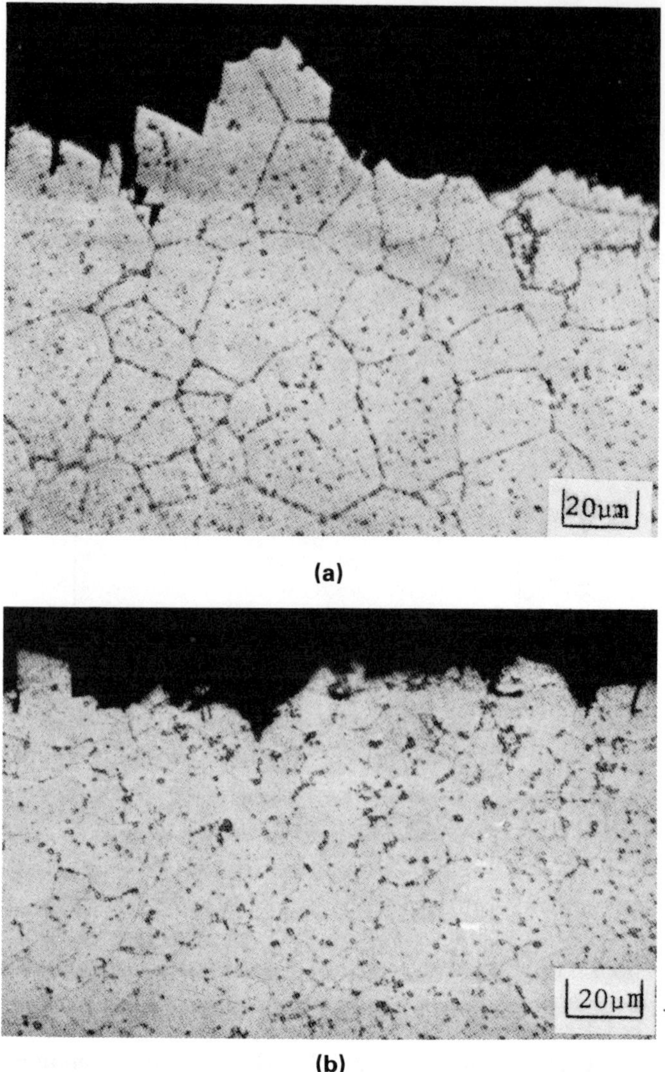

In (a) the material had a coarse grain size, fine γ' precipitates (~0.1 μm) and a crystalline appearance. In (b) the material had a fine grain size and contained large γ' precipitates (~0.3 μm), and the fracture surface was flatter than that in (a). Both micrographs correspond to ΔK of 65 MPa·m$^{1/2}$ (60 ksi·in.$^{1/2}$).

Fig. 8.9. Fracture paths as viewed by optical microscopy in René 95 tested at 540°C (1000°F) and 0.34 Hz (Ref 8.43)

advancing crack are much lower than had been previously thought. At low plastic strains, all of the deformation can be carried in well-defined slip bands. If the material is relatively clean, as are most modern nickel-base superalloys, then the probability of a slip band coming near an impurity is relatively low. Because the stresses are very large only near the tip of the slip band, the probability of a microcrack being nucleated ahead of the main crack is also low.

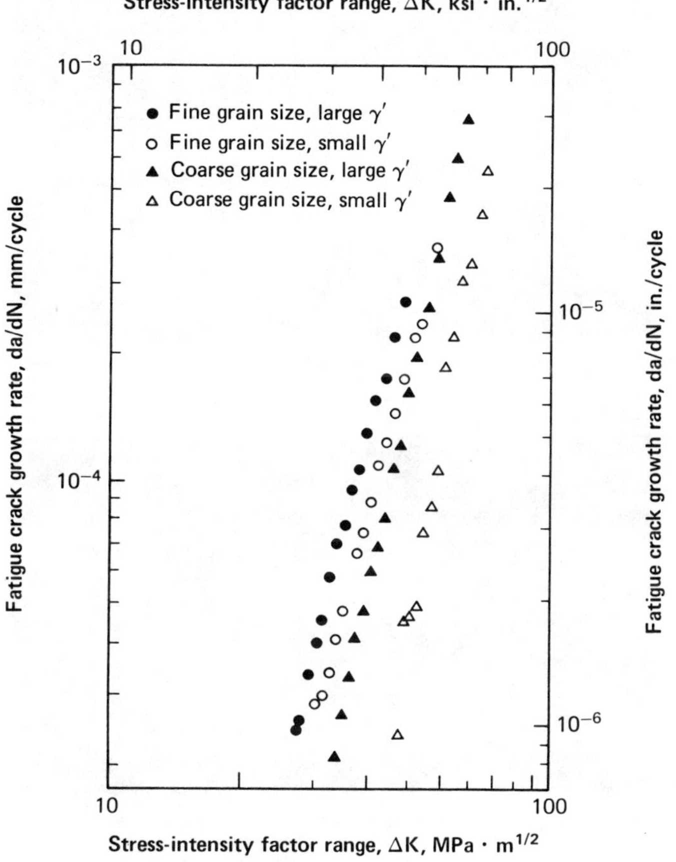

Fig. 8.10. Crack growth behavior of Waspaloy at room temperature as a function of heat treatment (Ref 8.20)

A further mechanism for enhancement of FCP performance in planar glide materials has been put forth in terms of a surface roughness effect (Ref 8.23). The basic idea is that, for a planar glide material, the fracture surface will show crystalline facets and, while individual facets are smooth, the combination of facets making up the surface will cause an over-all jagged topography. This is shown in Fig. 8.9 for René 95. In computation of the experimental FCP rate, the distance between the crack tip and some point ahead of the crack (measured along a straight line) is divided by the number of cycles required to propagate the crack this distance. If the surface is jagged, then the difference between the apparent distance traversed by the crack and the true distance can be substantial, leading to a lower apparent FCP rate. In addition to an increase in the crack path, the effective stress intensity may decrease as the crack deviates from a path normal to the applied load.* While these effects obviously can play a role, it is not clear

*It may be incorrect to think in terms of a local stress intensity parameter, because the stress intensity parameter is defined in terms of load transmission past the projected area of a crack and *local* variations in orientation have but a small effect. This problem has been reviewed by Kitagawa *et al* (Ref 8.24).

In (a), the heat treatment was such that there were large grains and small γ' particles (lowest crack growth rate in Fig. 8.10). In (b), the grains were small and the γ' particles were large (highest crack growth rate in Fig. 8.10). Note the distinctly crystalline features in (a).

Fig. 8.11. Typical fracture surface features of Waspaloy at a stress-intensity factor range of about 50 MPa·m$^{1/2}$ (45 ksi·in.$^{1/2}$) (Ref 8.20)

whether they can fully account for the large differences in behavior observed for Waspaloy (Fig. 8.10).

In a detailed investigation of Inconel 718, the effects of product form, heat-to-heat variation, and heat treatment were studied (Ref 8.25). For Heat I, the product was 1.5-by-4-cm (5⁄8-by-15⁄8-in.) bar stock melted by VIM-VAR* melt practice while for Heat II the product was 1.3-cm (½-in.) plate melted by VIM-EFR.† The compositions were within normal limits for Inconel 718. Each heat was given both

*Vacuum induction melted and vacuum arc remelted.

†Vacuum induction melted and electroflux remelted.

a conventional and a modified treatment designed to increase the fracture toughness. The heat treatments are listed below:

Conventional Treatment	*Modified Treatment*
(1) Heat to 950°C (1750°F) 1 h, AC to RT	(1) Heat to 1090°C (2000°F) 1 h, cool to 720°C (1325°F) at 55°C (100°F) per h
(2) Reheat to 720°C (1325°F) 8 h, FC to 620°C (1150°F)	(2) Hold at 720°C (1325°F) 4 h, cool to 620°C (1150°F) at 55°C (100°F) per h
(3) Hold at 620°C (1150°F) 18 h, AC to RT	(3) Hold at 620°C (1150°F) 16 h, AC to RT

The microstructures of Heat I for the conventional and modified heat treatments are shown in Fig. 8.12. In conventionally treated material, the grain size was ASTM 11½ and there were some coarse δ precipitates in the grain boundaries along with MC carbides. In material given the modified treatment, the grain size increased to ASTM 4, and, in addition to fine δ precipitates in the boundaries, there were also MC carbides which appeared to pin the boundaries. For the conventional heat treatment, Heat II had a grain size of ASTM 5 and showed banded regions in which δ phase platelets were localized in grain boundary regions. In addition to the normal γ'' and γ' precipitates there were also MC carbides distributed throughout the matrix. Modified treatment of Heat II produced a grain size of ASTM 2½ and significantly reduced the amount of δ phase, which appeared as a fine grain boundary precipitate. MC carbides were also observed in the boundary. The microstructures for both treatments are shown in Fig. 8.13.

The FCP test results are shown in Fig. 8.14 to 8.18 for five test temperatures. All testing was done at $R = 0.05$ and at a frequency of 0.67 Hz except for tests at room temperature, where the effect of frequency is minimal. It is noteworthy that for the conventional treatment, the Heat II material appears to be the most fatigue resistant even though it contains banded regions and significant amounts of δ phase in the grain boundaries of the banded regions. The authors ascribed these differences to differences in melting practice, but grain size cannot be ignored. As discussed previously, coarser grained materials tend to have markedly superior FCP characteristics. In response to the modified treatment, the FCP resistance of each heat increased with respect to the conventional treatment and the FCP properties of each heat tended to approach one another. The authors tentatively concluded that the improvements were due to a more homogeneous microstructure and cited the results of a previous study (Ref 8.26), indicated as Heat III in the figures, to reinforce that point. For Heat III, the austenitizing temperature was 1065°C (1950°F) and the results were very close to those for Heat II, in which the grain size was relatively large.

The basic fracture mechanisms were found to depend on testing temperature, ΔK, and to some extent on heat treatment. Representative fracture features for the standard heat treatment are shown in Fig. 8.19. While these fractographs were taken from specimens fractured at room temperature, they are representative of all testing temperatures except 540°C (1000°F) and 650°C (1200°F), where there was some evidence of intergranular fracture at low stress intensities. At low ΔK levels, there were numerous crystalline facets which are probably associated with intense deformation on $\{111\}$ planes. At the higher stress intensity levels, it was found that the striations corresponded closely to the macroscopic crack growth rate. This

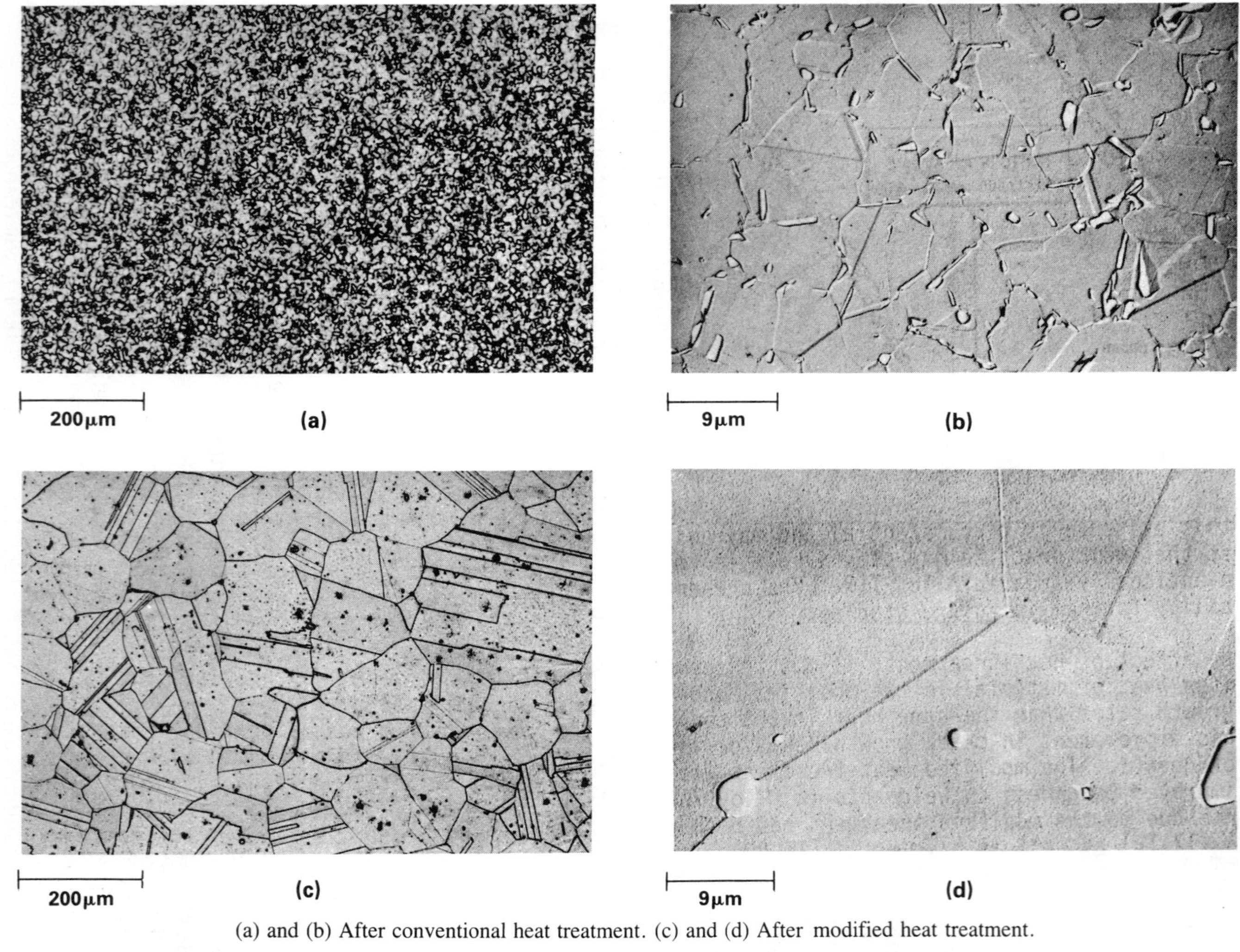

(a) and (b) After conventional heat treatment. (c) and (d) After modified heat treatment.

Fig. 8.12. Structures of Heat I of Inconel 718 (Ref 8.25)

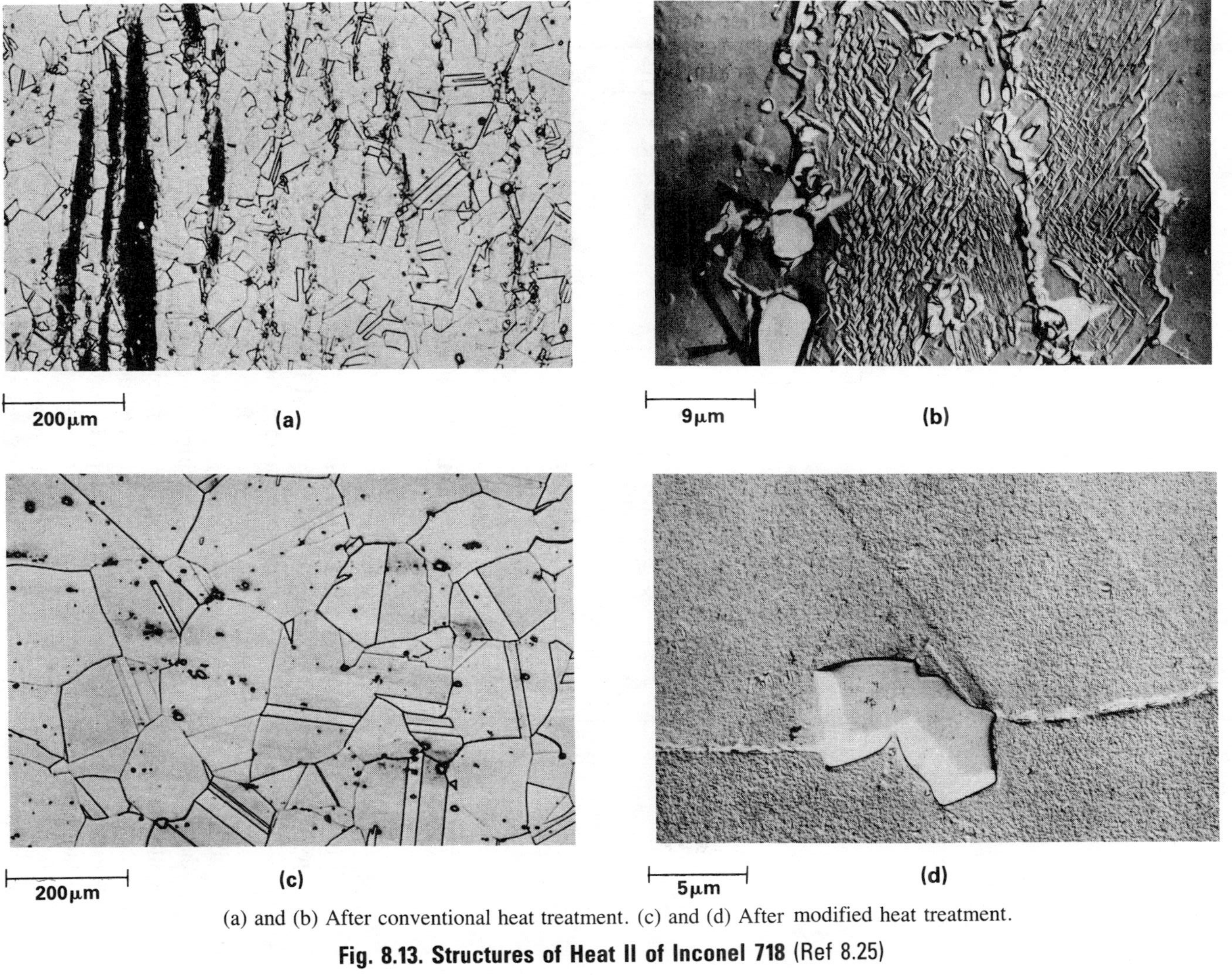

(a) and (b) After conventional heat treatment. (c) and (d) After modified heat treatment.

Fig. 8.13. Structures of Heat II of Inconel 718 (Ref 8.25)

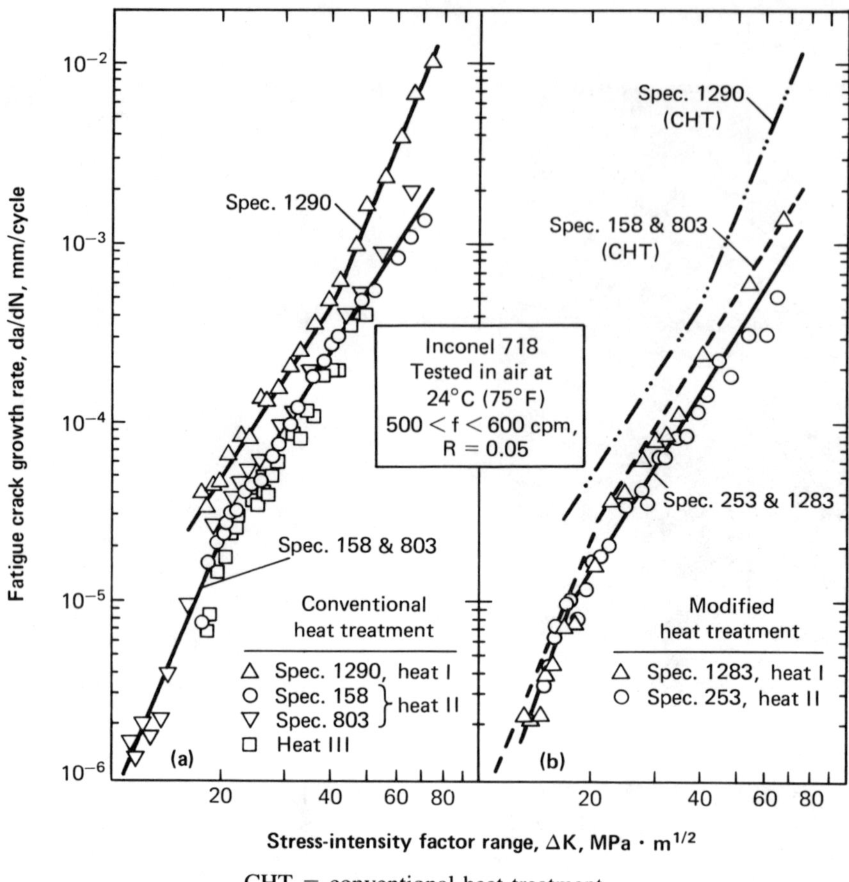

CHT = conventional heat treatment.

Fig. 8.14. Fatigue crack growth rate behavior of Inconel 718 tested in air at 24°C (75°F) (Ref 8.25)

one-to-one correspondence has been used by Popp and Coles (Ref 8.27) as an aid to failure analysis of components fabricated from Inconel 718.

For the modified treatment, the fracture surfaces tended to be similar to those of conventionally treated specimens with the additional observation that the crystalline fracture faces were less distinct, suggesting that a somewhat different deformation mode was operative. Unfortunately, no detailed observations were made on either the basic precipitate morphology or the nature of plastic deformation. It was suggested that there was a tendency for more intergranular fracture to occur following the modified treatment for tests at elevated temperatures. The latter observation is complicated by the fact that the junctions of large cleavage facets can have an intergranular appearance—especially at elevated temperatures, where oxidation plays an important role and where the oxidation products can obscure the features.

In summary, because of the complexity of most nickel-base alloys, variation in FCP performance cannot be attributed to a single microstructural effect, but it does appear to be amenable to metallurgical control through control of processing procedures, melting practice and/or heat treatment. FCP rates also appear to be

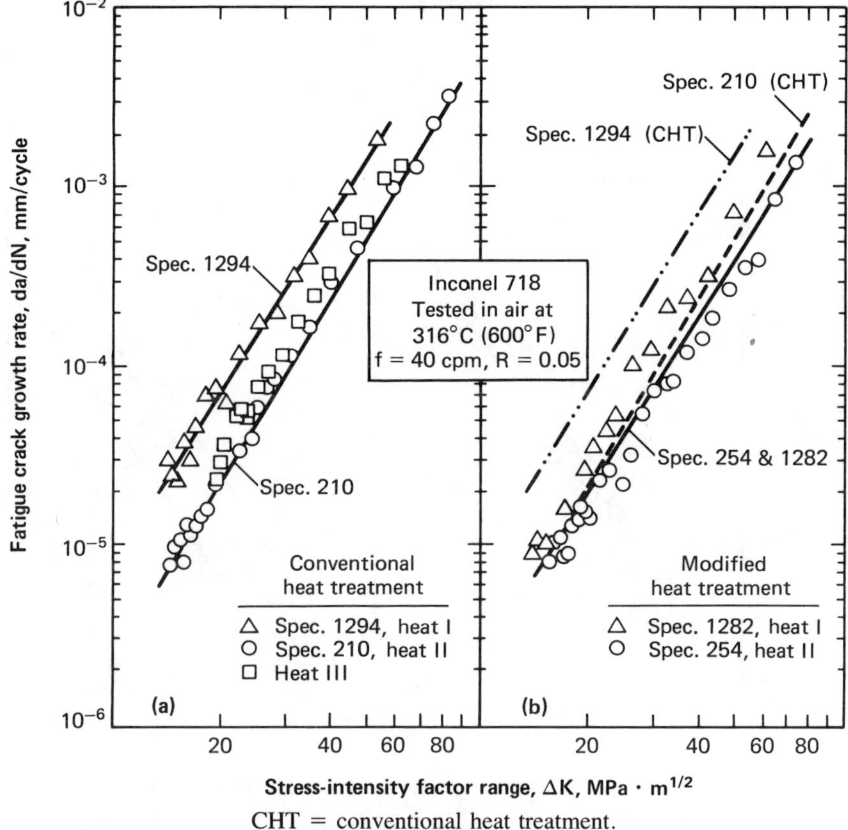

CHT = conventional heat treatment.

Fig. 8.15. Fatigue crack growth rate behavior of Inconel 718 tested in air at 316°C (600°F) (Ref 8.25)

lower when planar slip is promoted, which can be accomplished by compositional control (e.g., by decreasing the APB energy or by decreasing the degree of misfit) as well as by heat treatment (production of large grain sizes and small γ' sizes). Thus, when resistance to FCP is required in a particular application, the potential for enhancement is great, especially when one realizes that such enhancement is most pronounced in the low-to-mid-ΔK region, precisely the region of most engineering interest.

8.3.3. Effects of Testing Variables: Environment, Temperature, Frequency and Waveform

8.3.3.1. Environmental Factors

In many treatments of FCP, the effects of environmental factors are often ignored because testing generally is done in air at room temperature and at relatively high frequencies.* This is not true for nickel-base superalloys for most

*It should be noted that many investigators have observed frequency effects for steels, for titanium alloys and for aluminum alloys near the threshold at room temperature. These effects occur because the mechanical crack growth rate is low and because environmental influences may be quite pronounced.

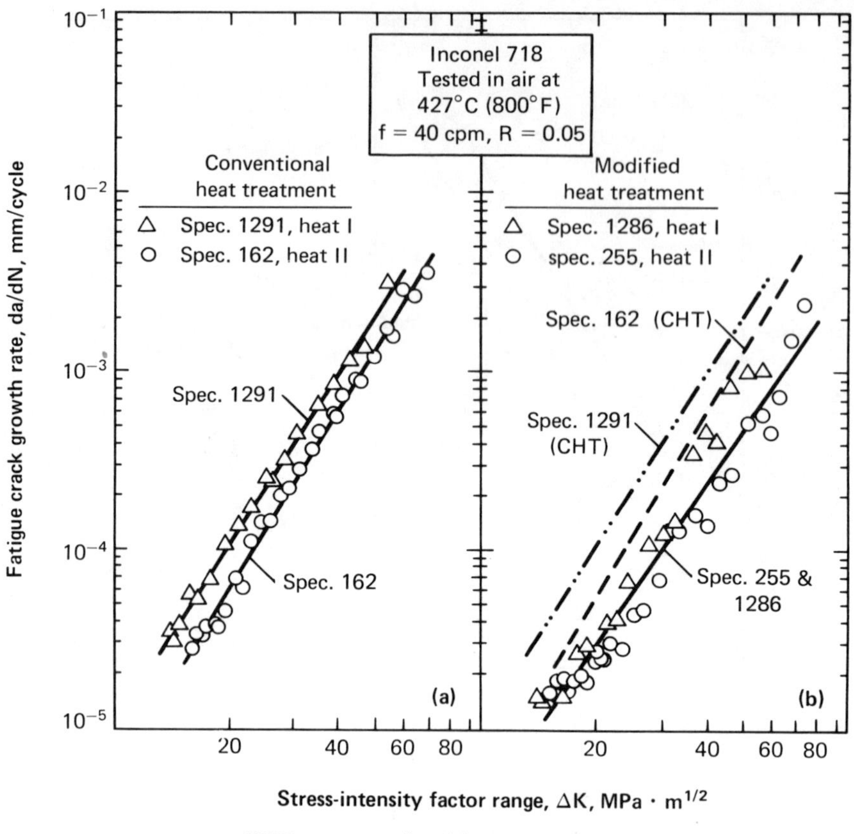

CHT = conventional heat treatment.

Fig. 8.16. Fatigue crack growth rate behavior of Inconel 718 tested in air at 427°C (800°F) (Ref 8.25)

applications of interest, because temperatures up to 1000°C (1832°F) may be encountered in turbine blades and because temperatures from 200 to 650°C (400 to 1200°F) occur in the discs. Thus, the effects of temperature, environment and waveform become very important.

That the effect of environment can be large may be inferred from some recent low-cycle fatigue studies of René 95 in which surface and subsurface cracking was observed at comparable strain ranges and defect sizes (Ref 8.28). As expected, the life of the subsurface crack was much greater than that of the surface crack, leading to the hypothesis of a strong environmental effect. This possibility is considered in more detail in an analysis of FCP properties of René 95 (Ref 8.29). The FCP rate was plotted as a function of temperature for a given ΔK range, as shown in Fig. 8.20. It is noteworthy that there is a minimum in the FCP rate at all ΔK levels except 22 MPa·$M^{1/2}$ (20 ksi·in.$^{1/2}$), where the data are at least suggestive of a minimum. Because any environmental interaction is thermally activated, the crack growth rate at a given ΔK level and frequency may be written as:

$$\frac{da}{dN} = A\exp - Q(\Delta K)/RT \qquad \text{(Eq 8.6)}$$

where A is a constant and Q(ΔK) is the apparent activation energy.

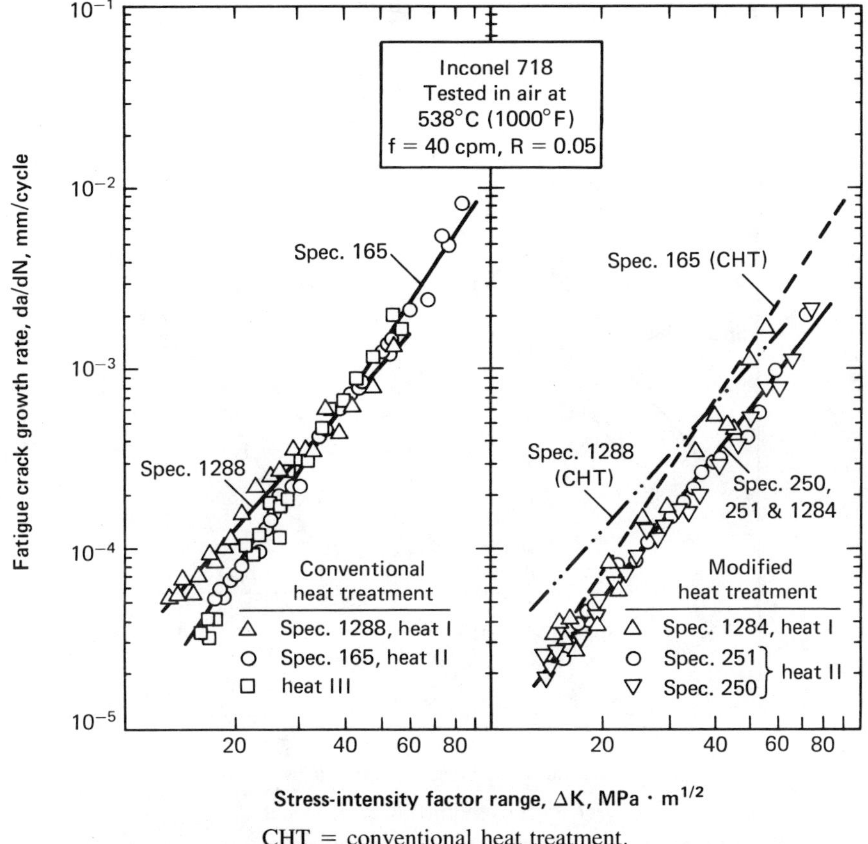

CHT = conventional heat treatment.

Fig. 8.17. Fatigue crack growth rate behavior of Inconel 718 tested in air at 538°C (1000°F) (Ref 8.25)

The dependence of Q on ΔK may be understood in several ways. For example, diffusion is facilitated by an expansion of the lattice ahead of a propagating crack (Ref 8.30)*, and the dilation is proportional to $(\Delta K)^2$. Consequently, the apparent activation energy is expected to be a linearly decreasing function of $(\Delta K)^2$. Such behavior has been observed for titanium (Ref 8.31) and for steel (Ref 8.32). In fact, extrapolation of the data for René 95 to ΔK = 0 gives a value for the characteristic activation energy of the process of about 30 k-cal/mol (Fig. 8.21). Values of 10 k-cal/mol have been observed by other investigators for IN 100 (Ref 8.33) and for Multimet (Ref 8.34), a solid solution iron-base superalloy that contains substantial amounts of nickel, cobalt and chromium. The authors of the latter report correctly point out that it is difficult to separate cracking due to a creep component from cracking associated with oxidation. Such a separation would require testing in vacuum or inert gas. On the other hand, creep need not always

*Other mechanisms are also possible and would be expected to cause the apparent activation energy to be proportional in some way to ΔK. For example at a high ΔK, there could be more oxide cracking which would account for a reduced apparent activation energy. There also would be an increased density of point defects and dislocations which would make diffusion easier and would reduce the apparent activation energy with increasing ΔK.

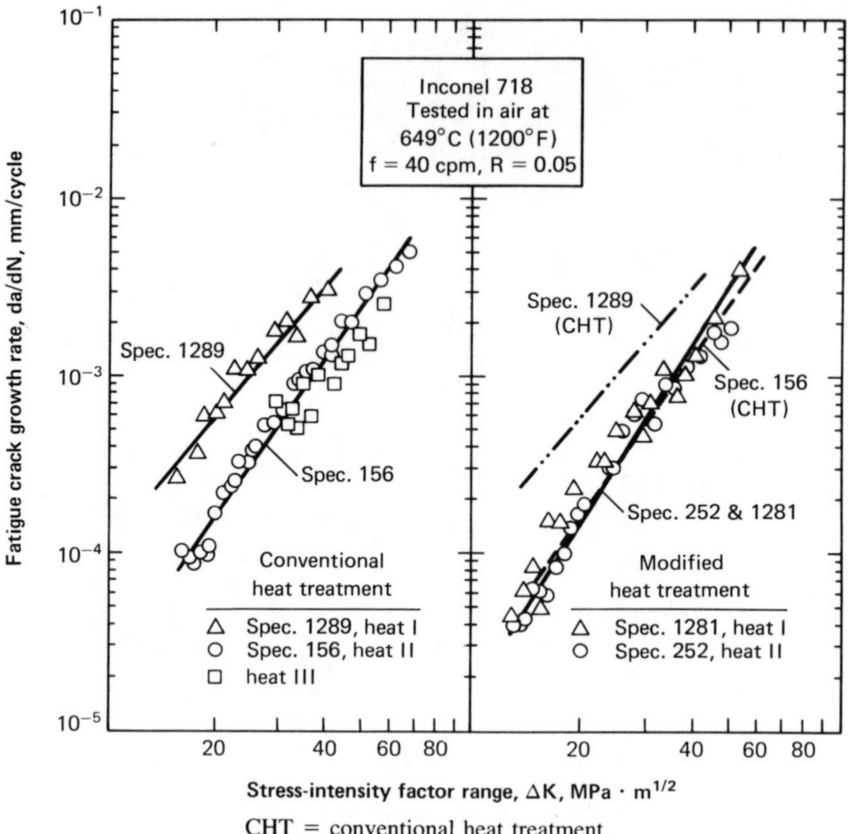

CHT = conventional heat treatment.

Fig. 8.18. Fatigue crack growth rate behavior of Inconel 718 tested in air at 649°C (1200°F) (Ref 8.25)

give rise to a more rapid rate of cracking. A microcreep effect at the crack tip could lead to crack tip blunting, which would effectively decrease the stresses in the region of the crack tip. It is conceivable that this mechanism could lead to a decreasing crack growth rate with increasing temperature. This effect would be counterbalanced by an environmental effect at the crack tip. Provided that the activation energy for environmental interaction is higher, there would come a point at which oxidation, or another crack tip environmental interaction, would become the dominant factor leading to an increase in FCP rate. The above discussion is intended to point out possibilities; such hypotheses are clearly in need of further detailed study.

Likewise, the effect of environment need not always lead to more rapid crack growth. It has been proposed that oxidation products could form in the crack tip region and prevent crack resharpening during the unloading portion of the cycle (Ref 8.35). If the stresses are sufficiently low, the oxidation products in the crack tip region will not be cracked and, in some systems, an elevation of the threshold might occur. Such effects would be pronounced at high temperatures and long hold times and have actually been observed in René 95, as shown in Fig. 8.22 (Ref 8.36). Once the stress intensity is high enough to crack the oxides, the rate

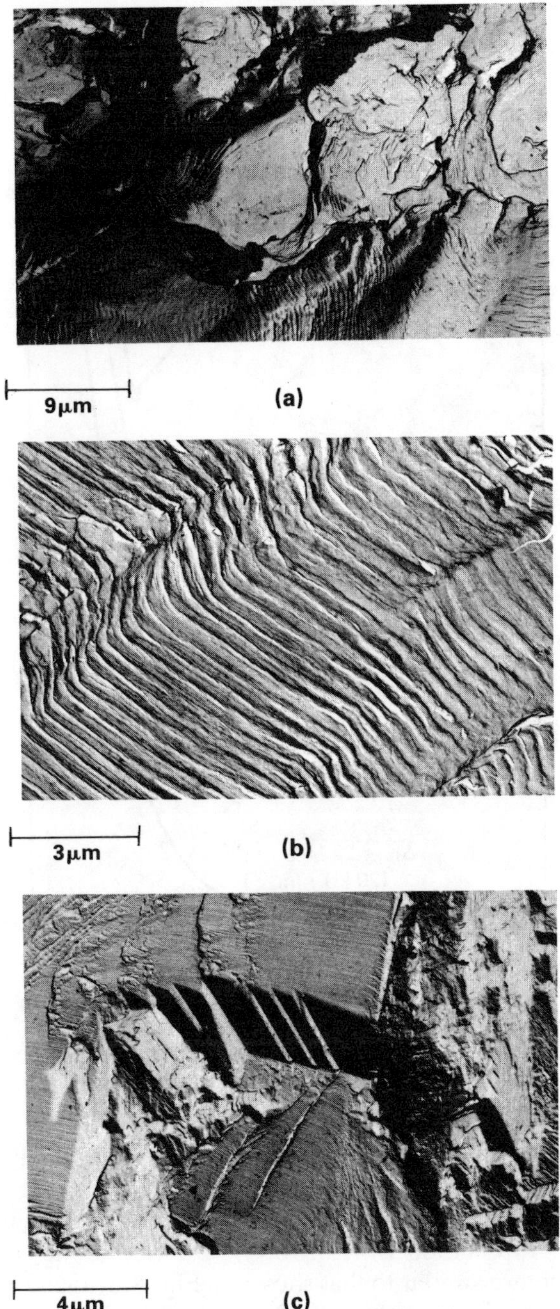

(a) Poorly defined microvoids and fatigue striations present at high ΔK levels. (b) Fatigue striations under intermediate ΔK conditions. (c) Cleavage-like faceted growth behavior in the low ΔK regime. Note the parallel lines superimposed on the faceted fracture surface.

Fig. 8.19. Typical electron fractographs showing room temperature fatigue fracture surface micromorphology of conventionally heat treated Inconel 718 (Ref 8.25)

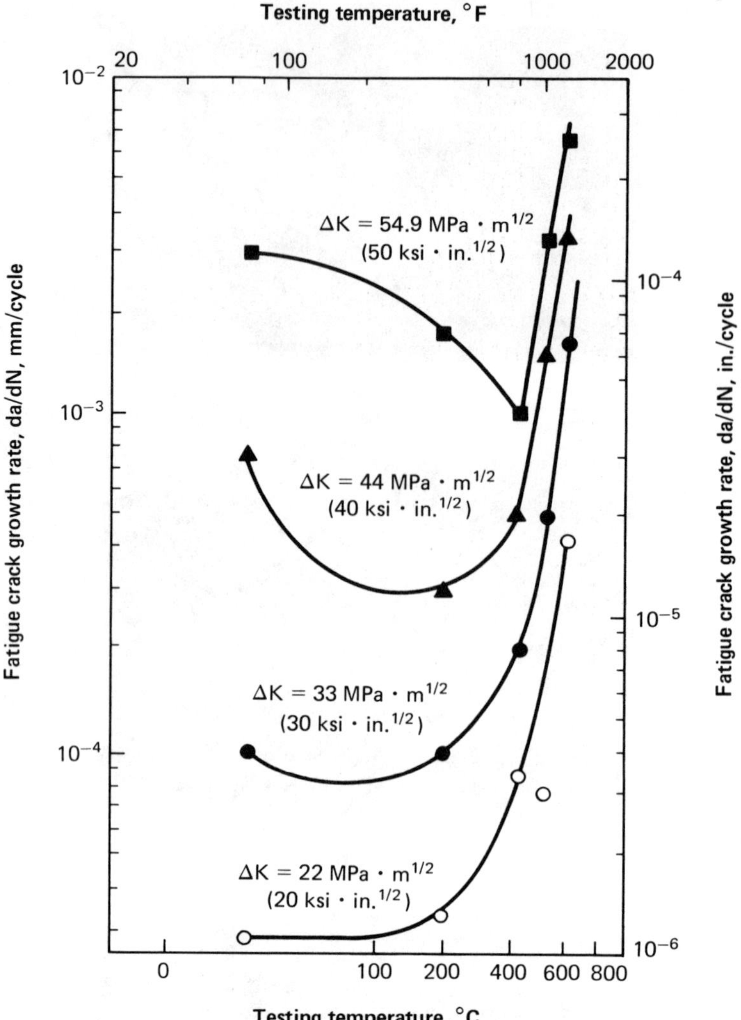

Fig. 8.20. Effect of temperature on fatigue crack growth rate at constant ΔK for René 95 (Ref 8.29)

of crack growth would be expected to increase due to the severely degraded region in the crack tip zone. Again, this is shown for René 95 in Fig. 8.22.

An environmental effect also was claimed to be operative during FCP studies of Udimet 700 at 850°C (1560°F) (Ref 8.37). Udimet 700 is similar to René 77 and has a microstructure similar to that shown in Fig. 8.1(a) and 8.3B. The fatigue crack growth rates are shown in Fig. 8.23, where they are compared with results obtained earlier by Gell *et al* (Ref 8.38) at room temperature. The crack growth rates were greatly accelerated by increases in temperature. The authors reported that, in stages I and II, crack propagation occurred by a cyclic cleavage mechanism resulting in fracture features similar to those shown in Fig. 8.11. In fact, fatigue fracture surfaces in most nickel-base alloys generally have cleavage-like features and, as has been seen, these features are even observed in tests carried out

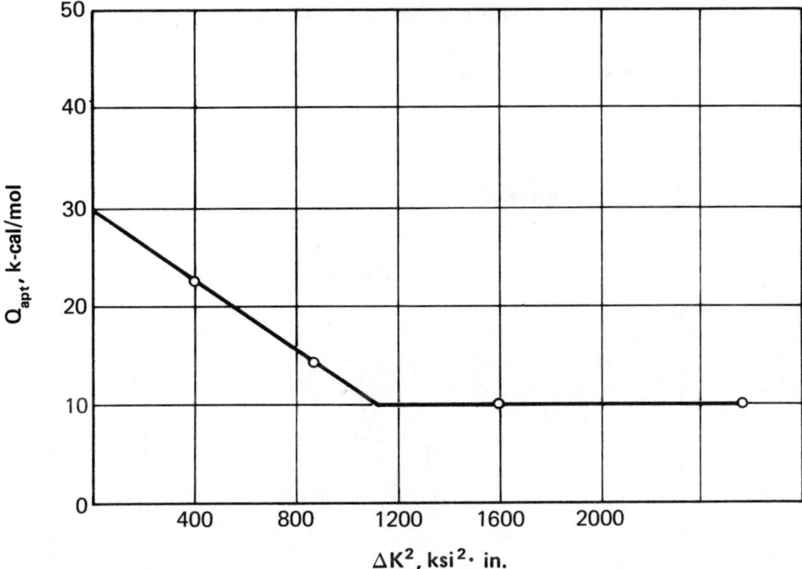

Fig. 8.21. Activation energy vs ΔK^2 for René 95 (Ref 8.29)

at room temperature. These features are probably related to the pronounced tendency of these materials to deform by planar glide for reasons cited previously. At the higher stress intensities, branch cracks started to form along grain boundaries, and intergranular cracking can be seen on the fractograph shown in Fig. 8.24, which the authors of this report attribute to a possible crack tip creep process.

The effect of environment has been studied extensively for Inconel 718 (Ref 8.39). Testing was done at 650°C (1200°F) at a frequency of 0.1 Hz using sinusoidal loading and R = 0. The heat treatment consisted of heating to 930°C (1700°F) and holding for 10 hours, air cooling to room temperature, reheating to 730°C (1350°F) and holding for 48 hours, and air cooling. This treatment develops a coarse accicular γ'' morphology and is more resistant to FCP than the conventional treatment, which develops fine γ'' plates with some small γ' spheres. It was found that the presence of small amounts of oxygen- or sulfur-bearing gases resulted in large increases in crack growth rate. The separate effects of air and SO_2 on FCP rates are shown in Fig. 8.25 and 8.26 while the combined effect is shown in Fig. 8.27. (The effect of air plus 0.5% SO_2 was similar to the effect of air alone.) It was observed that in the helium atmosphere, which was used to establish a baseline, cracking was generally transgranular with well-defined striations. In the air, oxygen-bearing and sulfur-bearing environments, the crack path changed from transgranular to intergranular, indicating that an important effect of the environment was to degrade the boundary strength by mechanisms that were not clearly defined. It was suggested that oxygen diffusion along grain boundaries and localized oxidation may have occurred. Another very important observation was that the effect of a given environment on FCP could not be predicted on the basis of unstressed exposure tests. The attack on the surfaces of unstressed specimens in aggressive SO_2 environments was minimal, but the SO_2 environments caused substantial increases in FCP.

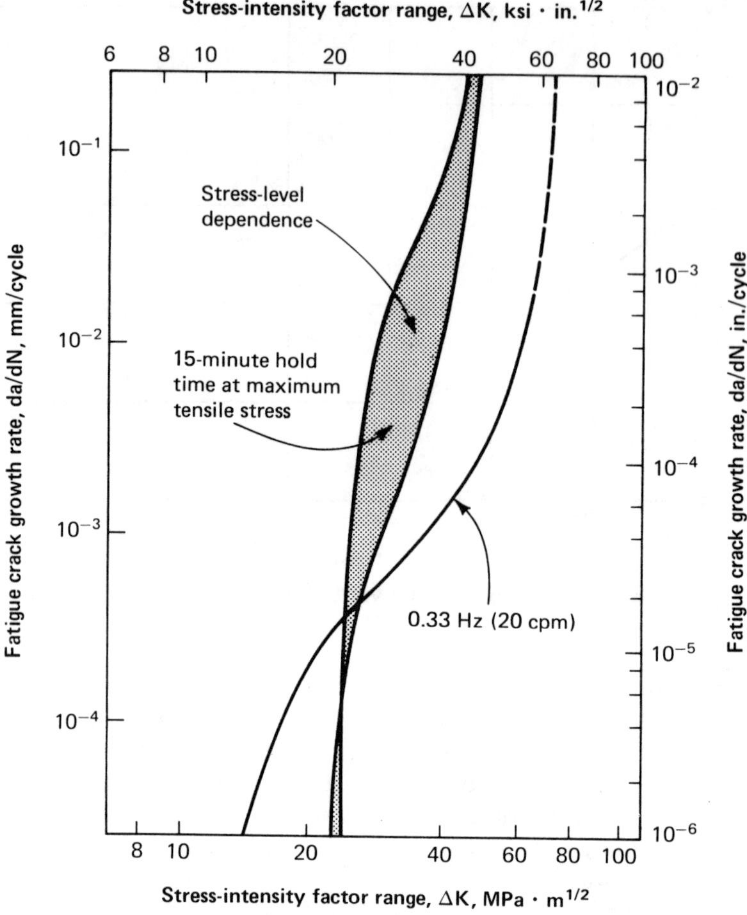

Fig. 8.22. Cyclic crack growth behavior for as-HIP René 95 under both continuous and hold time conditions at 650°C (1200°F) (Ref 8.36)

Similar observations have been reported recently for low-cycle fatigue specimens of René 80 that were stressed to about ⅓ of the yield stress at 982°C (1800°F) for 100 hours prior to testing (Ref 8.40). The prestressed specimens exhibited drastic decreases in fatigue life compared with specimens that were exposed but not stressed. The mechanism of degradation was suggested to be enhancement of oxygen diffusion along grain boundaries in the stressed specimens.

8.3.3.2. Temperature, Frequency and Waveform Effects on FCP at Elevated Temperatures

The effects of frequency and waveform have already been mentioned. For example, the effects of environment and creep tend to be minimized as the test frequency is increased. Also, if a hold time is imposed at maximum load, the FCP response may be different than in a continuous cycling test with the same over-all period. Such effects could be caused by an increased ramp rate, which could

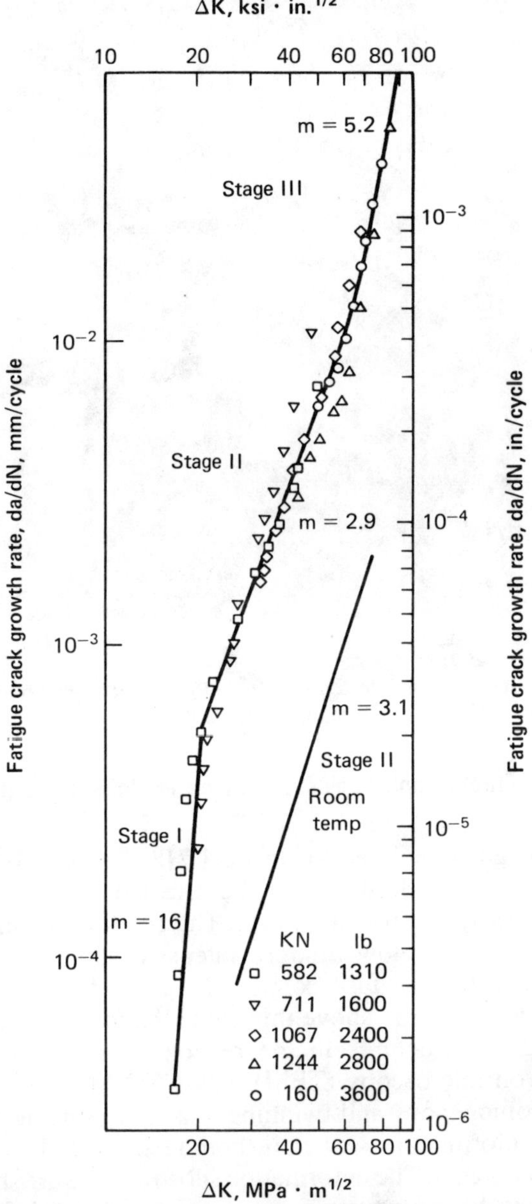

Fig. 8.23. Crack growth rates in terms of stress-intensity factor range for Udimet 700 at 850°C (1560°F) (Ref 8.37 and 8.38)

change the deformation mode and by an increased time at the maximum load for environmental interactions to occur. An additional effect that is frequently overlooked is the effect of the hydrostatic state of stress in the crack tip region, which can assist any diffusional processes that occur (Ref 8.30). In two cycles having the same period and the same maximum load, a dwell time at maximum load can lead to a reduction in the effective activation energy for diffusional processes.

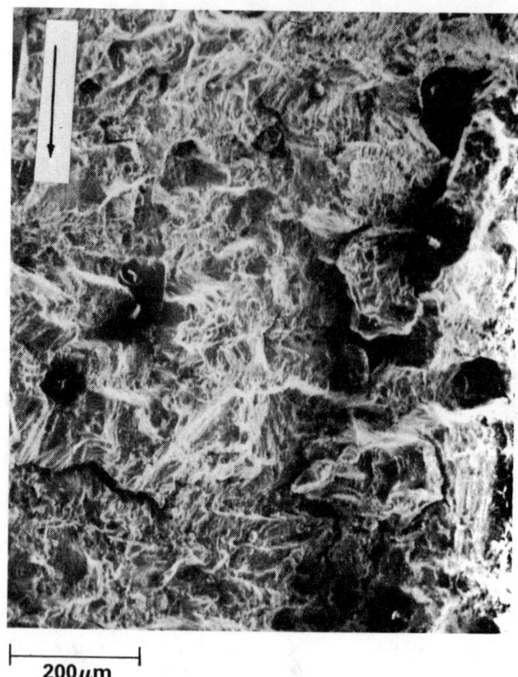

The specimen was tested at 850°C (1560°F) at a maximum load of 16 KN (3600 lb) at stage III region ($\Delta K = 68$ MPa · m$^{1/2}$, or 62 ksi · in.$^{1/2}$).

Fig. 8.24. Fractograph of FCP specimen of Udimet 700 (Ref 8.37)

These ideas have been considered for Inconel 718 (Ref 8.41). The FCP response at 20 Hz is shown in Fig. 8.28, where the effect of temperature is to increase the FCP rate, especially at low ΔK levels. The fracture features at 25°C (77 °F) are shown in Fig. 8.29. Below a stress intensity range of about 22 MPa·m$^{1/2}$ (20 ksi·in.$^{1/2}$), the fracture surface contained a high density of crystallographic facets on {111} planes whereas above this value the fracture surface was striated. The corresponding substructures in the reverse plastic zone were studied by transmission electron microscopy (TEM): below 22 MPa·m$^{1/2}$ (20 ksi·in.$^{1/2}$), deformation was inhomogenous and twinning was a prominent deformation mode; above this level, deformation was more homogenous and consisted of particle shearing and planar defects. The deformation substructures are shown in Fig. 8.30A and 8.30B. Both the fracture surface appearance and the deformation substructures were similar at 550°C (1025°F) and 20 Hz. The deformation substructures at 550°C were only studied at crack growth rates above 5×10^{-8} m/cycle (2×10^{-6} in./cycle).

The effect of frequency at 550°C (1025°F) was studied using a sine wave; the results are shown in Fig. 8.31. Below 0.5 Hz, the FCP rate was more rapid, and the crack surfaces showed an increased amount of intergranular fracture with decreasing frequency, with the crack path following the boundaries of the largest grains. One may be inclined to attribute the increase in FCP to either creep or environment, but this may not be the case, because different modes of deformation may have occurred at different strain rates.

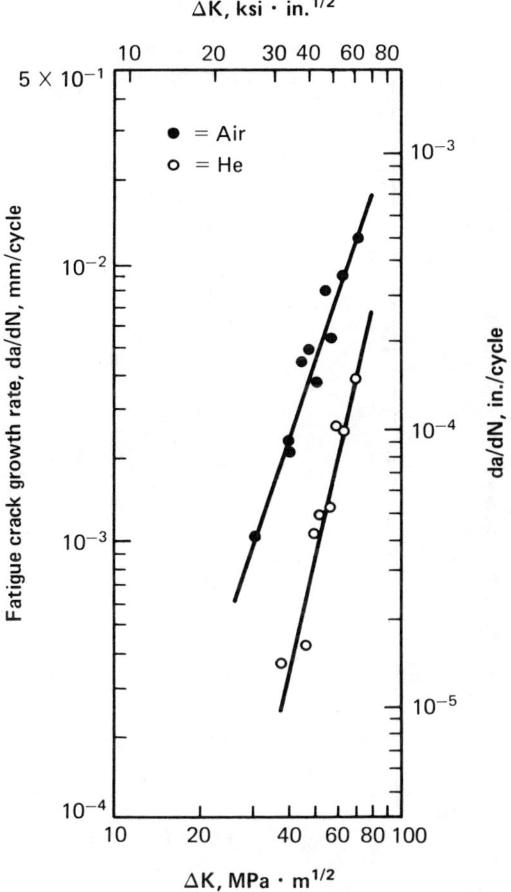

Frequency, 0.1 Hz. Temperature, 650°C (1200°F).

Fig. 8.25. Fatigue crack growth rate data for Inconel 718 in air and in helium (Ref 8.39)

To separate out the possible effects of creep or environment from deformation mode, the authors used triangular and square wave shapes, as shown in Fig. 8.32(a). The data obtained using the triangular wave at 2 Hz were the same as the data obtained in other tests using the sine wave at the same frequency which resulted in the lowest FCP rate. The effect of loading at the same rate but imposing a 10-second hold time at maximum load was to increase the FCP rate only slightly, as shown in Fig. 8.32(b). It is noteworthy that the deformation mode was similar in both cases, consisting of homogeneously distributed faults in the γ'' phase. Thus the small increase in the FCP rate appears to be unambiguously associated with an environmental (or creep) effect. For the low-frequency triangular wave, the amount of intergranular fracture increased, as shown in Fig. 8.33, and there were bands of intense deformation which were identified as twins. In addition, there were planar faults between the intense deformation bands, the density of which decreased with decreasing frequency. The conclusions to be drawn from these studies are that the change in deformation mode is more sensitive to strain rate than

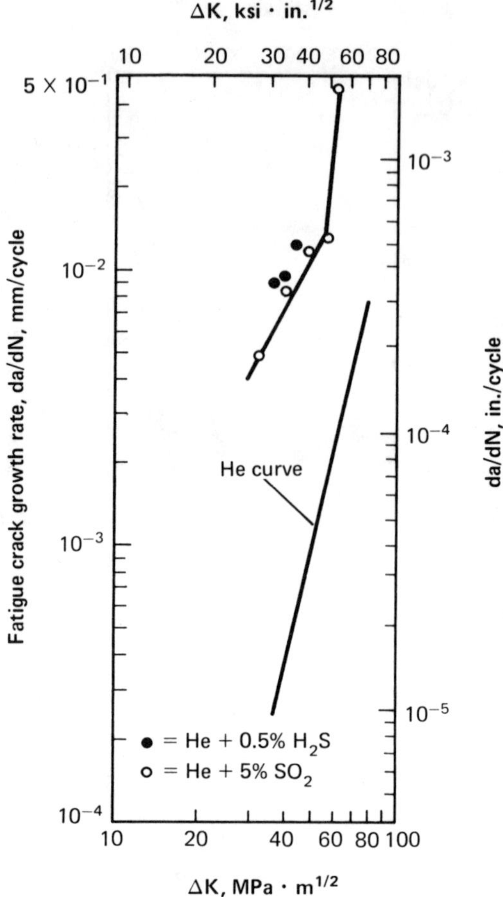

Frequency, 0.1 Hz. Temperature, 650°C (1200°F).

Fig. 8.26. Fatigue crack growth rate data for Inconel 718 in helium + 0.5% hydrogen sulfide and helium + 5% sulfur dioxide (Ref 8.39)

to frequency and that changes in deformation mode can be associated with large changes in FCP performance. This does not mean that, in the study just cited, twinning represents a severe form of *intrinsic* damage. Indeed it appears that the major effect of twinning at high temperatures is to promote environmental damage, because, as noted, the amount of grain boundary fracture increased at 550°C (1025°F) when twinning occurred. A further conclusion that can be drawn is that, because the big difference between the square wave and the triangular wave was associated with a change in deformation mode, creep was not an important factor in these experiments.

This study points out that great care must be exercised in analyzing and extrapolating the results of FCP tests. In particular, it is misleading to propose a general mechanism of life prediction at high temperature and low frequency, because, depending on testing conditions, different damage accumulation mechanisms may operate.

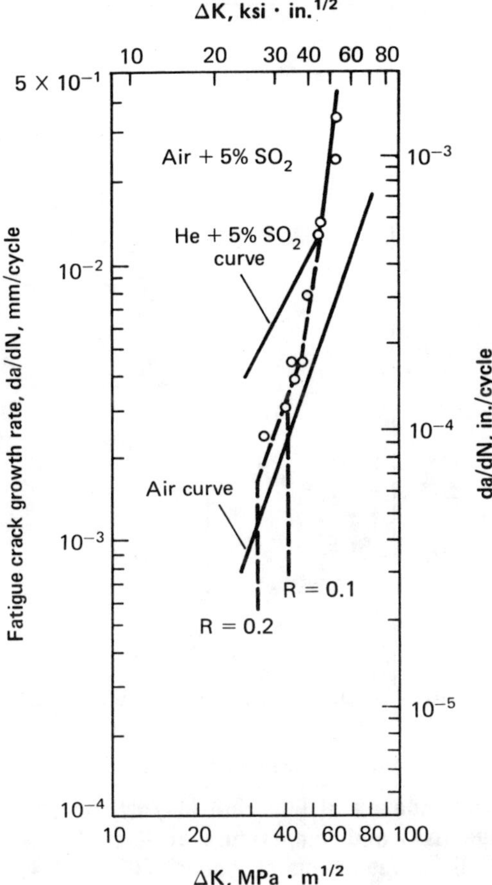

Fig. 8.27. Fatigue crack growth rate data for Inconel 718 in air + 5% sulfur dioxide (Ref 8.39)

8.3.3.3. Temperature Effects on Fracture Toughness and FCP of Nickel-Base Superalloys at Subzero Temperatures

As noted in previous chapters, metals with FCC crystalline structures generally retain good ductility and toughness at subzero temperatures. Many of the nickel-base superalloys not only retain good ductility and toughness but also have relatively high yield and ultimate tensile strengths at room temperature and at subzero temperatures. Among these alloys, Inconel 718 and Inconel X-750 have been selected for critical structural components of prototype superconducting motors and generators. Results of tests on Inconel 706 indicate that it also has suitable properties for certain applications at subzero temperatures.

Even at subzero temperatures, plane-strain fracture toughness (K_{Ic}) properties of nickel-base superalloys usually are too high to permit valid measurements to be made on specimens of reasonable size. However, valid K_{Ic} values for double-aged Inconel 718 have been determined on specimens 2.54 cm (1 in.) thick at room temperature and at cryogenic temperatures. Typical results for these tests and for

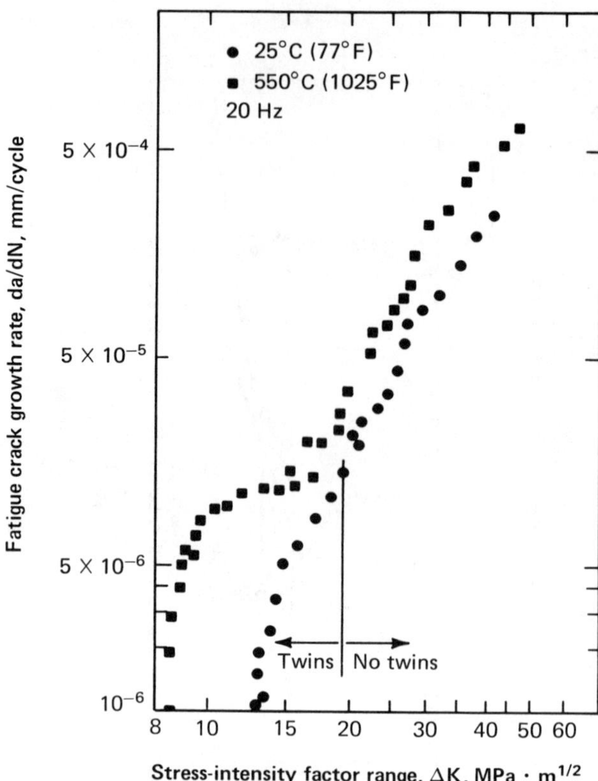

Fig. 8.28. Dependence of FCP rate (da/dN) on stress-intensity factor range (ΔK) and temperature at 20 Hz (sinusoidal wave shape signal) for specimens of Inconel 718 (Ref 8.41)

tension tests are shown in Table 8.2 for testing temperatures as low as −269°C (−452°F) (Ref 8.44). Yield and ultimate strengths increase as testing temperature is decreased, and ductility is higher at −269°C (−452°F) than at room temperature. Plane-strain fracture toughness also is higher at −269°C (−452°F) than at room temperature. Estimates of fracture toughness for Inconel 706 and Inconel X-750 alloys based on J_{Ic} values indicate that the fracture toughness of these alloys also is higher at cryogenic temperatures than at room temperature (Ref 8.45).

Results of FCP tests at room temperature and at temperatures as low as −269°C (−452°F) for Inconel 706, Inconel 718 and Inconel X-750 are shown in Fig. 8.34, 8.35 and 8.36. At equivalent ΔK values, the fatigue crack growth rates for these alloys are slightly lower at subzero temperatures than at room temperature (Ref 8.44 to 8.48). A comparison of FCP values from room temperature to −269°C (−452°F) for Inconel 718 and Inconel X-750 is shown in Fig. 8.37 (Ref 8.44) along with room temperature FCP data for Inconel 718 from Shahinian *et al* (Ref 8.49). The FCP data for these two alloys overlap in the ΔK range shown. Under some conditions, the FCP rate for Inconel 706 is slightly less than those for Inconel 718 and Inconel X-750 (Ref 8.48) at corresponding temperatures and ΔK levels. However, results of FCP tests are dependent on both melting practice and thermomechanical processing. Therefore, to ensure optimum properties

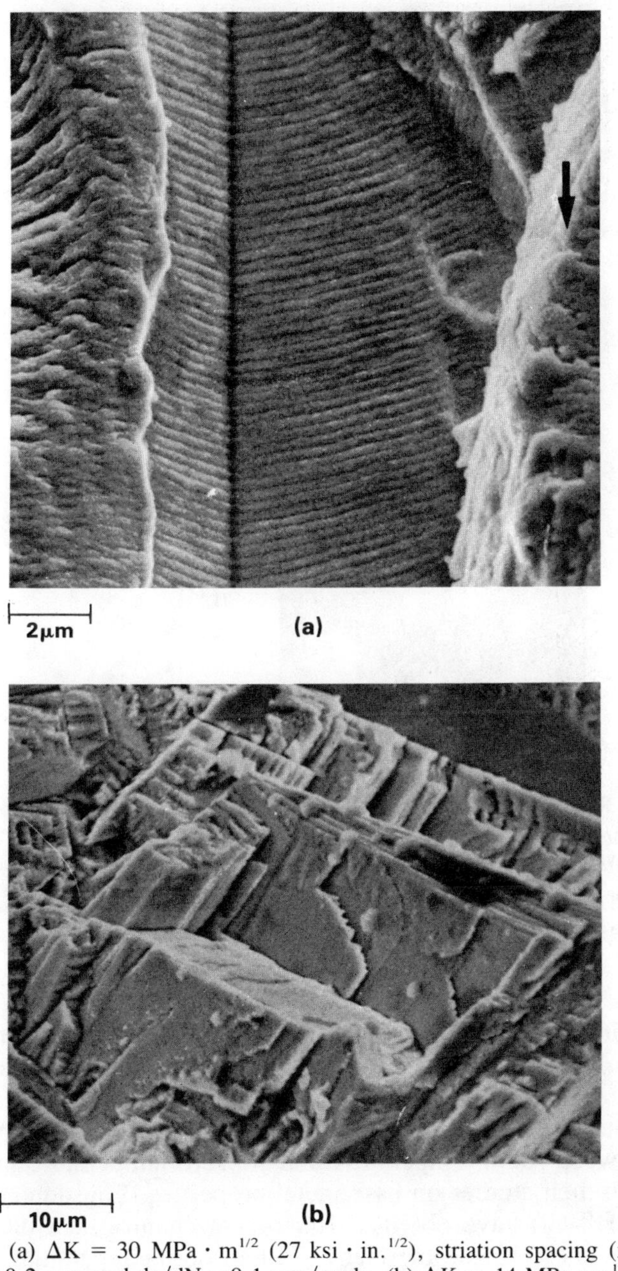

(a) $\Delta K = 30$ MPa · $m^{1/2}$ (27 ksi · in.$^{1/2}$), striation spacing (i) ≈ 0.2 μm, and da/dN ≈ 0.1 μm/cycle. (b) $\Delta K = 14$ MPa · $m^{1/2}$ (13 ksi · in.$^{1/2}$) and da/dN $= 2 \times 10^{-3}$ μm/cycle. The direction of crack propagation is indicated by an arrow in (a).

Fig. 8.29. Scanning electron micrographs of fracture surfaces of specimens of Inconel 718 at room temperature (Ref 8.41)

for critical components, information on melting practice, ingot reduction procedures and heat treatments are needed along with results of metallographic examinations and mechanical property tests at the service temperatures on heats to

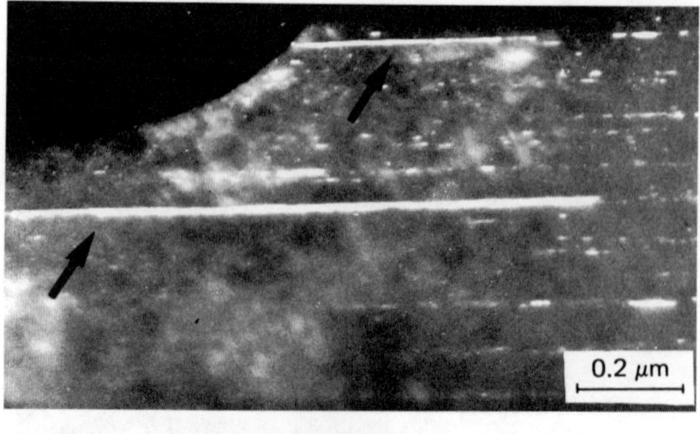

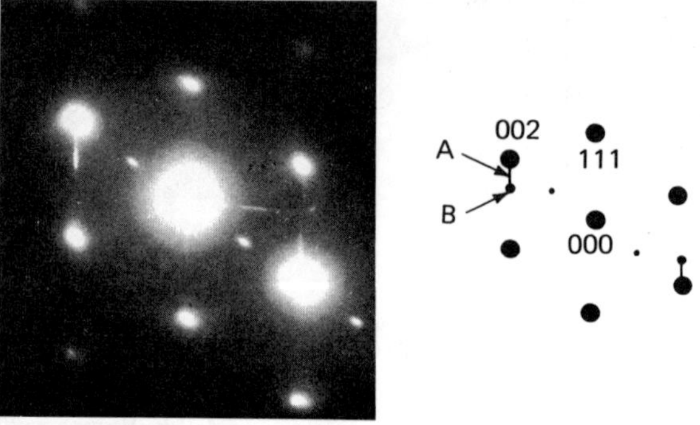

Room temperature; $\Delta K = 14$ MPa · m$^{1/2}$ (13 ksi · in.$^{1/2}$); da/dN = 0.002 μm/cycle. Dark field, and diffraction pattern. Twins are marked by arrows.

Fig. 8.30A. Identification of deformation twins in the threshold regime in specimens of Inconel 718 (Ref 8.41)

be used for critical components. Because the base metal properties of Inconel alloys improve as testing temperature is decreased from room temperature into the cryogenic range, one may use room temperature properties in design, and for qualification of these alloys for subzero service, if a relationship has been established between room temperature and subzero temperature properties.

In addition to their studies on base metal properties, Logsdon *et al* (Ref 8.45) and Wells (Ref 8.50) have obtained fracture mechanics data on weldments of Inconel 706, Inconel 718 and Inconel X-750 at room temperature and at subzero temperatures. Double-V butt welds were produced in Inconel 706 plate by the manual gas tungsten arc process using F-718 filler metal. The same filler metal was used in producing welds in Inconel 718 plate. The same welding process was used in producing welds in Inconel X-750 plate except that F-69 filler metal was used. Electron beam welding also was used in producing Inconel X-750 weldments. Tests were made on welded specimens in the as-welded condition and in the welded-plus-solution-treated and double-aged conditions. For the welded and fully heat treated specimens of Inconel 706 tested at −269°C (−452°F), the yield

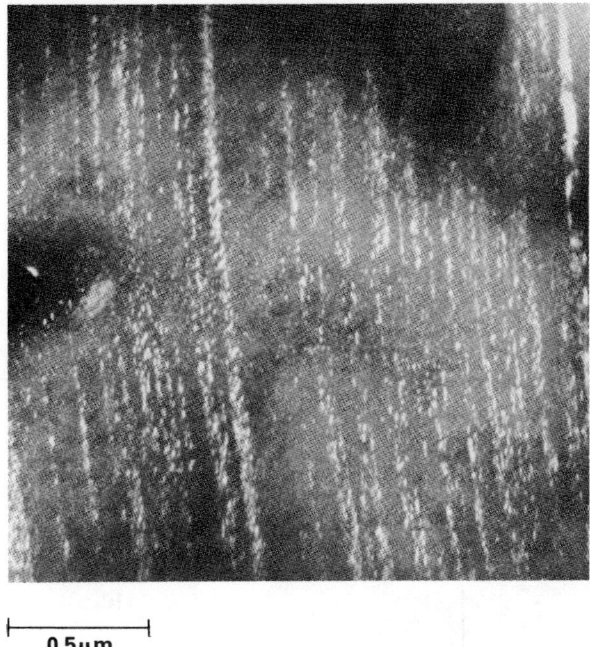

Temperature, 550°C (1025°F); $\Delta K = 25\ MPa \cdot m^{1/2}$ (23 ksi · $in.^{1/2}$); da/dN ≈0.10 μm/cycle; frequency, 20 Hz (sinusoidal load). Dark field obtained using a diffraction spot similar to that shown in Fig. 8.30A.

Fig. 8.30B. Fatigue plastic zone substructure in specimens of Inconel 718 (Ref 8.41)

strength of the weld was slightly lower than that of the base metal and the fracture toughness of the weld was much lower than that of the base metal. The same trend was observed in tests on welded specimens of Inconel 718. For welded specimens of Inconel X-750 in the fully heat treated condition, however, the yield strength at −269°C (−452°F) was slightly higher than that of the base metal at the same temperature, and the fracture toughness (from J_{Ic} tests) was substantially higher for the welded specimens than for the base metal at the same temperature.

Fatigue crack growth rates at room temperature and at −269°C (−452°F) for welded specimens of Inconel 706 and Inconel 718 in the fully heat treated conditions were substantially higher than those for corresponding specimens of base metal at the same ΔK values. Fatigue crack growth rate data for Inconel X-750 at −269°C (−452°F) were the same for all conditions of base metal and welded specimens. Therefore, when welds are required in superalloy structural components for room temperature or cryogenic temperature applications, fracture mechanics data indicate that Inconel X-750 is a reasonable candidate for those components. Because of the variability of welds produced in different shops, the welding practice to be used on any critical superalloy component should be qualified before fabrication is started.

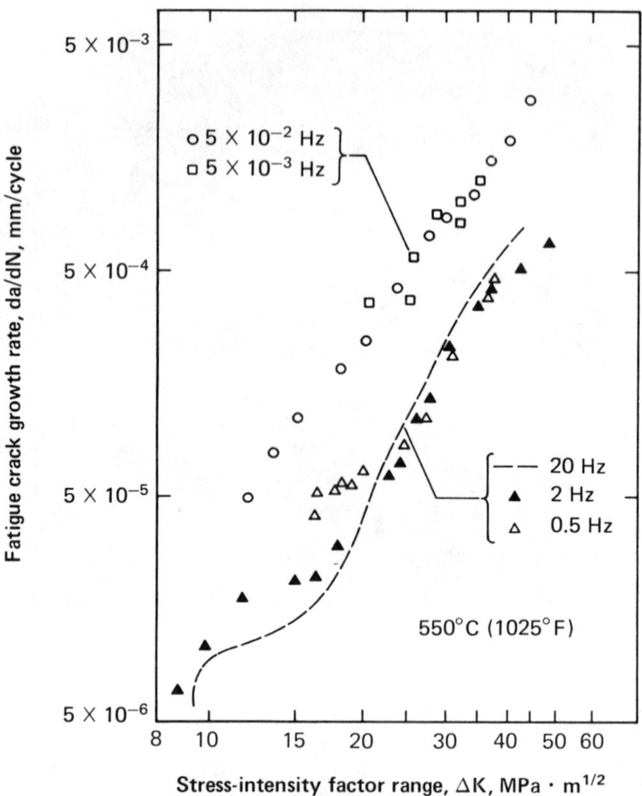

Fig. 8.31. Variation of FCP rate (da/dN) with stress-intensity factor range (ΔK) and frequency at 550°C (1025°F) (sinusoidal load) for specimens of Inconel 718 (Ref 8.41)

8.4. CREEP CRACK GROWTH

8.4.1. Applicability of Stress Intensity Parameter to Creep Crack Growth

When superalloys are used under conditions of high temperature and sustained loads, creep deformation is an important consideration in their application. As is the case for fatigue applications, the presence of cracks arising during manufacture or fabrication cannot be ignored and creep crack growth (CCG) is an important design consideration.

A question that naturally arises is whether or not this process can be characterized by the stress-intensity parameter. Floreen has investigated this question in a study of Inconel 718 in the temperature range from 595 to 704°C (1100 to 1300°F) (Ref 8.51). In this study, center-notched specimens were used, and specimen thickness was varied from 0.5 to 3.1 mm (0.02 to 0.12 in.); thickness had no apparent effect on the CCG characteristics. It was concluded that creep cracking occurred with only a small amount of creep strain at the crack tip and that stress intensity adequately characterizes stresses and strains in the region of the crack

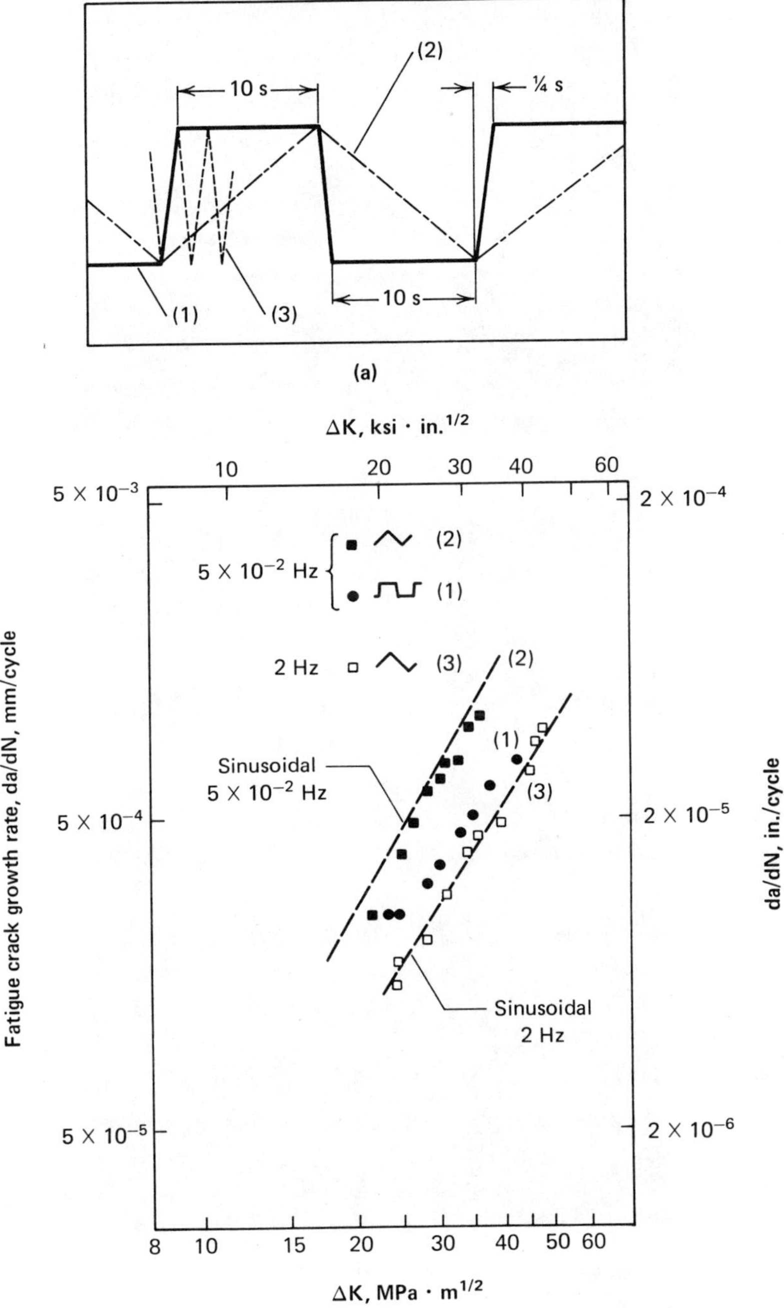

(a) Various forms of cyclic stress fluctuations used at 550°C (1025°F) at a frequency of 5×10^{-2} Hz. (b) FCP rates at 550°C under sinusoidal, triangular and square loads.

Fig. 8.32. Load/time waveforms and FCP rates for specimens of Inconel 718 (Ref 8.41)

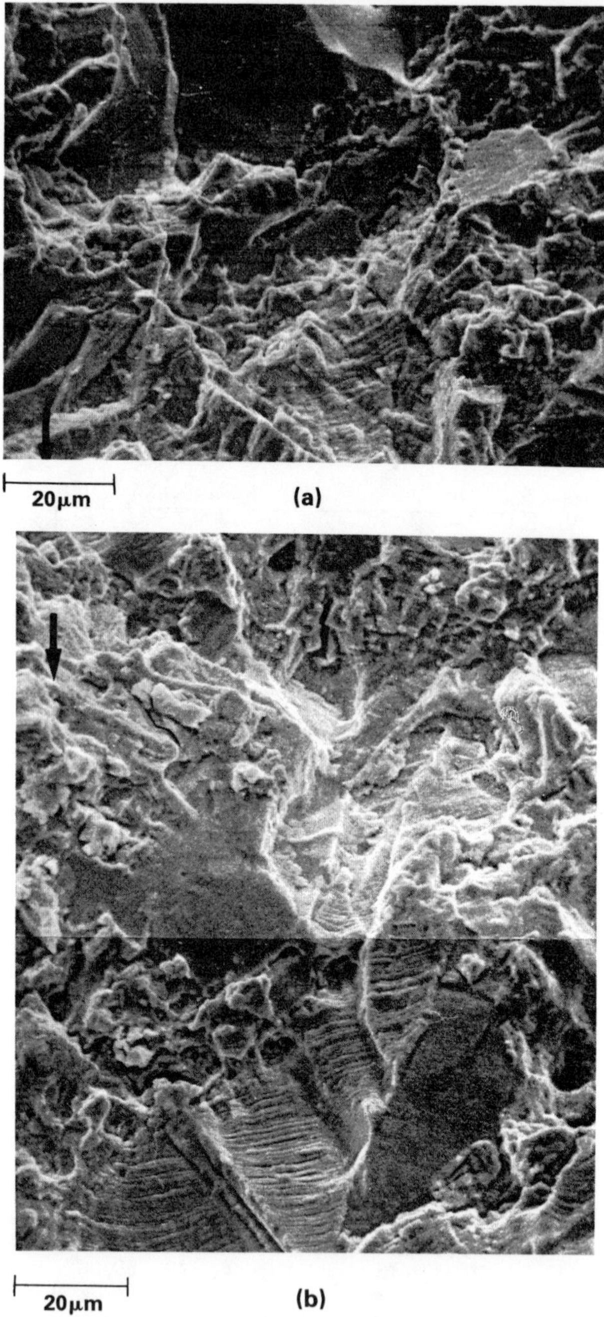

For definition of cycles, see Fig. 8.32. (a) Continuous triangular wave at a frequency of 5×10^{-2} Hz with $\Delta K = 35$ MPa · $m^{1/2}$ (32 ksi · $in.^{1/2}$) and da/dN = 0.8 μm/cycle. (b) Square wave with hold time of 10 s and with $\Delta K = 35$ MPa · $m^{1/2}$ and da/dN = 0.4 μm/cycle.

Fig. 8.33. Fracture surface in IN-718 at 550°C (1025°F) after cycling with a triangle wave (cycle 2) and a square wave (cycle 1) (Ref 8.41)

Table 8.2. Tensile and fracture toughness properties of Inconel 718(a) at room and subzero temperatures (Ref 8.44)

Testing temperature °C	Testing temperature °F	Yield strength MPa	Yield strength ksi	Tensile strength MPa	Tensile strength ksi	Elongation %	Reduction, in area %	Fracture toughness MPa·m$^{1/2}$	Fracture toughness ksi·in.$^{1/2}$
22	72	1172	170	1404	204	15.4	18.2	96.3	87.8
−196	−320	1342	197	1649	239	20.6	19.8	103	94
−269	−452	1408	204	1816	263	20.6	20.2	112	102

(a) Heat treatment: 980 °C (1800 °F) ¾ h, AC; double age 720 °C (1325 °F) 8 h, FC to 620 °C (1150 °F), hold 10 h; AC.

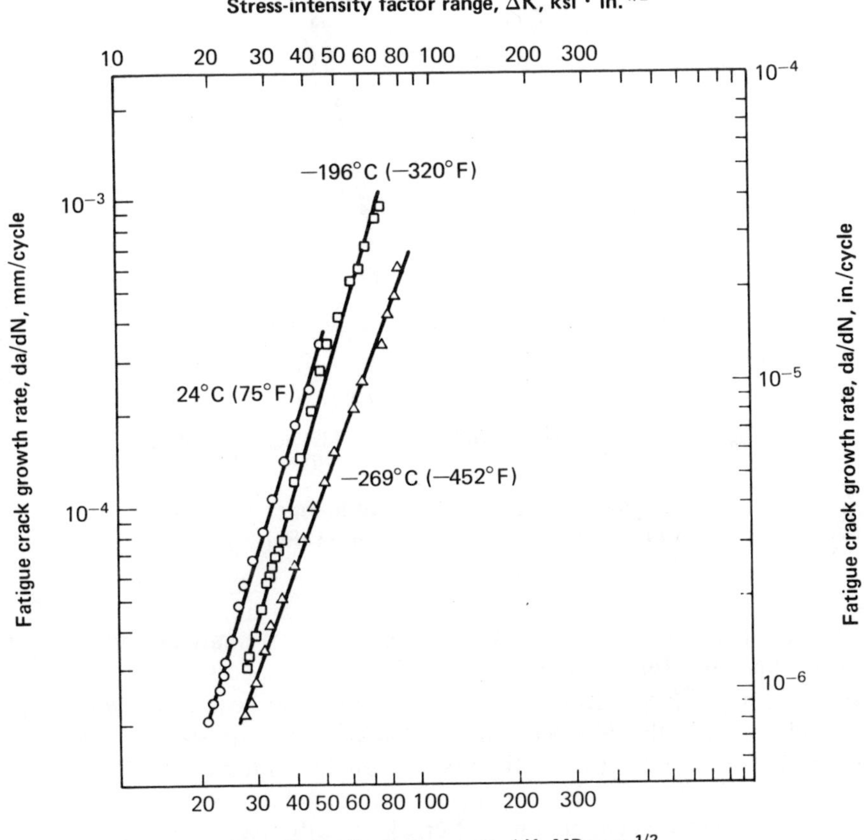

Heat treatment: 980°C (1800°F) 1 h, AC; double aged 730°C (1350°F) 8 h, FC to 620°C (1150°F), hold 8 h, AC.

Fig. 8.34. Fatigue crack growth rates of Inconel 706 forged billet (vacuum induction melted/vacuum are remelted) at an R ratio of 0.1 and a frequency of 10 Hz (Ref 8.45)

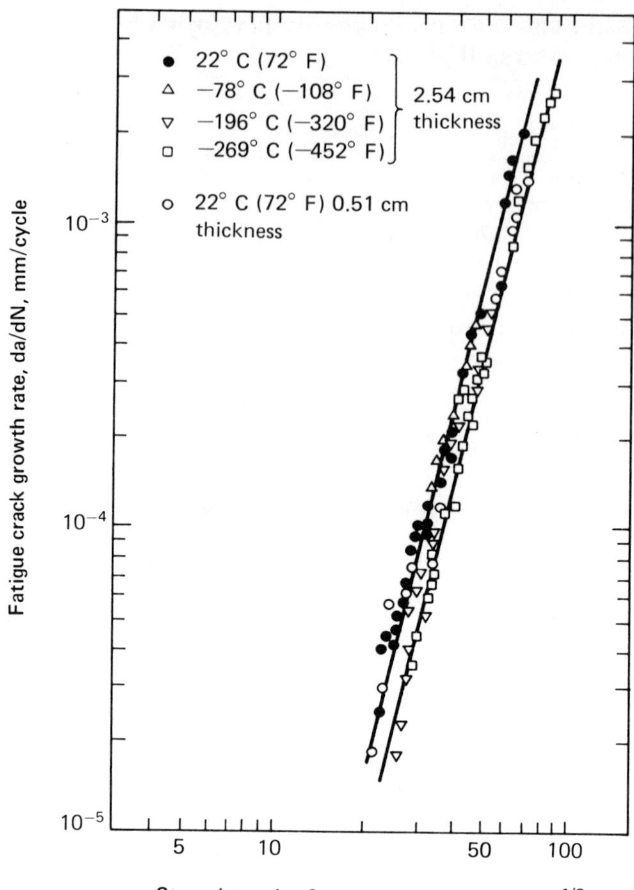

Heat treatment: 980°C (1800°F) ¾ h, AC; double aged 720°C (1325°F) 8 h, FC to 620°C (1150°F), hold 10 h, AC.

Fig. 8.35. Fatigue crack growth rates of Inconel 718 forged bar at an R ratio of 0.1 and a frequency of 20 Hz (Ref 8.44)

tip. On the other hand, it has been shown both experimentally and theoretically (Ref 8.52) that, for Type 304 stainless steel in the temperature range from 650 to 800°C (1200 to 1475°F), the crack growth rate can be best described in terms of net section stress. It also has been shown that when the stress-intensity parameter is used as a correlating variable, the results depend on the geometry of the specimen being used. Double-edge-notch and center-notch specimens tend to show approximately the same dependence of crack growth rate on K whereas a surface-flawed specimen shows much more rapid crack growth kinetics. This is especially significant because in practice most flaws are of the surface type. The reason that has been suggested for this behavior is that for Type 304 stainless steel there is a component of creep strain across the net section and considerable crack tip stress relaxation. The results of the latter study should be viewed as effectively defining the limits of applicability of the stress-intensity parameter approach to CCG. For

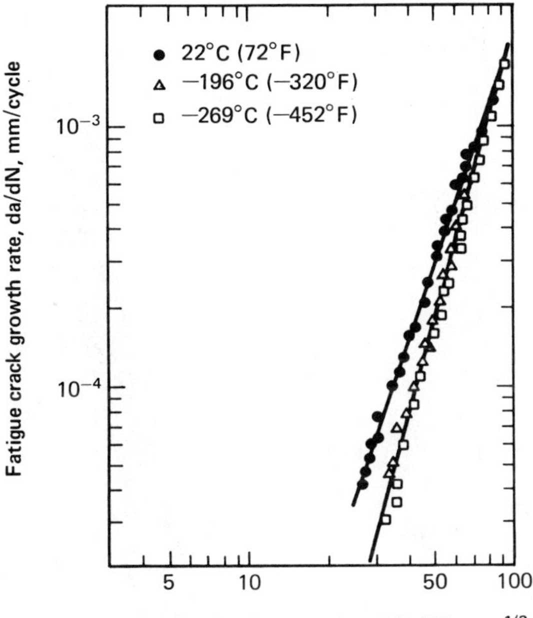

Heat treatment: solution treated and double aged.

Fig. 8.36. Fatigue crack growth rates of Inconel X-750 at an R ratio of 0.1 and at frequencies of 20 to 28 Hz (Ref 8.47)

superalloys that are used at high temperatures after being heat treated to high toughness and ductility, it must be demonstrated experimentally that K adequately describes the CCG process.

8.4.2. Mechanisms and Models for CCG

The CCG behavior of a series of typical superalloys has been studied at 704°C (1300°F), and the results are shown in Fig. 8.38. Compact tension specimens 9.2 mm (0.36 in.) thick were used in these experiments. The concept of a critical strain at some distance ahead of the crack tip has been advanced as a criterion for creep crack growth (Ref 8.54). Based on that idea, a formula for the initial stress-intensity factor required to produce failure in 100 hours has been derived:

$$K = [3\delta_o\sigma_{ys}E + 0.3\sigma_{ys}ED\varepsilon^*]^{1/2} \quad \text{(Eq 8.7)}$$

where δ_o is the initial crack tip displacement, σ_{ys} is the yield strength, E is Young's modulus, ε^* is the critical strain and D is the grain diameter. It has been found that the materials represented in Fig. 8.38 correlate very well with Eq 8.7 when ε^* is taken to be 0.1. Equation 8.7 is probably more indicative of a trend than predictive in nature. However, it does identify those parameters that strongly influence creep cracking. In particular, it is noteworthy that creep cracking is retarded by large grains.

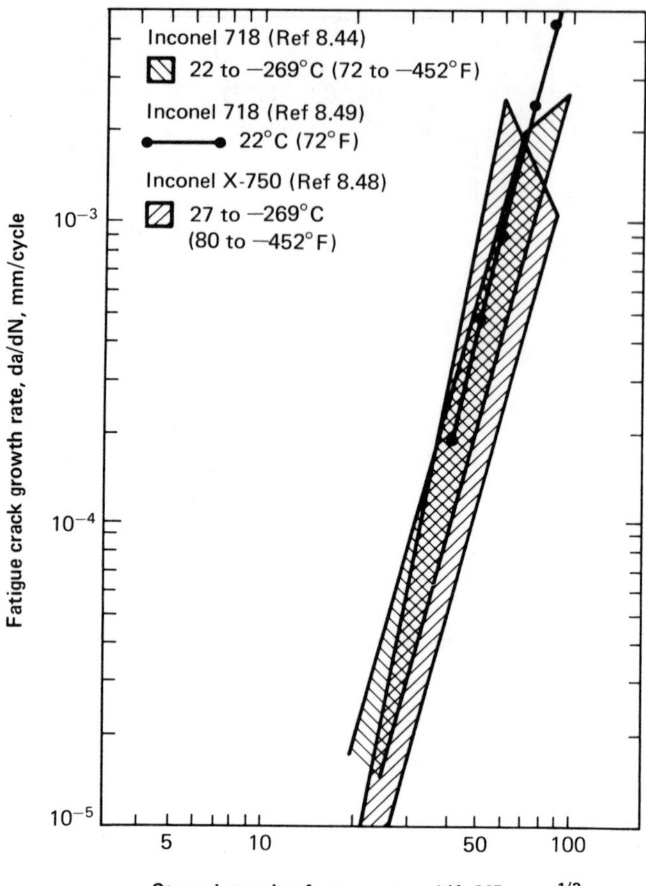

Fig. 8.37. Fatigue crack growth rates for Inconel 718 and Inconel X-750 in the subzero temperature range (Ref 8.44, 8.48 and 8.49)

A relationship for the rate of growth of a spherical cavity ahead of a crack has been derived by DiMelfi and Nix (Ref 8.55). The result of their calculations is:

$$\frac{1}{C}\frac{dC}{dt} = A\left(\frac{2}{\sqrt{3}}\right)^{n-1}\left(\frac{3K}{2n\sqrt{2\pi}}\right)^{n} x^{-n/2} \qquad \text{(Eq 8.8)}$$

where A is a constant, C is the cavity diameter, n is the exponent in the creep equation, x is the distance ahead of the crack and K is the stress-intensity factor.

If it is assumed that the CCG is the result of a cavitation process and that the CCG rate is proportional to the right-hand side of Eq 8.8, and if x is identified with grain size, then Eq 8.8 implies the following:

$$\frac{da}{dt} = B\left(\frac{K}{D}\right)^{n} \qquad \text{(Eq 8.9)}$$

where B is a constant and D is grain diameter. The data presented in Fig. 8.38 have

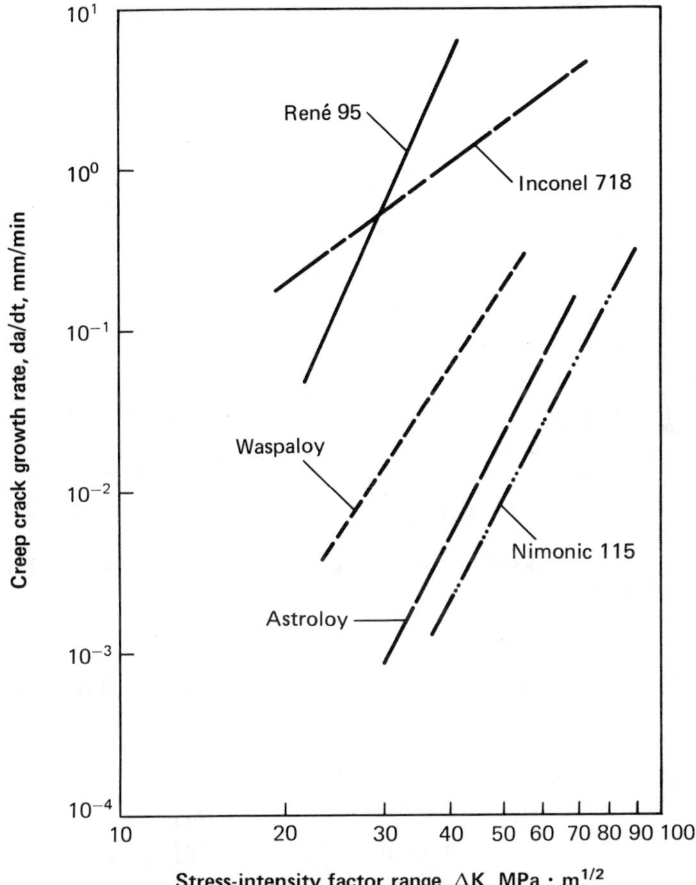

Fig. 8.38. Creep crack growth rate data for various superalloys at 704°C (1300°F) (Ref 8.53)

been replotted using this correlation (Ref 8.54), and the results are illustrated in Fig. 8.39. The fact that a good correlation is obtained for this wide range of materials has been interpreted to imply that grain size is the primary variable in determining the CCG rate (Ref 8.56).

8.4.3. Metallurgical Variables

It has already been pointed out that grain size is apparently of primary importance in controlling the CCG rate. It is also reasonable to expect that the shape of the grain boundary would also be important in CCG. In superalloys, it is possible to produce serrated grains by slowly cooling a warm worked alloy from above the γ' solvus. As the temperature falls below the solvus, γ' particles will nucleate heterogeneously on the grain boundaries, locally pinning an advancing boundary by a Zener-type mechanism. In other areas, where there are no boundary precipitates, the boundary will continue to advance into the warm worked matrix until it too either becomes pinned by precipitates or impinges on another boundary.

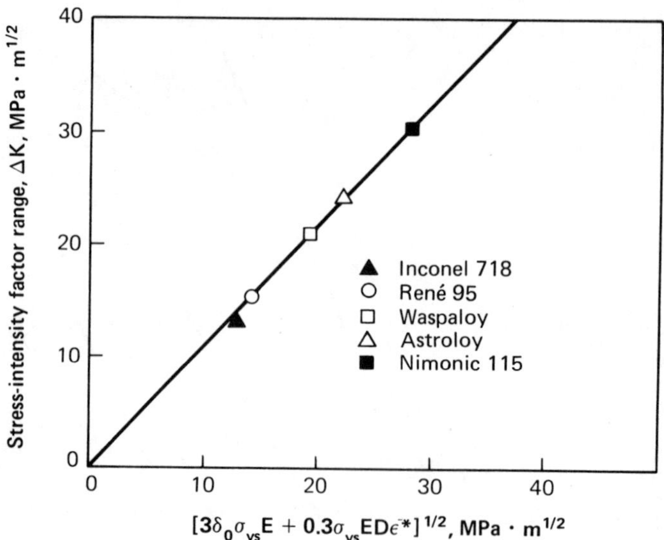

Fig. 8.39. Stress-intensity factors to produce failure in 100 h vs predicted values (Ref 8.54)

A serrated structure for IN-792 is shown in Fig. 8.40. It has been shown theoretically by Raj and Ashby (Ref 8.57) that the rate of boundary sliding (i.e., the creep rate) for a serrated boundary should be given by the relationship:

$$\dot{U} = \frac{8\,\tau_a \Omega\,\lambda}{\pi\,kT\,h^2} D_V \left\{1 + \frac{\pi \delta\, D_B}{\lambda\, D_V}\right\} \qquad \text{(Eq 8.10)}$$

where $\dot{U}$ is the sliding rate of the boundary, τ_a is the applied shear, k is Boltzman's constant, T is absolute temperature, λ is the periodicity of the boundary perturbation, h is the double amplitude of boundary perturbation, D_B is the boundary diffusion coefficient, D_V is the volume diffusion coefficient, Ω is the atomic volume and δ is the thickness of the grain boundary diffusion zone. The CCG rate is shown in Fig. 8.41 for both smooth and serrated boundaries of IN-792 (Ref 8.58), and the results are in qualitative agreement with the calculations of Raj and Ashby.

It is also known that overaging tends to improve the CCG characteristics of many alloys, and typical results are shown in Fig. 8.42 for Inconel 718 at 649°C (1200°F). These data can be explained in two ways. First, they can be considered to have resulted from the presence of large particles in the grain boundaries (Ref 8.57). Expressed mathematically, this becomes:

$$\dot{U} = \frac{1.6\tau_a \Omega\,\lambda^2}{kT\quad a^3}\left\{1 + 5\frac{\delta\, D_B}{a\, D_V}\right\} \qquad \text{(Eq 8.11)}$$

where a is the characteristic dimension of the particles in the grain boundaries and all other terms have been previously defined. It can be seen that the larger the particle, the larger the reduction in the sliding rate. In some instances, however,

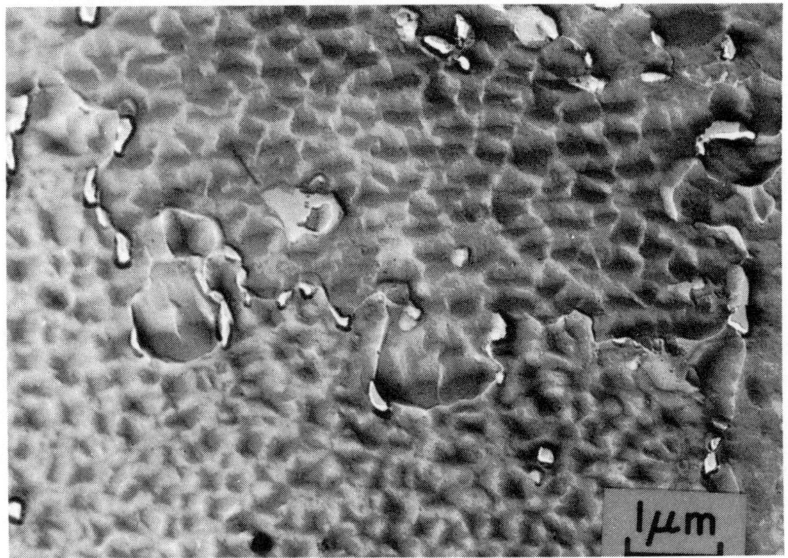

Fig. 8.40. Serrated grain boundary microstructure of IN-792 (Ref 8.58)

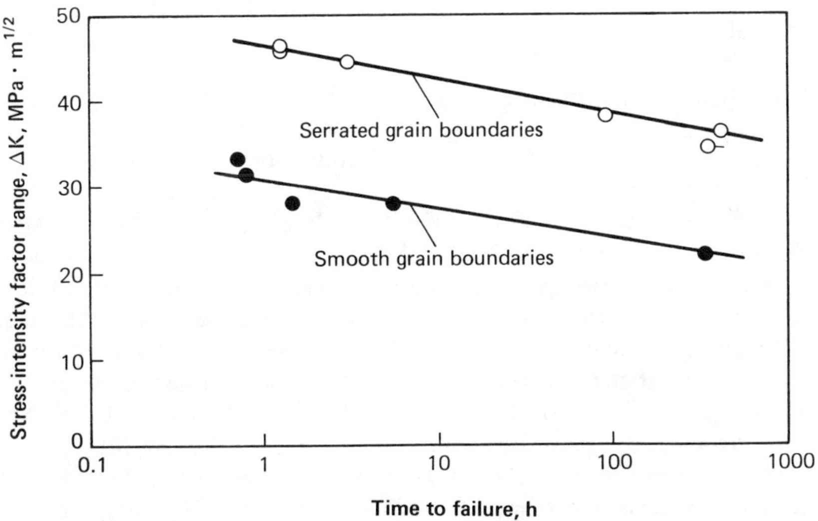

Fig. 8.41. Time to failure vs stress-intensity factor with either smooth or serrated grain boundaries for IN-792 at 704°C (1300°F) (Ref 8.58)

large precipitates in the grain boundaries can lead to rapid cracking (Ref 8.59). These particles could be large and brittle and could initiate cracking, or they could alter the local grain boundary structure and chemistry in such a way as to decrease the creep resistance. Another explanation (Ref 8.60) is that overaging tends to

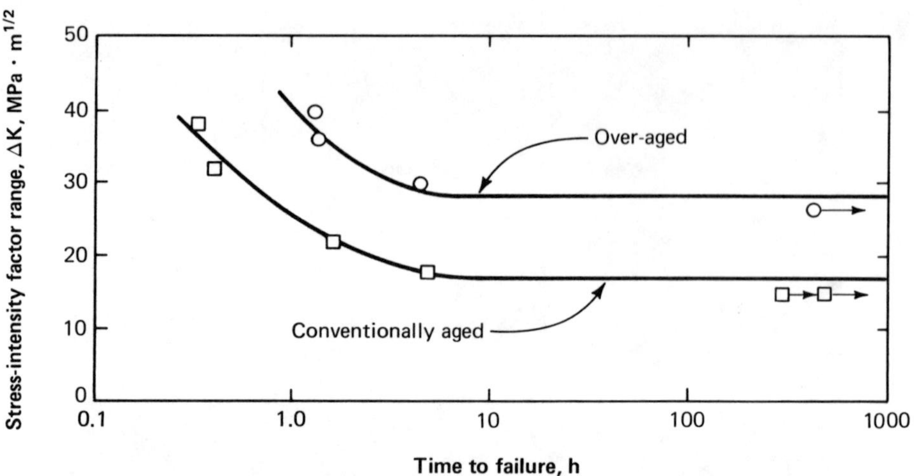

Fig. 8.42. Time to failure vs stress-intensity factor with various heat treatments for Inconel 718 at 649°C (1200°F) (Ref 8.53)

change the slip mode from dislocation shearing to looping. Because the dislocation looping mode promotes slip homogeneity, maximum stresses associated with slip bands are reduced at grain boundaries, and cracking should be less likely. Thus it is possible that the major benefit of overaging is primarily due to slip homogenization.

8.4.4. Environmental Effects

To this point, the effects of environment have not been considered. However, it is obvious that creep crack growth at elevated temperatures is in many ways analogous to stress corrosion cracking with air being the aggressive environment. The effects of environment on CCG in IN-718 have been studied by Floreen and Kane (Ref 8.61), and the results are shown in Fig. 8.43. Although cavity formation and transgranular cracking were observed for both the air and helium environments, the crack growth rate was almost 100 times higher in air than it was in helium, indicating that the effect of air may be to accelerate the cavity formation process. Creep cracking was also studied in IN-718 by Sadananda and Shahinian (Ref 8.62) over the temperature range 538 to 760°C (1000 to 1400°F), and their results are shown in Fig. 8.44. It is noteworthy that the CCG rate increased as the temperature was increased from 538 to 649°C (1000 to 1200°F), after which the rate either increased only slightly or decreased. This behavior has been interpreted as being due to a balance between two competing factors: vacancy diffusion to the crack tip, which accelerates cracking; and creep deformation, which causes the crack tip to become a less effective sink for point defects—presumably by a plastic blunting process. The effect of environment was not considered by these authors; however, the more recent results of Floreen and Kane (Ref 8.61) imply that the environment should be considered at least as a modifier of the various processes that occur at the crack tip.

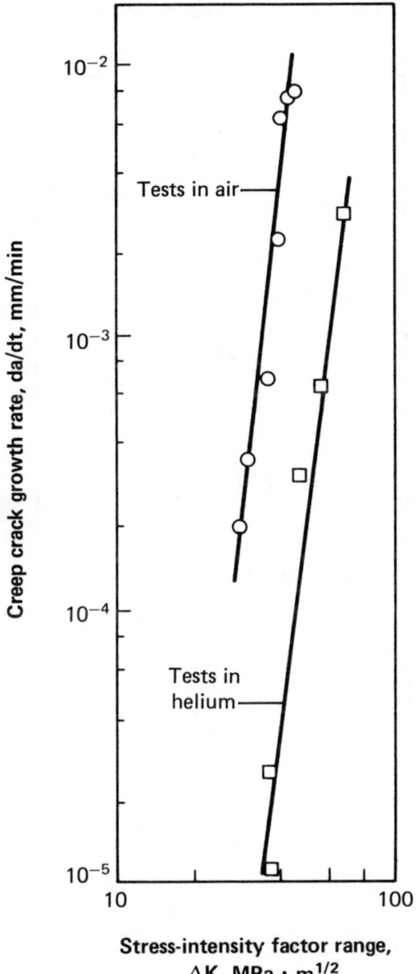

Fig. 8.43. Creep crack growth rates in air and in helium for Inconel 718 at 650°C (1200°F) (Ref 8.61)

8.5. SUMMARY

The FCP behavior of nickel-base superalloys is affected by factors such as temperature, environment and hold time. The precise effects of these variables depend on the composition and heat treatment of the system being considered. Furthermore, metallurgical changes — such as precipitate coarsening, changes in carbide composition and morphology, and twinning — can occur during the course of a test. All of these changes depend to a greater or lesser degree on temperature, time and deformation. In some cases, the same phenomenon can be either beneficial or detrimental. For example, for specimens of René 95, a long hold period at maximum load increased the threshold stress-intensity factor range for initiation

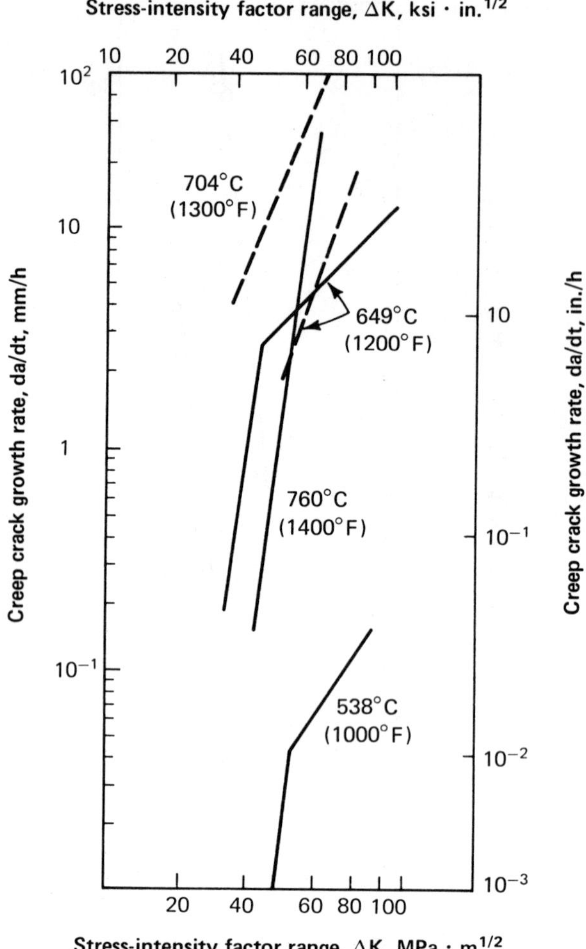

Fig. 8.44. Effect of temperature on creep crack growth behavior in Inconel 718 (Ref 8.62)

of crack growth, but once the crack began to propagate, the rate was significantly accelerated over that observed for other comparable specimens of René 95. This was interpreted in terms of oxidation, which prevented crack tip sharpening at low stress-intensity levels, but at higher stress intensities, where the oxide could be cracked, an oxidized region around the crack tip failed and the crack advance per cycle was increased.

The effect of microstructure is significant in both the γ'- and γ''-strengthened systems and is especially important at temperatures at which the microstructure is stable and at which environmental/creep mechanisms are not primary damage modes. The most fatigue resistant microstructures were those that enhanced slip planarity (i.e., microstructures with coarse grains and fine precipitates). With implementation of the retirement-for-cause design philosophy, crack propagation considerations are likely to become more important, and the potential for life extension of components such as jet engine discs by use of heat treatments designed for maximum fatigue crack growth resistance appears to have great promise.

Certain nickel-base superalloys have high fracture toughness and good resistance to FCP at cryogenic temperatures. At these temperatures, the microstructures are stable and the environment generally does not influence the properties. Of the nickel-base alloys that have been evaluated at cryogenic temperatures, Inconel 706, Inconel 718 and Inconel X-750 have greater fracture toughness and greater resistance to FCP at −269°C (−452°F) than at room temperature. Limited results of FCP tests on superalloy weldments at −269°C indicate that FCP rates in welds are the same as those in the base metal for Inconel X-750 but higher than those in the base metal for the other alloys studied.

It has been shown that CCG is affected by temperature, environment, microstructure and composition. The CCG rate could be reasonably well represented by the stress-intensity parameter for most nickel-base alloys of practical interest. However, care must be exercised in using this approach for alloys and heat treatments having high ductility.

The effect of temperature is generally to increase the CCG rate, although there may be competing crack tip mechanisms that can result in decreasing CCG rates with increasing temperature. It has also been shown that environmental factors can greatly accelerate CCG rates at elevated temperatures, even when failure occurs by purely creep-type processes such as cavity formation at the crack tip.

The CCG rate generally can be reduced in a given alloy by increasing the grain size, by heat treating so as to produce serrated grain boundaries, or by overaging. Serrated grain boundaries reduce the creep rate by retarding grain boundary sliding, whereas overaging provides large particles on the boundaries and homogenizes the slip. In some cases, very large particles in the grain boundaries increase the CCG rate, and thus some care must be taken in attempting to reduce the CCG rate by heat treatments that produce large particles in the grain boundaries.

8.6. REFERENCES

8.1. Analysis of Microstructures in Nickel Base Alloys: Implications for Strength and Alloy Design, by J. M. Oblak and B. H. Kear: in *Electron Microscopy and the Structure of Materials,* edited by G. Thomas, R. M. Fulrath and R. M. Fisher, University of California Press, Berkeley, 1972, p 565-616

8.2. Strengthening Mechanisms in Ni Base Superalloys, by R. F. Decker: Climax Molybdenum Co. Symposium, Zurich, May 5-6, 1969

8.3. The Metallurgy of Ni Base Alloys, by R. F. Decker and C. T. Sims: in *The Superalloys,* edited by C. T. Sims and W. C. Hagel, John Wiley and Sons, New York, 1972, p 33-77

8.4. A Dynamic Theory of Coherent Precipitation Hardening with Application to Ni Base Superalloys, by S. M. Copley and B. H. Kear: *Transactions of AIME,* Vol 239, 1967, p 984-992

8.5. The Temperature Dependence of the Flow Stress of the γ' Phase Based Upon Ni_3Al, by P. H. Thornton, R. G. Davies and T. L. Johnston: *Metallurgical Transactions,* Vol 1, No. 1, Jan 1970, p 207-224

8.6. Coherency Strains of γ' Hardened Nickel Alloys, by R. F. Decker and J. R. Mihalisin: *ASM Transactions Quarterly,* Vol 62, No. 2, June 1969, p 481-489

8.7. The Effect of Orientation and Sense of Applied Uniaxial Stress on the Morphology of Coherent γ' Precipitates in Stress Annealed Ni Base Superalloy Crystals, by J. K. Tien and S. M. Copley: *Metallurgical Transactions,* Vol 2, No. 2, 1971, p 543-553

8.8. Effects of Stress Coarsening on Coherent Particle Strengthening, by J. K. Tien and R. P. Gamble: *Metallurgical Transactions,* Vol 3, No. 8, Aug 1972, p 2157-2162

8.9. Tensile Embrittlement of Turbine Blade Alloys After High Temperature Exposure, by W. H. Chang: in *Superalloys Processing,* Section V, MCIC-72-10, Metals and Ceramics Information Center, Battelle, Columbus, OH, 1972

8.10. Prediction of Sigma Type Phase Occurrence from Compositions in Austenitic Superalloys, by L. R. Woodyatt, C. T. Sims and H. J. Beattie, Jr.: *Transactions of AIME,* Vol 236, No. 4, April 1966, p 519-527

8.11. Identification of the Strengthening Phase in Inconel Alloy 718, by P. S. Kotval: *Transactions of AIME,* Vol 242, No. 8, Aug 1968, p 1764-1765

8.12. The Precipitation of Ni_3Nb Phases in a Fe-Ni-Cr-Nb Alloy, by I. Kirman and D. H. Warrington: *Metallurgical Transactions,* Vol 1, No. 10, Oct 1970, p 2667-2675

8.13. The Metallurgy of Ni-Fe Alloys, by D. R. Muzyka: in *The Superalloys,* edited by C. T. Sims and W. C. Hagel, John Wiley and Sons, New York, 1972, p 113-143

8.14. Fatigue Crack Growth at High Temperatures, by M. O. Speidel: in *High Temperature Materials in Gas Turbine Engines,* edited by P. R. Sahm and M. O. Speidel, Elsevier Scientific Publishing Co., 1974, p 208-255

8.15. Application of Fracture Mechanics at Elevated Temperatures, by R. M. Wallace, C. G. Annis, Jr., and D. Sims: Report AFML-TR-76-176, Part II, 1976

8.16. A Three Component Model for Representing Wide Range Fatigue Crack Growth Data, by A. Saxena, S. J. Hudak, Jr., and G. M. Jouris: *Engineering Fracture Mechanics,* Vol 12, No. 1, 1979, p 103-115

8.17. A Critical Analysis of Crack Propagation Laws, by P. C. Paris and F. Erdogan: *Journal of Basic Engineering, Transactions of ASME,* Series D, Vol 85, Dec 1963, p 528-534

8.18. The Effect of Microstructure on Elevated Temperature Crack Growth in Ni Base Alloys, by H. F. Merrick and S. Floreen: *Metallurgical Transactions,* Vol 9A, No. 2, Feb 1978, p 231-233

8.19. Effect of Grain Size and γ' Size on FCP in René 95, by J. Bartos and S. D. Antolovich: *Fracture 1977,* Vol 2, 1977, p 996-1006

8.20. The Effect of Microstructure on the FCP Properties of Waspaloy, by S. D. Antolovich, C. Bathias, B. Lawless and B. Boursier: (paper in preparation)

8.21. Dislocation Precipitate Interaction and Cyclic Stress Strain Behavior of a γ' Strengthened Superalloy, by R. Stolz and A. Pineau: *Materials Science and Engineering,* Vol 34, No. 3, Aug 1978, p 275-284

8.22. Low Cycle Fatigue, Fatigue Crack Propagation and Substructures in a Series of Polycrystalline Cu-Al Alloys, by A. Saxena and S. D. Antolovich: *Metallurgical Transactions,* Vol 6A, No. 9, Sept 1975, p 1809-1828

8.23. Influence of Micromechanisms of Cyclic Deformation at Elevated Temperature on Fatigue Behavior, by M. Clavel, C. Levaillant and A. Pineau: in *Creep-Fatigue-Environment Interactions,* edited by R. M. Pelloux and N. S. Stoloff, American Institute of Mining, Metallurgical and Petroleum Engineers, New York, 1980, p 24-45

8.24. Crack-Morphological Aspects in Fracture Mechanics, by H. Kitagawa, R. Yuuki and T. Ohira: *Engineering Fracture Mechanics,* Vol 7, 1975, p 515-529

8.25. Effect of Heat Treatment on Elevated Temperature Fatigue Crack Growth Behavior of Two Heats of Alloy 718, by W. J. Mills and L. A. James: ASME Publication 7-WA/PUP-3, 1979

8.26. The Effect of Temperature on the Fatigue Crack Growth Behavior of Two Ni Base Alloys, by L. A. James: *Journal of Engineering Materials and Technology, Transactions of ASME,* Series H, Vol 95, No. 4, Oct 1973, p 254-256
8.27. Subcritical Crack Growth for IN 718 at Elevated Temperatures, by H. G. Popp and A. Coles: Proceedings of Air Force Conference on Fatigue and Fracture of Air Force Structures and Materials, 1969, p 71-86
8.28. Low Cycle Fatigue of As-HIP and HIP+Forged René 95, by S. Bashir, Ph. Taupin and S. D. Antolovich: *Metallurgical Transactions,* Vol 10A, No. 10, Oct 1979, p 1481-1490
8.29. Crack Propagation Under Thermal Mechanical Cycling, by P. Domas: an interim progress report to the Air Force Materials Laboratory, Contract F33615-77-C-5193, Period 9/1/77-1/15/79 (section written by S. D. Antolovich)
8.30. *Point Defects and Diffusion in Strained Metals,* by L. A. Girifalco and R. O. Welch: Gordon and Breach, New York, 1967, p 124
8.31. Activation Energy Dependence on Stress Intensity in SCC and Corrosion Fatigue, by P. Bania and S. D. Antolovich: STP 610, American Society for Testing and Materials, Philadelphia, 1976, p 157-175
8.32. Environment Assisted Fatigue Crack Growth in TRIP Steels, by L. C. Jea: M. S. Thesis, University of Cincinnati, 1974
8.33. An Interpolative Model for Elevated Temperature Fatigue Crack Propagation, by C. G. Annis, R. M. Wallace and D. L. Sims: Report AFML-TR-76-176, 1976
8.34. Fatigue Crack Propagation at Elevated Temperatures in Solid Solution Strengthened Superalloys, by D. A. Jablonski, J. V. Carisella and R. M. Pelloux: *Metallurgical Transactions,* Vol 8A, No. 12, Dec 1977, p 1893-1900
8.35. Mechanisms of High Temperature Fatigue, by M. Gell and G. R. Leverant: STP 520, American Society for Testing and Materials, Philadelphia, 1973, p 37-67
8.36. Evaluation of Cyclic Behavior of Aircraft Turbine Disk Alloys, by V. Shahani and H. G. Popp: NASA Report NASA-CR-159433, 1978
8.37. A Fracture Mechanics Approach to High Temperature Fatigue Crack Growth in Udimet 700, by K. Sadananda and P. Shahinian: *Engineering Fracture Mechanics,* Vol 11, No. 1, 1979, p 73-86
8.38. The Fatigue Strength of Nickel-Base Superalloys, by M. Gell, G. R. Leverant and C. H. Wells: STP 467, American Society for Testing and Materials, Philadelphia, 1970, p 113-153
8.39. Effects of Environment on High Temperature Fatigue Crack Growth in a Superalloy, by S. Floreen and H. Kane: *Metallurgical Transactions,* Vol 10A, No. 11, Nov 1979, p 1745-1751
8.40. Low Cycle Fatigue of René 80 as Affected by Prior Exposure, by S. D. Antolovich, P. Domas and J. L. Strudel: *Metallurgical Transactions,* Vol 10A, No. 12, Dec 1979, p 1859-1868
8.41. Frequency and Waveform Effects on the Fatigue Crack Growth Behavior of Alloy 718 at 298°K and 823°K, by M. Clavel and A. Pineau: *Metallurgical Transactions,* Vol 9A, No. 4, April 1978, p 471-480
8.42. Metallurgical Aspects of High Temperature Fatigue, by S. D. Antolovich: in *La Fatigue des Matériaux et des Structures,* edited by C. Bathias and J. P. Bailon, Maloine S. A., Paris, 1980, p 465-496
8.43. Effect of Microstructure on the Fatigue Crack Growth of a Powder Metallurgy Nickel Base Superalloy, by J. Bartos: Ph.D. Dissertation, University of Cincinnati, 1976
8.44. Low Temperature Effects on the Fracture Behavior of a Nickel Base Superalloy, by R. L. Tobler: *Cryogenics,* Vol 16, No. 11, Nov 1976, p 669-674
8.45. Cryogenic Fracture Toughness and Fatigue Crack Growth Rate Properties of

Inconel 706 Base Material and Gas Tungsten Arc Weldments, by W. A. Logsdon, J. M. Wells and R. Kossowsky: paper presented at the International Cryogenic Materials Conference, University of Wisconsin, Madison, WI, Aug 21-24, 1979

8.46. The Influence of Processing and Heat Treatment on the Cryogenic Fracture Mechanics Properties of Inconel 718, by W. A. Logsdon, R. Kossowsky and J. M. Wells: in *Advances in Cryogenic Engineering,* Vol 24, edited by K. D. Timmerhaus, Plenum Press, 1978, p 179-209

8.47. Fatigue Crack Growth Resistance of Structural Alloys at Cryogenic Temperatures, by R. L. Tobler and R. P. Reed: in *Advances in Cryogenic Engineering,* Vol 24, edited by K. D. Timmerhaus, Plenum Press, 1978, p 82-90

8.48. Cryogenic Fracture Mechanics Properties of Several Manufacturing Process/Heat Treatment Combinations of Inconel X-750, by W. A. Logsdon: in *Advances in Cryogenic Engineering,* Vol 22, edited by K. D. Timmerhaus, R. P. Reed and A. F. Clark, Plenum Press, 1977, p 47-57

8.49. Fatigue Crack Growth in Selected Alloys for Reactor Applications, by P. Shahinian, H. E. Watson and H. H. Smith: *Journal of Materials,* Vol 7, No. 4, April 1972, p 527-535

8.50. Evaluation of Inconel X-750 Weldments for Cryogenic Applications, by J. M. Wells: in *Advances in Cryogenic Engineering,* Vol 22, edited by K. D. Timmerhaus, R. P. Reed and A. F. Clark, Plenum Press, 1977, p 80-90

8.51. The Creep Fracture Characteristics of Nickel Base Superalloy Sheet Samples, by S. Floreen: *Engineering Fracture Mechanics,* Vol 11, No. 1, 1979, p 55-60

8.52. Characterization of the Elevated Temperature Static Load Crack Extension Behavior of Type 304 Stainless Steel, by P. L. Jones and A. S. Tetelman: *Engineering Fracture Mechanics,* Vol 12, No. 1, 1979, p 79-97

8.53. The Creep Fracture of Wrought Nickel-Base Alloys by a Fracture Mechanics Approach, by S. Floreen: *Metallurgical Transactions,* Vol 6A, No. 9, Sept 1975, p 1741-1749

8.54. A Critical Strain Model for the Creep Fracture of Nickel Base Superalloys, by S. Floreen and R. H. Kane: *Metallurgical Transactions,* Vol 7A, No. 8, Aug 1976, p 1157-1160

8.55. The Stress Dependence of the Crack Growth Rate During Creep, by R. I. DiMelfi and W. D. Nix: *International Journal of Fracture,* Vol 13, No. 3, 1977, p 341-348

8.56. High Temperature Crack Growth Structure-Property Relationships in Nickel Base Superalloys, by S. Floreen: in *Creep-Fatigue-Environment Interactions,* edited by R. M. Pelloux and N. S. Stolloff, AIME, New York, 1980, p 112-128

8.57. On Grain Boundary Sliding and Diffusional Creep, by R. Raj and M. F. Ashby: *Metallurgical Transactions,* Vol 2, No. 4, p 1113-1127

8.58. Metallurgical Factors Affecting the Crack Growth Resistance of a Superalloy, by J. M. Larson and S. Floreen: *Metallurgical Transactions,* Vol 8A, No. 1, Jan 1977, p 51-55

8.59. Tensile Behavior of René 95 in the Thermomechanically Processed and Conventionally Processed Forms, by M. N. Menon and W. H. Reimann: *Metallurgical Transactions,* Vol 6A, No. 5, May 1975, p 1075-1085

8.60. The Dependence of the Notch Sensitivity of Waspaloy at 100°-1400°F on the Gamma Prime Phase, by D. J. Wilson: *Journal of Engineering Materials and Technology,* Vol 95, Series H, No. 1, 1973, p 15-20

8.61. An Investigation of the Creep/Fatigue Environment Interaction in a Nickel Base Superalloy, by S. Floreen and R. H. Kane: *Fatigue of Engineering Materials and Structures,* Vol 2, No. 4, 1979, p 401-412

8.62. Creep Crack Growth in Alloy 718, by K. Sadananda and P. Shahinian: *Metallurgical Transactions,* Vol 8A, No. 3, March 1977, p 439-449

Chapter 9

Design, Materials Selection and Failure Analysis

W. W. Gerberich and A. W. Gunderson

9.0. INTRODUCTION

If the fracture mechanics principles and material properties outlined in Chapters 1 to 8 are to be used in a meaningful way, the materials engineer must be cognizant of available design procedures. This chapter will reiterate the basic principles of fracture toughness and subcritical crack growth approaches and then will discuss several procedures that are required when these principles are used in designing cost-effective and reliable high-performance components. These procedures include nondestructive evaluation (NDE) methods, failure analysis and materials selection. The last section, which is on materials selection, draws heavily on the case study approach wherein fracture mechanics principles have been utilized.

A summary of applications of relevant fracture mechanics principles is given in a comprehensive survey of 226 leaders in the field conducted in 1975 by Rich, Tracy and Cartwright (Ref 9.1). From the over-all responses listed in Table 9.1, it can be seen that at least 44 general areas of design, including materials selection for those designs, are represented. Since 1975, the number of areas open to materials selection approaches has further expanded; some of these will be discussed in this chapter. With regard to trends in usage, that same survey indicated that the application of fracture mechanics approaches will continue to expand, as noted in Fig. 9.1, and that the fracture toughness approach will be used more widely than the Charpy impact test and other correlation tests for materials selection, as noted in Fig. 9.2. Cyclic growth was noted as the most frequently used design aspect, with fracture toughness a close second followed by environmentally induced slow crack growth or corrosion fatigue.

With the fracture toughness concepts and methods outlined in the first three chapters, design methodologies can be invoked to minimize the risk of failure of structures containing flaws or cracks. As with any risk/cost/benefit approach, it is possible to overdesign a component so that it will never fail but will become

Table 9.1. Results of survey of application of fracture mechanics to design (Ref 9.1)

Type of design	No.(a)	Type of design	No.(a)
1. Adhesive joints	29	23. Mechanical joints & fasteners	35
2. Aerospace vehicles	52	24. Metal forming & casting	13
3. Aircraft engines	15	25. Microstructures	29
4. Aircraft structures/airframes	77	26. Missile structures	25
5. Analytical methods/mechanics	89	27. NDT	45
6. Bearings	7	28. Nuclear structures	25
7. Bridges and buildings	13	29. Passenger land vehicles	2
8. Cargo land vehicles	2	30. Piping	24
9. Control systems	4	31. Pressure vessels/gases	35
10. Dental structures	4	32. Pressure vessels/launch tubes	15
11. Electronics equipment	2	33. Pressure vessels/submarines	15
12. Fatigue prediction	115	34. Reliability analysis	35
13. Fracture control planning	85	35. Rigs – offshore structures	4
14. General structures	54	36. Rocket motors	15
15. Geological applications	6	37. Rotating machinery/electric	8
16. Helicopter structures	17	38. Rotating machinery/turbines	22
17. High-temperature applications	42	39. Shells	19
18. Internal combustion engines	6	40. Ship structures	29
19. Materials development	61	41. Standards development	33
20. Materials fracture characterization	119	42. Structures testing	55
21. Materials production	13	43. Surface coatings	10
22. Materials testing/evaluation	127	44. Welding/fabrication	31

(a) Number of participants identified with each design type.

noncompetitive. The other extreme is to underdesign a critical component, with the result that injury or death occurs, causing sufficient liability losses or consumer distrust to cripple an organization or company. Clearly, engineering input into management decision making is required with regard to where fracture mechanics design methodologies belong in this spectrum of cost/benefit considerations. The following sections describe the fracture toughness and subcritical slow crack growth approaches. Implementation of these approaches often requires information on sizes of cracks in critical components, a quantitative description of the quality of material used, and input from failure analyses on how the components failed in service. Because of the importance of these subjects, two additional sections, covering quality control and failure analysis, are presented in this chapter.

At the end of this chapter, several case studies using either fracture toughness or subcritical crack growth approaches to materials selection are presented. Different types of materials selection are required in response to varying circumstances, such as (*a*) an unanticipated material/stress/environment combination which might lead to failure in service; (*b*) a relatively noncritical component, such as a tool bit or a gear, which might be improved for longer life; or (*c*) an intensive design/material evaluation approach which would be required prior to selection of

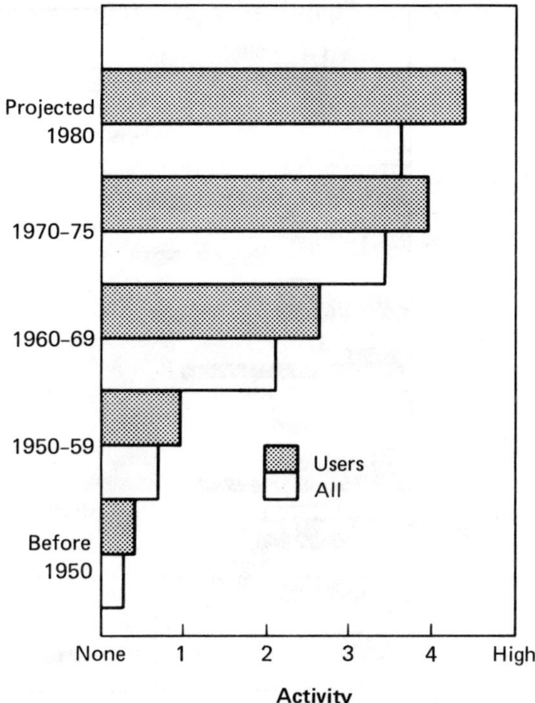

Fig. 9.1. Trend in application of fracture mechanics in design (Ref 9.1)

materials for highly critical components. Thus, the several cases presented will cover postfailure analyses of what happened and of how to avoid recurrence, fracture mechanics approaches for selection of materials to provide increased service life, and fracture control methodologies implemented to select safe materials and designs for critical components.

9.1. PRINCIPLES OF THE FRACTURE TOUGHNESS APPROACH

Fracture resistant design is most simply applied where a known material with fracture toughness, K_c or K_{Ic}, and yield strength, σ_{ys}, must survive at a given design stress, σ, due to a monotonically increasing load. This is clearly presented by Hertzberg (Ref 9.2) for plane stress conditions where a crack of length 2a exists in an infinitely large plate, giving:

$$K = K_c = \sigma \sqrt{\pi a} \qquad \text{(Eq 9.1)}$$

K_c → Material selection; σ → Design stress; $\sqrt{\pi a}$ → Allowable crack size or NDE crack detection level

Defect tolerances for monotonically loaded structures have been developed for

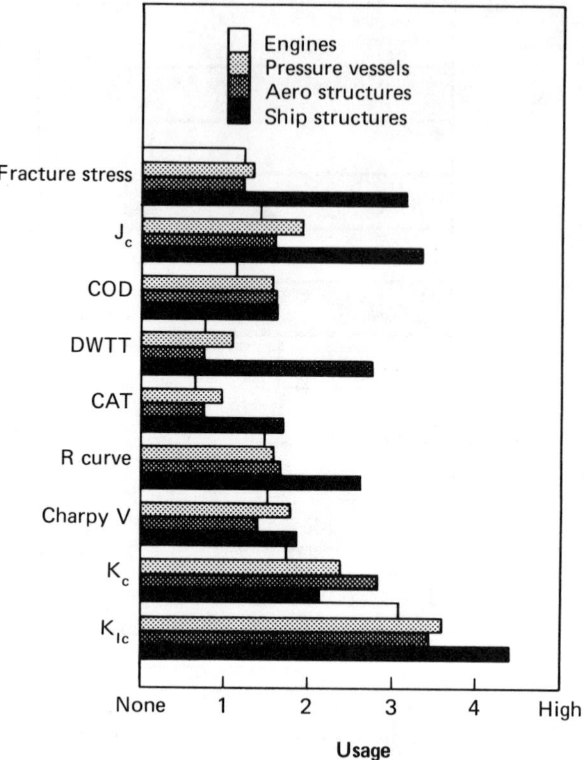

Fig. 9.2. Relative usage of toughness tests in materials selection (Ref 9.1)

pressure vessels, piping and other structures. Three such approaches will now be described in some detail.

9.1.1. Pressure Vessel Leak-Before-Burst Concept

If monotonically loaded thin-walled pressure vessels were only expected to last a short time, then fracture toughness could be chosen to allow for a leak-before-burst condition—that is, if a vessel containing pressurized gas or liquid contains a growing crack, the toughness should be sufficiently high to tolerate a defect size which will allow the contents to leak out before it grows catastrophically. Because in order to cause a leak the crack must grow through the cylinder wall thickness, t, and to a distance along the axis of about $2a \simeq 2t$, then the fracture toughness and design stress should be chosen so that, using Eq 9.1:

$$(K_c/\sigma)^2 \geqslant \pi t \qquad \text{(Eq 9.2)}$$

This example is illustrated by the sketch in Fig. 9.3. This is somewhat of an oversimplification because it does not take into account the plastic zone or curvature correction factors; these are detailed later in Section 9.5. One of the most conservative uses of the leak-before-burst criterion has been adopted by the United States Coast Guard where it recommends that the critical crack size be ten times

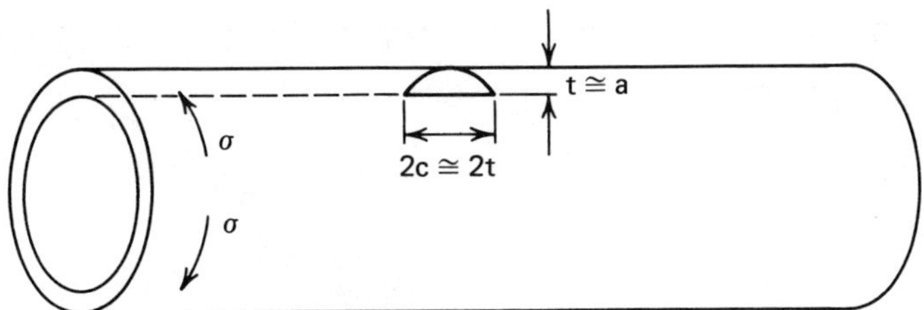

Fig. 9.3. Sketch illustrating a leak-before-burst condition in a thin-wall pressure vessel

greater than the minimum value which could be detected by evidence of escaping gas. Further applications of such approaches to shipping containers for liquified natural gas are discussed in detail by Burns *et al* (Ref 9.3) and by Tenge *et al* (Ref 9.4).

9.1.2. Overpressurization of Thick-Walled Vessels

Quite often present in thick-wall pressure vessels are part-through cracks which may not grow through the thickness to cause a reduction in pressure before unstable growth occurs. For these situations, an alternative but less reliable procedure has been used. Consider first a single pressurization to the design stress of the vessel. Any critical-size flaw in the vessel would cause failure. The problem with applying this type of pressurization cycle as a design or test procedure is that subcritical crack growth in service could allow a marginal crack which had just escaped detection to grow to critical size during service. To avoid this, a proof-pressure cycle to higher than design stress is sometimes applied to allow for a "safety factor" with regard to operational time. Subsequent nondestructive testing and proof pressure testing at well-defined intervals may then be used to ensure additional time intervals for safe operation.

An example would be a welded box-beam component which had already been manufactured and which was found to have a marginal fracture toughness of 55 MPa $\cdot$ $m^{1/2}$ (50 ksi $\cdot$ $in.^{1/2}$). Should it be rejected, or could it be safely employed by subjecting it to a stress of 1.5 times its applied stress, σ_A, via an overpressurization cycle? Note that, although the component here is loaded as a beam in service, the proof loading was done by internal pressurization. Any convenient proof loading can be used, provided that the stresses due to the proof load are known and are of the same type—e.g., tension. The component has walls 40 mm (1.6 in.) thick and, assuming that a surface crack 40 mm (1.6 in.) in length could be easily detected by nondestructive testing, then the crack depth, a, for a semicircular crack* would be half the wall thickness, or 20 mm (0.8 in.). The applicable fracture toughness equation would be:

$$K_{Ic} \simeq 1.1\sigma_A \left[\frac{\pi a}{Q}\right]^{1/2} \qquad \text{(Eq 9.3)}$$

where σ_A is the applied stress of 300 MPa (43.5 ksi), which is 50% of the yield

*A semicircular crack is picked here because this is the most conservative case.

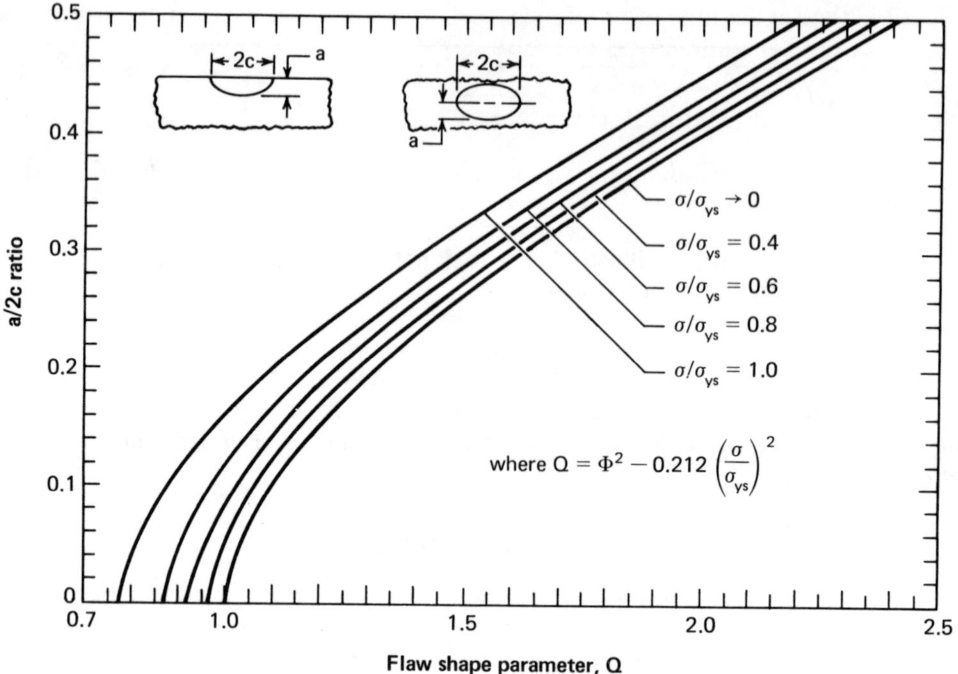

Fig. 9.4. Flaw shape parameter curves for surface and internal cracks (Ref 9.5)

strength in this case. Q is a flaw-shape parameter for elliptically shaped cracks; Q can be approximated by the standard values of the elliptical integral, Φ, and by the ratio of applied stress to yield stress (see Fig. 9.4: Ref 9.5). For $\sigma_A/\sigma_{ys} = 0.5$ and $a/2c = 0.5$, $Q = 2.28$; from Eq 9.3, calculation shows that use of a material with $K_{Ic} = 55$ MPa · $m^{1/2}$ (50 ksi · $in.^{1/2}$) would lead to component failure if a crack with a depth of 20 mm (0.8 in.) or greater were present. Fortunately, nondestructive testing could detect such a crack. However, because of suspected service conditions, it was anticipated that cracks might grow 10 mm/year (0.4 in./year) in depth.* According to Eq 9.3, a 1.5 × overpressurization to a stress of 450 MPa (65.2 ksi) would ensure that any crack present would be less than 9.3 mm (0.37 in.) deep. Cracks deeper than this would have caused failure. Simply stated, the requirement would be that the difference between the critical crack, a_c, at the applied stress and the maximum crack detected by overpressurization, a_{op}, be greater than the growth in the time interval, t_i, between proof tests, or:

$$a_c - a_{op} > \overline{(da/dt)}t_i \qquad \text{(Eq 9.4)}$$

where $\overline{da/dt}$ is the average crack growth rate. From this example, it would appear that yearly overpressurizations would be necessary to ensure that the maximum

*Such a growth rate of 1.6×10^{-10} m/sec (6.3×10^{-9} in./sec) might easily occur even near threshold stress intensities under cyclic or environmental conditions.

possible flaw at the time of proof overpressurization had not grown to critical size. This example is summarized in Fig. 9.5.

9.1.3. General Design/Performance Model

Most structural components are not pressurized, in which case a leak-before-burst criterion cannot be utilized. Furthermore, many components cannot be proof tested once they have been put into service. For these cases, a general design/performance model is useful. Although they did not address a specific materials selection problem, Jackson and Wright (Ref 9.6) presented an overview of how the fracture toughness approach may be used to select a material for steel castings. They reviewed both the classical "factor of safety" design approach as well as the fracture mechanics approach. Thus, reference to this paper is recommended for the materials selection engineer who is considering fracture mechanics for the first time. Jackson and Wright first illustrated a flow diagram, as shown in Fig. 9.6, which deals with the duality of the stress analysis and materials

Given: σ_A = applied stress = 300 MPa (43.5 ksi)

K_{Ic} = fracture toughness = 55 MPa · $m^{1/2}$ (50 ksi · $in.^{1/2}$)

σ_{ys} = yield strength = 600 MPa (87 ksi)

t = wall thickness = 0.04 m (1.5 in.)

a_c = critical crack depth $\geqslant$ NDE detection capability $\approx$ 0.02 m (0.75 in.)

Procedure:

i) Determine that Q = 2.28 for σ_A/σ_{ys} = 0.5 from Fig. 9.3 and 9.4.

ii) Establish the critical flaw size from Eq 9.3:

$$a_c = \frac{K_{Ic}^2}{1.21\sigma_A^2}\left(\frac{Q}{\pi}\right) = \left(\frac{1}{1.21}\right)\left(\frac{55}{300}\right)^2\left(\frac{2.28}{\pi}\right) = 0.02 \text{ m (0.75 in.)}$$

iii) Establish nondestructive testing for 0.02-m crack depths ($\approx$ 0.04-m surface flaws).

iv) Estimate the average slow crack growth rate to be $\overline{da/dt} \approx 0.01$ m/year (0.39 in./year) in depth.*

v) Establish the maximum crack depth, a_{op}, that could exist at a 1.5 X overpressurization. Q = 2.37 for σ_{op}/σ_{ys} = 0.75.

$$a_{op} = \frac{1}{1.21}\left(\frac{55}{450}\right)^2\left(\frac{2.37}{\pi}\right) = 0.0093 \text{ m (0.37 in.)}$$

vi) The following must be satisfied for any time interval, t_i:

$$(a_c - a_{op}) > (\overline{da/dt})t_i$$

$$(0.02 \text{ m} - 0.0093 \text{ m}) > (0.01 \text{ m/year}) \text{ 1 year}$$

*It is emphasized that components used in aggressive environments may exhibit much faster growth rates, in which case this approach cannot be recommended.

Fig. 9.5. Overpressurization scheme for providing safer operation of a structural component

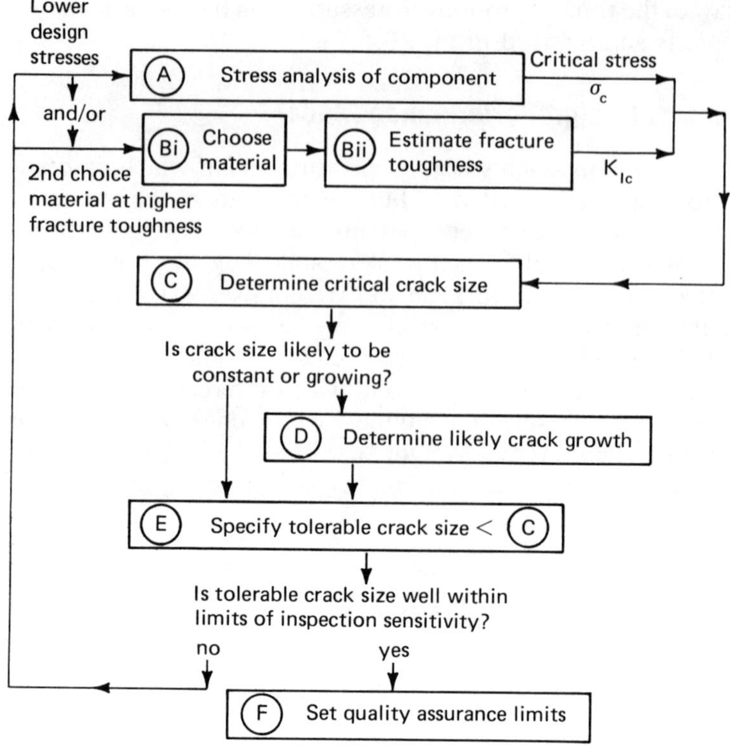

Fig. 9.6. Design/performance model based on fracture toughness criteria (Ref 9.6)

selection process. In the design/performance model, it is noted that either design stresses for a fixed materials selection or new materials selection for a fixed design must be compatible with available nondestructive evaluation (NDE) and quality assurance limits.

Jackson and Wright described three types of cast steels, with yield strengths of 480, 740 and 1280 MPa (70, 110 and 185 ksi) and K_{Ic} values of 104, 86 and 46 MPa · $m^{1/2}$ (95, 78 and 42 ksi · in.$^{1/2}$), respectively. They assumed that surface cracks of depth a are present in components made from these cast steels. Using the available solutions for surface cracks, the final critical defect sizes were determined from plots such as those in Fig. 9.7. As one follows the flow diagram in Fig. 9.6, decisions with regard to appropriate working stresses, defect sizes and available NDE methods may lead to a second or third materials choice. Also discussed in this overview are ramifications of subcritical crack growth by fatigue and stress corrosion as well as the accuracy of current NDE technology.

9.2. SUBCRITICAL CRACK GROWTH APPROACH

If cracks grow in service due to environmental or fatigue effects, then the time to failure will be governed by stress and environmental history. The simplest case is depicted in Fig. 9.8(a), where crack length increases with time at a constant

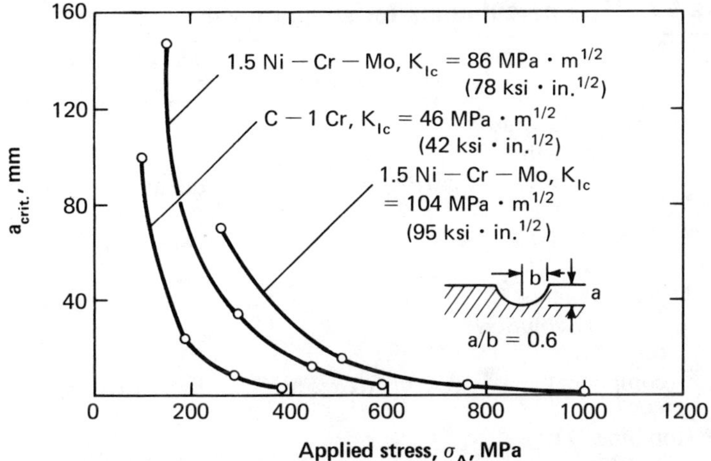

Fig. 9.7. Effect of fracture toughness, K_{Ic}, and applied stress on critical size of surface cracks (Ref 9.6)

stress — for example σ_A or σ_B. However, if the stress is sufficiently low, such as σ_C, no growth occurs. Differentiation of the curves in Fig. 9.8(a) at a number of points along the time scale gives the log plot shown in Fig. 9.8(b). A single curve results as long as the three stress levels represent essentially linear elastic behavior. Integrating such a relationship from a known defect size, a_i, to a critical defect size a_c, will then allow determination of the time to failure, t_f, or the number of cycles to failure, N_f, as outlined in Table 9.2. For an environmentally induced slow crack growth or fatigue crack growth, a general life expression is:

$$N_f,\ t_f = \int_{a_i}^{a_c} \text{const. } f(\sigma, a)da \qquad \text{(Eq 9.5)}$$

The procedure for determining component life in extremely simplified cases is outlined in Table 9.2. In the fatigue case, one would have to determine C and n for the material and environment of interest and assume that environment and temperature were constant. In the sustained load case, it is assumed that a single set of constants, c_o and n_o, and a single activation energy, ΔH, govern over a temperature range $T_1 < T < T_2$, where one mechanism controls. For the special case where cyclic stress, cyclic frequency and temperature are constant, Hertzberg (Ref 9.2) has shown that the cyclic life is:

$$N_f = \frac{2}{(n - 2)CY^n\Delta\sigma^n}\left[\frac{1}{a_i^{[(n/2)-1]}} - \frac{1}{a_{cr}^{[(n/2)-1]}}\right];\ n \neq 2 \qquad \text{(Eq 9.6)}$$

where Y is the geometrical factor which accounts for flaw and structure shape in the stress-intensity solution and n is the exponent of the fatigue power law expression (See Eq 2.16). More complicated cyclic load histories are discussed in Section 9.2.1.

If the applied stress is low enough, as for σ_c in Fig. 9.8, no growth will occur from some initial small flaw (e.g., a corrosion pit, an inclusion or a weld defect).

Table 9.2. Life calculations for simple cases

Step		Simplified fatigue	Simplified sustained load cracking
1	Determine laboratory crack growth relationship	$da/dN = C\Delta K^n$	$da/dt = C_o K_I^{n_o} \exp\left(-\frac{\Delta H}{RT}\right)$
2	Determine stress-intensity solution for service component	$\Delta K = f_1(\Delta\sigma, a)$	$K_I = f_2(\sigma, a)$
3	Combine (1) and (2)	$da/dN = f_3(\Delta\sigma, a)$	$da/dt = f_4(\sigma, a, T)$
4	Integrate for life	$N_f = \int_{a_i}^{a_c} \frac{da}{f_3(\Delta\sigma, a)}$	$t_f = \int_{a_i}^{a_c} \frac{da}{f_4(\sigma, a, T)}$

This is a second line of defense in that if defects no greater than a_{th} can be guaranteed, then a threshold design stress may be specified by:

$$\sigma_{th} = \frac{K_{th}}{Y[\pi a_{th}]^{1/2}} \quad \text{or} \quad \Delta\sigma_{th} = \frac{\Delta K_{th}}{Y[\pi a_{th}]^{1/2}} \qquad \text{(Eq 9.7)}$$

where Y is the factor taking into account flaw and structure shape.* Equations 9.7 summarize the threshold design approach for subcritical and fatigue crack growth, respectively. These and other design approaches are considered in the following sections.

9.2.1. Fatigue Methodology

For service loading that causes fatigue crack growth, the stress-intensity factor varies continuously because of load fluctuations and crack growth. The justification for applying the stress-intensity factor to the modeling of fatigue crack growth is based on the observation that the geometric component of the stress-intensity factor changes only negligibly for the small crack movements that occur during an increment of service loading. Attention is thus focused on characterization of the service loading because, again, it is assumed that the crack length is quasistationary and that a one-to-one relationship exists between structural loading and crack tip stresses.

For a fatigue loading history, the example of the cracked structure shown in Fig. 9.9(a) will exhibit crack growth behavior as described in Fig. 9.9(b). Most structural cracks will respond similarly when the structure is not under a

*One should be very careful in the use of Eq 9.7 for very small flaws, $a < 1$ mm, since recent evidence suggests that ΔK_{th} for the small crack situation may be less than ΔK_{th} for the long crack situation.

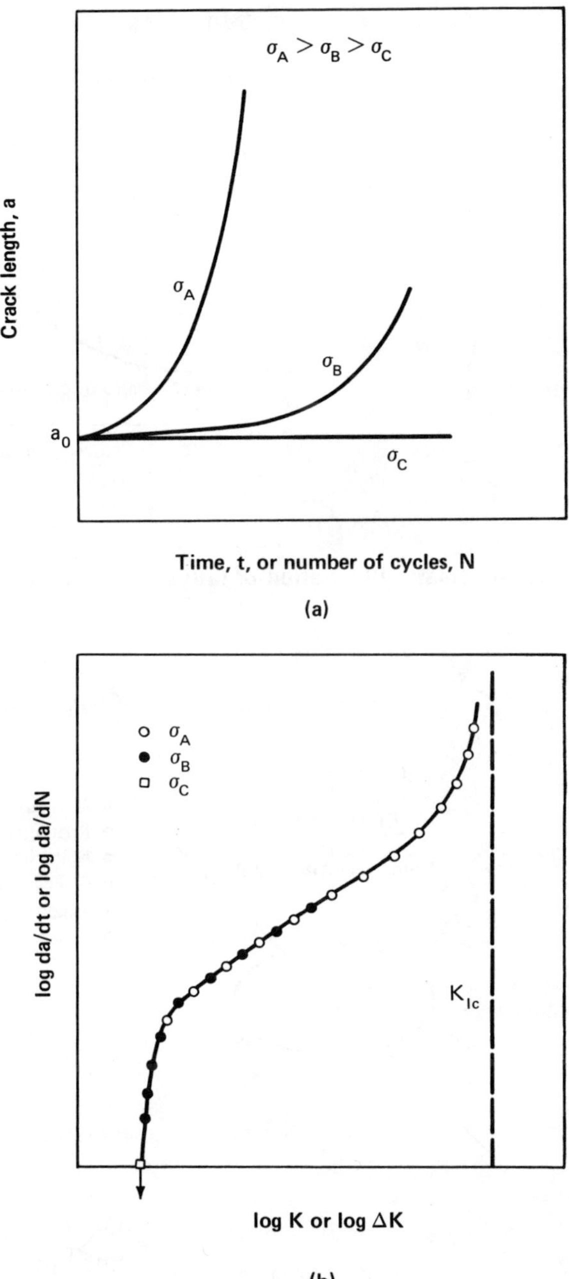

Fig. 9.8. Schematic representation of subcritical crack growth

displacement-controlled loading situation. Normally, then, the longer the crack the faster the rate of crack growth. Mechanical loading and other parameters which have been noted to have important impacts on the fatigue crack initiation life of structures also produce similar effects on the fatigue crack growth life. Butler (Ref 9.7) assembled Fig. 9.10 to display those factors which must be taken into account in design.

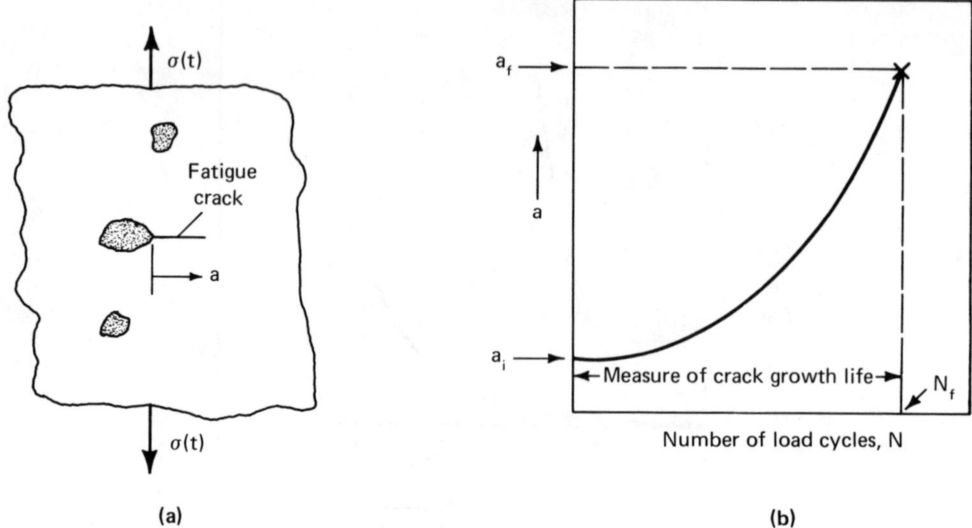

Fig. 9.9. Schematic illustration of fatigue crack growth

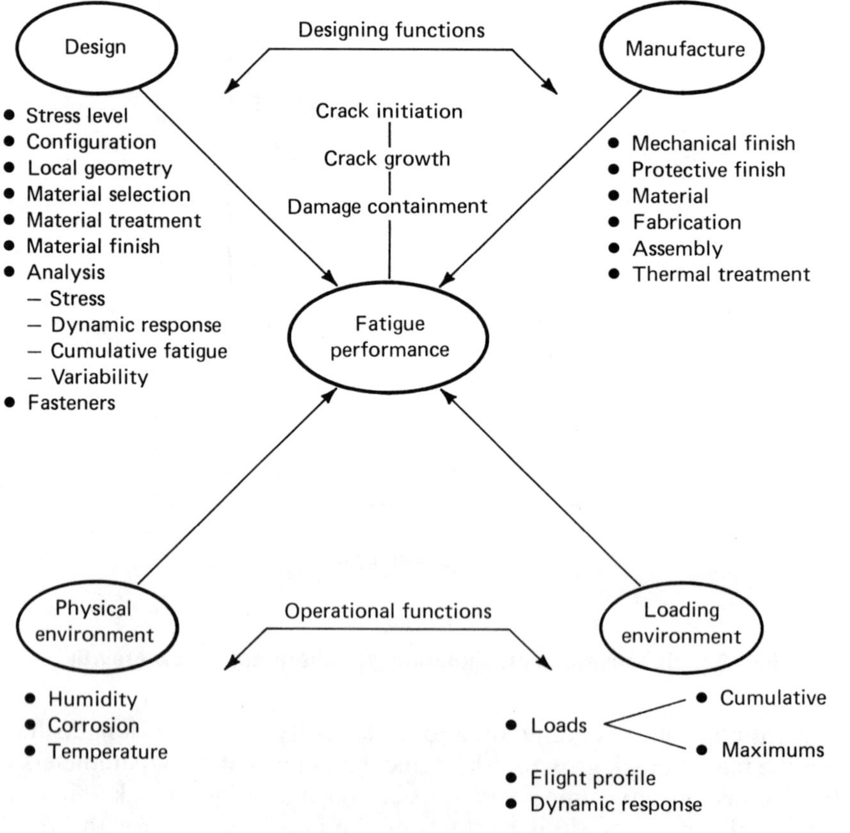

The input functions to fatigue performance.

Fig. 9.10. The elements of fatigue performance (Ref 9.7)

In the design approach based on fracture mechanics methodology, the focus is on the behavior of cracks. The following are typical questions that the designer might address: How big an initial flaw can the various manufacturing processes induce into the part? Will such a crack grow to critical size and cause catastrophic failure during the expected life of the part? The term fracture control design is normally used to describe the design philosophy in which the emphasis is on the control of potential cracking problems. Because of the unknowns from a processing and inspection standpoint, fracture control and design efforts start with an assumed initial crack size.

The basic design procedure is to assume some initial flaw size, analyze the growth of that flaw under the applied service loads until the point where an inspection would detect the growing flaw, and thus facilitate repair before catastrophic failure. If undetected, the crack would grow until the stressed member failed. Further definitions of these procedures follow:

9.2.1.1. Initial Flaw Size

The *initial flaw size* (a_i: see Fig. 9.9b) on which the analysis is based can be established in several ways:

1. By inspection capability: The initial crack size can be defined by the inspection size-detection limits and reliability. In such cases, a demonstration program is usually required to define the reliability and confidence of a specific inspection system.
2. By proof test: In some materials, an overload or overpressure can be used to define an initial flaw size (see Section 9.1.2.). The fracture mechanics methodology is used to calculate an initial flaw size based on the fracture toughness of the material and the applied stress that did not cause failure during the proof test.
3. By fatigue initiation: In some instances where exceptional inspection techniques are used (such as for certain rotating engine parts), the penalty for assuming an initial flaw is too high from a stress standpoint. For these cases, the time for a crack to initiate and grow to a certain fixed size (often about 1.0 mm, or 0.04 in.) is taken into consideration; both fatigue initiation and fatigue crack growth analyses are merged into a common procedure to predict the total life of the part.
4. By fiat or legislation: The size is set by either industry-wide design codes or government regulation based on historical data.

9.2.1.2. Final Crack Size

The *final crack size* (a_f in Fig. 9.9b) is also established in the fracture control design philosophy. Several choices are available:

1. The size or length at which the crack becomes readily inspectable by visual inspection or with NDE equipment.
2. The size or length at which impending failure is obvious and visually noticeable, such as a leaking tank or larger-than-normal vibrations or deflections.
3. The size or length at which the crack can be calculated via fracture mechanics methodology to be at a critical size for abrupt component fracture.

9.2.1.3. Crack Growth Curve

The *crack growth curve* (or life capacity curve) is described in Fig. 9.9b. The essential purpose of this section is to explain the design analysis required to establish the crack growth behavior and life exhibited by cracked structures under service-type loads and environments. Without fracture mechanics methodology, establishing this behavior would require duplication of all mechanical, structural, environmental and material features that affect crack growth behavior. Fracture mechanics methodology reduces the task to duplication of essential mechanical (loading), environmental and material features during the study of crack behavior in small-scale laboratory tests. Thus, the methodology provides the structural transfer function and a new method of relating laboratory data to service life analysis (Fig. 9.10).

Shown in Fig. 9.11 are sketches which describe the steps by which basic constant-amplitude data on fatigue crack growth rates are generated from laboratory specimens (Ref 9.8). The stress-intensity factor range (ΔK) for the specimen provides the transfer function required to develop crack growth behavior in the

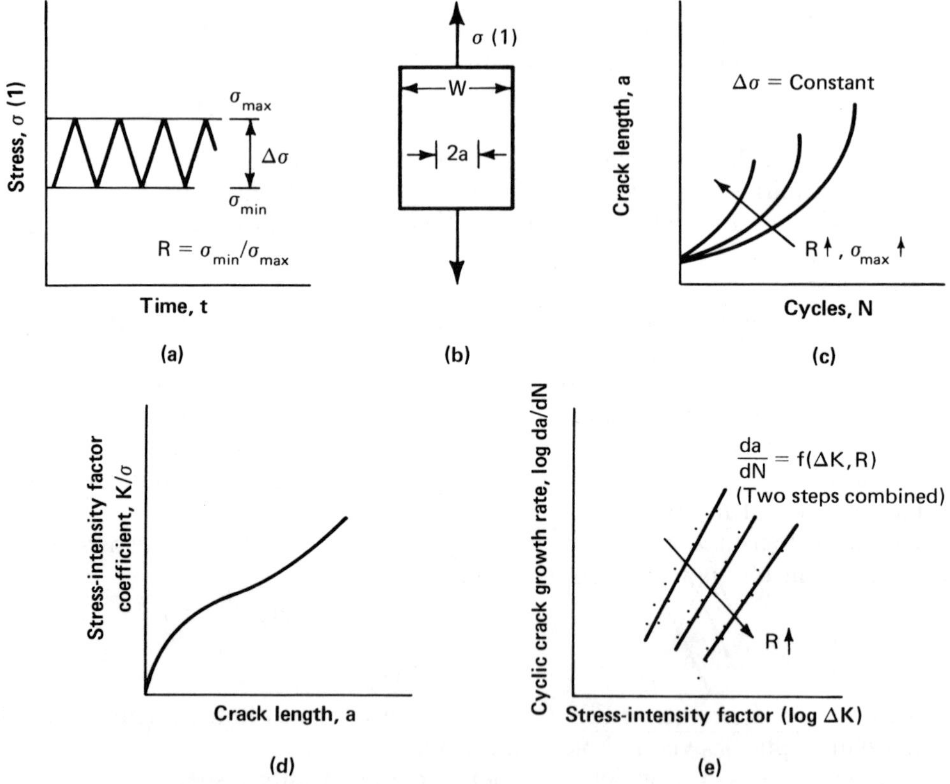

(a) Constant-amplitude loading. (b) Laboratory specimen. (c) Observed crack growth behavior (several cases). (d) Stress-intensity factor solution. (e) Differentiated a vs N behavior as a function of ΔK.

Fig. 9.11. Development of constant-amplitude crack growth rate data from laboratory tests (Ref 9.8)

laboratory, to evaluate the behavior, and to subsequently utilize the laboratory-generated data in making life predictions for structures subjected to identical loading and environmental conditions. Typically, life predictions are made by an integration process—that is, by inverting the differentiation scheme illustrated in Fig. 9.11. In fact, a common method for demonstrating the accuracy of the methodology associated with both (*a*) the growth rate data-reduction scheme, and (*b*) subsequent utilization of the crack growth rates in fatigue life prediction, requires that the laboratory-generated a vs N curves be regenerated by the integration procedure.

The general procedure for predicting fatigue life is outlined by the sketches in Fig. 9.12 (Ref 9.8). For relatively simple geometries and other service conditions, a closed form expression for fatigue life, such as Eq 9.6, can be developed and used with success to predict fatigue life. More often, an incremental damage integration model, as sketched in Fig. 9.12, is required to obtain useful predictions of fatigue crack growth and life. In the damage integration model, an iteration is conducted between the initial crack length, a_i, and any intermediate crack length, a_k, where $a_i < a_k < a_f$. This results in:

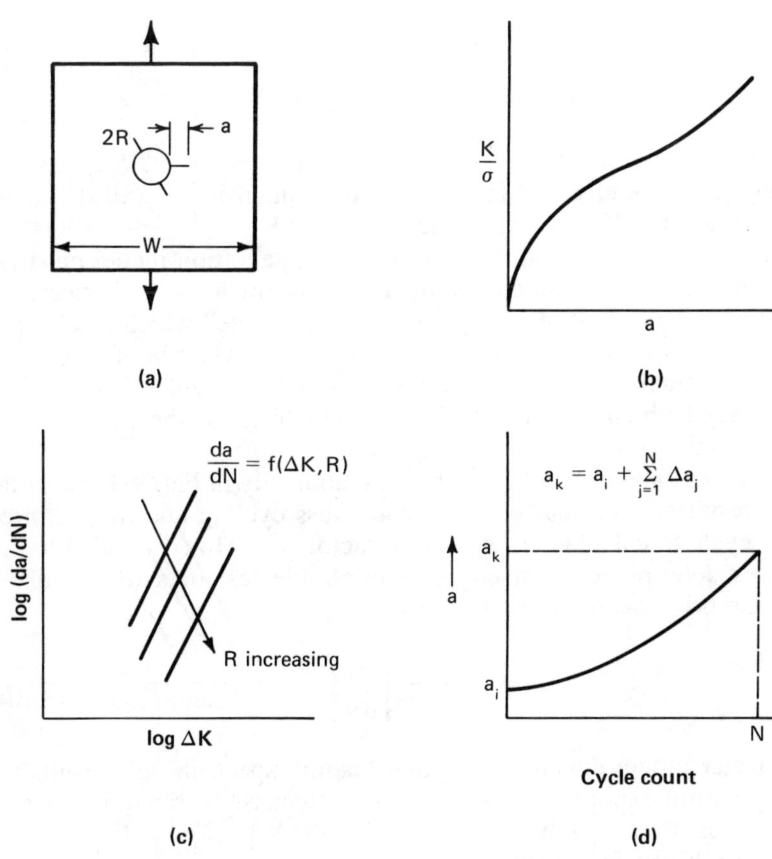

(a) Geometry of interest. (b) Stress-intensity factor solution. (c) Crack growth rate behavior. (d) Integration of da/dN expression.

Fig. 9.12. Procedure for predicting fatigue life (Ref 9.8)

$$a_k = a_i + \sum_{j=1}^{N} \Delta a_j \qquad \text{(Eq 9.8)}$$

where N is the number of cycles corresponding to the intermediate crack length a_k (see Fig. 9.12d). The next cycle of the stress history induces a crack length growth increment, Δa_{N+1}. The damage integration model provides the analysis capability to determine this crack growth increment. The growth increment Δa_{N+1} is expressed in terms of the cycle increment, ΔN, and is set equal to the constant-amplitude crack growth rate which, in turn, is expressed as a function of stress-intensity factor range, ΔK, and stress ratio, R, as:

$$\Delta a_{N+1} = \left.\frac{da}{dN}\right|_{N+1} \Delta N = f\left(\Delta K\Big|_{N+1}, R_{N+1}, \Delta N\right) \qquad \text{(Eq 9.9)}$$

The ΔK and R factors in Eq 9.9 are determined using the maximum and minimum stresses in the N + 1 cycle of the given load history and by using the stress-intensity factor coefficients associated with the given structural geometry at the crack length a_k. A crack growth rate equation that interrelates ΔK and R is useful in relation to the damage integration model. Such an equation is the Forman equation (Eq 3.5), which is repeated here:

$$\frac{da}{dN} = \frac{C\Delta K^n}{(1-R)K_c - \Delta K} \qquad \text{(Eq 9.10)}$$

This equation also models crack growth region 3 (see Section 2.5.2 and Fig. 2.10) by accounting for the effect of the critical stress intensity, K_c, of the material on crack growth at high K levels approaching K_c or K_{Ic}.

For variable amplitude loadings, the damage integration model must take into account a "retardation" effect caused by a high-to-low load level interaction. The most popular concept has been to predict a "yield zone" which has been created by a prior high-load level and to reduce the crack growth rate of subsequent load cycles until the crack grows out from the high-load yield zone. Two basic models will be discussed which use this principle but which vary in the method of reducing the crack growth rates.

The Wheeler model (Ref 9.9), as shown graphically in Fig. 9.13, uses the yield zone sizes resulting from current low-load stress cycling and from a prior high-load stress cycle to calculate a retardation factor, C_p. This is applied to the crack growth rates determined from constant-amplitude test data to account for the effects of the prior overload, as follows:

$$a_k = a_i + \sum_{j=1}^{N} C_{p_j}\left[\frac{da}{dN}\right] \qquad \text{(Eq 9.11)}$$

The Wheeler model depends on actual loading spectrum test results to determine the spectrum exponent "m" (Fig. 9.13). Because of the uniqueness of "m" to a particular material/spectrum combination, the Wheeler model is limited in its predictive capability for other loading spectra and materials. Another model, devised by Willenborg (Ref 9.10), uses an effective stress concept to reduce the growth rates following an overload.

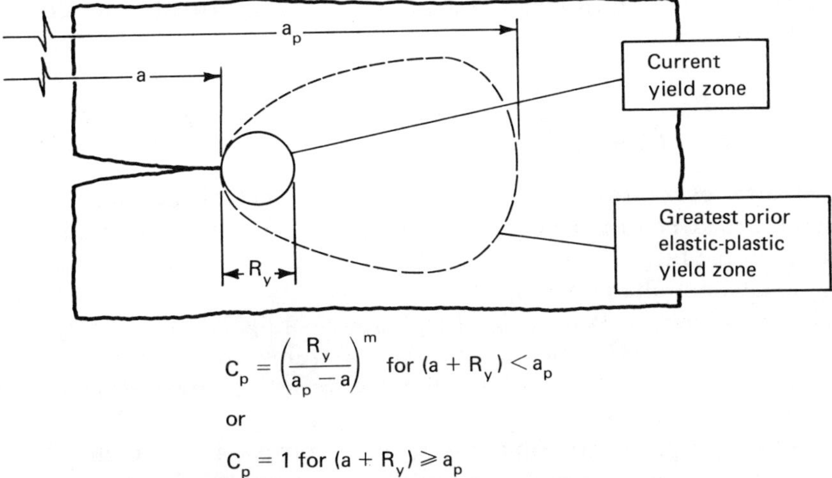

$$C_p = \left(\frac{R_y}{a_p - a}\right)^m \text{ for } (a + R_y) < a_p$$

or

$$C_p = 1 \text{ for } (a + R_y) \geq a_p$$

Fig. 9.13. Wheeler model for fatigue crack growth retardation through crack tip yield zones (Ref 9.9)

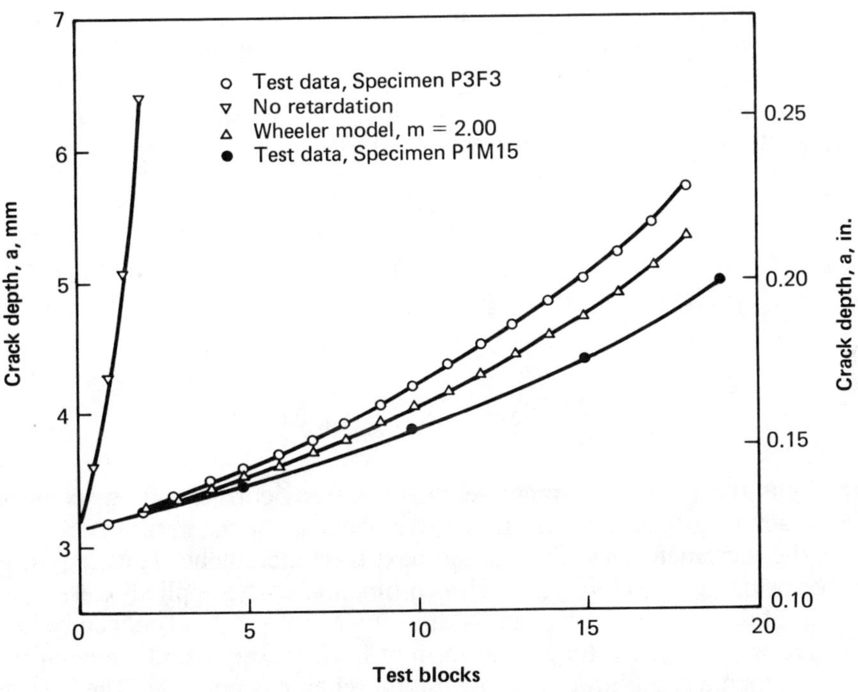

Fig. 9.14. Comparison of laboratory test data with damage-integration fatigue crack growth models (Ref 9.11)

A comparison of laboratory test data with results from damage integration models with and without retardation is shown in Fig. 9.14 (Ref 9.11). The spectrum used was from a military aircraft in which the overloads are expected to produce considerable crack growth retardation. This is verified by the comparison of the laboratory and model results.

9.2.2. Stress Corrosion Cracking, Hydrogen Embrittlement and Corrosion Fatigue

The use of stress corrosion cracking or hydrogen embrittlement data for analysis of sustained-load environmental cracking is analogous to the iterative scheme illustrated for the fatigue case in Fig. 9.11 and 9.12. Two obvious differences are that the load is not alternating and that there would not be crack closure effects. This simplifies the analysis and often allows straightforward integration for making lifetime predictions. The other extreme of complexity involves corrosion/fatigue interactions and is still in the formative stages of understanding. Nevertheless, a few analytical approaches have been proposed which offer some insight as to the separate contributions of stress corrosion and fatigue. These two environmentally controlled processes will be discussed briefly with an analytical representation of the sustained-growth process.

In many medium- to ultrahigh-strength steels and titanium alloys, stress corrosion cracking (SCC) and hydrogen embrittlement may be the same phenomenon, with the source of the hydrogen being different. For illustrative purposes, a SCC analysis will be presented with the understanding that a similar approach may suffice for hydrogen embrittlement. The example will actually be for a high-strength aluminum alloy, 7075-T651, which has a known susceptibility to SCC, particularly in the short transverse orientation. The easiest approach is to predict the crack at any time, t_i, in the growth process so that analogs to Eq 9.8 and 9.9 are:

$$a_k = a_i + \sum_{j=1}^{N_{\Delta t}} \Delta a_j \qquad \text{(Eq 9.12a)}$$

and

$$\Delta a_{k+1} = \left.\frac{da}{dt}\right|_{k+1} \Delta t = f(K_{I_k}, T, C_o, \Delta t) \qquad \text{(Eq 9.12b)}$$

where $N_{\Delta t}$ is the number of incremental time steps corresponding to the intermediate crack length, a_k; Δt is the time increment used in the iterative scheme; and Δa_{k+1} is the increment of growth in the next time increment. Thus, da/dt is the stress corrosion crack velocity, which is a function of the applied stress intensity at the start of the increment, K_{I_k}, and of the temperature, T, and concentration, C_o, of the environment. If the functional form of $f(K_I)$ is simple and the environment is constant, then a straightforward integration scheme is possible. The first step is to represent $f(K_I)$ for the growth rate, as illustrated in Fig. 9.15(a), which shows crack growth rate as a function of applied stress intensity for a specimen of aluminum alloy 7075-T651 tested in a 3.5% aqueous solution of sodium chloride at 21°C (70°F) (Ref 9.9). The analytical representation was previously suggested by two investigations (Ref 9.12 and 9.13) which indicated a similar equation for either SCC or fatigue, giving:

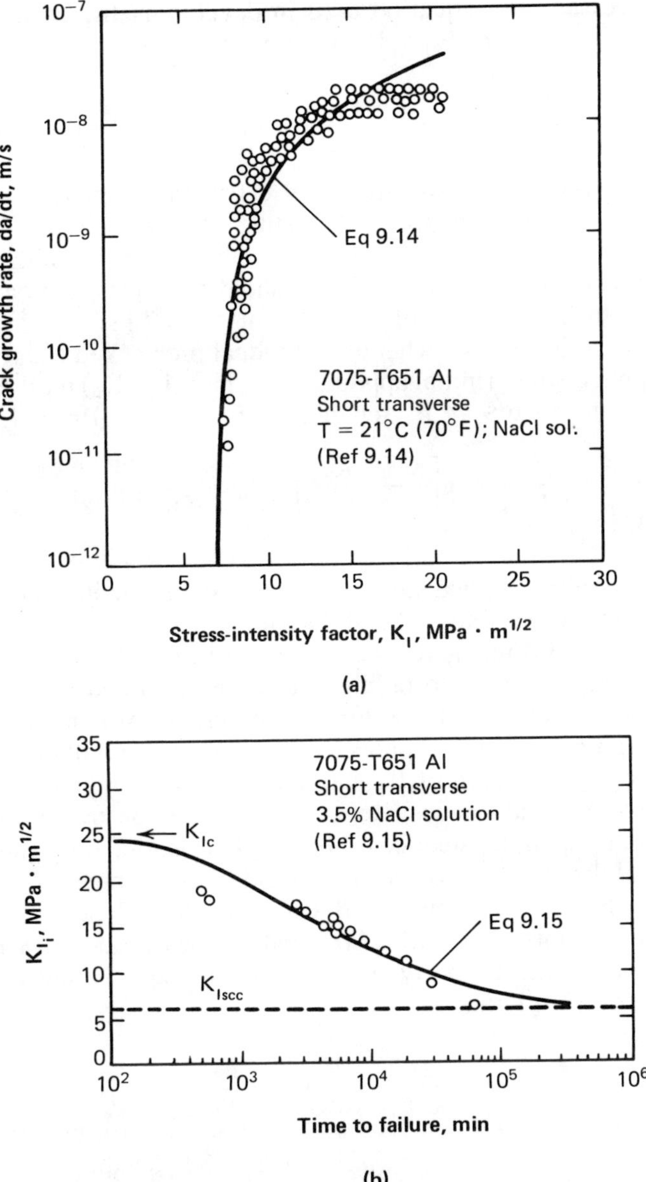

Fig. 9.15. Results of room temperature stress corrosion tests on precracked specimens of aluminum alloy 7075-T651 (Ref 9.13, 9.14 and 9.15)

$$\frac{da}{dt} = \alpha(T)(K_I - K_{Iscc})^{n_o} \; ; \quad \frac{da}{dN} = C(\Delta K - \Delta K_{th})^n \qquad \text{(Eq 9.13)}$$

where $\alpha(T)$ is equivalent to $C_o \exp(-\Delta H/RT)$ in Table 9.2, and K_{Iscc} and ΔK_{th} correspond to threshold values in the corresponding stress corrosion and fatigue cracking processes. The equation used in developing the curve in Fig. 9.15(a) is a slight modification of Eq 9.13, with $n_o = 2$, giving:

$$\frac{da}{dt} = C_o \left\{\frac{K_I - K_{Iscc}}{K_{Iscc}}\right\}^2 \exp\left(-\frac{\Delta H}{RT}\right) \qquad \text{(Eq 9.14)}$$

where $C_o \exp(-\Delta H/RT) \simeq 10^{-8}$ m/s (3.9×10^{-7} in./s) and $K_{Iscc} = 6.7$ MPa·m$^{1/2}$ (6.1 ksi · in.$^{1/2}$) for room temperature testing of aluminum alloy 7075-T651. In Fig. 9.15(a), it can be seen that this representation is reasonably good for levels of applied stress intensity, K_I, from threshold to values approaching the plane strain fracture toughness, K_{Ic}, of 25 MPa · m$^{1/2}$ (22.7 ksi · in.$^{1/2}$). This may then be applied to a series of tests wherein individual precracked coupons were held at a constant applied stress (initial applied stress intensity, K_{Ii}) until failure occurred. First, integration of Eq 9.14 by separation of variables (Ref 9.13) leads to:

$$t_f = \frac{2K_{Iscc}{}^2 a_i}{C_o \exp\left(-\frac{\Delta H}{RT}\right) K_{Ii}^2} \left\{ \ln\left[\frac{K_{Ic} - K_{Iscc}}{K_{Ii} - K_{Iscc}}\right] - \frac{K_{Iscc}}{K_{Ic} - K_{Iscc}} + \frac{K_{Iscc}}{K_{Ii} - K_{Iscc}} \right\} \qquad \text{(Eq 9.15)}$$

where K_{Ii} is the initial applied stress intensity and a_i is the initial half crack length. Using the values for $C_o \exp(-\Delta H/RT)$ and K_{Iscc} above and an initial half crack length of 0.02 m (0.8 in.) gives the prediction in Fig. 9.15(b). Considering that the crack velocity and time to failure data were generated in two different heats of material in two different laboratories, the agreement is realistic.

A similar approach may be applied to corrosion fatigue if the failure process is one of simple superposition of the fatigue and SCC components. This was first pointed out by Wei and Landes (Ref 9.16), who demonstrated that the corrosion fatigue rate in a maraging steel could be roughly predicted by superimposing the contribution of the stress corrosion component (time averaged over a given fatigue cycle) on the fatigue contribution. Because da/dt is dependent on the applied K_I level (for example, Eq 9.14), Wei and Landes proposed a numerical integration method similar to the type of iteration in Eq 9.12 but integrated over a given fatigue cycle, giving:

$$\frac{da}{dN} = \left(\frac{da}{dN}\right)_f + \left(\frac{da}{dN}\right)_{scc}$$

$$= \left(\frac{da}{dN}\right)_f + \int \frac{da}{dt}\{K(t)\}\,dt \qquad \text{(Eq 9.16)}$$

where da/dt {K(t)} is the SCC growth rate, which is a function of both (*a*) the maximum stress level in the fatigue cycle and (*b*) crack length, which increases with time. Thus, da/dt {K(t)} is considered to be the variable sustained-load growth rate, which is a function of a time-varying stress-intensity factor.

In another study, this superposition model was considered in an analytical way wherein the sustained growth rate was given by Eq 9.14. This was accomplished by defining a stress corrosion cracking threshold value which is lower under fatigue conditions than it is under static conditions. Such a fatigue-modified SCC threshold, defined as $K_{Iscc(f)}$, has been observed in several investigations (Ref 9.13 to 9.18). With the realization that $K_{Iscc(f)} < K_{Iscc}$, a superposition model can be analytically developed (Ref 9.13) for SCC systems which follow Eq 9.14:

$$da/dN = \left(\frac{da}{dN}\right)_f + \frac{C_o \exp\left(-\frac{\Delta H}{RT}\right)}{3\omega\ K^2_{Iscc(f)}} \left[\frac{\left\{\Delta K\left(\frac{1}{1-R}\right) - K_{Iscc(f)}\right\}^3}{\Delta K}\right] \qquad \text{(Eq 9.17)}$$

where ω is the cyclic test frequency and R is the fatigue load ratio. This particular analysis applies to situations where the minimum stress-intensity values are low—i.e., $K_{min} < K_{Iscc(f)}$. By applying this to Speidel's data (Ref 9.12) on 12% chromium stainless steel, one can first utilize his fit to the fatigue data as indicated in Fig. 9.16(a) for $(da/dN)_f$. For the SCC component, a reasonable representation of the sustained-load data in Fig. 9.16(b) is given by Eq 9.14, with $C_o \exp(-\Delta H/RT)$ taken to be 2.8×10^{-7} m/s (1.1×10^{-5} in./s) and K_{Iscc} taken to be 20 MPa · $m^{1/2}$ (18.2 ksi · $in.^{1/2}$). Because the threshold decreases under fatigue conditions, $K_{Iscc(f)}$ for this material is chosen as 8 MPa · $m^{1/2}$ (7 ksi · $in.^{1/2}$). With these values in Eq 9.17, the data are predicted reasonably well in Fig. 9.16(c) for the 2.3 Hz frequency tests in distilled water but are overpredicted by a factor of about four for the lowest cyclic frequencies. This is not surprising in light of the fact that Speidel (Ref 9.12) demonstrated that corrosion fatigue at these low frequencies involved growth for only about 20% of the load cycle at high ΔK values, even though $K_{Iscc(f)}$ was exceeded for about 70% of the load cycle. Thus, although superposition might give a qualitative estimate of growth rates, it can be seen that other factors might greatly affect growth rates—in this case, perhaps crack closure effects.

Corrosion fatigue descriptions are further complicated by the fact that the environment may produce multiple effects. For example, Suresh *et al* (Ref 9.17) demonstrated that dry hydrogen may produce a *frequency-sensitive* environmental effect analogous to SCC at intermediate ΔK values and a *frequency-insensitive* environmental effect near the threshold. This is illustrated in Fig. 9.17 for 2¼ Cr-1Mo steel tested in air and in 138-kPa hydrogen gas. Because the sustained-load threshold for this steel is on the order of 90 MPa · $m^{1/2}$ (82 ksi · $in.^{1/2}$), the $K_{th(f)}$ of about 22 MPa · $m^{1/2}$ (20 ksi · $in.^{1/2}$) gives $K_{th(f)} \ll K_{th}$.

It can be seen for ΔK values greater than $K_{th(f)}$ that there is a large increase in growth rate for the low test frequencies but not for the higher ones. Therefore, this regime may be considered to be one where superposition might apply. In addition, however, there is a true threshold, ΔK_{th}, which appears to be frequency-insensitive but which nevertheless decreased by about 30% to 5.4 MPa · $m^{1/2}$ (4.9 ksi · $in.^{1/2}$) because of the hydrogen environment. Such multiple effects are poorly understood and are clearly possible in a large number of material/environment systems.

Thus, the reader is cautioned not to apply simple superposition or extend it beyond known temperature/environment/stress combinations unless it is certain that the growth mechanisms and frequency responses are the same over the entire spectrum of variables.

With this overview of the design procedures, it is useful to describe the other necessary components—that is, quality control and failure analysis—which may often be deciding factors in judging past, present or future performance.

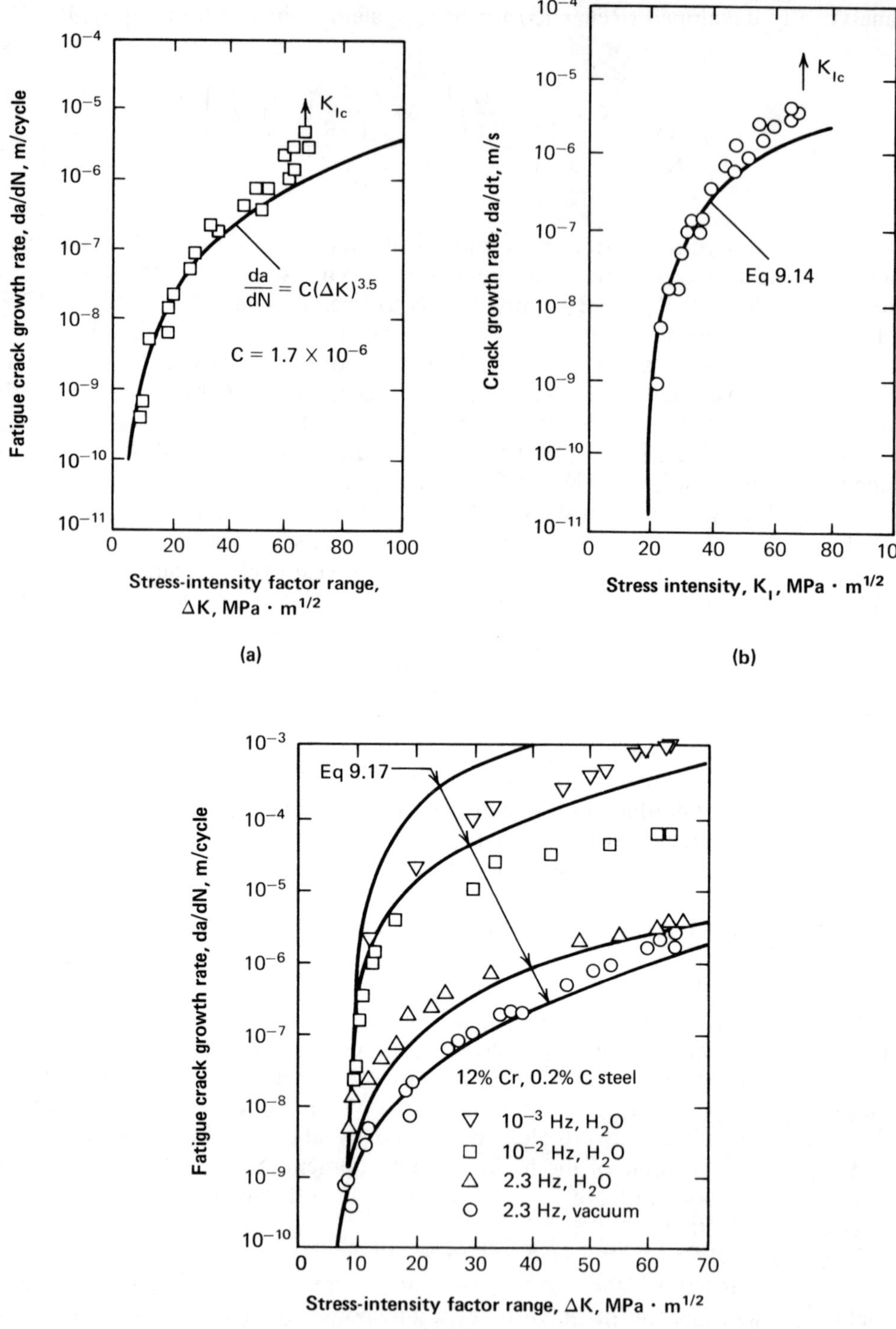

Fig. 9.16. Fatigue crack growth rate data for precracked specimens of 12% chromium stainless steel (Ref 9.12 and 9.13)

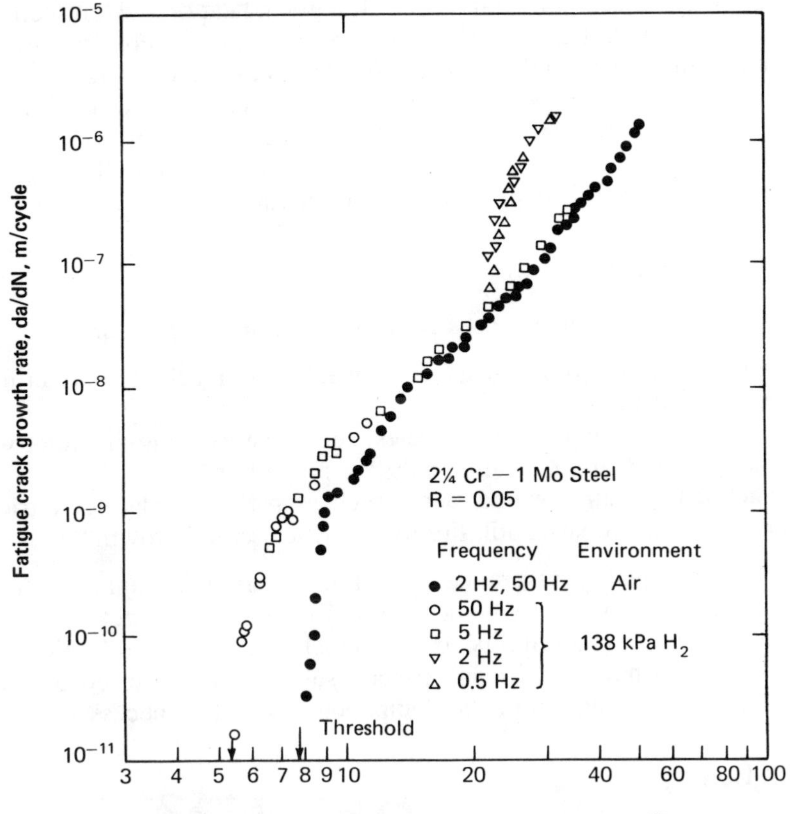

Fig. 9.17. Fatigue crack growth rates in 2¼Cr-1Mo steel in air and in hydrogen (Ref 9.17)

9.3. QUALITY CONTROL

There are essentially three stages of quality control: control of *incoming material*, control of *in-plant fabrication* and control of *exit finished products* — all of which may be crucial to a particular fracture design methodology.

9.3.1. Incoming Quality

Some of the questions that may be asked in reference to incoming quality are:

1. How uniform in properties is the material as purchased from the vendor?
2. What minimum level of toughness is guaranteed?
3. What NDE methods are used to accept or reject material lots?

For incoming quality, the types of data spread that one might expect, given the testing and materials variability associated with fracture measurements, were illustrated in Chapter 4. For example, even though the nominal value of K_{Ic} for a quenched and tempered steel with a yield strength of 1400 MPa (200 ksi) could be 90 MPa · $m^{1/2}$ (82 ksi · in.$^{1/2}$) in Fig. 4.3, a given heat or a given section from the same heat might fail at 60 MPa · $m^{1/2}$ (55 ksi · in.$^{1/2}$). For fracture-critical components, then, some vendor or user specifications are often required. These may be material fracture properties such as K_{Ic} or screening tests for which results have been shown to correlate with fracture properties. However, in using screening tests such as precracked Charpy tests for steels or titanium alloys, great care must be taken with regard to statistical variations, as discussed in Ref 9.19.

9.3.2. In-Plant Control

Fabrication of a component raises additional questions about quality, such as:

1. How do fabrication processes such as welding or machining affect material properties and performance?
2. Do additional required thermal cycles, such as stress relieving after welding or baking after plating, change material properties?
3. Do chemical or surface treatments such as pickling, plating, nitriding or carburizing change susceptibility to subcritical crack growth?

As part of any quality assurance program, a materials engineer should be involved in the assessment of whether any fabrication step affects the incoming fracture toughness or susceptibility to slow crack growth. If the fabrication procedure is known to have significant effects, such as those caused by welding, fracture mechanics evaluation of the components would be necessary.

9.3.3. Exit Quality

As the finished component exits the plant, the final question is: What NDE or design methods are used to guarantee safe defect sizes as the components are put into service? It is beyond the scope of this book to educate the reader in NDE methods. Some useful information is presented in Ref 9.20. In addition, it is important to note that different degrees of sensitivity exist for the different NDE techniques with regard to the character of the flaw (that is, surface versus subsurface, orientation, and so on). One early study revealed the uncertainty with which NDE methods could detect even surface cracks (Ref 9.21). For example, x-ray methods were not sensitive to cracks 7.5 mm (0.3 in.) long.

By 1975, it appeared that laboratory detection methods for such flaws had improved significantly but that in-service capabilities had not (Ref 9.1). Although methods may have been improved since then, the reader is cautioned that utilization of NDE for quality assurance should only follow a program demonstrating that such flaws may be routinely detected.

9.4. FAILURE ANALYSIS

Even careful fracture design and quality control implementation cannot guarantee that a component will not fail in service. There are many reasons for this, such as (*a*) unanticipated loading or environmental factors; (*b*) incorrect usage of the component, such as might result from incorrect maintenance procedures; and (*c*) unknown synergistic effects that lead to failure processes not considered in design or laboratory simulations. Two examples of unanticipated environmental factors that may cause failure are severe vibrations during shipment by truck or rail and combinations of corrosive environments and residual stresses that produce stress corrosion cracking of stored spare parts. Synergistic effects could arise from the spectra of service environments and loading conditions which could produce unexpected interactions, such as creep/fatigue/environment interactions that may be difficult to simulate in the laboratory.

Because of the possibility of unanticipated failures, good failure analysis is often an essential part of fracture design. Briefly, the components of failure analysis are fractography, metallography and spectroscopy combined with a thorough knowledge of all service conditions. Fractography might include both scanning electron microscopy and replica microscopy with a transmission electron microscope, depending on the topology of the fracture surface involved. Probably the most important feature of a failure analysis is identification of the failure mode, and this invariably involves scanning electron microscopy. In addition, extraction techniques of particles with a scanning transmission electron microscope (STEM) or nondispersive x-ray techniques might identify second phases. Metallography could provide identification of crack paths, microstructures and appropriate surfaces for microhardness traverses (across welds, for example). Little more will be said about these standard failure analysis techniques, because they are adequately covered in Ref 9.20 and 9.22.

With regard to spectroscopy in failure analysis, this is often done at three levels, as indicated in Ref 9.23: (*a*) macroscopic level for bulk composition; (*b*) microscopic level for second phase identification; and (*c*) atomistic level for grain boundary segregants. Spectroscopic examination of bulk samples for determination of chemical composition is a standard technique and is a routine technique for determination of whether a given component is within acceptable limits. Nondispersive x-ray identification of second phases on the micrometer level is often possible by scanning electron microscopy, an example of which would be identification of large MnS stringers or silicide inclusions in steel. Smaller second phases can be identified by scanning transmission electron microscopy. At the atomistic level, it is sometimes important to establish whether tramp element segregants, which may only cover the top 100 Å of a fractured grain boundary, have contributed to subcritical crack growth or brittle fracture. This may be accomplished by an Auger spectroscopy technique and is described in detail in Ref 9.24.

9.5. MATERIALS SELECTION

Over the last two decades it has often proven difficult if not impossible to anticipate all possible combinations of macroscopic and microscopic failure mechanisms which a new design may experience. It is therefore not surprising that many materials selection processes are actually re-selection processes that are undertaken after failure analysis has demonstrated a particular material/loading/environment combination to be unacceptable. Nevertheless, an attempt will be made to provide material selection data where prior thought has gone into the selection process and where predictive attempts based on laboratory data have been made. Because this type of design/material selection procedure is still in its infancy, there will necessarily be information gaps. Where possible, selection based on the following three considerations in various material systems will be presented:

1. K_{Ic}, fracture toughness
2. ΔK, fatigue crack propagation
3. K_{Iscc}, stress corrosion cracking threshold.

Applications to steel, aluminum alloy and titanium alloy components will be given.

9.5.1. Case Studies Using K_{Ic} Considerations

9.5.1.1. Pressure Vessels

Application of fracture mechanics technology to pressure vessel design and material selection dates back to the missile motor cases of the early 1960's in the aerospace industry, later to brittle fractures in petrochemical plants. Two striking examples of postfailure analysis for assessment of what went wrong and how to avoid future mishaps are the explosion of an ammonia pressure vessel (Ref 9.25) and the failure of a second-stage missile motor case (Ref 9.26), both of which failed during hydrostatic proof pressure testing.

As indicated in Fig. 9.18, the large ammonia vessel, 18.2 m (716.5 in.) high by 2 m (78.7 in.) in diameter by 0.15 m (5.9 in.) in wall thickness, suffered considerable damage when it failed within 2% of the intended operating pressure of 35 MPa (5.1 ksi). The proof pressure was intended to be 40% higher. The failure origin was traced back to hairline surface cracks on the order of 8 mm

Fig. 9.18. Failed ammonia pressure vessel (Ref 9.25)

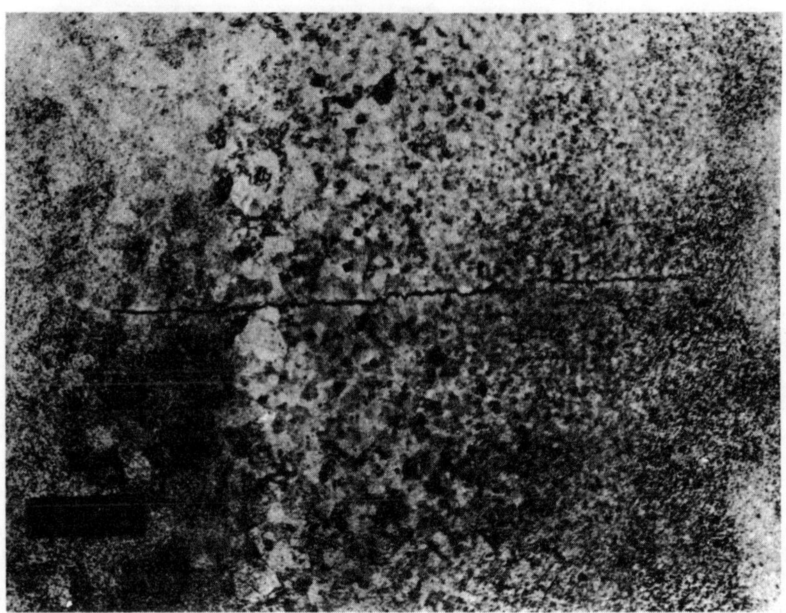

Fig. 9.19. Welding crack in the heat-affected zone of the vessel shown in Fig. 9.18 (Ref 9.25)

(0.3 in.) in length in the weld heat-affected zone, one of which is shown in Fig. 9.19. From the fracture toughness in the weld (58 MPa · $m^{1/2}$; 53 ksi · $in.^{1/2}$), which had been lowered from that of the base forging (120 MPa · $m^{1/2}$; 109 ksi · $in.^{1/2}$) due to improper stress relief, it was calculated that the burst pressure was reasonable only if the total stress that produced fracture was a summation of applied membrane stress and residual stress. By use of this calculation, along with consideration of the excessive hardness and low toughness in the heat-affected zone, the cause of the failure was traced back to a cold spot in the stress-relieving furnace. It was ascertained that this vessel could have been repaired and that avoidance of cold spots in the stress-relieving furnace could prevent future mishaps in similar vessels without a change in material.

Failure on hydrotesting of the second-stage missile motor case originated at a crack at the edge of a seam weld (Ref 9.26). The case had been produced from AISI 4340 steel and had been heat treated to a yield strength of 1450 MPa (210 ksi). The forward dome and aft funnel had been attached to the cylindrical section by circumferential seam welds following heat treatment. Cracks had occurred in one of these welds. Better NDE control could prevent the larger defects from seeing service. In addition, use of higher tempering temperatures could increase the toughness with little sacrifice in strength level. Thus, in both cases a heat treatment solution was possible.

A different conclusion was reached by Srawley and Esgar (Ref 9.21) in assessment of the hydrotest failure of a 6.6-m- (260-in.-) diam rocket motor case produced from 18Ni(250) maraging steel. In this instance, failure resulted not from incorrect heat treatment but from incorrect materials selection for the chosen design. The maximum scheduled hydrotest pressure was 6.6 MPa (960 psi),

Origin of fracture is indicated by arrows.

Fig. 9.20. Failed motor case pieces laid out in approximately the proper relation to each other (Ref 9.21)

which was 10% higher than the maximum design pressure of the motor case. During pressurization, accelerometers located on the vessel indicated that stress waves (perhaps from a crack source) were emitted at pressures greater than 1.7 MPa (242 psi). As the pressure was increased to 3.8 MPa (542 psi), another stress wave was detected and a loud report from the test tower indicated catastrophic failure. With the pieces arranged properly in regard to their positions in the as-fabricated vessel (Fig. 9.20), the failure was traced back to the origin using chevron markings on the fracture surfaces. This is indicated in Fig. 9.21A, where it can be seen that the stress-wave emission triangulation did approximately trace back to the failure origin, as indicated in Fig. 9.21B. Here, it was found that at the clean origin of weld W7/302 there was a defect large enough to cause failure at a stress of 660 MPa (96 ksi). This corresponded to an applied stress intensity of 60 MPa · $m^{1/2}$ (55 ksi · $in.^{1/2}$), which, although considerably below the fracture toughness of the base metal, was equal to or greater than the toughness in the weld zone or in the heat-affected zone.

The details of material selection, design and failure analysis for this rocket motor case are given in Fig. 9.22. Aside from the fact that the safety factor was nonconservative, the alarming feature of this analysis (as verified by postfailure

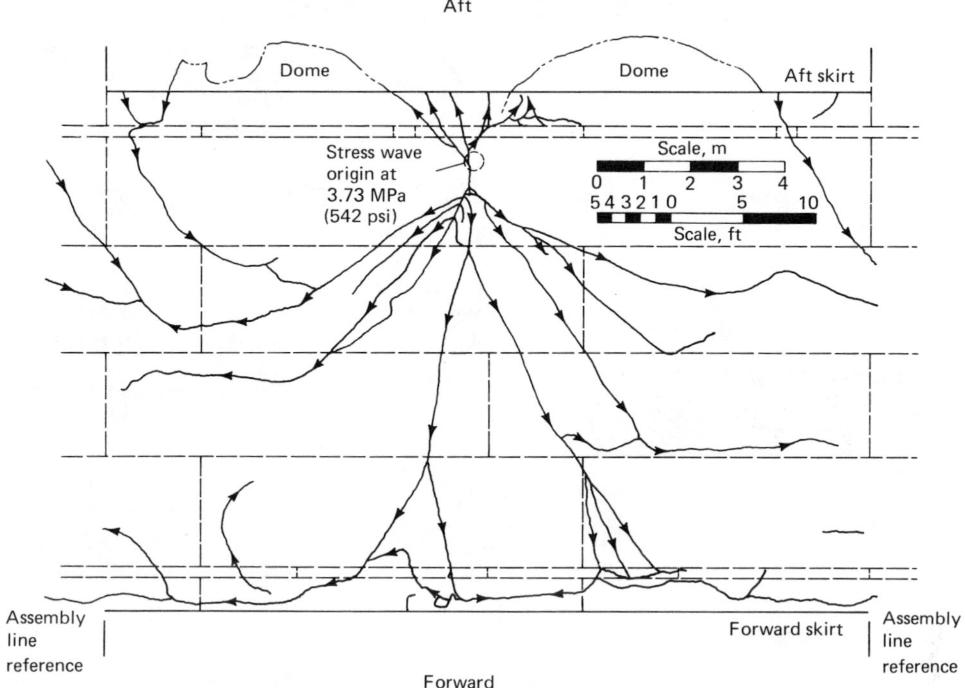

Arrowheads indicate directions of fracture propagation. Dashed lines indicate welds.

Fig. 9.21A. Map of fracture paths as viewed when looking at inside surface developed onto plane obtained by unrolling cylinder (Ref 9.21)

round robin NDE testing) was that subsurface tight cracks of large size could easily escape detection. That is, although cracks 1.15-mm (0.045-in.) deep should have been a cause for rejection, failure was caused by a crack 5.6 mm (0.22 in.) deep by 35 mm (1.4 in.) long.* A macroscopic view of the defect in question is given in Fig. 9.23, and an oxidized large flaw that clearly went through heat treatment is shown in Fig. 9.24. It is interesting that a different producer of the same rocket motor chamber had chosen a Grade 200 maraging steel that was 10% lower in yield strength but more than 100% greater in weld fracture toughness than the Grade 250 steel. Selection of this alternative material resulted in successful hydrotests and in successful motor firing of chambers 6.6 m (260 in.) in diameter with no failures.

9.5.1.2. Line Pipe Steels

An extensive series of tests and quality assurance evaluations of low-strength steel line pipe have been reported by Duffy, Eiber and Maxey (Ref 9.27) and by Hahn and Sarrate (Ref 9.28). Full-scale tests on precracked 0.76-m- (30-in.-) diam line pipe with a nominal yield strength of 414 MPa (60 ksi) were conducted to burst. The hoop stresses (circumferential direction) at failure were measured

*The short and long dimensions are 2a and 2c, respectively, as in Fig. 9-4.

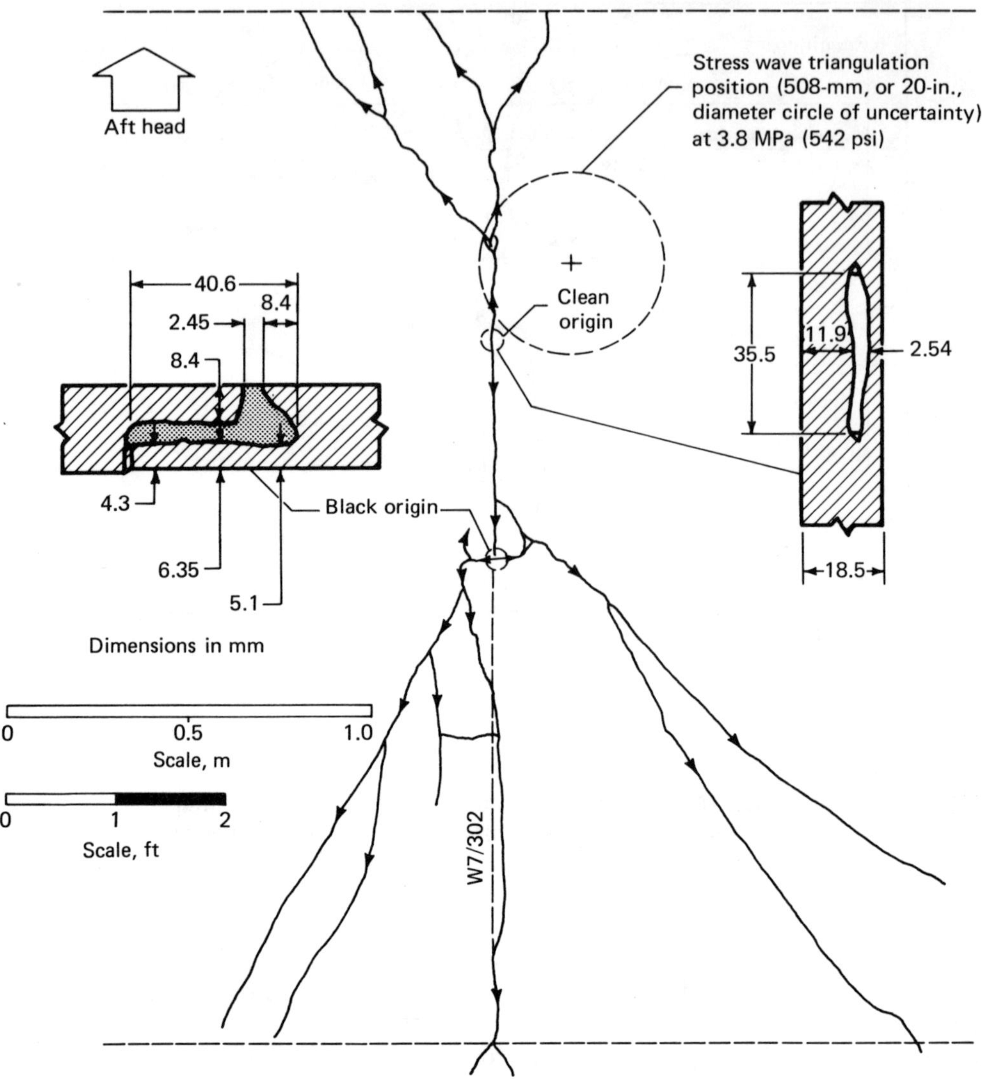

Inset sketches show shapes and dimensions of the two origins.

Fig. 9.21B. Detail in immediate vicinity of primary and secondary origins (Ref 9.21)

and compared with predictions based on two approaches. The first involved a plane-stress fracture toughness approach based on a Dugdale plastic zone correction (Ref 9.29) to the Folias solution (Ref 9.30) for a longitudinal crack in a cylindrical vessel, as given by:

$$K_c^2 = \frac{\pi c \sigma_h^2}{\cos\theta}\left(1 + 1.61\frac{c^2}{Rt}\right)\left(\frac{4-\kappa}{2}\right) \quad \text{(Eq 9.18)}$$

where K_c = critical stress-intensity factor, in MPa · $m^{1/2}$;
c = half crack length of axial through-wall cracks, in meters;

Requirement	6.0 MPa (875 psi) design pressure. Welded cylindrical structure 6.6 m (260 in.) in diameter
Materials selection	Grade 250 Maraging Steel σ_{ys} (plate) $\cong$ 1590 MPa (230 ksi) 90% weld efficiency $\geqslant$ 1420 MPa (207 ksi) 1.33 safety factor $\geqslant$ 1070 MPa (156 ksi) K_{Ic} (base plate) = 87 MPa · $m^{1/2}$ (79 ksi · $in.^{1/2}$) K_{Ic} (weld) = 58 MPa · $m^{1/2}$ (52.9 ksi · $in.^{1/2}$)
Design (vessel thickness)	Membrane stress $\sigma = \frac{PR}{t}$; t $\cong$ 18.5 mm (0.73 in.)
NDE	Ultrasonic, radiographic, magnetic particle: $>$ 1.8 mm (0.071 in.) in plate $>$ 1.15 mm (0.045 in.) in weld reject
Hydrotest	110% of design = 6.6 MPa (960 psi) internal pressure
Premature failure	56% of desired hydrotest = 3.8 MPa (542 psi)
Failure analysis	Origin 5.6 mm (0.22 in.) deep (2a) by 35 mm (1.4 in.) long (2c) caused failure at a membrane stress of 660 MPa (96 ksi) a/2c = 0.07 Q $\cong$ 1 (see Fig. 9.4) a_c = 2.8 mm (0.110 in.), σ = 660 MPa (96 ksi) $K_I = \sigma \left(\frac{\pi a}{Q}\right)^{1/2}$ = 62 MPa · $m^{1/2}$ (55 ksi · $in.^{1/2}$) $K_I > K_{Ic}$ (weld) at failure.

Fig. 9.22. Incorrect approach to design of 260-in. rocket motor case (Ref 9.21)

R =average radius of vessel, in meters;
t =wall thickness, in meters;
$\theta = (\pi/2)(\sigma_h/\sigma_c)$; σ_h=nominal hoop failure stress, in MPa;
σ_c=failure stress for an unflawed vessel, in MPa; and
$\kappa = (3-\nu)/(1+\nu)$ plane stress.

The second criterion was essentially a flow or plastic instability model where the average of the yield and ultimate strengths σ_{flow}, was used to predict burst pressure. These two approaches apparently fit the burst hoop stress results quite well, as shown in Fig. 9.25 and 9.26, which represent through-crack lengths from 0.025 to 0.5 m. The fracture toughness approach is slightly superior for long cracks. On the other hand, for a large number of pressure vessels made of low-strength steels with varying radius-to-thickness ratios, the observed burst strengths could be predicted reasonably well using the plastic instability approach. With the Folias correction (Ref 9.30), the burst strength is given by:

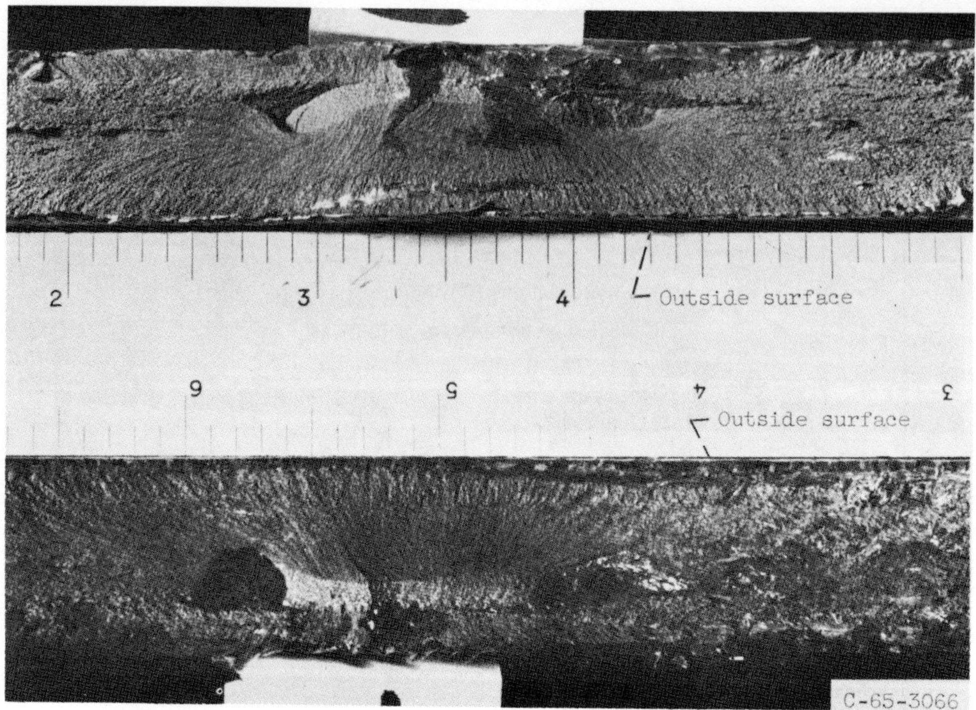

Fig. 9.23. Mating surfaces of clean origin on W7/302 weld (Ref 9.21)

Fig. 9.24. Typical chevron markings on fracture surface, showing directions of fracture propagation away from black origin (Ref 9.21)

$$\sigma_h = \overline{\sigma}_{flow}\left[1 + 1.61\frac{c^2}{Rt}\right]^{-1/2} \quad \text{(Eq 9.19)}$$

where, for low-strength structural steel, Hahn and Sarrate (Ref 9.28) took the average flow stress in MPa to be:

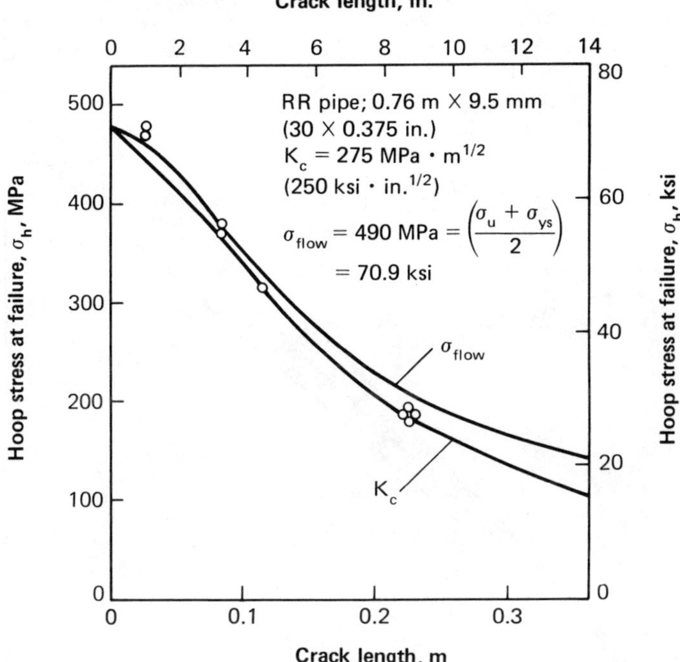

Fig. 9.25. Hoop stress at failure vs crack length for RR pipe (Ref 9.27)

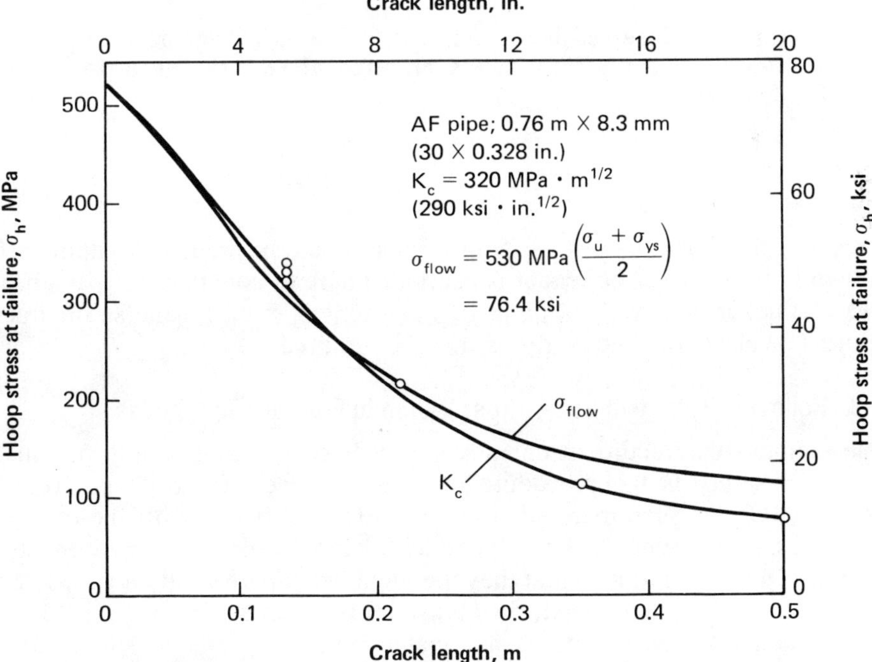

Fig. 9.26. Hoop stress vs crack length at failure for AF pipe (Ref 9.27)

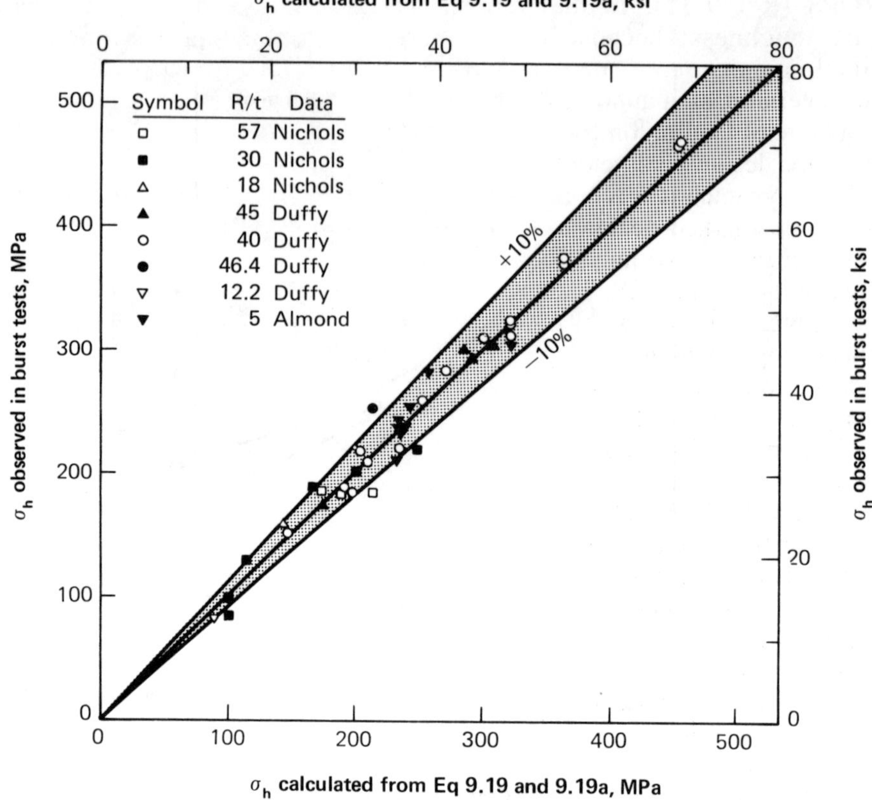

Fig. 9.27. Observed hoop stress at failure vs calculated hoop stress using the flow stress criterion of Eq 9.19 and 9.19a (Ref 9.28)

$$\overline{\sigma}_{\text{flow}} = 1.04\ \sigma_{ys} + 69\ \text{MPa} \qquad \text{(Eq 9.19a)}$$

The result is that burst hoop stress from 45 tests could be predicted within ±10%, as shown in Fig. 9.27. The reader is cautioned here to note that, for part-through cracks in thicker-wall vessels or in higher-strength steels, plane-strain fracture toughness or elastic-plastic analyses may be required.

9.5.1.3. Rollover Protective Structures: Dynamic Fracture Toughness

One example of a fracture mechanics approach to a material/design combination selected for safety is the procedure of selecting steels for rollover protective structures (ROPS). Such materials to be used for roll bars for off-the-road vehicles, for example, must satisfy dynamic fracture resistance requirements. In addition, ROPS are unique in that they are intended to perform their function only once, and, while they may bend, they must not break in performing this function. Although material considerations associated with bend and tensile elongation behavior are also important, only the resistance to fracture is discussed here. For other design considerations, the reader is referred to the SAE-ISTC report on ROPS (Ref 9.31). In that report, the emphasis is placed on dynamic fracture toughness, because this property is of paramount importance to structural integrity.

The material/design approach was to evaluate the effects of temperature, thickness and rate of loading on fracture resistance. This information was then utilized to arrive at a design parameter based on a simple correlation between dynamic K_{Ic} and Charpy V-notch (CVN) impact energy. For any given loading rate, structural steels exhibit marked decreases in toughness with decreases in temperature and increases in rate of loading. This is shown by a sketch of typical CVN impact energy results in Fig. 9.28. This transition from plastic (plane-stress) to relatively elastic (plane-strain) behavior of notched specimens occurs over temperature and thickness ranges which are dependent on loading rate. From Fig. 9.29, which considers one thickness, it can be seen that the transition region occurs at lower temperatures for static loading than for impact loading. For steels having yield strengths between 345 and 485 MPa (50 and 70 ksi), the transition temperature shifts due to static versus CVN impact loading are typically 56 to 78°C (133 to 172°F), with the larger shift being typical of the lower-yield-strength material. The loading time identified as "ROPS" in this figure is based on a recording of strain versus time during an actual rollover of a large earthmoving vehicle.

In addition to the interaction between temperature and loading rate, there is also an effect of thickness on fracture resistance. This can be seen in Fig. 9.30, where the plane-strain constraint is sufficiently large in the 25-mm- (1-in.-) thick dynamic tear test (DT) specimen to cause an increase of 10°C (18°F) in the ductile-to-brittle transition temperature over that of the 16-mm- (0.62-in.-) thick specimen. A criterion of 50% shear on the fracture surface was used to determine the transition temperature. A slow bend test performed on a specimen of the same type indicated a 20°C (36°F) increase in transition temperature for the thicker specimen. Thus, an increase in either thickness or loading rate will cause an increase in the transition temperature. Considering the effects of thickness, loading rate and test temperature on toughness, the final materials selection pro-

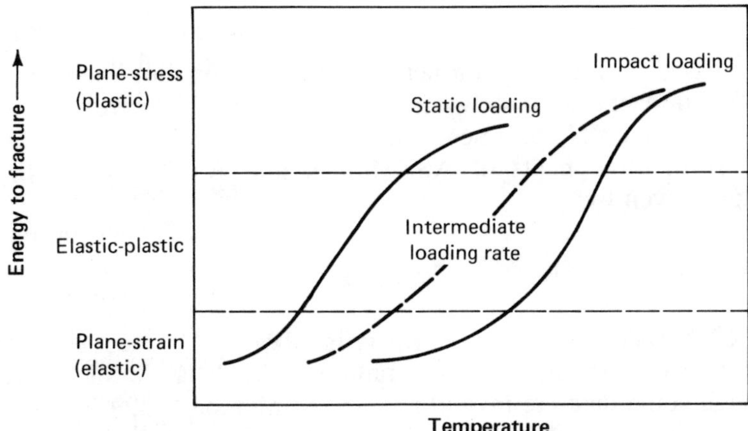

Fig. 9.28. Sketch of Charpy V-notch energy results for structural steels (Ref 9.31)

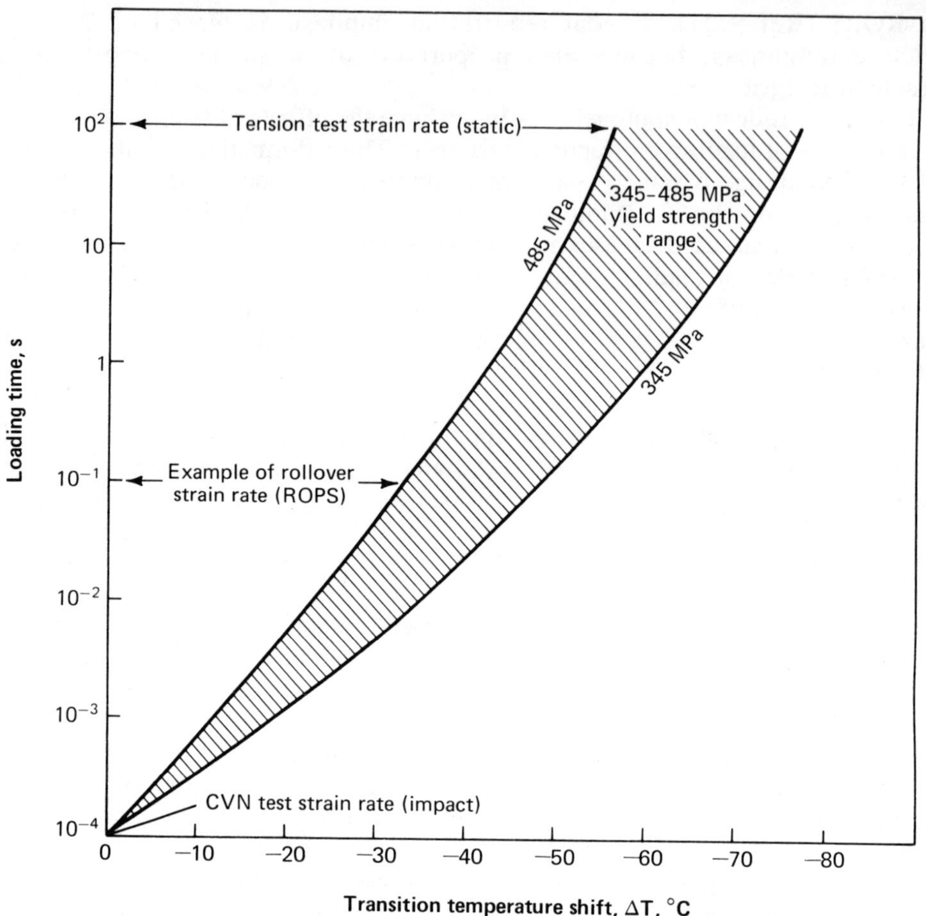

Fig. 9.29. Loading time vs transition temperature shift for structural steels (Ref 9.31)

cedure was based on a correlation between dynamic fracture toughness and the standard CVN impact test, as shown in Fig. 9.31. Here, the correlation represents dynamic testing of CVN specimens and 12.7-mm- (0.5-in.-) thick compact K_{Ic} specimens from three heats of ASTM A36 and A572 steels. The log-log relationship is given by:

$$K_{Ic_{(0.01\ s)}} = 26(CVN)^{0.43} \qquad \text{(Eq 9.20)}$$

where the CVN data are in joules, K_{Ic} is in MPa · $m^{1/2}$, and the parenthetical subscript represents time in seconds to failure in the dynamic fracture toughness test. Although some data are invalid per ASTM Method E-399, data points 1 to 4 meet all requirements of E-399 for plane-strain fracture. Because the time to failure is one-tenth that for a ROPS, this relation may be used as a conservative design criterion for materials selection. It was considered that if through-thickness

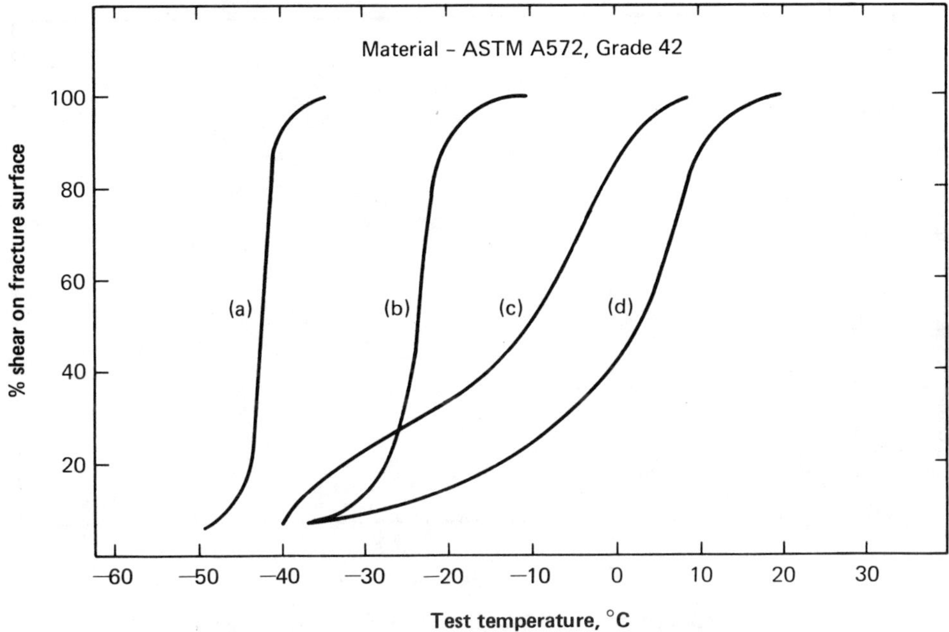

(a) Slow bend; 16-mm- (0.62-in.-) thick DT specimen. (b) Slow bend; 25-mm- (1-in.-) thick DT specimen. (c) Impact; 16-mm- (0.62-in.-) thick DT specimen. (d) Impact; 25-mm- (1-in.-) thick DT specimen.

Fig. 9.30. Percent shear on fracture surface vs testing temperature for dynamic tear test specimens of A572, Grade 42, steel (Ref 9.31)

yielding at the crack tip could be developed, a ROPS structure would bend before it would break. The criterion used for through-thickness yielding was taken as:

$$K_{Ic_{(0.01\ s)}} > 1.0\ \sigma_{ys}\ \sqrt{B} \qquad \text{(Eq 9.21)}$$

where B is material section thickness in meters and σ_{ys} is dynamic yield strength. This, along with the dynamic rate effect on yield and the CVN-$K_{Ic(0.01\ s)}$ relationship (Eq 9.20), gave the material (yield strength and toughness requirement) and design (thickness requirement) necessary to provide safe operation at the lowest service temperature of −30°C or −22°F (see Fig. 9.32). Because of the correlation in Fig. 9.31, the materials selection and quality control process could be provided by relatively inexpensive CVN impact testing. For example, a steel with a static yield strength of 485 MPa (70 ksi) in sections 15 mm (0.60 in.) thick would require a CVN impact energy of 15 J at a temperature of −30°C (−22°F).

9.5.2. Case Studies Using ΔK Considerations

9.5.2.1. Hydraulic Equipment at Fatigue Threshold

In the manufacture of port blocks for a ship's hydraulic equipment, it was common to machine the blocks from heavy plate. Then, because failure of these

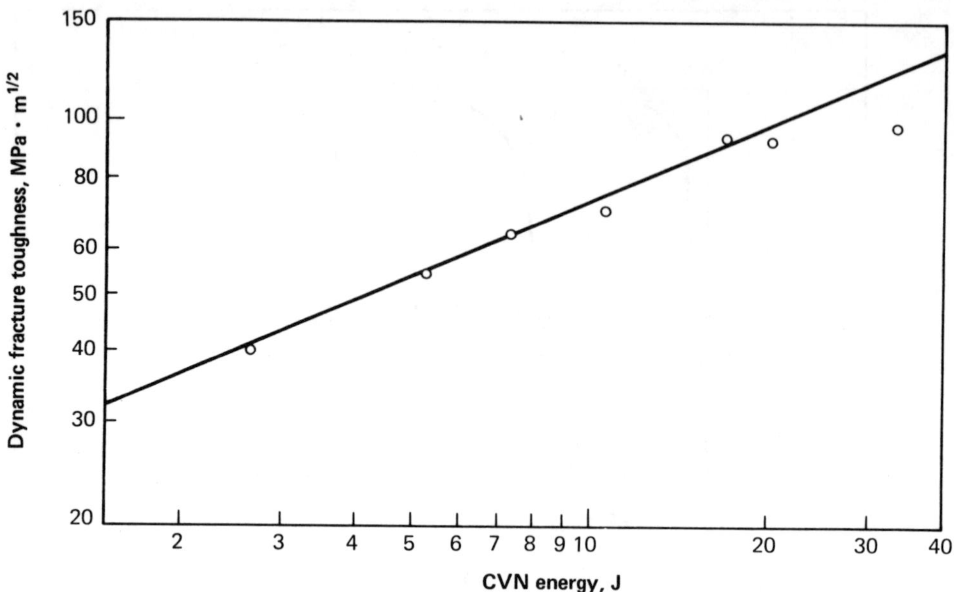

Fig. 9.31. Dynamic fracture toughness vs CVN impact energy for A36 and A572 steels (Ref 9.31)

blocks could disable the ship, the blocks were 100% ultrasonically examined for potential defects. A larger than normal port block was required and an ultrasonic examination of the 100-mm- (3.9-in.-) thick plate gave the indications illustrated in Fig. 9.33. At the National Engineering Laboratory in East Kilbride, Scotland, Pook (Ref 9.32) was given the task of analyzing the situation. It was clear that the laminations were natural defects and could grow to failure under the on-off pressure of the hydraulic equipment (maximum pressure: 34.5 MPa, or 5 ksi). The questions then were: "How dangerous are such laminations?" and "If very dangerous, could they not be avoided during drilling of the ports?". With little time for decision, fatigue thresholds for a number of steels were collected and compared, and a value of $\Delta K_{th} \leq 7.3$ MPa · $m^{1/2}$ (6.6 ksi · $in.^{1/2}$) was determined for the steel in question. In this application, calculations indicated a peak hoop stress of 34.5 MPa (5 ksi) at the ID of the port. In addition, the oil flowing through the port could pressurize a lamination and cause an equivalent additional stress of 34.5 MPa (5 ksi), so that a total $\Delta\sigma$ of 69 MPa (10 ksi) was used. With a ΔK_{th} of 7.3 MPa · $m^{1/2}$ (6.6 ksi · $in.^{1/2}$), the maximum permissible defect size was calculated from:

$$\Delta K_{th} = \Delta\sigma(\pi a)^{1/2}\alpha \qquad \text{(Eq 9.22)}$$

where α is a crack geometry factor on the order of unity. The resulting defect that was required to be detected was 3.6 mm (0.14 in.) since any larger lamination would have eventually grown to failure. The investigators decided that no ultrasonic method available to them was sufficient for detection of such flaws with 100% confidence. Based on the above calculations and on a conservative stance with regard to NDE, they recommended scrapping the plate. Furthermore, be-

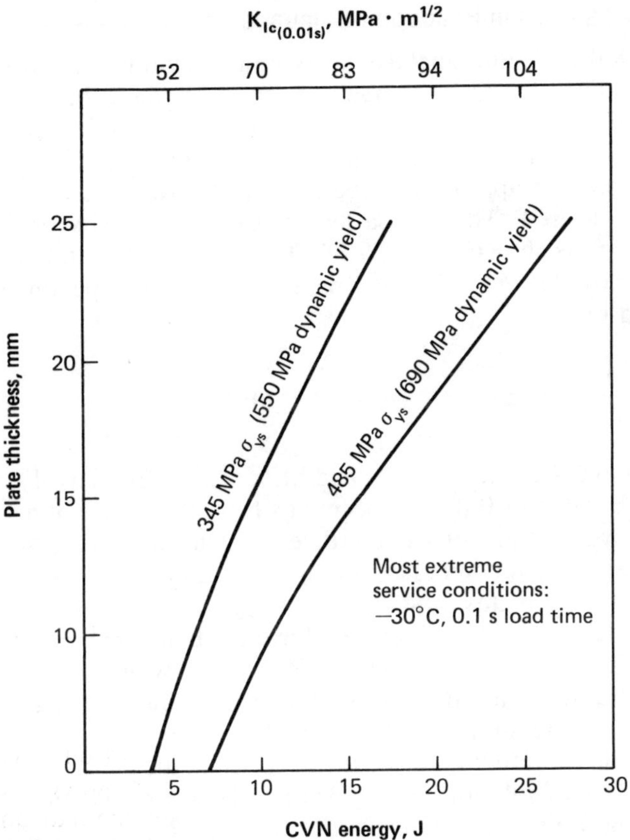

Fig. 9.32. Material properties and thickness required for most severe service conditions for rollover protective structures (Ref 9.31)

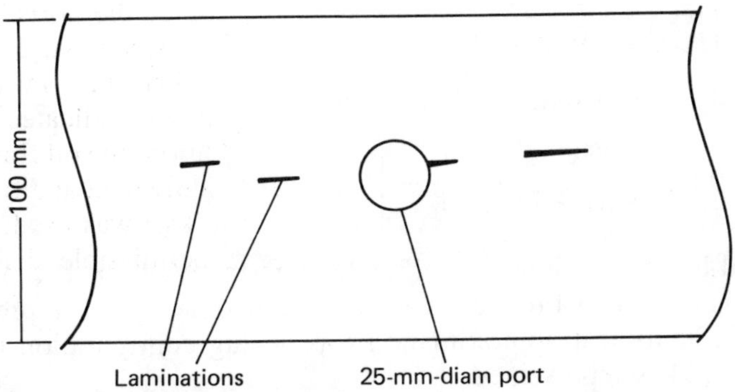

Fig. 9.33. Schematic section through 100-mm (4-in.) plate for a hydraulic port block (Ref 9.32)

cause other available commercial materials had no higher thresholds, the solution was to machine the blocks from specially produced forgings that would be free of laminations.

9.5.2.2. Fatigue Failures in Helicopter Components

Since the 1960's, failure analyses of titanium, aluminum and steel helicopter components undergoing fatigue crack propagation have been fairly common. However, relatively few studies in the open literature have quantified the observations using fracture mechanics principles. In one recent case study by Zola (Ref 9.33), a safe-life approach to the design of a rotor hub for the Heavy Lift Helicopter was desired. This was a laboratory design study of fretting-fatigue-induced cracking, as shown in Fig. 9.34. The study utilized fatigue crack growth rate data on alloy Ti-6Al-4V from fracture mechanics specimens. These data are represented by

$$\frac{da}{dN} = C(\Delta K)^4;\ C = 1 \times 10^{-12}\ \text{MPa}^{-4} - \text{m}^{-1} \qquad \text{(Eq 9.23)}$$

where da/dN and ΔK are in m/cycle and MPa $\cdot$ $m^{1/2}$, respectively. The data were obtained at an R value of 0.43 on specimens from the center of a solution treated and overaged forging and were then utilized to determine the effects of spectrum loading on the simulated component failure shown in Fig. 9.34. Although Zola's work used several estimates of stress intensity, and came to several conclusions based on possible defect sizes, it did not have the benefit of a more recent study. In that study, Schijve and Hoeymakers (Ref 9.34) evaluated fretting-corrosion-fatigue-induced crack initiation on aluminum alloy lugs, using procedures that may be applied to titanium alloys. The over-all approach is to demonstrate an analytical solution for fatigue life and then apply it to the previous failure analyses. Newman's analytical solution (Ref 9.35) has been experimentally verified and is a good starting point. It is given by:

$$\Delta K = \Delta\sigma_n \sqrt{\pi a}\, f_w\, f_b \left(\sec\frac{\pi D}{2W}\right)^{1/2} G \qquad \text{(Eq 9.24)}$$

where: $f_w = \left[\sec\left(\frac{\pi}{2} \cdot \frac{D + a}{W - a}\right)\right]^{1/2}$,

$f_b = 0.707 - 0.18\lambda + 6.55\lambda^2 - 10.54\lambda^3 + 6.84\lambda^4;\ \lambda = \frac{1}{1 + (2a/r)}$ and

$G = 0.5 + \frac{W}{\pi(D + a)}\left[\frac{D}{D + 2a}\right]^{1/2}$.

The crack and lug dimensions as well as loading (a, W, D, r, t, σ_n) are noted in Fig. 9.35. It is clear that Eq 9.24 is very cumbersome for a life prediction, and thus a suitable analytical approximation for the lug configuration of interest is useful. If Eq 9.24 is represented by:

$$\Delta K = \Delta\sigma_n \sqrt{\pi a} \cdot Y \qquad \text{(Eq 9.25)}$$

with Y representing the last four terms in Eq 9.24, then Y may be calculated for $0 \leq a/r \leq 0.16$ and for the specific dimensions of the lug as indicated in Table 9.3. As shown there, the agreement between this calculated Y and a simpler, convenient expression given by:

Fig. 9.34. Rotor hub lower plate lug cracks (Ref 9.33)

$$Y \simeq \frac{8.159}{[1 + 20\,(a/r)]^{1/4}} \qquad \text{(Eq 9.26)}$$

is very good. For the fatigue crack sizes of interest, the agreement for most of the life, to $a/r \simeq 0.12$, is within 1.3% and is still within 5% to $a/r = 0.16$, where-

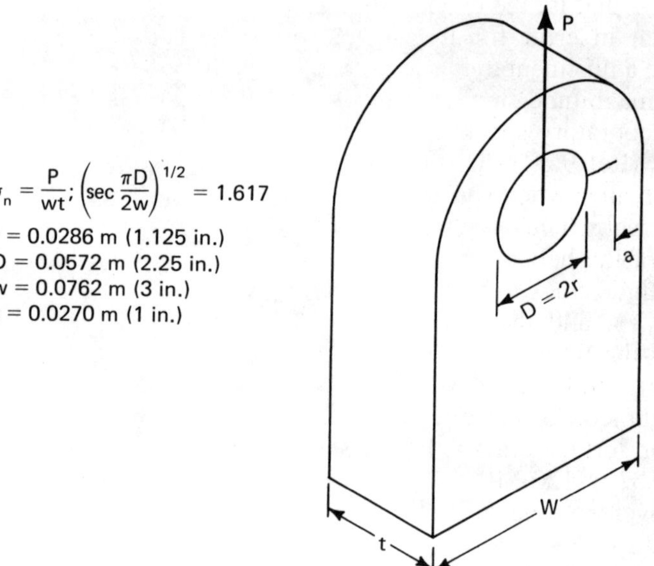

Fig. 9.35. Sketch of a cracked helicopter rotor lug

Table 9.3. Determination of Y from Newman's analysis (Eq 9.24 and 9.25) compared to Y from Eq 9.26 (Ref 9.34)

a/r	λ	f_b	f_w	G	Y	Y(Eq 9.26)
0	1	3.377	1.617	0.924	8.159	8.159
0.02	0.9615	3.066	1.659	0.916	7.534	7.501
0.04	0.9259	2.816	1.706	0.908	7.054	7.044
0.06	0.8929	2.613	1.758	0.900	6.685	6.699
0.08	0.8621	2.445	1.817	0.893	6.415	6.425
0.10	0.833	2.305	1.884	0.885	6.214	6.199
0.12	0.8065	2.187	1.961	0.878	6.089	6.009
0.16	0.7565	2.000	2.151	0.865	6.017	5.700

upon catastrophic failure occurred in the actual component. The convenient feature is that when Y is now used in the fatigue crack propagation equation, it is easily integrable—that is, combining Eq 9.23, 9.25 and 9.26 gives:

$$da/dN = C\Delta\sigma_n^4 \pi^2 a^2 \cdot \frac{4431}{[1 + 20(a/r)]} \qquad \text{(Eq 9.27)}$$

Separating the variables and integrating from an initial flaw, a_o, to a final flaw, a_f, gives:

$$N_f = \frac{2.3 \times 10^{-5}}{C\Delta\sigma_n{}^4}\left[\frac{1}{a_o} - \frac{1}{a_f} + \frac{20}{r}\ln\left(\frac{a_f}{a_o}\right)\right] \qquad \text{(Eq 9.28)}$$

Note that these equations assume that all cycles are involved in fatigue crack propagation—that is, the crack initiates on the first load cycle. This is probably not too much in error for fretting corrosion fatigue at relatively high cyclic stresses. One adjustment to this approach is that the value of C in Eq 9.28 is about an order of magnitude smaller than the value in Eq 9.23, for fatigue growth in precracked laboratory specimens. This is due to the observation of Schijve and Hoeymakers (Ref 9.34) that the growth rate at low ΔK was nearly an order of magnitude smaller when starting with fretting fatigue cracks than when starting with starter cracks. Thus, with $C_{eff} \simeq 1 \times 10^{-13}$ m/cycle (3.9×10^{-2} in./cycle), an analysis of the helicopter lug failure was made. By taking the initial fretting corrosion fatigue crack to be of grain or colony size ($a_o \approx 0.03$ mm, or 0.001 in., for Ti-6Al-4V), and the size of the final observed crack, a_f, to be 4.78 mm (0.19 in.) at the time of catastrophic failure, a calculation was derived as summarized in Fig. 9.36. This estimate should be conservative, because the analysis is for straight cracks through the thickness dimension of the lug rather than for corner or surface cracks. In fact, the calculation gives 4.1×10^4 cycles to failure for $\Delta\sigma_n = 120$ MPa (17 ksi) whereas the observed numbers of cycles at this stress level were between 3.2×10^4 and 13.6×10^4 cycles after the fatigue crack started.

Such an approach also may be applied to the observed fatigue lives of Ti-6Al-4V lug configurations evaluated by Salkind and Lucas (Ref 9.37). For various net stress levels, cyclic fatigue lives were predicted from Eq 9.28 keeping all other

Requirement of 3600 flight hours	Peak loads represent $\Delta\sigma \cong 120$ MPa (17 ksi) stresses in the lug.
Materials selection	Ti-6Al-4V solution treated and over-aged. $da/dN \cong C(\Delta K)^4$; $K_{Ic} = 70$ MPa · m$^{1/2}$ (63 ksi · in.$^{1/2}$) $C = 1 \times 10^{-12}$ MPa^{-4}-m^{-1}-cycle^{-1}
Design	Laboratory test full size components with spectrum loading.
Original analysis	Mission profile suggested between 3.2×10^4 and 13.6×10^4 cycles of fatigue cracking at peak loads. A mission life of only 2645 h was achieved, 73% of requirement.
Failure analysis	Fractography identifies fretting corrosion fatigue as the initiating mechanism.
New analysis	$N_f = \frac{2.3 \times 10^{-5}}{C\Delta\sigma_n^4}\left[\frac{1}{a_o} - \frac{1}{a_f} + \frac{20}{r}\ln\left(\frac{a_f}{a_o}\right)\right]$ $C_{eff} \cong 1 \times 10^{-13}$ for fretting-induced surface flaws. $a_o \cong 0.03$ mm (0.001 in); $r = 0.0286$ m (1.125 in.) $a_f = 4.8$ mm (0.19 in.); $\Delta\sigma_n = 120$ MPa (17 ksi) A predicted life of 4.1×10^4 cycles results.

Fig. 9.36. Helicopter lug fatigue case: a retrospective analysis (Ref 9.33)

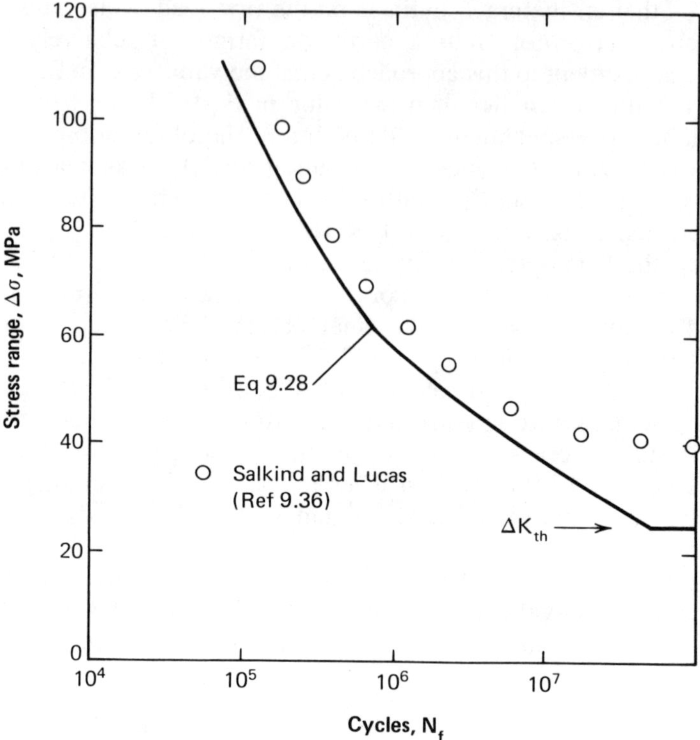

Fig. 9.37. Predicted and observed fatigue lives for Ti-6Al-4V helicopter lugs (Ref 9.36)

parameters the same. The stress range used in the predictions may not exactly correspond to that of the experiments, because the reported stresses are for prototype configurations (Ref 9.36). Thus, the close proximity of the predicted curve and experimental data in Fig. 9.37 may be fortuitous. What is important is that the predicted curve almost exactly matches the shape of the experimental cyclic fatigue curve in the range from 10^4 to 10^6 cycles, where fatigue crack propagation is controlling. At lower applied stresses, threshold considerations become more important and the simple power law approach breaks down.

It was clear to the investigators who studied helicopter lug problems that if fretting corrosion fatigue could be eliminated then the initiation stage could be delayed and an extended life could be achieved. Both Zola (Ref 9.34) and Salkind and Lucas (Ref 9.36) recommended sacrificial liners and/or fretting inhibitors. With regard to the former, Fig. 9.38, from Salkind and Lucas, shows that the fatigue limit at 10^6 cycles may be doubled by the use of bonded liners.

9.5.3. Case Studies Using K_{Iscc} Considerations

9.5.3.1. Hydrogen Embrittlement of Plated Fasteners

One of the most prevalent types of delayed subcritical crack failure occurs in high-strength fasteners where either corrosion, electrochemical corrosion or electroplating processes evolve hydrogen, which degrades the load-carrying capacity

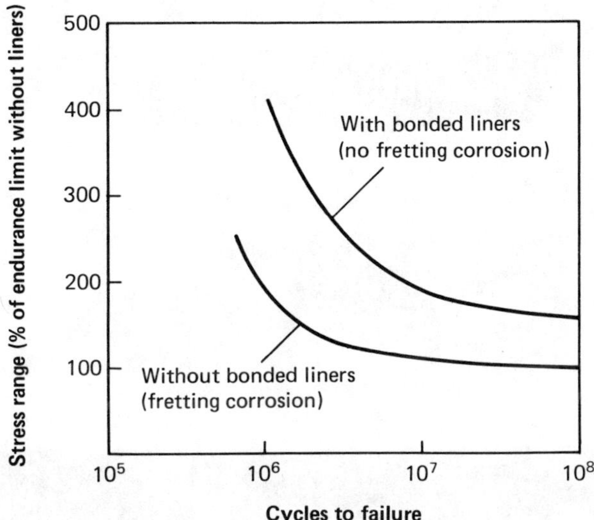

Fig. 9.38. Effect of liners on fatigue lives of helicopter lugs (Ref 9.36)

of the fastener. One such case occurred in a large construction project where workers were putting up fluorescent light fixtures with zinc-plated nails driven into concrete by a pneumatic gun. The resulting residual tensile stresses in the nails, along with the hydrogen content from the plating operation, represented a dangerous combination. The next day, as the workers were putting up additional fixtures, the previous day's work started crashing down about them. Analysis showed the delayed failure mechanism to be hydrogen embrittlement.

A well-documented case of the same type of phenomenon comes from Thornton and Colangelo (Ref 9.37), who analyzed a number of failed threaded fasteners from a 152-mm cannon. The screws, made of AISI 4037 steel and heat treated to 42 to 43 HRC, were supposed to be able to withstand design loads of 25,900 N (5820 lb) and had been rated at 32,500 N (7300 lb) in tension. For additional protection against corrosion, the fasteners were electroplated with cadmium. Based on a root diameter of about 5 mm (0.20 in.), the design load for the fasteners was about 80% of the ultimate strength level. However, a large number of premature failures occurred. Failures in two of these fasteners are shown in Fig. 9.39(a), and the metallographic section in Fig. 9.39(b) clearly demonstrates the damage near the plated surface. Furthermore, a scanning electron micrograph of the fracture surface, shown in Fig. 9.40, reveals intergranular facets typical of hydrogen embrittlement. Thornton and Colangelo surmised that hydrogen had entered the steel during the plating operation and, because of an insufficient or nonexistent baking operation, had not been removed.

A fracture mechanics calculation can show that premature failure would be expected if such fasteners containing hydrogen were used near design loads. By taking the applied stress intensity associated with a circumferential notch (Ref 9.38) and assuming that the thread roots were sufficiently sharp to represent a natural crack, the calculated stress intensity can be obtained from:

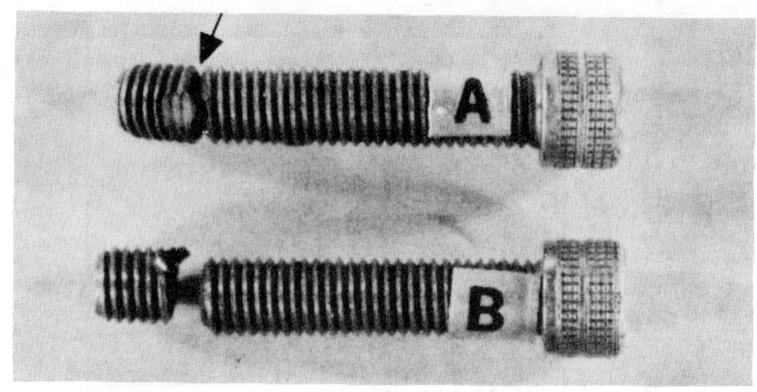

(a)

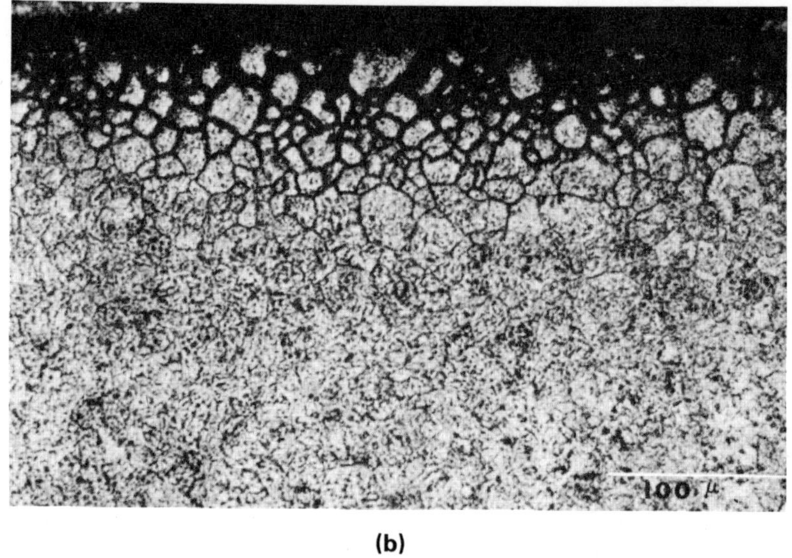

(b)

(a) Cracked fasteners. Arrow indicates crack. (b) Photomicrograph showing extensive grain boundary degradation at surface of shank portion of a fastener.

Fig. 9.39. Photograph and metallographic section of cadmium-plated steel fasteners (Ref 9.37)

$$K_I = (0.9)\,\frac{P}{D^{3/2}} \qquad \text{(Eq 9.29)}$$

where P is the applied load in meganewtons, D is the outer diameter in meters, and the 0.9 factor applies for the ratio of thread root diameter to outer diameter in this problem. For the applied design load, this gives an applied stress intensity of 45 MPa $\cdot$ m$^{1/2}$ (41 ksi $\cdot$ in.$^{1/2}$), which exceeds by far the stress intensity that a steel with a strength of 1500 MPa (215 ksi) can withstand when exposed to stress corrosion cracking or hydrogen embrittlement conditions for long periods of time (see Fig. 4.45). However, if the hydrogen could be removed, the fracture tough-

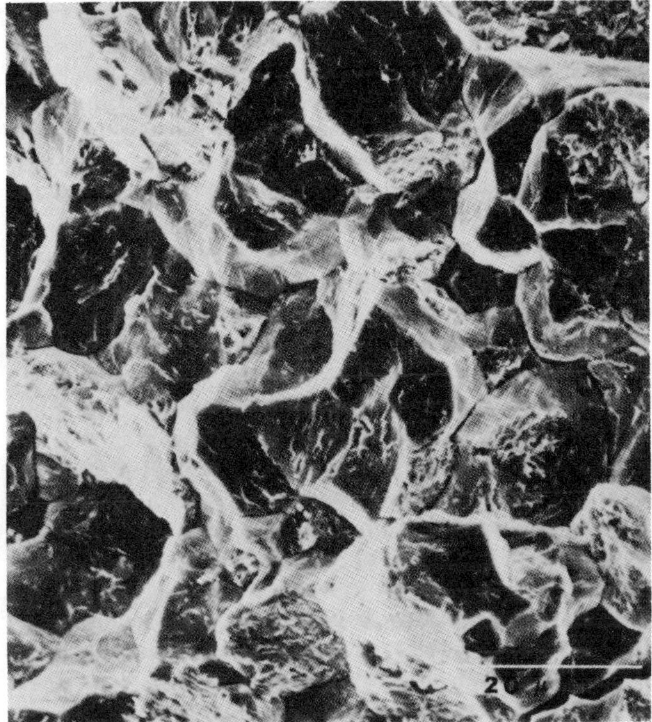

Fig. 9.40. SEM micrograph showing intergranular fracture and secondary cracking of cadmium-plated fasteners (Ref 9.37)

ness should be on the order of 70 MPa · $m^{1/2}$ (63.7 ksi · $in.^{1/2}$), which is enough to sustain working loads. Following this investigation, Thornton and Colangelo recommended a bake-out at 190 to 204°C (375 to 400°F) for 6 hours to remove hydrogen from all cadmium-plated fasteners. No failures were observed in the baked-out fasteners. This case is summarized in Fig. 9.41.

9.5.3.2. Corrosion-Related Failures in Petroleum Processing

Because of their high resistance to corrosion, titanium and titanium alloys are being employed to an increasing extent in the petrochemical industry (Ref 9.39). A U-bend bundle of titanium tubing for a heat exchanger is shown in Fig. 9.42. Although considerable success has been achieved with titanium in this type of application, problems have been encountered in situations where the TiO_2 film has been disrupted, resulting in crevice or pitting corrosion. In addition, under severe conditions, hydrogen cracking may result. Examples of three such failures are shown in Fig. 9.43, where crevice corrosion, pitting and cracking can be observed. The pitting (Fig. 9.43b) was traced to an iron scratch in the surface, which disrupted the TiO_2 film and established an electrochemical cell.

The cracking problem, shown in Fig. 9.43(c), was identified in a titanium tube for a heat exchanger, the requirements of which are given in Fig. 9.44 along with a summary of the analysis. The premature failure was associated with titanium hydrides produced by a galvanic couple between Fe and Ti and with the fact that the H_2S in the solution would increase the corrosion current. Hydrogen diffusion

Requirements	25,900-N (5820-lb) design load on fasteners; Steel fasteners electroplated for corrosion protection.
Materials selection	AISI 4037 steel heat treated to 42–43 HRC ≈ 1500 MPa (215 ksi) ultimate strength.
Design	Cyclindrical-head cap screws with 1/4" - 28 UNF threads manufactured to Mil. Spec. MS 21262 (AGS) and plated per QQ-P-416.
Premature failure	Many fasteners failed even though rated at 32,500 N in tension.
Fractography	Intergranular fracture (Fig. 9.40) was characteristic of hydrogen embrittlement.
Failure analysis	$K_{applied} \cong (0.9)\dfrac{P}{D^{3/2}} \cong 45$ MPa · m$^{1/2}$ (41 ksi · in.$^{1/2}$) $K_{applied} \geq K_{th}$ (Fig. 4.45) for hydrogen embrittlement of 1500 MPa (215 ksi) steel
Recommendation and result	All Cd-plated fasteners were baked for 4–6 h at 190–204°C (375–400°F) with no recurrence of cracking.

Fig. 9.41. High-strength fasteners from a 152-mm cannon: an error in electroplating (Ref 9.37)

in the temperature range from 20 to 110°C (68 to 230°F) was considered sufficiently rapid to produce the hydrides that are visible in the photomicrograph in Fig. 9.45. In an attempt to reconstruct the failure process, the time to failure was considered in terms of a "stress corrosion" cracking mechanism provided by hydrides. Previous data (Ref 9.40) on a commercially pure Ti75A material had shown that cracking in 3.5% NaCl solutions could be represented by:

$$\frac{da}{dt} \simeq \alpha K^n \qquad \text{(Eq 9.30)}$$

where $\alpha = 2.4 \times 10^{-16}$ (m/s units) and $n = 5.4$. To determine the stresses in the tube, calculations based on internal pressures and temperature gradients were made. It was reported (Ref 9.39) that there were no residual stresses. The important stress is the thermal stress, as given by:

$$\sigma \simeq E_{Ti}\,(\overline{\alpha}_{Fe}\Delta T_{Fe} - \overline{\alpha}_{Ti}\Delta T_{Ti}) \simeq 124 \text{ MPa (18 ksi)} \qquad \text{(Eq 9.31)}$$

where the $\overline{\alpha}$ values are the average coefficients of thermal expansion and the ΔT values represent the temperature changes at the inlet side. This thermal stress value is slightly less than the minimum value for fracture of TiH_2, which had been determined by Irving and Beevers (Ref 9.41) to be 170 MPa (24.6 ksi). Because other stresses from concentration gradients, or some slight residual stresses, could

Fig. 9.42. U-bend bundle of titanium tubing for heat exchanger in petroleum refining plant (Ref 9.39)

have been present, this minimum value will be used—that is:

$$\sigma \geq 170 \text{ MPa (24.6 ksi)} \qquad \text{(Eq 9.32)}$$

The discontinuous cracking of large grains in Fig. 9.45 suggests that hydrides of grain-diameter length can initiate cracking, so that a reasonable value of a_o is a large grain size of about 135 μm (0.005 in.). Considering the minimum stress of 170 MPa and an opening mode stress intensity, then:

$$K_{th} \simeq 1.1\sigma\,(\pi a_o)^{1/2} \simeq 3.8 \text{ MPa} \cdot \text{m}^{1/2} \text{ (3.5 ksi} \cdot \text{in.}^{1/2}) \qquad \text{(Eq 9.33)}$$

This value is close to the threshold value observed by Simpson and Puls (Ref 9.42) for zirconium hydride embrittlement of Zr-2.5Nb. To determine the final crack size, the minimum stress of 170 MPa (24.6 ksi) is combined with the fracture toughness of 135 MPa · $m^{1/2}$ (123 ksi · in.$^{1/2}$), from Ref 9.40, to give:

$$a_f \simeq \left(\frac{K_c}{\sigma}\right)^2 \frac{1}{\pi} \simeq 0.20 \text{ m (8 in.)} \qquad \text{(Eq 9.34)}$$

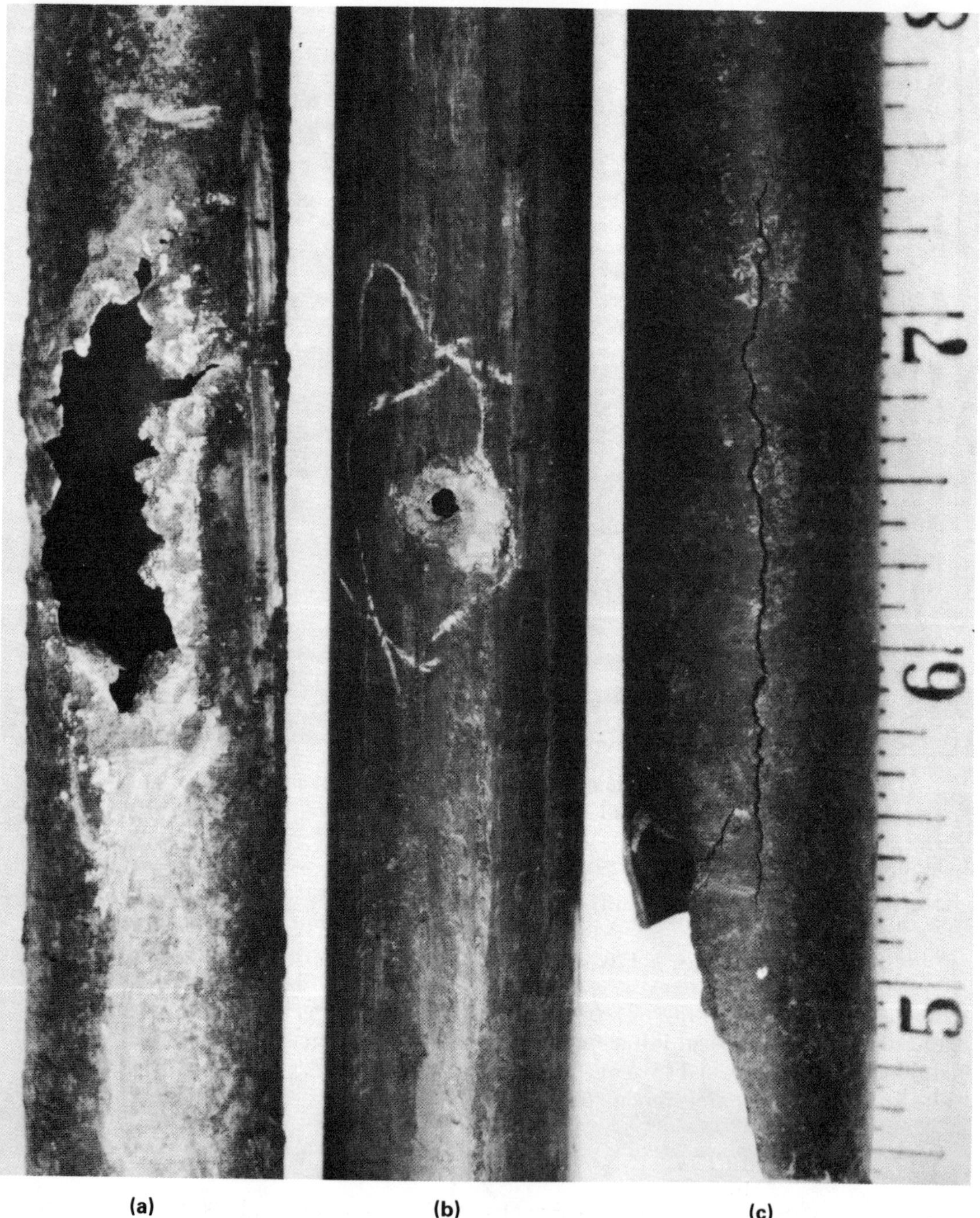

(a) (b) (c)

Fig. 9.43. Examples of (a) crevice corrosion, (b) pitting and (c) stress corrosion cracking in titanium tubing (Ref 9.39)

Requirements	Inlet: H_2O at 20°C (68°F) for cooling tube; stream at 110°C (230°F) at shell. Outlet: H_2O at 40°C (104°F); stream at 45°C (113°F).
Materials selection	Titanium tubing with a carbon steel shell.
Design	With 2% moisture, autoclave tests show TiO_2 resists hydrogen absorption. Use welded construction with 1.8-mm (0.070-in.) wall, 19-mm (0.75-in.) diameter tubing.
Premature failure	One tube cracked in 2½ years. A crack 0.61 m (24 in.) long was found parallel to the weld in the heat-affected zone.
Metallography	Titanium hydrides were found near the outside diameter of the tubes with grain length discontinuous cracks.
Failure analysis	Thermal stresses: $\sigma \cong E_{Ti}(\bar{\alpha}_{Fe}\Delta T_{Fe} - \bar{\alpha}_{Ti}\Delta T_{Ti})$ $\cong$ 124 MPa (18 ksi) $t_f \cong \frac{2}{(n-2)\alpha(1.1\sigma)^n\pi^{n/2}}\left[\frac{1}{a_o^{(\frac{n}{2}-1)}} - \frac{1}{a_f^{(\frac{n}{2}-1)}}\right]$ $n = 5.4;\ \alpha = 2.4 \times 10^{-16}$ (m/s units) $\sigma \geqslant 170$ MPa (24.6 ksi); $a_o \cong 135\ \mu$m (0.005 in.); $a_f \cong 0.3$ m (12 in.) $t_f \cong$ 7 years
Recommendation	Eliminate the Fe-Ti couple.

Fig. 9.44. Titanium tube from a heat exchanger (Ref 9.39)

The observed half-crack length of the longitudinal through-crack in the tube was 0.3 m (12 in.). This is sufficiently close to the calculated value to permit further use of this stress with some degree of confidence. Combining Eq 9.30 and 9.33 and integrating gives:

$$\int_0^{t_f} \simeq \frac{1}{\alpha(1.1\sigma)^n\pi^{n/2}}\int_{a_o}^{a_f} a^{-n/2}\,da \qquad \text{(Eq 9.35)}$$

which reduces to:

$$t_f \simeq \frac{2}{(n-2)\alpha(1.1\sigma)^n\pi^{n/2}}\left[\frac{1}{a_o^{(n/2-1)}} - \frac{1}{a_f^{(n/2-1)}}\right] \qquad \text{(Eq 9.36)}$$

The approximation in the expressions above is due to the direct extrapolation of Eq 9.30 to threshold. As can be seen in Fig. 9.44, this analysis gives a calculated value of 7 years for time to failure rather than the 2½ years observed. However, in this type of calculation, it is unrealistic to expect better agreement. In the first place, although the laboratory test was performed in a 3.5% NaCl solution, the

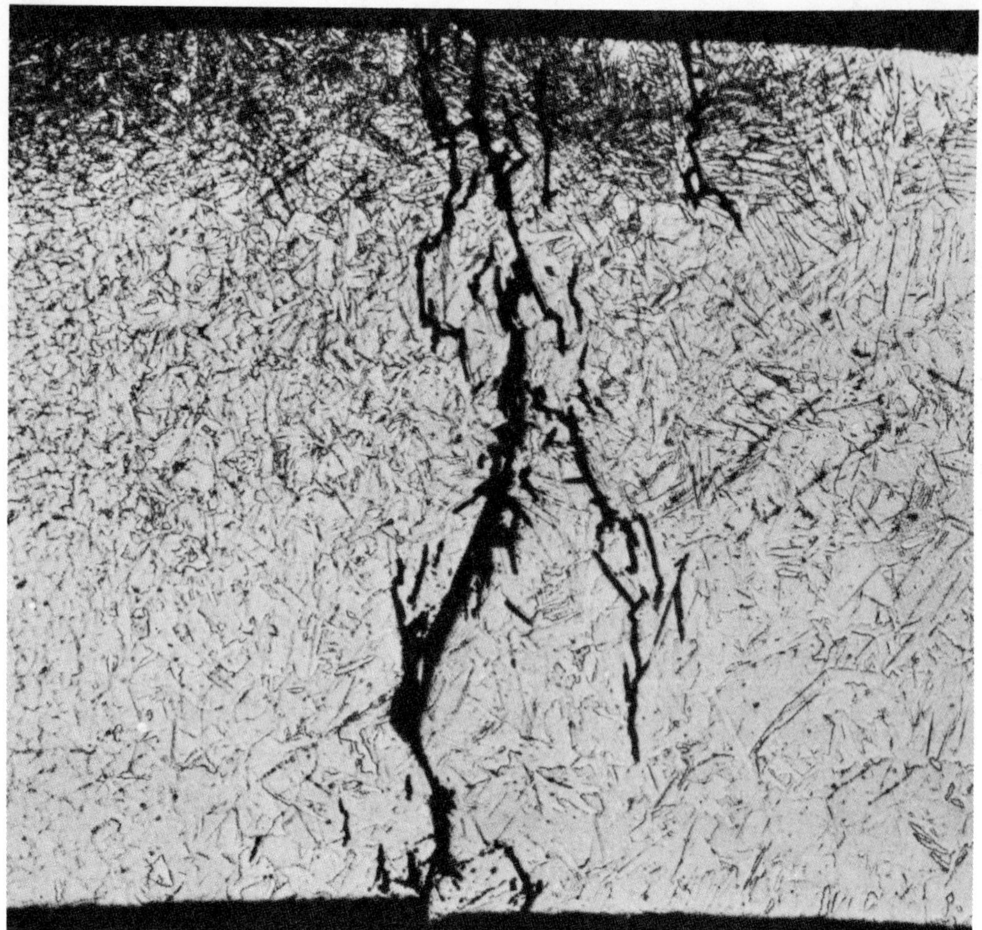

The crack apparently was caused by galvanic action between the tube and the heat exchanger's shell of carbon steel. Note heavy concentration of titanium hydrides at top (OD of tube).

Fig. 9.45. Section of titanium tube from a heat exchanger (Ref 9.39)

service failure environment contained H_2S, which is known to be a greater promoter of cathodic reactions. Secondly, there was a galvanic cell between Fe and Ti and it is known that greater cathodic overpotentials will increase crack velocities if the hydride mechanism predominates. Thus, such a failure might have been anticipated and avoided by eliminating the Fe-Ti couple, which not only produced thermal stresses but also exacerbated the hydrogen cracking problem.

For an addendum to this chapter, the reader should refer to Ref 9.43, which is an overview article on material problems in the hydrocarbon processing industries and which cites several case histories on hydrogen embrittlement and temper embrittlement. Although little work has been done on combined effects of hydrogen and temper embrittlement, at least one review paper (Ref 9.44) indicates that substantial progress has been made, primarily by McMahon and coworkers (Ref 9.45 and 9.46), in analyzing these effects.

9.6. REFERENCES

9.1. A Survey of Fracture Mechanics Applications in the United States, by T. P. Rich, P. G. Tracy and D. J. Cartwright: AMMRC MS 76-1, Army Materials and Mechanics Research Center, Watertown, MA, Feb 1976

9.2. *Deformation and Fracture Mechanics of Engineering Materials,* by R. W. Hertzberg: John Wiley and Sons, New York, 1976, p 270

9.3. The Design and Failure Analysis of a Cargo Containment System for a Liquid Natural Gas Ship, by D. J. Burns, R. J. Pick and T. Brown: in *Fracture 1977,* Vol 1, ICF4, Waterloo, Canada, 1977, p 173-191

9.4. Significance of Defects in Liquified Natural Gas Tanks in Ships, by P. Tenge, O. Solli and O. Forli: STP 579, American Society for Testing and Materials, Philadelphia, 1975, p 10-41

9.5. Plane Strain Crack Toughness Testing of High Strength Metallic Materials, by W. F. Brown, Jr., and J. E. Srawley: STP 410, American Society for Testing and Materials, Philadelphia, 1966

9.6. Fracture Toughness Approach to Steel Castings Quality Assurance, by W. J. Jackson and J. C. Wright: *Metals Technology,* Vol 4, Part 9, Sept 1977, p 425-433

9.7. Reliability Analysis of Transport Aircraft, by J. P. Butler: paper presented at Conference on Structural Safety and Reliability, 1969

9.8. Estimating Fatigue Crack Lives for Aircraft: Techniques, by J. P. Gallagher: *Experimental Mechanics,* Vol 16, No. 11, Nov 1976, p 425-433

9.9. Spectrum Loading and Crack Growth, by D. E. Wheeler: *Journal of Basic Engineering, Transactions of ASME,* Series D, Vol 94, No. 1, March 1972, p 181-186

9.10. J. Willenborg, R. M. Engle and H. A. Wood: Technical Memorandum 71-1-FBR, 1971

9.11. H. A. Wood: Private communication, AFFDL, Wright-Patterson Air Force Base, Dayton, OH, 1972

9.12. Corrosion Fatigue in Fe-Ni-Co Alloys, by M. D. Speidel: in *Stress Corrosion Cracking and Hydrogen Embrittlement of Iron Base Alloys,* National Association of Corrosion Engineers, Vilieux, Firminy, France, June 1973, p 1071-1094

9.13. On the Superposition Model for Environmentally-Assisted Fatigue Crack Propagation, by W. W. Gerberich, J. P. Birat and V. F. Zackay: in *Corrosion Fatigue,* NACE-2, National Association of Corrosion Engineers, Houston, 1972, p 396-408

9.14. SCC in High Strength Aluminum Alloys, by M. V. Hyatt and M. D. Speidel: in *SCC in High Strength Steels and in Titanium and Aluminum Alloys,* edited by B. F. Brown, Naval Research Laboratory, 1972, p 147-244

9.15. Influence of Environment on Crack Propagation Characteristics of High Strength Aluminum Alloys, by J. H. Mulherin: STP 425, American Society for Testing and Materials, Philadelphia, 1967, p 66

9.16. Correlation Between Sustained-Load and Fatigue Crack Growth in High-Strength Steels, by R. P. Wei and J. D. Landes: *Materials Research and Standards,* Vol 9, No. 7, July 1969, p 25-27

9.17. Hydrogen-Assisted Fatigue Crack Growth in 2¼Cr-1Mo Low Strength Steels, by S. Suresh, C. M. Moss and R. O. Ritchie: Proceedings of the 2nd International Symposium on Hydrogen, Japan Institute of Metals, Minikami Spa, Japan, Nov 1979

9.18. Effects of the Environment on Stable Crack Growth in High-Strength Steels Under Static and Cyclic Loading, by L. Hyspecka, M. Tvrdy and K. Mazanec: in *Advances in Fracture Research,* Vol 2, edited by D. Francois, 5th International Conference on Fracture, Cannes, April 1981, p 973-980

9.19. Rapid Inexpensive Tests for Determining Fracture Toughness: Report of Committee, National Materials Advisory Board, National Academy of Sciences, Washington, DC, 1976

9.20. *Failure Analysis and Prevention:* Vol 10 of *ASM Metals Handbook* (8th Ed.), American Society for Metals, Metals Park, OH, 1975
9.21. Investigation of Hydrotest Failure of Thiokol Chemical Corporation 260-Inch-Diameter S1-1 Motor Case, by J. E. Srawley and J. B. Esgar: NASA TM X-1194, NASA Lewis Laboratories, Jan 1966
9.22. *Fractography and Atlas of Fractographs:* Vol 9 of *ASM Metals Handbook* (8th Ed.), American Society for Metals, Metals Park, OH, 1974
9.23. Corrosion Failure Analysis, by W. G. Dobson: *Metal Progress,* Vol 116, No. 3, Aug 1979, p 57-62
9.24. An Auger Electron Spectroscopic Study of Phosphorous Segregation in the Grain Boundaries of Nickel Base Alloy 600, by M. Guttmann *et al: Corrosion,* Vol 37, No. 7, July 1981, p 416-425
9.25. Explosion of an Ammonia Pressure Vessel, by A. R. Rosenfield and C. N. Reid: in *Case Studies in Fracture Mechanics,* edited by T. P. Rich and D. J. Cartwright, AMMRC MS 77-5, Army Materials and Mechanics Research Center, Watertown, MA, June 1977, p 3.5.1-3.5.12
9.26. Metallurgical Evaluation of Proof Test Failure of a Second Stage Missile Motor Case, by E. B. Kula and A. A. Anctil: in *Case Studies in Fracture Mechanics,* edited by T. P. Rich and D. J. Cartwright, AMMRC MS 77-5, Army Materials and Mechanics Research Center, Watertown, MA, June 1977, p 1.7.1-1.7.14
9.27. Recent Work on Flaw Behavior in Pressure Vessels, by A. R. Duffy, R. J. Eiber and W. A. Maxey: in *Practical Fracture Mechanics for Structural Steel,* edited by M. O. Dobson, Chapman and Hall Ltd., London, 1969, p M1-M34
9.28. Failure Criteria for Through-Cracked Vessels, by G. T. Hahn and M. Sarrate: in *Practical Fracture Mechanics for Structural Steel,* edited by M. O. Dobson, Chapman and Hall Ltd., London, 1969, p P1-P15
9.29. Yielding of Steel Sheets Containing Slits, by D. S. Dugdale: *Journal of Mechanics and Physics of Solids,* Vol 8, No. 2, 1960, p 100-104
9.30. An Axial Crack in a Pressurized Cylindrical Shell, by E. S. Folias: *International Journal of Fracture Mechanics,* Vol 1, 1965, p 104-113
9.31. Steel Products for Rollover Protective Structures (ROPS) and Falling Object Protective Structures (FOPS): SAE-ISTC Ad Hoc Committee, Dec 7, 1978
9.32. Avoidance of Fatigue Crack Growth in a Ship's Hydraulic Equipment, by L. P. Pook: in *Case Studies in Fracture Mechanics,* edited by T. P. Rich and D. J. Cartwright, AMMRC MS 77-5, Army Materials and Mechanics Research Center, Watertown, MA, June 1977, p 4.3.1-4.3.4
9.33. Analysis of Heavy Lift Helicopter (HLH) Rotor Hub Lower Lug, by J. C. Zola: in *Case Studies in Fracture Mechanics,* edited by T. P. Rich and D. J. Cartwright, AMMRC MS 77-5, Army Materials and Mechanics Research Center, Watertown, MA, June 1977, p 2.2.1-2.2.15
9.34. Fatigue Crack Growth in Lugs, by J. Schijve and A. H. W. Hoeymakers: *Fatigue of Engineering Materials and Structures,* Vol 1, 1979, p 185-201
9.35. Predicting Failure of Specimens with Either Surface Cracks or Corner Cracks at Holes, by J. C. Newman, Jr.: NASA TN D-8244, June 1976
9.36. Fretting Fatigue in Titanium Helicopter Components, by M. J. Salkind and J. J. Lucas: in *Corrosion Fatigue,* NACE-2, National Association of Corrosion Engineers, Houston, 1972, p 627-630
9.37. Environmentally Assisted Failures in Ordnance Components, by P. A. Thornton and V. J. Colangelo: in *Risk and Failure Analysis for Improved Performance and Reliability,* edited by J. J. Burke and V. Weiss, Plenum Press, New York, 1980, p 203-224

9.38. *The Stress Analysis of Cracks Handbook,* by H. Tada, P. C. Paris and G. R. Irwin: Del Research Corp., Hellertown, PA, 1973, p 27.1
9.39. Titanium Solves Corrosion Problems in Petroleum Processing, by L. C. Covington: *Metal Progress,* Vol 111, No. 2, Feb 1977, p 38-45
9.40. Micromechanical Approach to Stress Corrosion Cracking in Titanium Alloys, by Y. Katz: Ph.D. Thesis, University of California, Berkeley, Sept 1969
9.41. Some Observations on the Deformation Characteristics of Titanium Hydride, by P. E. Irving and C. J. Beevers: *Journal of Materials Science,* Vol 7, No. 1, Jan 1972, p 23-30
9.42. The Effects of Stress, Temperature and Hydrogen Content on Hydride-Induced Crack Growth in Zr-2.5% Nb, by L. A. Simpson and M. P. Puls: *Metallurgical Transactions A,* Vol 10A, No. 8, Aug 1979, p 1093-1105
9.43. Material Problems in the Hydrocarbon Processing Industries, by G. R. Prescott: *Metal Progress,* Vol 120, No. 2, July 1981, p 24-30
9.44. Recent Studies of Grain Size and Hydrogen Concentration Effects on Hydrogen Embrittlement in BCC Metals, by W. W. Gerberich and A. G. Wright: 2nd Conference on Environmental Degradation, VPI, Blacksburg, VA, Sept 1981, p 183-206
9.45. Embrittlement of a 5 Pct. Nickel High Strength Steel by Impurities and Their Effects on Hydrogen-Induced Cracking, by C. L. Briant, H. C. Feng and C. J. McMahon, Jr.: *Metallurgical Transactions A,* Vol 9A, No. 5, May 1978, p 625-633
9.46. Solute Segregation and Intergranular Brittle Fracture in Steels, by M. L. Jokl, J. Kameda, C. J. McMahon, Jr., and V. Vitek: *Metal Science,* Vol 14, No. 8-9, Aug-Sept 1980, p 375-384

Index

Alloy steels. *See* Carbon and alloy steels
Alpha stabilizers, description of, 214
Aluminum alloy 7075-T6, sustained-load tests on, 19, 20
Aluminum alloys, fatigue crack propagation, 189-196
composition, microstructure and thermal temperature, effects of, 190-192
alloy 7075, dispersoid type, 191, 192
alloy 7475, high and intermediate stress intensity, 190
particle sizes, comparison of, 191
exposure, temperature and humidity, effects of, 192-193, 195, 196
fatigue crack growth, summary of, 189
load ratio, effect of, 195-196, 198
load-time history, effect of, 193-195, 197
product form and orientation, effects of, 192, 193, 194
fatigue crack growth rates, 192, 194
orientation, effect of, 192, 193
Aluminum alloys, fracture properties of, 169-208
fatigue crack propagation, 189-196
fracture toughness, 171-189
introduction, 169
sustained-load crack growth, 196-208
Aluminum alloys, fracture toughness, 171-189
critical stress intensity factor measurements, 171, 175
mechanical processing and orientation, effects of, 184-187
orientation, 184-185, 186
product form, 185
product size, 185-187
microstructure, composition and heat treatment, effects of, 173-184, 185
2000 and 7000 alloy series, composition of, 177-178, 180, 182
2024, 2124, 7075, and 7475, fracture toughness comparisons, 175, 180
2024-T851, cracked particles in, 175, 179
2024-T851, fractograph of, 175, 179
7075 and 7175, plane strain fracture toughness of, 177-178, 182
ancillary element additions, effect of, 176, 181
crack growth resistance curves, 184
crack length, gross section stress vs., 184, 185
crack propagation, effect of grain size, 182, 183
fracture modes, 174
iron content, effect of, 175, 180
tear strength-yield strength, effect of quench water temperature, 182, 183
toughness-yield strength, effects of solute content and over-aging, 179-180, 182
plane strain fracture toughness, 171-177
crack length, influence of, 171, 173
measurements of, 171, 174
specimen thickness, influence of, 171, 172
unit propagation energy, relationship between, 173, 177
values for, 172, 176
temperature, effect of, 187-189
plane strain fracture toughness vs., 187, 188
toughness, variation with yield strength, 173, 178
Aluminum alloys, high-strength
ASTM Method E338, 33
ASTM Method E602, 33
crack-mouth-opening displacement, load vs., 25, 26
Aluminum alloys, major systems, 169-171
Aluminum alloys, sustained-load crack growth, 196-208
composition, microstructure and thermal processing, effects of, 200-204
individual elements, effects of, 201
strength and stress-corrosion resistance, variations in, 202, 203
stress-corrosion ratings, 204, 205-206
tensile yield strength, stress-corrosion cracking threshold stress vs., 202, 203
product form and orientation, effects of, 204
stress and stress intensity threshold characterizations, 199, 200, 202
stress-corrosion crack velocity curves, 196, 201, 202
stress-corrosion cracking, effect of humidity and stress intensity, 196, 199
stress-corrosion cracking, effect of outdoor exposure and stress intensity, 196, 200
temperature and environment, effects of, 204-208
creep crack growth, alloy 2219-T851, 204, 207
Aluminum alloys, types
2024/7075, 175, 177, 180
2024-T851, 179, 190-191
2124, 177-178, 185
2124/7475, 175, 180

Aluminum alloys, types *(continued)*
  2124-T851, 190
  2219-T851, 189
  5083-0, 193
  7050, 177
  7075, 184, 190, 192
  7075-T6, 184, 195
  7075-T651, 328, 329
  7075-T7651, 36
  7475, 177-178, 184, 185, 190
  7475-T761, 184
  Alclad 7075-TC, 193
  Ti-6Al-4V, 350, 353, 354
Applied stress, 5, 6, 8, 11
Assumptions, conceptual, 9-14
  fracture toughness, 12-14
  material flaws and nondestructive inspection, 9-10
  stress intensity factor, 10-12
ASTM Methods
  B645, 171
  E23, 32
  E338, 33
  E399, 13, 23-27, 171, 185, 346
    crack-mouth-opening displacement, 25
    fatigue precracking process, 25
    fracture specimens, orientation of, 23, 25
    high-strength aluminum alloys, 25, 26
    requirements, 23-25
    specimen types usable, 23, 24
  E561, 29, 30, 171
  E602, 33
  E604, 32-33
    compared to other tests, 33
  E647, 19, 34-35, 189
Auger spectroscopy technique, 55, 335
Austenite matrix phase, 255
Austenitic stainless steels, composition of, 106-107, 108
  Kromarc 58, 107, 108
  Type 304, 107, 108
  Type 310S, 106, 108
  Type 316, 107, 108
Austenitic stainless steels, fracture properties of, 112-142
  alloys 21-6-9 and 22-13-5, 137-140
    annealed 21-6-9, fatigue crack growth rates in, 137-138, 140
    gas tungsten arc welding process, 139, 141
    Kromarc 58, 140-141, 142
    Pyromet 538 weld metal, fatigue crack growth rates, 139, 141
    shielded metal arc welding process, 139, 141
  summary, 141-142
  Type 301, 112-114
    effective stress intensity factor, 113
    fatigue crack growth rates, 112, 113
    fatigue crack growth rates, ½-hard type steel, 113, 114
  Types 304 and 304L, 114-124
    aging, effect on fatigue crack growth rates, 119, 121
    base metal and welds, Type 304LN, 123-124
    cyclic variation, effect of, 116
    fatigue crack growth rates, annealed and cold worked, 119, 120
    fatigue crack growth rates, cryogenic temperatures, 123
    fatigue crack growth rates over time, 117, 118
    fatigue crack growth rates, room and elevated temperatures, 115-123
    fatigue crack growth rates, scatter band of, 116, 117
    humidity, effect on fatigue crack growth rates, 119, 122
    testing temperature, effect of, 115, 115-116
    Type 304, Type 308 welds in, 124, 125
  Types 309S and 310S, 124-127
    fatigue crack growth rates, 309S, 125
    fatigue crack growth rates, 310S, 126, 127
    fracture toughness data, 310S, 126-127
  Types 316 and 316N, 127-137
    air exposure, effect on fatigue crack growth rates, 128, 130
    cold worked 316, fatigue crack growth rates of, 129-130, 132
    cyclic crack growth data, 136
    cyclic frequency variation, effect on fatigue crack growth rate, 128, 131
    gas environments, effect on fatigue crack growth rates, 134-136
    holding times, effect on fatigue crack growth rates, 130-131, 133
    static crack growth data, 136
    temperature, effect on fatigue crack growth rate, 128, 129
    Type 316 base and weld metal, fatigue crack growth rates, 131-132, 134
  Types 321 and 348, 135, 137, 138, 139
    fatigue crack growth rates, room temperature, 137, 138
    fatigue crack growth rates, 321 aged and unaged, 137, 139

Basic concepts, fracture mechanics, 5-20
  additional applications and limitations, 16-20
  assumptions, 9-14
  basic quantities, 5
  energy-release rate, 5-9
  plastic zone, 14-16
Bend specimens
  ASTM Method E604, use in, 32-33
  crack-opening displacement testing methods, use in, 32
Beta stabilizers, description of, 214

Calcium treatment, 73
Cantilever specimens
  martensitic stainless steels, Type 420, 148
Carbides, superalloy, 257-262
Carbon and alloy steels, fatigue crack propagation, 66-86
  environments, aggressive, 78-86
    corrosion fatigue behavior, types of, 78-79, 84
    corrosion fatigue crack growth rates, effect of composition, 79, 87
    corrosion fatigue crack growth, selected data, 81-82, 88
    corrosion fatigue, effect of cyclic frequency, 79, 85, 86
    stress intensity factor rate, fatigue crack growth rate vs., 83, 89
  fracture modes, microstructural, 66-68

crack growth, graphs of, 66
fatigue failure modes, microstructural, 67
microstructure and heat treatment, effects of, 68-73
ASTM Method A533B, fatigue crack growth rates for, 73, 76
fatigue crack growth rates, effect of strength level, 68, 69
fatigue crack growth rates, upper limits of, 68, 70
stress intensity factor range, crack growth rate vs., 68, 71
stress intensity factor range, crack propagation rate vs., 68, 72
threshold, effect of grain size, 68-71, 73
threshold stress intensity range, effect of yield strength, 71-72, 74
threshold stress, variation with crack size, 73,75
variables, 73-78
fatigue crack growth rates, effect of elevated temperatures, 77, 82
fatigue crack growth rates, effect of low temperatures, 77, 83
fatigue crack growth rate testing, load sequences for, 77, 80, 81
fatigue crack propagation, influence of temperature, 77-78, 84
fatigue testing, nomenclature used in, 73-74, 77
fatigue threshold stress intensity factor range, effect of stress ratio, 74, 78, 79
Carbon and alloy steels, fracture properties of, 41-99
fatigue crack propagation, 66-86
fracture toughness, 46-66
introduction, 41-42
metallurgy of, 42-46
sustained-load crack propagation, 86-96
Carbon and alloy steels, fracture toughness, 46-66
alloy chemistry, effects of, 49-56
carbon content and testing temperature, effect of, 51, 52
fracture toughness and yield strength, relationship between, 49, 50
nickel content, effect of, 50-51
sulfide inclusions, effect on ferrite steel toughness, 53-55
sulfur content, effect on fracture toughness, 52, 53
transition temperature shift, grain boundary concentration vs., 55
fracture modes, microstructural, 47-49
cleavage, 47-49
dimpled rupture, 48, 49
intergranular fracture, 48, 49
microstructure and heat treatment, effects of, 56-62
fracture toughness, effect of carbon content and tempering temperature, 56, 58
fracture toughness, effect of grain size, 56, 59
fracture toughness, effect of particles, 60-61
fracture toughness, relationship to inverse square root of grain size, 56-60
fracture toughness, variation with tempering temperature, 56, 57
variables, 62-66
fracture toughness, effect of temperature, 62, 63
fracture toughness, effect of temperature and strain rate, 62, 64
fracture toughness, influence of temperature and loading rate, 63, 65
transition temperature shift, effect of yield strength, 62-63, 64
Carbon and alloy steels, nominal compositions of, 43-44
Carbon and alloy steels, sustained-load crack propagation, 88-99
alloy chemistry, effects of, 88-92
static stress-corrosion threshold, effect of alloying element parameter, 89, 91
static stress-corrosion threshold, effect of molybdenum content, 90, 92
fracture modes, microstructural, 88
microstructure and heat treatment, effects of, 92-93
aging temperature, effects of, 92, 95
crack resistance, influence on, 92-93, 96
stress-corrosion cracking, effects of tempering, 92, 94
tensile strength and stress intensity factor, correlation between, 92, 93
temperature and environment, effects of, 93-99
crack growth rate, dependence of, 93-95, 97
crack growth rate, effect of inhibitors, 96, 97
stress-corrosion cracking, observation of, 95, 98
Center-cracked specimens
ASTM Method E647, use in, 34
Charpy impact test (*See also* ASTM Method E23), 13, 311
Charpy specimens
illustrations of, 32
tests, use in, 34
Charpy V-notch impact energy, 345-347
Compact specimens
ASTM Method E647, use in, 34
martensitic stainless steels, Type 403, 147
martensitic stainless steels, Type 431, 149
sustained-load crack growth testing, use in, 36, 37
Constant amplitude loading. *See* Loading
Conventional melting, 73
Corrosion fatigue, 328-333
Crack deformation, modes of, 11-12
Crack-face displacement. *See* Crack deformation; Stress intensity factor
Crack growth
elastic-plastic, 16-17
fatigue, 17-19
shear-lip type, 16
stress-corrosion, aluminum alloy 7075-T6, 19, 20
testing environment, effects of, 19-20
Crack length, 5, 8, 11
Crack-mouth-opening displacement, 32, 36
Crack-opening displacement test method, 32
Crack-tip-opening displacement, 52
Critical plane strain energy release rate, 5-9
Cryogenic temperature
effect on fatigue crack growth rate, Types 301 and 304 stainless steels, 123

Cyclic loading. *See* Loading
Cyclic loading tests
  Types 316 and 316N stainless steels, 136

Design, fracture toughness approach, 313-318, 319
  design/performance model, 317-318, 319
  pressure vessel leak-before-burst concept, 314-315
  vessels, overpressurization of, 315-317
    cracks, flaw shape parameter curves, 316
    scheme, example of, 317
Design, fracture mechanics of
  application, survey of, 311, 312
  application, trend in, 311, 313
Design, material selection and failure analysis, 311-362
  failure analysis, 334-335
  fracture toughness approach, 313-318, 319
  introduction, 311-313, 314
  materials selection, 335-362
  quality control, 333-334
  subcritical crack growth approach, 318-333
Design, quality control, 333-334
  exit quality, 334
  incoming quality, 333-334
  in-plant control, 334
Design, subcritical crack growth approach, 318-333
  cyclic life calculation, 319
  fatigue methodology, 320-328
    crack growth curve, 322, 324-328
    fatigue crack growth, illustration of, 320, 322
    fatigue crack growth retardation, Wheeler model for, 326, 327
    fatigue life, prediction of, 325-326
    fatigue performance, elements of, 321, 322
    initial flaw size, 323
    laboratory test data, comparison of, 327-328
  general life calculations, 319, 320
  schematic representation, 318-319, 321
  stress-corrosion cracking, hydrogen embrittlement and corrosion fatigue, 328-333
    fatigue crack growth rate data, 331, 332
    stress-corrosion tests, results of, 328-330
  threshold design stress calculation, 320
Distortional strain energy criterion, 14
Double-beam specimens
  sustained-load crack growth testing, use in, 36, 37
Dynamic tear test, 33, 345

Elastic modulus. *See* Young's modulus
Elastic-plastic crack growth, 16-17
  elastic strain energy release rate, 16
  J-integral concept, 16-17
  resistance-curve concept, 16
Elastic-plastic fracture toughness tests, 30-32
  advantage of, 32
  J value, 31
  results, typical, 31
Elastic strain energy release rate, 16
Electron microscopy, scanning transmission, 286, 335
Electroslag remelting, 52, 73
Embrittlement
  temper, 46
  tempered martensite, 46, 56
Energy release rate, 5-9
  center crack, loaded plate, 7
  center crack, wide plate, 6
  cyclic crack growth, 9
  Griffith's energy balance criterion, 7
  stress and crack length, relationship between, 8
  subcritical crack growth, 9
  Young's modulus, 6
Eutectic decomposition, 175

Failure analysis (*See also* Design, material selection and failure analysis), 334-335
Fatigue crack growth
  A533 steel, behavior of, 17-18
  ASTM Method E647, 19
  steel pressure vessel, 10
Fatigue loading, 9
Fatigue precracking process, 25
Ferritic stainless steels, composition of, 107
Forman equation, 35, 196, 326
Fracture test methods, 23-38
  correlation fracture tests, 29-34
  fatigue crack growth rate tests, 34-35
  non-plane strain toughness tests, 29-32
    crack-opening displacement methods, 32
    elastic-plastic fracture, 30-32
    R-curve determination, 29-30
  plane strain fracture toughness tests, 23-29
    ASTM Method E399, 23-27
    stress intensity factor tests, new, 27-29
  sustained-loading crack growth tests, 35-36
  test results, use of, 36-38
Fracture toughness, 12-14
  ASTM Method E399, 13
  critical K value, effect of specimen thickness, 12, 13
  notch toughness, comparison to, 13
  plane strain fracture toughness, variation with yield strength, 13-14

Gamma prime precipitate, 255-257, 258, 259
Gas tungsten arc welding process. *See* Welding process
Grain boundaries, superalloy, 262-263
Griffith's energy balance criterion, 7

Historical developments, fracture mechanics, 1-3
Hydrogen embrittlement, 328-333

J-integral concept (*See also* Elastic-plastic fracture toughness tests), 16-17
J-integral method
  martensitic stainless steels, fracture properties of, 143

Kahn-type tear test, 172

Load-carrying capabilities, 5
Loading
  cantilever, used in sustained-load crack growth testing, 36, 37

constant amplitude, 34, 35
cyclic, 34
random, 34
spectrum, 34

Martensitic stainless steels, composition of, 107-110
Type 403, 107, 109
Martensitic stainless steels, fracture properties of, 142-149
Type 403 standard and modified, 143-148
compositions, 143
fatigue crack growth rate scatter band, 143-145
fatigue crack growth rates, various environments, 146-148
fracture toughness, range of temperatures, 143, 144
fracture toughness scatter bands, 143, 144
Type 420, 148-149
Type 422, 149, 150
Type 431, 149
Material flaws, 9-10
flaw size, maximum calculation, 10
steel pressure vessel, fatigue surface of, 9, 10
Materials selection (*See also* Design, material selection and failure analysis), 335-362
Materials selection case studies, fatigue crack propagation, 347-354, 355
helicopter components, 350-354, 355
lug fatigue, 353
lug fatigue lives, 354
lug fatigue lives, effect of liners, 354, 355
rotor hub cracks, 350, 351, 352
Ys, comparison of, 350-352, 352
hydraulic equipment, 347-350
illustration of, 348, 349
Materials selection case studies, fracture toughness, 336-347, 348, 349
line pipe steels, 339-344
burst hoop stress results, 341-344
pressure vessels, 336-339, 340, 341, 342
ammonia, 336
design approach, incorrect, 338, 341
failure origins, 338, 340
fracture paths, maps of, 338, 339
fracture surface, chevron markings on, 339, 342
motor case pieces, 338
welding crack, 336-337
weld, mating surfaces on, 339, 342
rollover protective structures, 344-347, 348, 349
Charpy V-notch energy results, 345
impact energy, dynamic fracture toughness vs., 345-346, 348
material properties and thickness, 347
testing temperature, percent shear vs., 345, 347
transition temperature, loading time vs., 345, 346
Materials selection case studies, stress-corrosion cracking threshold, 354-362
petroleum processing, corrosion-related failures in, 357-362
crevice corrosion, pitting and cracking, 357, 360
titanium tube, 357, 361
titanium tube, photomicrograph of, 358-359, 362
plated fasteners, hydrogen embrittlement of, 354-357, 358
high-strength fasteners, 357, 358
micrograph of, 355, 357
photograph and metallograph of, 355, 356

Nondestructive inspection methods. *See* Material flaws

Orowan criterion, 57

Paris equation, 35, 234, 265
Phases, superalloy, 255-263
Plane strain fracture toughness factor. *See* Fracture toughness
Plane strain fracture toughness, measurement of. *See* Fracture test methods
Plane strain fracture toughness, stainless steels, 111
Plastic deformation, 10, 11, 14, 32
Plastic zone
plastic deformation, 14
stress distribution, illustration with, 14, 15
Poisson's ratio, 12-13, 15, 16, 221
Precipitation hardening stainless steels, composition of, 109, 110-111
A-286, 109, 111
classes, 110, 111
Precipitation hardening stainless steels, fracture properties of, 149-163
austenitic, 159-163
fatigue crack growth rates, A-286 steel, 161, 163
fracture toughness data, A-286 steel, 161-163
martensitic, 150-156
fatigue crack growth rate, 15-5PH stainless steel, 151, 155
fatigue crack growth rate, 17-4PH stainless steel, 151, 156
fatigue crack growth rate, PH13-8Mo cantilever beam specimens, 156, 159
fatigue crack growth rate, PH13-8Mo steel, 155-156, 158
fatigue crack growth rate scatter band, PH13-8Mo steel, 151-154, 157
fracture toughness data, 151, 152, 153
stress-corrosion threshold data, 151, 154
semiaustenitic, 156-159, 160
fracture toughness and threshold stress-corrosion cracking data, 159, 160

Random loading. *See* Loading
R-curve concept. *See* Resistance-curve concept
R-curve determination, 29-30
ASTM Method E561, 29
ASTM Method E561, test results, 29, 30
specimens, center-cracked tensile, 29
specimens, compact, 29
Resistance-curve concept, 16

Scanning electron microscope fractograph
A533B steel, cleavage fracture in, 47, 48
HY130 steel, intergranular fracture in, 48, 49
Scanning transmission electron microscopy, 286, 335
Shear deformation (*See also* Stress intensity factor), 12

Shielded metal arc welding process. *See* Welding process
Specimens, types of, 23, 27
Spectrum loading. *See* Loading
Stainless steel alloys, types (*See also* Austenitic stainless steels, fracture properties of)
  21-6-9, 137-140
  22-13-5, 137-140
Stainless steels, types (*See also* Austenitic, Martensitic and Precipitation hardening stainless steels)
  15-5PH, 150-156
  17-4PH, 150-156
  17-7PH, 156-159, 160
  300M, 68, 73, 75, 96, 99
  301, 106, 112-114
  304, 107, 111, 114-124, 125, 298
  304L, 114-124, 125
  304LN, 123-124
  308, 124, 125
  309S, 124-127
  310S, 106, 124-127
  316, 107, 111, 127-137
  316N, 127-137
  321, 134-137, 138, 139
  348, 137, 138, 139
  403, 107, 111, 112, 142, 143-148
  410, 142
  420, 148-149
  422, 149, 150
  431, 149
  A286, 111, 159-163
  AM355, 156-159, 160
  Custom 455, 151-156
  Kromarc 58, 107, 140-141, 142
  list of, 108-109, 110-111
  PH13-8Mo, 151-156, 157, 158, 159
  PH15-7Mo, 156-159, 160
Stainless steels, wrought. *See* Wrought stainless steels
Steels, AISI types
  3340, 46, 55
  4335, 9-10
  4340, 12, 48, 49, 56, 61, 62, 68, 79, 83, 93-95
  4340M, 79
  4345, 53
Steels, ASTM specification
  A36, 62, 64, 80, 346, 348
  A212B, 77
  A514F, 80
  A516, 81
  A517E, 77
  A533B, 47, 48, 59, 73, 76, 77, 81, 83
  A533B-1, 17, 18, 74, 78
  A572, 346, 348
  A588A, 80
Steels, classes of, 42
Steels, commercial grades
  HP-9-4-30, 68
  HP-9-4-35, 77
  HY 130, 48, 49
  X65 pipeline, 53, 54, 79
Steels, high-strength low-alloy, 42, 45, 59
Steels, maraging type
  10Ni-Cr-Mo-Co, 42, 79
  12Ni-5Cr-3Mo, 42, 79, 86
  18Ni, 89, 91, 92, 95, 95-96
Steels, metallurgy of, 42-46
  embrittlement, special considerations, 46
  phase relationships and chemistries, 42-46
  carbon and alloy steels, composition of, 43-44
  ferrite and carbide phases, distribution of alloying elements, 45
  steels, ferrite-pearlite, 45
Stress-corrosion crack growth. *See* Crack growth
Stress-corrosion crack growth rate, measurement of. *See* Sustained-load crack growth tests
Stress-corrosion cracking, 95-96, 328-333
Stress-corrosion cracking resistance, 91
Stress intensity factor, 5, 10-12, 26
  crack deformation, modes of, 11-12
  linear-elastic theory, use of, 10-11
  shear deformation, 12
  through-thickness crack, illustration of, 11
Stress intensity factor fracture tests, 27-29
  application of, 29
  disk specimen, 27, 28
  rod specimen, 27-29
Stress-strain curve, 6
Superalloys, creep crack growth, 294-304, 305, 306
  environmental effects, 304, 305, 306
  mechanisms and models, 299-301, 302
  metallurgical variables, 301-304
    grain boundary microstructure, serrated, 302, 303
    stress intensity factor, time to failure vs., 302, 303, 304
    stress intensity parameter, applicability of, 294-299
Superalloys, fatigue crack propagation, 265-294
  empirical data, 265
  environmental factors, 277-284, 285, 286, 287, 288, 289
    crack growth rates, stress intensity factor range, 282, 285
    cyclic crack growth behavior, continuous and time hold conditions, 280-282, 284
    ΔK^2, activation energy vs., 279, 283
    fatigue crack growth rate data, 283, 287, 288, 289
    fatigue crack growth rate, effect of temperature, 278, 282
    Udimet 700, fractograph of, 283, 286
  Inconel 718, electron fractographs of, 273, 281
  microstructure, effects of, 265-277, 278, 279, 280, 281
    cyclic crack growth rate, effect of gamma prime precipitate size, 265, 266
    cyclic crack growth rate, effect of grain size, 266, 267
    heat treatment, effects of, 273, 274, 275
    heat treatment results, 273, 276, 277, 278, 279, 280
    René 95, fracture paths of, 267, 270, 271
    René 95, micrographs of, 266-267, 268, 269
    residual cycle life, effect of particle and grain sizes, 267, 269
    Waspaloy, crack growth behavior of, 267, 271, 271-272
    Waspaloy, fracture surface, 272
  temperature, frequency and waveform effects, elevated temperatures, 284-289, 290, 291, 292, 293, 294, 295, 296
    fatigue plastic zone substructure, 286, 293
    fracture surfaces, scanning electron micrographs

of, 286, 291
frequency, effect of, 286, 294
IN-718, fracture surface in, 287, 296
Inconel 718, deformation twins in, 286, 292
stress intensity factor range and temperature, effect of, 286, 290
time/load waveforms, 287, 295
temperature, subzero effects, 289-294, 297, 298, 299, 300
fatigue crack growth rates, 290, 297, 298, 299, 300
Superalloys, fracture properties of, 253-307
creep crack growth, 294-305, 306
environmental effects, 304-305
metallurgical variables, 301-304
models and mechanisms, 299-301, 302
stress intensity parameter, applicability of, 294-299
fatigue crack propagation, 265-294, 295, 296, 297, 298, 299, 300
empirical models, 265
microstructure, effects of, 265-277, 278, 279, 280, 281
testing variables, effect of, 277-294, 295, 296, 297, 298, 299, 300
introduction, 253-254
nickel-base, physical metallurgy of, 254-265
gamma double phase, 263-265
gamma prime phase, 254-263
summary, 305-307
Superalloys, list of, 256
Superalloys, nickel-base, 254-265
gamma double prime, 263-265
gamma prime phase, 254-263
carbide structure, René 95, 260
composition, 255, 256
general comments, 254-255
grain boundary structure, René 77, 260, 261
phases, 255-263
precipitate structures, 257, 259
time, temperature and stress, effects of, 257, 259
Superalloys, types
IN-792, 302, 303
Inconel 706, 289, 290, 292, 293, 297, 307
Inconel 718, 256, 263-264, 272-283, 286-307
Inconel X-750, 289, 290, 292, 293, 299 300, 307
Multimet, 279
René 77, 256, 258, 260, 261, 282
René 80, 256, 258, 259, 284
René 95, 256, 260, 266-271, 278-283, 305, 306
Udimet 500, 256, 258
Udimet 700, 282, 285, 286
Waspaloy, 266, 271, 272
Sustained-load crack growth tests, 35-36
test specimens and loading, illustration of, 36, 37
Subcritical crack growth, 9

Temper embrittlement, 46, 49, 55
Tempered martensite embrittlement, 56
Titanium alloys, fatigue crack propagation, 232-241, 242, 243, 244, 245
alloy chemistry, effects of, 235-237
alloy type, effect on fatigue crack propagation resistance, 235
oxygen, effect of, 236
crystallographic texture, effects of, 238
environmental effects, 238-241
mechanical factors, 238-240
thermal and chemical environments, 240-241, 242, 243, 244, 245
microstructures, effect of, 237-238
scatter, illustration of, 233-234, 235
yield strength and other variables, variation with, 233
Titanium alloys, fracture properties of, 213-247
conclusions, 244-247
fatigue crack propagation, 232-241, 242, 243, 244, 245
fracture toughness, 224-232
introduction, 213-214
metallurgy of, 214-224
sustained-load crack propagation, 241-244, 246, 247
tensile properties, 224
Titanium alloys, fracture toughness, 224-232
alloy chemistry, effects of, 225-227
fracture toughness, influence of oxygen content, 226, 226-227, 230
high-strength alloys, fracture toughness of, 225
plane strain fracture toughness, effect of hydrogen content, 227
environment, effects of, 231-232
microstructure, effects of, 227-230
alloy Ti-6Al-4V, fracture toughness in welds, 229, 230
fracture toughness, effect of forging procedure, 229
plane strain fracture toughness and transformed structure, relationship between, 228
plane strain fracture toughness, effect of primary alpha dispersion, 228, 229
texture, effects of, 230-231
Titanium alloys, metallurgy of, 214-223
alpha-beta alloys, high-strength, 222-223
phase relationships and chemistry, 214
super alpha alloys, 223
Ti-6Al-4V, 214-222
composition ranges, 215
deformation modes, 219-222
grade, extra low interstitial, 214-215
microstructure and transformation behavior, 215-219, 220
Titanium alloys, sustained-load crack propagation, 241-244, 246, 247
microstructure and temperature, effect of, 242, 247
Ti-6Al-4V, 242-246
Titanium alloys, tensile properties, 224, 225
specifications for, 225
Titanium alloys, types
Ti-4Al-3Mo-1V, 232, 243
Ti-4Al-2Sn-4Mo-0.5Si, 228, 229
Ti-4.5Al-5Mo-1.5Cr, 229
Ti-5Al-2.5Sn, 244
Ti-6Al-2Sn-4Zr-2Mo, 223, 235, 239

Titanium alloy, types *(continued)*
  Ti-6Al-2Sn-4Zr-6Mo, 222-223, 229, 230, 231, 235, 238, 239
  Ti-6Al-4V, 213, 214-222, 223, 226, 228-230, 232, 236, 237, 240, 246
  Ti-6Al-4V ELI, 232, 238, 240
  Ti-6Al-6V-2Sn, 222-223, 229, 231, 232, 238, 241
  Ti-8Al-1Mo-1V, 223, 235, 239

Uniform tensile stress, 5

Vacuum arc remelting, 52
Vacuum induction remelting, 52
von Mises strain energy criterion. *See* Distortional strain energy criterion

Welding process, gas tungsten arc, 139, 141
Welding process, shielded metal arc, 139, 141
Wheeler model, 326, 327
Wrought stainless steels, fracture properties of (*See also* Austenitic, Martensitic and Precipitation hardening stainless steels), 105-163
  compositions, 106-111
    austenitic, 107-109
    ferritic, 107
    martensitic, 107-110
    precipitation hardening, 110-111
  fracture mechanics data, 111-163
    austenitic, 112-142
    martensitic, 142-149
    precipitation hardening, 149-163
  introduction, 105-106

Yield strength, variation with plane strain fracture toughness, 13-14
Young's modulus, 6, 53

Zener mechanism, 262, 301

C. H. U.
6/82